Benchmark Papers in Optics

Series Editor: Stanley S. Ballard
University of Florida

Published Volumes and Volumes in Preparation

POLARIZED LIGHT
William Swindell
INFRARED DETECTORS
Richard D. Hudson, Jr., and Jacqueline Wordsworth Hudson
INTEGRATED OPTICS
John F. Ebersole and Charles M. Wolfe
REMOTE SENSING
Richard D. Hudson, Jr., and Jacqueline Wordsworth Hudson
LIGHT IN THE SEA
John E. Tyler

A *BENCHMARK* ® Books Series

POLARIZED LIGHT

Edited By
WILLIAM SWINDELL
Optical Sciences Center
The University of Arizona

Stroudsburg, Pennsylvania

Distributed by
HALSTED PRESS *A Division of John Wiley & Sons, Inc.*

Benchmark Papers in Optics, Volume 1
Library of Congress Catalog Card Number: 74-26881
ISBN: 0-470-83997-X

77 76 75 1 2 3 4 5

Manufactured in the United States of America.

LIBRARY OF CONGRESS CATALOGING IN PUBLICATION DATA

Swindell, William, comp.
Polarized light.

(Benchmark papers in optics ; v. 1)
Includes bibliographical references.
1. Polarization (Light)--Addresses, essays, lectures.
I. Title.
QC442.S94 535.5'2 74-26881
ISBN 0-470-83997-X

Exclusive Distributor: **Halsted Press**
A Division of John Wiley & Sons, Inc.

Acknowledgments and Permissions

ACKNOWLEDGMENT

DOVER PUBLICATIONS, INC.—*Treatise on Light*

PERMISSIONS

The following articles have been reprinted with the permission of the authors and the copyright holders.

AMERICAN INSTITUTE OF PHYSICS—*The Journal of Chemical Physics*
Polarization of Light Scattered by Isotropic Opalescent Media

INSTITUT MITTAG-LEFFLER, SWEDEN—*Acta Mathematica*
Generalized Harmonic Analysis

MASSACHUSETTS INSTITUTE OF TECHNOLOGY—*Journal of Mathematics and Physics*
Optical Algebra

MASSON ET CIE, ÉDITEURS—*Nouvelle Revue d'Optique*
A Unified Formalism for Polarization Optics by Using Group Theory

OPTICAL SOCIETY OF AMERICA—*Journal of the Optical Society of America*
Some Aspects of the Development of Sheet Polarizers

PRESIDENT AND FELLOWS OF HARVARD COLLEGE—*Polarized Light*
Bibliography

SOCIETA ITALIANA DI FISICA—*Il Nuovo Cimento*
Coherence Properties of Partially Polarized Electromagnetic Radiation
On the Matrix Formulation of the Theory of Partial Polarization in Terms of Observables

VAN NOSTRAND REINHOLD COMPANY—*Colloid Chemistry*
Dichroism and Dichroic Polarizers

Series Editor's Preface

Optics, pure and applied, comprises a very broad field indeed. Its roots are ancient; its modern applications are novel and exciting. A difficult and far-reaching task is hence faced by the series editor of the Benchmark Papers in Optics. He must consider the classical fields, so rich in history and so necessary for the understanding of all optical phenomena, and he must bring to the scientific audience current information on the burgeoning areas of applied optics. And what should be done in the several inter- or multidisciplinary fields of which optics forms a part?

For all subjects, the classical or benchmark papers must be included, be they centuries or only a decade old, be they in English or another language. In addition, the papers must be placed in proper perspective by commentaries written by knowledgeable and skillful volume editors. Thus the impact of the book will be satisfyingly greater than the sum of the original papers that are presented in facsimile. These volumes should be of value to scholars, teachers, applied physicists, engineers, and practical opticists. All this constitutes a courageous goal; we can only hope that it is being at least partially achieved.

It seems appropriate that the first volume in the Benchmark Papers in Optics series deals with a fundamental property of light: polarization. It was, of course, polarization phenomena that established the fact that light is a transverse wave motion. A full treatment of the history, theory, and applications of polarized light would require many volumes. Therefore, the editor, Professor Swindell, wisely restricted his coverage chiefly to those papers that have contributed most to our current understanding of the nature of polarized light. The 36 papers cover a time span of three centuries; they include important early contributions from Europe, some of which have been translated into English, and modern topics.

Although the scientific literature contains many references to polarized light and its applications, there are only a very few books devoted entirely to this subject. The present volume should prove to be an important addition because of its coverage, the careful selection of papers, and the illuminating commentaries written by Professor Swindell.

Stanley S. Ballard

Contents

II. FOUNDATIONS OF MODERN METHODS

III. UNPOLARIZED LIGHT

IV. THE JONES CALCULUS

V. OTHER DESCRIPTIONS OF POLARIZATION

VI. SYNTHETIC POLARIZERS

VII. BIBLIOGRAPHY

Contents by Author

Introduction

Despite the numerous flyby missions, and hard and soft landings on Venus, nearly all the information we have on the droplets that form the cloud structures covering Venus have been obtained from earth-based observations. We know, for example, that the mean droplet diameter is 1.1 ± 0.1 micrometers and that the refractive index of the droplet is 1.46 ± 0.5. These results have been obtained from the measurements on the state of polarization of the sunlight that is scattered from the planet.[1]

Let us consider some additional uses of polarized light. There are security systems that activate an alarm when a beam of light is broken. By using a polarized beam of light with suitable optical components at the detector, it becomes virtually impossible to defeat the system by substituting a secondary beam of light while the primary beam is interrupted. The system can be made even harder to defeat by temporally modulating the polarization form and using the appropriate demodulation at the detector.

Many organic dyes fluoresce with a decay time that is of the order of 1 nanosecond. An ingenious way to measure these decay periods is described by Ravilious et al.[2] If the dye, in solution, is excited by polarized radiation, then the reradiation is depolarized by an amount that depends in part on the decay time. Depolarization is caused by Brownian rotation of the excited dipoles. The elegance of the method stems from the fact that a fast event is measured with essentially a static experiment. What one observes is the time-averaged effect of a continuing interaction between the decaying molecules and the calculable Brownian motion.

In situations where radar screens must be viewed in the presence of ambient room lighting, there is the possibility that weak "blips" may be missed because of the low visual contrast. The simple expedient of placing a sheet of circularly polarizing material in front of the radar tube face can largely eliminate this problem. The extraneous ambient illumination is polarized by the filter and, after reflection by the screen face, is still circularly polarized but with the opposite handedness. Even allowing for nonideal conditions, nearly all of it is absorbed by the filter and never reaches the observer. Only half of the wanted light from the cathode-ray tube is lost to the filter, and the contrast ratio and the performance of the system as a whole is significantly increased.

A final example is the Kerr-cell shutter. The device consists of a pair of crossed polarizing elements surrounding a cell filled with nitrobenzene. Normally no light can pass through the combination. The application of a high voltage across the cell induces birefringence in the nitrobenzene and light is transmitted. If the electric field is pulsed, extremely short shutter opening times, as short as 10^{-7} second, can be achieved. This arrangement permits synchronized photography of high-speed events.

These few examples have been chosen to illustrate the diverse and adroit applications of polarized light. Hundreds, or even thousands more, exist and are described in the journal and textbook literature. (For a thorough overview, see Shurcliff[3] and Shurcliff and Ballard.[4])

In preparing this book I have taken the standpoint that polarized light is a tool, a tool that has enabled many interesting and valuable devices to be developed. We have already looked at a few of them. As such, I feel that it is more important to review man's understanding of the nature of polarized light than to reprint selected articles from recent literature. To arbitrarily select some of these articles and endorse them with "Benchmark" status is not realistically possible. Any selection resulting from such an undertaking would give only an incomplete representation of the present state of the art and would unavoidably reflect the editor's personal interests and prejudices.

Thus I have selected mainly those papers that have contributed to our knowledge and understanding of the nature of polarized light. To accomplish this it was necessary to delve back into history, and the first part of this book represents my attempt to provide a more or less continuous coverage of the (mainly theoretical) evolution of the subject matter. In omitting some of the worthwhile modern articles, I am somewhat consoled by the fact that they are readily accessible to the interested reader in other places.

Our understanding has evolved slowly over three hundred years, and until the twentieth century all major advances took place in Europe. As a result, some of the chosen articles are in a foreign language and many are drawn from rather obscure sources and are difficult to obtain. I welcome this opportunity to draw these works together; they have long been widely scattered. In the Benchmark context, these earlier papers should not be dismissed as being of purely historical interest. They are a cogent part of our scientific development and may be neglected only at the expense of impaired scholarship.

Just about every textbook on optics contains a chapter or two on polarized light. However, very few books have been written exclusively on the subject. In a sense these specialized books are also benchmark papers and deserve mention. These books include the following:

1. *The Depolarization of Electromagnetic Waves,*[5] by P. Beckmann. Depolarization in this book denotes the change in polarization from one state to another as a result of scattering or propagation in anisotropic media. Following the first two chapters of introductory theory, Beckmann discusses depolarization by scattering from solid objects, anisotropic media, rough surfaces, random scatterers, and in random media. The book is written with emphasis on radio waves but also contains sections on optics. The theory, of course, is applicable to either.

2. *Polarized Light and Optical Measurement,* [6] by D. Clarke and J. F. Grainger. This book contains a review of modern descriptions and chapters on the interaction of light and matter, optical elements used in polarization studies, measurement of the state of polarization, and the role of polarization in optical instrumentation.

3. *Polarized Light,* [3] by W. A. Shurcliff. The following appears on the dust jacket: "This book fulfills a long standing need for a compact, scholarly account of the theory, production, and application of polarized light. It is the first text devoted exclusively to this subject and is intended for use by experimentalists and engineers in nearly every branch of science and technology that employs light." No other book in the English language has supplanted this one in terms of diversity and breadth of coverage.

4. *Polarized Light,* [4] by W. A. Shurcliff and S. S. Ballard. This is Momentum Book No. 7, written at the request of the Commission on College Physics. In a concise and readable fashion, it explains the fundamentals of polarized light, indicates the power of shortcut theoretical methods, surveys the time-honored applications of polarized light, and describes some of the newest and most dramatic applications.

There are also many books that deal specifically with one aspect of polarized light, e.g., photoelasticity,[7] crystal optics,[8] and polarization interferometers.[9]

What of the future? Clearly there are many avenues still to be explored. Here we can look at a very few of the current lines of research.

One application of polarized light is to be found in the rapidly developing field of picosecond (time resolved) spectroscopy. Consider an ensemble of atoms (say hydrogen atoms) initially in their ground state. If these atoms are sequentially pumped by picosecond pulses having different polarization states, the relative transmission of the ensemble, measured as a function of time delay, will yield information concerning the very short relaxation times between different magnetic states.[10]

The all-optical computer, still pretty much of a pipe dream, will use polarized light in several different ways. The binary logic of electronic computers is represented by different voltage levels. With optical signals the binary states can be orthogonal polarization forms. Integrated optics and fiber optics will also play an important role in the development of such devices. Electronic-optical hybrids will first appear with optical devices gradually taking over from their electronic counterparts. Optical read-and-write memories are presently available using photochromics. Large read-only, random-access memories using the Faraday and Kerr effects have also been developed. Curie-point writing, which permits rapid read–write capability, has also been demonstrated. There is an enormous research effort in this area presently underway. Without doubt computer technology will lean more and more heavily on optical systems in general and polarizing optics in particular.

Earth-resource remote-sensing programs provide pictures of the earth taken from high-altitude aircraft or orbiting satellites. Multiband sensing dramatically increases the amount of information that can be extracted from such imaging systems. Comparatively little attention has been paid to studying the polarization content of the scene. There is additional information that may be useful in the polarization structure of image-forming light. We can expect to see further research along these lines.

This introduction ends with a look at something that might have been. As early as

1920 a scheme was proposed for equipping car headlights and car drivers with suitably oriented polarizing visors.[11] Thus drivers would see readily by the lights of their own cars and yet not be dazzled by oncoming vehicles. Despite extensive campaigning by a chief proponent, E. H. Land, the system was never adopted. The loss of ability to drive with full headlights at all times, with the concomitant increase in safety and comfort, is perhaps something we should all mourn.

I apologize to those readers who have little time for history, and to those who would rather have seen a fuller treatment of the phenomenology and applications of polarized light. I hope also not to have offended those of my colleagues whose work, but for the laws of chance, would also have been included in the latter stages of this book. The selection of articles is entirely mine.

References

1. J. E. Hansen and A. Arking, *Science,* **171,** 669–672 (1971).
2. C. F. Ravilious, R. T. Farrar, and S. H. Liebson, "Measurement of Organic Fluorescence Decay Times," *J. Opt. Soc. Am.,* **44,** 600 (1954).
3. W. A. Shurcliff, *Polarized Light,* Harvard University Press, Cambridge, Mass., 1962.
4. W. A. Shurcliff and S. S. Ballard, *Polarized Light,* Van Nostrand Reinhold, New York, 1964.
5. P. Beckmann, *The Depolarization of Electromagnetic Waves,* Golem Press, Boulder, Colo., 1968.
6. D. Clarke and J. F. Grainger, *Polarized Light and Optical Measurement,* Pergamon, Oxford, 1971.
7. H. T. Jessop and F. C. Harris, *Photoelasticity,* Dover, New York, 1960. (See for an example.)
8. N. H. Hartshorne and A. Stuart, *Crystals and the Polarizing Microscope,* Arnold Press, London, 2nd ed., 1950. (See for an example.)
9. M. Françon and S. Mallick, *Polarization Interferometers, Application in Microscopy and Macroscopy,* Wiley-Interscience, New York, 1971. (See for an example.)
10. Work of this kind is currently being pursued at the Optical Sciences Center, University of Arizona, Tucson, Ariz.
11. One of the several papers written by E. H. Land on this subject is: E. H. Land, "The Use of Polarized Headlights for Safe Night Driving," *Traffic Quart.,* 330–339 (Oct. 1948).

I
Early Studies

Editor's Comments on Papers 1 Through 8

Erasmus Bartholinus (1625–1698), a Danish physician and naturalist, provides us with the earliest account of a phenomenon directly related to polarized light. The transparent "Iceland Crystal," or Iceland spar as it is now known, was brought to his attention because of its unusual property of creating double images when nearby objects were viewed through it. In this sixty-page book *Experimenta crystalli Islandici disdiaclastici, quibus mira et insolita refractio detegitur,* he describes seventeen experiments, some chemical, but mainly optical, that were performed on the substance.

Experiment 7, reproduced in Paper 1, is the first of several in which this remarkable property of double refraction is investigated and is the one included here. Later experiments (16 and 17) showed that one of the two images was formed according to the already well known Snell's Law, and Bartholinus is to be credited with the still-used terminology "ordinary" and "extraordinary" as applied to rays and images.

Apart from observing that the two images appeared equally bright, he did not investigate any of the intrinsic properties of the refracted light. The fundamental differences between the light rays that formed the separate images were not known to him. It is apparent that he regarded double refraction as a property of the crystal only and that the two images and the light that formed them were, apart from geometrical considerations, equal in all respects. Thus it cannot be said that he discovered the phenomenon of "polarized light."

Christiaan Huygens (1629–1695) made one of the most significant contributions to the field of optics when he proposed his theory concerning the wave nature of light. *Traité de la lumière* was completed in 1678. It was read before the Académie des Science in 1679 but not published until 1690. This brilliant work accounted for reflection and ordinary refraction phenomena by means of a wave theory of light in which spherical waves played the central role.

Huygens developed the theory using the analogy of the propagation of sound, but there was great difficulty in understanding the nature of the waves themselves, which were thought to be longitudinal ether waves.

One triumphant success for his theory was that it accounted perfectly for the phenomenon of extraordinary refraction as seen in Iceland spar. It was necessary only to replace the spherical wave by a spheroidal wave to solve the mystery of predicting the direction of the extraordinary ray. Chapter V, "On the Strange Refraction of Iceland Spar," deals with this extension to his theory.

Paper 2 (sections 15 through 35 of Chapter V) is a concise and elegant exposition and proof of his theory. That this theory has stood the test of time and still stands today is a tribute to the genius of this philosopher.

Huygens discovered polarized light. The extract reproduced here (from section 43 of Chapter V) describes an experiment, the results of which showed that there is an intrinsic difference between the light that is refracted into the two beams. Because Huygens believed that the waves, whatever their nature, were longitudinal, he was unable to explain the difference and left the subject for others to investigate.

Space forbids that the whole treatise be reproduced here. However, the Preface to Huygens' work is an interesting part of the whole and should be read. Thus it is included. These extracts are from *Treatise on Light* by Christiaan Huygens; translated by Silvanus P. Thompson.

Isaac Newton (1642–1727) was vehemently opposed to the wave theory of light and despite the success of Huygens, the wave theory lay almost dormant during the eighteenth century because of the shadow cast by this great philosopher. Yet, in Part III of his book *Opticks,* Newton admits that his corpuscular theory is deficient in being able to explain, or agree with, certain observed optical phenomena. He does this by posing a number of questions: ". . . queries, in order to a farther search to be made by others."

We include in Paper 3 Queries 25 and 26, in which Newton is grasping for an explanation of the differences between the usually and unusually refracted beams of light from a crystal of Iceland spar.

In Query 25 he concludes that polarization is an intrinsic property of light and not an effect that is imparted to it by the crystal. In Query 26 he concludes that some kind of transverse asymmetry must exist: "Have not the rays several sides. . . ." Blinded by the corpuscular theory, he progressed no further.

In Query 29 Newton draws an analogy between the poles of a magnet and the "sides" of a ray. This is apparently the origin of the expression "polarized light" used by later authors, although not by Newton himself.

The next significant advance to the understanding of polarized light came from E. Louis Malus (1775–1812). Over a century had passed since the publication of Newton's *Opticks* and Huygens' *Treatise on Light* when the Paris Académie des Sciences in 1808 offered a prize for a mathematical theory of double refraction. During his studies, for which he eventually won the prize in 1810, Malus made several important discoveries, of which three were particularly significant. First, he showed conclusively that Newton's suggestion (Query 25) was correct—that polarization is an intrinsic property of light itself and not something that is added by passing through the crystal. He determined this following the chance observation of the light of the setting sun reflected off a glass window, viewed through a crystal of Iceland spar. Second, he formulated the now well-known "cosine-squared" law. Third, he showed that polarization is produced also by transmission through transparent bodies. Incidentally, his mathematical theory was based on Newton's ideas, which were soon to completely fall from favor. Paper 4 is one of several in which Malus reports his remarkable discoveries.

Augustin Fresnel (1788–1827) perhaps contributed more than anyone else to the science of optics. With the support of overwhelming experimental evidence, and awareness of earlier suggestions of Thomas Young, he proposed that light be considered as a transverse wave motion. This astonishing pronouncement was met with hostility and rejection by fellow scientists. The ether was thought to be a fluid and thus incapable of propagating transverse waves. However, the model worked supremely well. Hitherto observed but unexplained facts were readily explained in terms of transverse waves. These included the phenomena of double refraction, the observations of Arago on chromatic polarization and interference of polarized light, and Malus's cosine-squared law.

Fresnel applied the theory to partial reflection, a problem that the corpuscular theory had been completely unable to explain in any reasonable manner. Not only was partial reflection explained qualitativcly, but also quantitatively. The celebrated Fresnel reflection formulas are still in use today. Perhaps Fresnel was lucky to have obtained the same formulas as Drude, who in the twentieth century rederived them using the electromagnetic theory.

Papers 5 and 6 are extracts from *Oeuvres complètes de Fresnel,* Volume I. The first, "Considérations mécaniques sur la polarisation de la lumière," is taken from a discussion on the chromatic polarization in thin crystals and includes the pronouncement of the transverse wave theory. The second-extract, "Mémoire sur la réflexion de la lumière polarisée," contains Fresnel's formulation of the reflection coefficient formulas.

Sir David Brewster, K. H., D. C. L., F. R. S. Edinburgh (1781–1868), contributed widely to the science of optics. He discovered photoelasticity (1816), and the Brewster telescope and Brewster fringes are named after him. He authored several articles dealing with polarized light. In the article reproduced here, he extends the work of Malus and discovers the relationship between the polarizing angle (now known as the Brewster angle) and the refractive power of dielectric materials.

Paper 7 reproduces only the first few and last two pages of his article "On the Laws Which Regulate the Polarization of Light by Reflexion from Transparent Bodies." The remaining twenty-eight pages describe twenty-nine propositions, all of which are based on laws enunciated in the pages shown here.

In a later series of articles, published anonymously in the *Edinburgh Philosophical Journal,* Brewster gives a historical account of polarization and double refraction. At the beginning he writes:

> One of the principal objects of the present series of papers is to correct these absurd misapprehensions; and we have no doubt that we shall be able to render the subject intelligible to such of our readers as have but a very slender portion either of physical or mathematical knowledge; and to convince those whose attainments are of a higher order, that almost all the phenomena of double refraction and polarisation, intricate and capricious as they appear to be, have been brought under the dominion of general laws, and can be calculated with as much accuracy as that with which the astronomer can compute the motions and positions of the heavenly bodies.

William Nicol (1768–1851) announced the discovery of a new polarizing prism in 1828. One can see from the title of Paper 6 that the significant feature was not that the device produced two beams of polarized light, but that the two beams were widely separated. Rochon [*Nova Acta Petropol.,* **6,** 37 (1788)] and Wollaston [*Phil. Trans.,* **110,** 126 (1820)] had both described prisms generating two diverging polarized beams. However, at the exit face the beams almost completely overlapped each other, thus limiting the usefulness of the devices.

Nicol's prism used the phenomenon of total internal reflection to deviate one of the beams completely to one side, thus permitting, for the first time, a large-area polarizing system with no overlap. There have been many improvements in detail on Nicol's original design, which now is seldom used. However, the principle of operation is still embodied in polarizing prisms, where highest purity of polarization and wide spectral bandwidth are required.

1

ERASMI BARTHOLINI

EXPERIMENTA CRYSTALLI ISLANDICI DISDIACLASTICI

Qvibus mira & insolita

REFRACTIO

detegitur.

ANNO

1670.

HAFNIÆ,

Sumptibus DANIELIS PAULLI Reg. Bibl.

QVOD NE DIVINARET OLIM
GRÆCIA
IN ISLANDIA SEPULTUM,
NUNC PRIMO DETECTUM:
UT
MAXIMO DANIÆ MONARCHÆ,
DEBERET NON MINUS OPTICA
SUA INCREMENTA,
QVAM ASTRONOMIA;
ET QVI OMNEM COLLIGUNT EX
SENSIBUS DOCTRINAM,
APPREHENDANT OCULIS VERI-
TATEM,
QVAM MENTIBUS NON POSSUNT;
ATQVE IN SEPTENTRIONE
NON REMITTI FRIGORE,
SED INTENDI EXPERIANTUR
LUMINIS RADIOS.

ERASMI

ERASMI BARTHOLINI

EXPERIMENTA CRYSTALLI ISLANDICI DISDIACLASTICI.

QVòd ſi tam celebris eſt apud omnes gloria Adamantis, atqve varia iſta opum gaudia, gemmæ unionesq́ve, ad oſtentationem tantum placent, ut digitis colloq́ve circumferantur; non minori afficiendos ſperaverim gaudio eos, qvibus curioſitatis conſcientia qvàm deliciarum eſt potior, novitate corporis alicujus, inſtar cryſtalli translucidi, qvod ex Islandia nuper ad nos perlatum eſt; cujus tam mira eſt conſtitutio, ut haud ſciam, num alias magis naturæ apparuerit gratia. Mihi eqvidem corporis hujus admiratione diu fatigato, varia ſeſe obtulerunt experimenta, qvæ naturæ induſtriis, & lautioribus ingeniis, in promovenda cognitione humana,

A cum

cum viderentur nonnihil vel utilitatis, vel voluptatis esse allatura, haud gravatè publico communicare volui.

EXPERIMENTUM I.

PRincipio, non minus jucunda mihi fuit eximia atqve rara corporis hujus figura, exterius conspicua, qvàm olim vel Nivis, vel Salium, vel

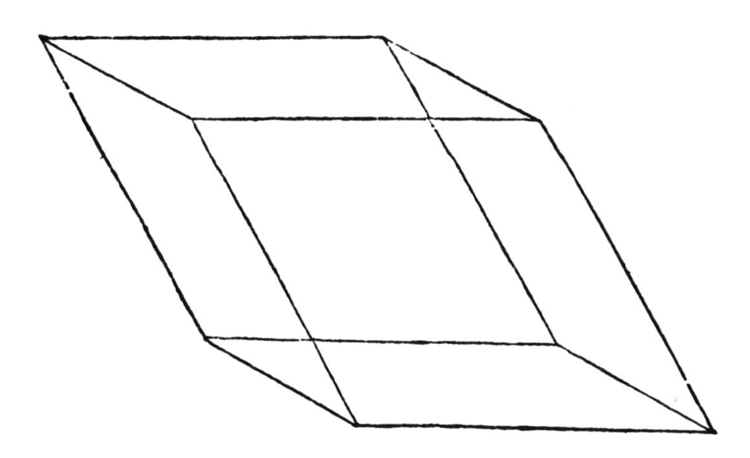

lapidum aliorum & crystallorum: qvippe qvadrangulis qvidem nascitur superficiebus planis, lateri-

lateribusqve æqvidiſtantibus, ſed disparibus angulorum intervallis, qvalem figuram planam, ſi Geometrarum lingva uti placuerit, appellabimus Rhombum vel Rhomboides; adeo ut totum corpus repræſentet Priſma vel Rhombicum, vel Rhomboides. Neqve tantum integrum corpus talem oſtendit figuram, ſed partes qvoqve omnes, ſi minutim diffringatur, conſtantiſſime eandem ſervant; qvam, tam adjecta figura, qvàm ſeqventes ferè demonſtrant; excepto aliqvo caſu, ubi Trigonicam Pyramidalem figuram nativum exhibuit ſolum.

EXPERIMENTUM II.

DEinde, aliqvid ſimile Electricis in eo deprehendi; nempe qvod attritu, acceptâ caloris animâ, trahat ad ſe paleas, aliaqve corpora leviora; qvæ ut propius veniunt, trahuntur, & complexu hærent. Id qvod non minus huic corpori accidere compertum habeo, qvàm ſuccino, etiam non affricto, dummodo calefacto: prout expertus ſum admotis plumis, pilis aliisqve levioribus, aliàs non ſuccedit experimentum. Et, cùm vitro & cryſtallo hæc ſit natura communis

A 2 cum

cum hoc corpore; non tantum candore & perſpicuitate, ſed etiam hac trahendi proprietate, videtur ad naturam cryſtalli qvàm Talci propius accedere.

EXPERIMENTUM III.

EST qvoqve non tanta duritie corporis, ut inexpugnabilis ferro ſit; ſed debiliori etiam frangentis vi divellantur minutiores partes. Non tamen ut Talcum in lamellas ſciſſile, ſed in exiguas partes, longitudine & latitudine ſimiles exhibentes figuras, fiſſili vena diſſolvitur, mortario commiſſum. Poliri qvoqve recuſat propter teneritatem. Igni præterea non facile conſumitur, aut in calcem redigitur; verùm ſi flammæ committatur, & igni reverberii, calcinari chymicâ lege expertus ſum. Qvippe, cum fruſtulum hujus cryſtalli, flammæ lampadis, per fiſtulam, qvâ vitra hermeticè occluduntur, animatæ, admoverem; mox animadverti redigi in calcem ſimilem calci vivæ, qvæ caloris ſenſu afficeret digitum admotum & leviter humectatum, atqve calorem excitaret; poſtremò, cum aqva fontana adſpergeretur, ebulliret, atqve in calcem communem verteretur.

EX-

EXPERIMENTUM IV.

POstea, ut certior evaderem an in liqvoribus Chymicis dissolvi posset; in ejus superficiem immisi aqvam fortem: qvam subitò, ingenti strepitu agere comperi, & ita corrodere lapidis externam faciem, ut bullas & motum excitaret, relicto qvoqve erosionis vestigio. Hinc in mortario comminutum, atqve in subtilem pulverem redactum aqvæ forti immisi: cum mox ebullire aqva fortis pergeret, donec totum qvod immissum erat dissolutum esset,& aqva fortis colore flavescente notaretur, qvæ prius limpida & perspicua erat. Thermoscopium postea, vitrea pila concava insignitum, pro immittendis humoribus,adhibebatur; qvo insignis qvalitatum caloris & frigoris differentia conspicua erat. Dissoluto in aqva forti pulvere, immisi Spiritum Vitrioli, qvi crassum à tenui rursus secerneret, & præcipitaret calcem albam, atqve in fundum remitteret.

EXPERIMENTUM V.

UT verò exactius nobis essent omnia perspecta, singulas figuræ, qvam possidet hoc corpus, partes examinandas suscepimus; atqve animad-

A 3

madvertimus lævorem laterum ita esse absolutum, ut nulla arte possit æqvari: eumqve tum facile obtineri, si tenuius frustum digitis subito diffringatur. Aliàs si malleo contundatur, non æqvali momento vis percussionis singulis partibus imprimitur, nec pari resistentia ubivis pugnatur; unde plana laterum crystalli sæpe exsurgunt scabrosa, adhærentibus passim exiguis fragmentis. Nitor toti corpori veriùs qvàm splendor, coloris limpidæ aqvæ; qvem aqva immersum exsiccatumqve mutat & hebetat. Hinc superiorem superficiem, & externam obscuriorem nativus locus præbet, cum imbres aut hyemes non patiatur. Apparent & qvandoqve

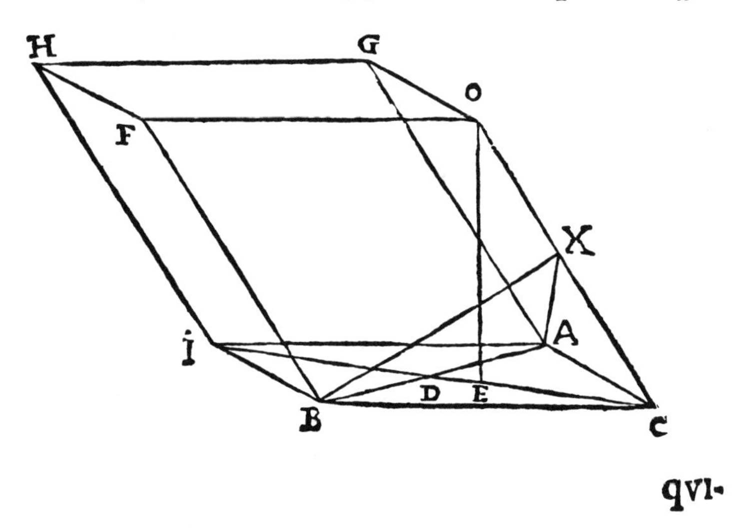

qvi-

qvidam colorum repercuſſus, qvales in cœleſti arcu ſpectantur. Mucronibus angulorum non eſt eadem ſpecies, obliqve ſibi invicem inclinatis omnibus lateribus planis. Plana oppoſita ſunt ſibi invicem parallela. Nempe planum OGHF æqvidiſtat plano ſeu Baſi CAIB, & ſuperficies OGAC, parallela eſt ſuperficiei FHIB. Hinc lineæ qvoqve OG & FH ſunt parallelæ & æqvales; nec non lineæ GH & OF. Deniqve anguli oppoſiti GOF & GHF, æqvantur, uti & anguli OFH & OGH. Expreſſi autem figuram hujus corporis, ſex Rhombis ſibi invicem

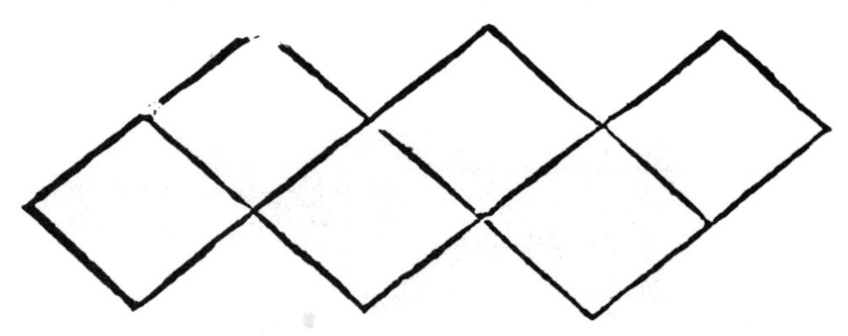

uno latere annexis, non tamen æqviangulis, ſed eâ menſurâ angulorum prædítis, qvam exhibet ipſum corpus; adeò ut, qvi ex charta, vel aliâ aliqvâ commodâ materiâ, exſciſſam hanc figuram complicaverit, obtineat facile ſimile corpus, qvod effingit ipſum Prisma, vera qvantitate

titate angulorum, qvam in ſchematibus ſeqventibus exprimere neqvivimus.

EXPERIMENTUM VI.

Cum verò verſemur heic in corporis ſolidi conſideratione, non tantum perpendendi erunt anguli plani, ſed etiam ſolidi. Eſt verò Angulus Solidus, qvi conſtat tribus angulis planis, qvalis in hoc Priſmate, eſt unus angulus ſolidus, qvi continetur angulis tribus planis OCB, OCA

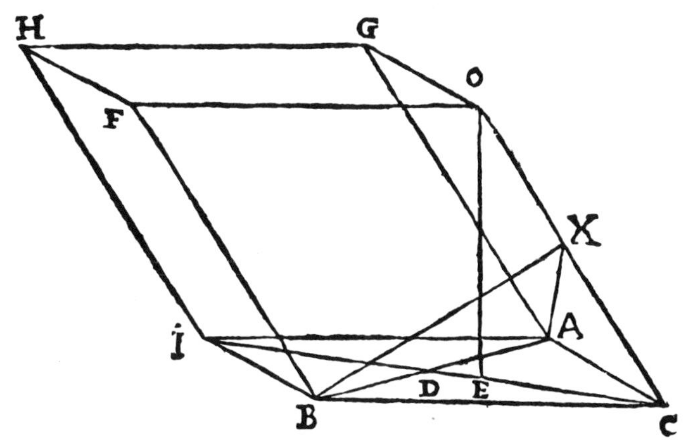

& ACB. Qvorum ſingulorum qvælibet ſuperficies, ut ACB & OCB, extremitate anguli ſui inclinatur ipſi OC, communi ſectioni reliqvorum duorum planorum OCA & OCB, ACO & BCO,

BCO, & efficit Angulum, qvi dicetur *Angulus conjunctus inclinationi anguli plani ad planum unde factus est.*

Et qvandoqvidem qvilibet angulorum planorum, ut ACB, hujus anguli solidi, factus est per communes sectiones duorum reliqvorum planorum OCA & OCB cum tertio plano angulo ACB; hi duo reliqvi anguli plani, continent angulum inclinationis inter se; qvi angulus inclinationis vocetur, *Angulus collateralis inclinationis ejusdem anguli plani* ACB, angulus scilicet inclinationis planorum angulorum OCA, & OCB. Qvare in hoc angulo solido, novem sunt nobis perpendenda; tres anguli plani, tres anguli collaterales inclinationum, & tres anguli conjuncti inclinationi.

In nostro Prismate crystallino, planorum angulorum duo sunt semper acuti, & duo reliqvi obtusi; & nullus unqvam æqvalis angulis collateralibus inclinationum. In præcedente figura, anguli obtusi sunt GOF & GHF; acuti verò OGH & OFH, licet in Schemate non ita oculis exhibeantur. Horum angulorum mensura mechanice inventa est; ipsius nempe anguli

B GOF

GOF vel ACB grad. 101. & ipsius OGH vel CAI vel ACO vel BCO grad. 79; qvod est complementum prioris anguli ad semicirculum. Qvibus datis, possunt inveniri haud difficulter &

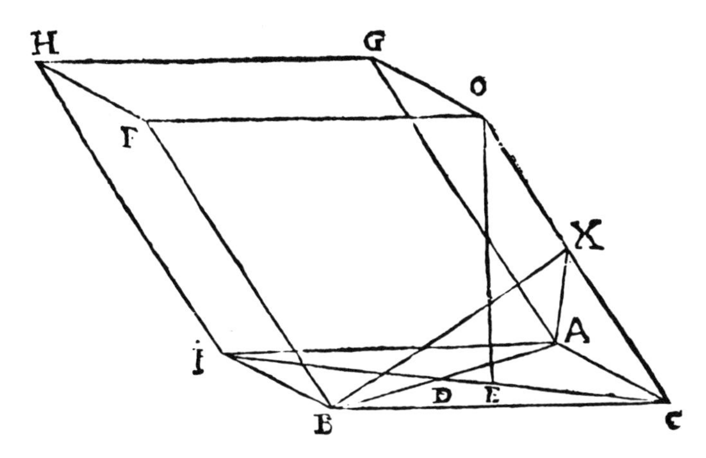

angulus collateralis inclinationis anguli plani ACB, & angulus conjunctus inclinationi anguli ACB, qvorum ille notatur in schemate per AXB, hic vero per XCD. Hinc ductâ perpendiculari BX, ex B in OC, ponatur BC 1000 part. eritqve BX Sin. Rect. anguli BCO 982, & XC, sin. compl. anguli ejusdem BCO 191, AX verò est æqvalis ipsi BX. Deinde, demittatur ex C perpendicularis CD in AD, qvæ dividat angulum ACB in duas partes æqvales, sicuti & lineam am AB; eritqve angulus BCD grad. 30. min. 30. cujus Sin. Rect. est 772, linea nempe BD: Siqvidem CB est communis utriqve Triangulo XCB & CDB. Qvare in Triangulo Plano AXB, ex datis tribus lateribus, qværatur angulus AXB, vel in hoc casu, cum Prisma est Rhombicum, facilius, in Triangulo XDB; ut XB ad Sin. Tot. ita DB, ad Sin. Rect. ang. DXB, 51. grad. 50. min. cujus duplum 103. grad. 40. min. est angulus AXB, collateralis inclinationis anguli plani ACB qvæsitus.

Jam, pro angulo conjuncto inclinationi, qvem efficiunt communes sectiones planorum ad plana reliqva, sint in Triangulo Rectangulo XCD, cognita duo latera, scilicet crus CX, 191, & Basis CD, 636, ex qvibus elicitur angulus qvæsitus DCX 72. grad. 34. min. qvi est Angulus conjunctus inclinationi anguli plani ad planum unde factus est.

EXPERIMENTUM VII.

HUjus crystalli examini cum ulterius incumberem, mirum & insolitum apparuit Phænomenon, qvo objecta per id conspecta, non sicuti

B 2 in

in aliis corporibus pellucidis, ſimplici imagine refracta exhiberentur, ſed Dupla. Detinuit diutius & oculos & mentem hoc experimentum, & ab aliis avocavit, qvæ perſeqvi ſuo tempore ſtatueram; cum animadverterem me Dioptrices qvædam fundamenta de Refractione attigiſſe. Et qvamvis, non ſatis advertentes, facile effugiat hoc experimentum, tamen ſeqventi ratione veritas Phænomeni conſtabit. In menſa, vel charta pura, ponatur objectum aliqvod, ut punctum, vel aliud aliqvid, ejus magnitudinis cujus eſt B, vel A, eiqve ſuperponatur Prismatis Rhom-

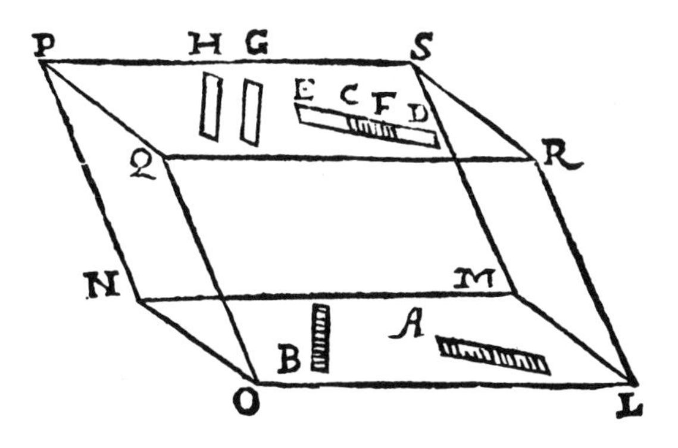

boidis infima ſuperficies LMNO. Tum per ſuperiorem ſuperficiem RSQP, conſpiciatur objectum

ctum B, vel A, radiis visus, per substantiam totius Prismatis RSPQOLMN directis, primò ad objectum B, deinde ad A. Jam, cum species objecti per corpora pellucida alia apparens, ut per vitrum, aqvam, aliaqve, eadem & simplex saltem repræsentetur; horum singula, dupla conspicientur imagine, B nempe in G, & H; & A in CD & FE, in superficie RSPQ, ut in adjectâ figurâ apparet. Notandum verò, distantiam, inter H & G species ab objecto B prodeuntes, esse nunc majorem, nunc minorem, prout Prisma crystallinum magnitudinis fuerit differentis, adeò ut in tenuioribus portionibus ferè evanescat illud duplicis speciei discrimen; sed eò remotiores ab invicem esse, qvo majoribus crystalli hujus frustis experimenta hæc tentaveris.

EXPERIMENTUM VIII.

NEqve tantum objecti B apparentia dupla exhibetur visui, sed & utraqve dilutiore colore: interdum ejusdem speciei pars una, obscurior altera. Nam, cum supra objectum A commovetur hinc inde Prisma; hæ apparentiæ, non eadem intentione coloris conspiciuntur, qva

B 3 ocu-

oculis nudis conſpiciuntur, vel per aliud corpus pellucidum videntur. Hinc ſpecierum, EF & CD, aliqvando apparebit una pars altera diluti-or. Ut, ſi in fig. præcedente, fuerit objecti lo-

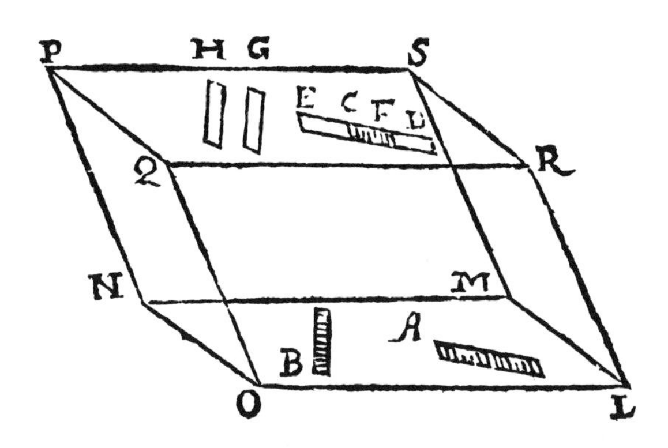

co linea aliqva denſior A; dum circumvolvitur Priſma ei incumbens, ſuperficie eadem deorſum vergente, animadvertemus in certo aliqvo ſitu, apparentiam objecti A, repræſentari in ſuperficie RSPQ per DE, ita ut pars FC ſit obſcuriore colore, qvàm extremitates DF & CE.

EXPERIMENTUM IX.

ANimum & aciem oculorum probe intendentibus apparet una ex duabus hiſce imaginiginibus, altera elevatior. Id qvod adjecta figura demonstrat, in qva HI sectio delineata est Prismatis, per oculum O, objectum A & imagines E & D, qvas ostendit in superficie, objectum A.

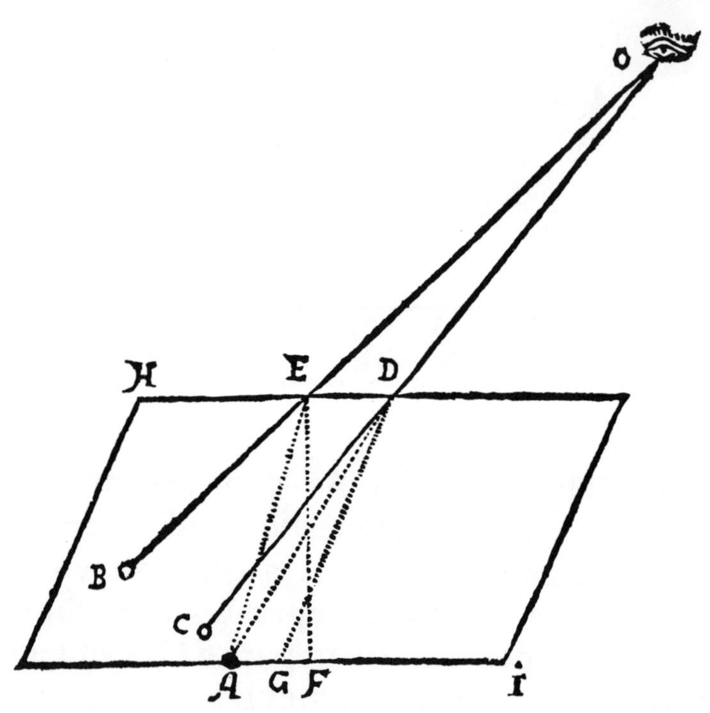

Jam, si productos ab oculo O imagineris singulorū imaginum radios, OD nempe in C, & OE in B, ubi oculus judicat intra Prisma occurrere imagines; facile constat B locum ex qvo provenire judicadicamus imaginem E, excelſiori eſſe loco qvàm C, ex qvo judicat viſus prodire imaginem alteram D.

EXPERIMENTUM X.

DAtur & locus aliqvis, ubi imago tantum apparet una & ſimplex, per Priſma hoc aſpicientibus, non ſecus ac per aliud aliqvod Diaphanum. Invenietur autem hic locus ſeqventi me-

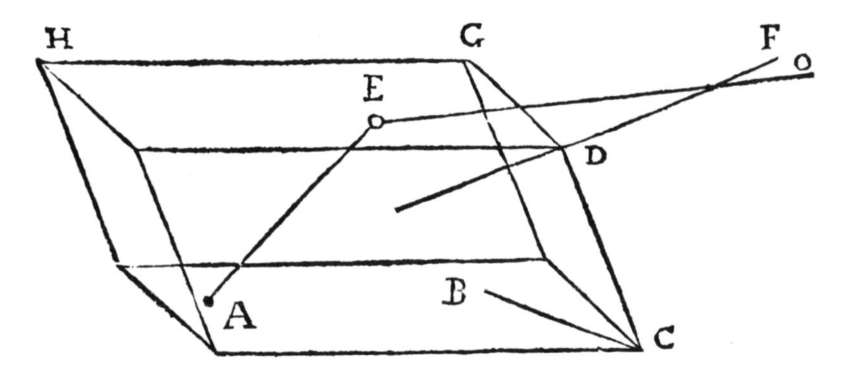

thodo. In figura adjuncta ſit objectum A, qvod in ſuperficie radiet, per lineam AE. Sitqve linea CB, ducta per unum angulum, dividens hunc angulum in duas partes æqvales. Hinc ſi ducatur FD, linea perpendicularis lineæ CD, lateris

teris plani Prismatis, in plano qvod per DCB imaginari possumus; & oculus O statuatur infra lineam perpendicularem FD; non videbitur species una, qvæ ab A emittitur, sed tantummodo altera in E; idqve donec oculus est supra planum DGH. Experimentum hoc aliqva difficultate absolvitur, propter exiguitatem anguli EOD; successum verò habebit feliciorem si assumatur Prismatis frustum amplius & perspicuum.

1

Experiment VII

ERASMUS BARTHOLINUS

This translation was prepared by Richard C. Jensen, University of Arizona, from Experimenta crystalli Islandici disdiaclastici, *Daniel Paul, 1670, pp. 1–17*

When I spent further time on this crystal, a strange and wonderful phenomenon appeared, according to which objects seen through it appeared not with single images, as in other transparent bodies, but with double images. This experiment held my attention for some time and distracted me from other experiments which I had intended to pursue in the same time, when I noticed that I had touched on certain fundamentals of Dioptrix concerning refraction. And although this experiment might easily escape those who do not pay sufficient attention, the truth of the phenomenon will be established by the following account:

On a table or on unmarked paper is placed a certain object, such as a stylus or something else, of the size of B or A, and upon it is placed LMNO, the lower surface of a rhomboid prism. Then through RSQP, the upper surface, object B or A is seen, as the rays of vision are directed through the substance of the whole prism RSPQOLMN first toward object B, then toward A. Now, although the shape of the object showing through other transparent objects (e.g., through glass, water, etc.) is represented as the same or at least single, the separate objects in this case are seen with double image, B with images G and H; A with CD and EF on surface RSPQ, as appears in the attached diagram. Note: The distance between images H and G coming from object B is sometimes greater, sometimes less, according to the differing size of the crystalline prism, so much so that in thinner sections the distinction between the two images almost disappears, but they appear more separated as you attempt the experiment with larger pieces of this crystal.

2

Reprinted from *Treatise on Light* (1690), translated by Silvanus P. Thompson, Dover Publications, Inc., New York, 1962, pp. v–vii, 60–76, 92–94, by permission of the publisher

Treatise on Light

CHRISTIAAN HUYGENS

PREFACE

WROTE this Treatise during my sojourn in France twelve years ago, and I communicated it in the year 1678 to the learned persons who then composed the Royal Academy of Science, to the membership of which the King had done me the honour of calling me. Several of that body who are still alive will remember having been present when I read it, and above the rest those amongst them who applied themselves particularly to the study of Mathematics; of whom I cannot cite more than the celebrated gentlemen Cassini, Römer, and De la Hire. And although I have since corrected and changed some parts, the copies which I had made of it at that time may serve for proof that I have yet added nothing to it save some conjectures touching the formation of Iceland Crystal, and a novel observation on the refraction of Rock Crystal. I have desired to relate these particulars to make known how long I have meditated the things which now I publish, and not for the purpose of detracting from the merit of those who, without having seen anything that I have written, may be found to have treated of

Editor's Note: A row of asterisks indicates that material has been omitted from the original article.

of like matters: as has in fact occurred to two eminent Geometricians, Messieurs Newton and Leibnitz, with respect to the Problem of the figure of glasses for collecting rays when one of the surfaces is given.

One may ask why I have so long delayed to bring this work to the light. The reason is that I wrote it rather carelessly in the Language in which it appears, with the intention of translating it into Latin, so doing in order to obtain greater attention to the thing. After which I proposed to myself to give it out along with another Treatise on Dioptrics, in which I explain the effects of Telescopes and those things which belong more to that Science. But the pleasure of novelty being past, I have put off from time to time the execution of this design, and I know not when I shall ever come to an end if it, being often turned aside either by business or by some new study. Considering which I have finally judged that it was better worth while to publish this writing, such as it is, than to let it run the risk, by waiting longer, of remaining lost.

There will be seen in it demonstrations of those kinds which do not produce as great a certitude as those of Geometry, and which even differ much therefrom, since whereas the Geometers prove their Propositions by fixed and incontestable Principles, here the Principles are verified by the conclusions to be drawn from them; the nature of these things not allowing of this being done otherwise. It is always possible to attain thereby to a degree of probability which very often is scarcely less than complete proof. To wit, when things which have been demonstrated by the Principles that have been assumed correspond perfectly to the phenomena which experiment has brought under observation; especially when there are a great number of them,

them, and further, principally, when one can imagine and foresee new phenomena which ought to follow from the hypotheses which one employs, and when one finds that therein the fact corresponds to our prevision. But if all these proofs of probability are met with in that which I propose to discuss, as it seems to me they are, this ought to be a very strong confirmation of the success of my inquiry; and it must be ill if the facts are not pretty much as I represent them. I would believe then that those who love to know the Causes of things and who are able to admire the marvels of Light, will find some satisfaction in these various speculations regarding it, and in the new explanation of its famous property which is the main foundation of the construction of our eyes and of those great inventions which extend so vastly the use of them. I hope also that there will be some who by following these beginnings will penetrate much further into this question than I have been able to do, since the subject must be far from being exhausted. This appears from the passages which I have indicated where I leave certain difficulties without having resolved them, and still more from matters which I have not touched at all, such as Luminous Bodies of several sorts, and all that concerns Colours; in which no one until now can boast of having succeeded. Finally, there remains much more to be investigated touching the nature of Light which I do not pretend to have disclosed, and I shall owe much in return to him who shall be able to supplement that which is here lacking to me in knowledge. The Hague. The 8 January 1690.

* * * * * * *

15. But I found in this refraction that the ratio of FR to RS was not constant, like the ordinary refraction, but that it varied with the varying obliquity of the incident ray.

16. I found also that when QRE made a straight line, that is, when the incident ray entered the Crystal without being refracted (as I ascertained by the circumstance that then the point E viewed by the extraordinary refraction appeared in the line CD, as seen without refraction) I found, I say, then that the angle QRG was 73 degrees 20 minutes, as has been already remarked; and so it is not the ray parallel to the edge of the Crystal, which crosses it in a straight line without being refracted, as Mr. Bartholinus believed, since that inclination is only 70 degrees 57 minutes, as was stated above. And this is to be noted, in order that no one may search in vain for the cause of the singular property of this ray in its parallelism to the edges mentioned.

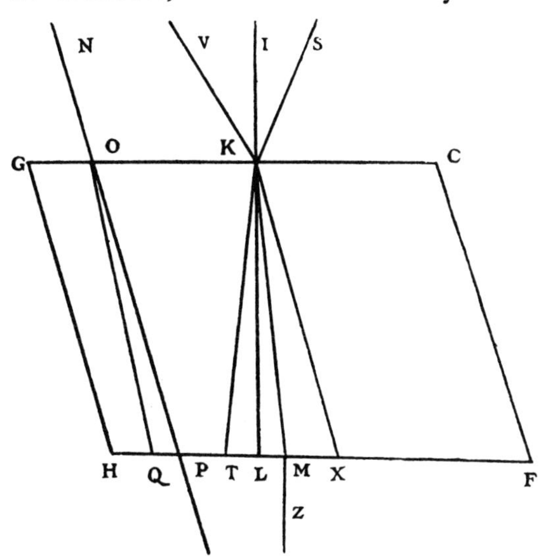

17. Finally, continuing my observations to discover the nature

nature of this refraction, I learned that it obeyed the following remarkable rule. Let the parallelogram GCFH, made by the principal section of the Crystal, as previously determined, be traced separately. I found then that always, when the inclinations of two rays which come from opposite sides, as VK, SK here, are equal, their refractions KX and KT meet the bottom line HF in such wise that points X and T are equally distant from the point M, where the refraction of the perpendicular ray IK falls; and this occurs also for refractions in other sections of this Crystal. But before speaking of those, which have also other particular properties, we will investigate the causes of the phenomena which I have already reported.

It was after having explained the refraction of ordinary transparent bodies by means of the spherical emanations of light, as above, that I resumed my examination of the nature of this Crystal, wherein I had previously been unable to discover anything.

18. As there were two different refractions, I conceived that there were also two different emanations of waves of light, and that one could occur in the ethereal matter extending through the body of the Crystal. Which matter, being present in much larger quantity than is that of the particles which compose it, was alone capable of causing transparency, according to what has been explained heretofore. I attributed to this emanation of waves the regular refraction which is observed in this stone, by supposing these waves to be ordinarily of spherical form, and having a slower progression within the Crystal than they have outside it; whence proceeds refraction as I have demonstrated.

19. As to the other emanation which should produce the

the irregular refraction, I wished to try what Elliptical waves, or rather spheroidal waves, would do; and these I supposed would spread indifferently both in the ethereal matter diffused throughout the crystal and in the particles of which it is composed, according to the last mode in which I have explained transparency. It seemed to me that the disposition or regular arrangement of these particles could contribute to form spheroidal waves (nothing more being required for this than that the successive movement of light should spread a little more quickly in one direction than in the other) and I scarcely doubted that there were in this crystal such an arrangement of equal and similar particles, because of its figure and of its angles with their determinate and invariable measure. Touching which particles, and their form and disposition, I shall, at the end of this Treatise, propound my conjectures and some experiments which confirm them.

20. The double emission of waves of light, which I had imagined, became more probable to me after I had observed a certain phenomenon in the ordinary [Rock] Crystal, which occurs in hexagonal form, and which, because of this regularity, seems also to be composed of particles, of definite figure, and ranged in order. This was, that this crystal, as well as that from Iceland, has a double refraction, though less evident. For having had cut from it some well polished Prisms of different sections, I remarked in all, in viewing through them the flame of a candle or the lead of window panes, that everything appeared double, though with images not very distant from one another. Whence I understood the reason why this substance, though so transparent, is useless for Telescopes, when they have ever so little length.

21. Now

21. Now this double refraction, according to my Theory hereinbefore established, seemed to demand a double emission of waves of light, both of them spherical (for both the refractions are regular) and those of one series a little slower only than the others. For thus the phenomenon is quite naturally explained, by postulating substances which serve as vehicle for these waves, as I have done in the case of Iceland Crystal. I had then less trouble after that in admitting two emissions of waves in one and the same body. And since it might have been objected that in composing these two kinds of crystal of equal particles of a certain figure, regularly piled, the interstices which these particles leave and which contain the ethereal matter would scarcely suffice to transmit the waves of light which I have localized there, I removed this difficulty by regarding these particles as being of a very rare texture, or rather as composed of other much smaller particles, between which the ethereal matter passes quite freely. This, moreover, necessarily follows from that which has been already demonstrated touching the small quantity of matter of which the bodies are built up.

22. Supposing then these spheroidal waves besides the spherical ones, I began to examine whether they could serve to explain the phenomena of the irregular refraction, and how by these same phenomena I could determine the figure and position of the spheroids: as to which I obtained at last the desired success, by proceeding as follows.

23. I considered first the effect of waves so formed, as respects the ray which falls perpendicularly on the flat surface of a transparent body in which they should spread in this manner. I took AB for the exposed region of the surface. And, since a ray perpendicular to a plane, and coming

coming from a very distant source of light, is nothing else, according to the precedent Theory, than the incidence of a portion of the wave parallel to that plane, I supposed the straight line RC, parallel and equal to AB, to be a portion of a wave of light, in which an infinitude of points such as RH*h*C come to meet the surface AB at the points AK*k*B. Then instead of the hemispherical partial waves which in a body of ordinary refraction would spread from each of these last points, as we have above explained in treating of refraction, these must here be hemispheroids. The axes (or rather the major diameters) of these I supposed to be oblique to the plane AB, as is AV the semi-axis or semi-major diameter of the spheroid SVT, which represents the partial wave coming from the point A, after the wave RC has reached AB. I say axis or major diameter, because the same ellipse SVT may be considered as the section of a spheroid of which the axis is AZ perpendicular to AV. But, for the present, without yet deciding one or other, we will consider these spheroids only in those sections of them which make ellipses in the plane of this figure. Now taking a certain space of time during which the wave SVT has spread from A, it would needs be that from all the other points K*k*B there should proceed, in the same time, waves similar to SVT and similarly situated. And the common tangent NQ of all these semi-ellipses would be the propagation of the wave RC which fell on AB, and would

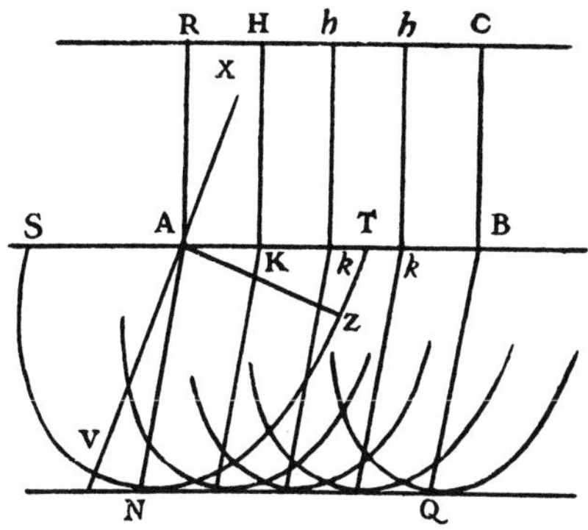

would be the place where this movement occurs in much greater amount than anywhere else, being made up of arcs of an infinity of ellipses, the centres of which are along the line AB.

24. Now it appeared that this common tangent NQ was parallel to AB, and of the same length, but that it was not directly opposite to it, since it was comprised between the lines AN, BQ, which are diameters of ellipses having A and B for centres, conjugate with respect to diameters which are not in the straight line AB. And in this way I comprehended, a matter which had seemed to me very difficult, how a ray perpendicular to a surface could suffer refraction on entering a transparent body; seeing that the wave RC, having come to the aperture AB, went on forward thence, spreading between the parallel lines AN, BQ, yet itself remaining always parallel to AB, so that here the light does not spread along lines perpendicular to its waves, as in ordinary refraction, but along lines cutting the waves obliquely.

25. Inquiring subsequently what might be the position and form of these spheroids in the crystal, I considered that all the six faces produced precisely the same refractions. Taking, then, the parallelopiped AFB, of which the obtuse solid angle C is contained between the three equal plane angles, and imagining in it the three principal sections, one of which is perpendicular to the face DC and passes through the edge CF, another perpendicular to the face BF passing through the edge CA,

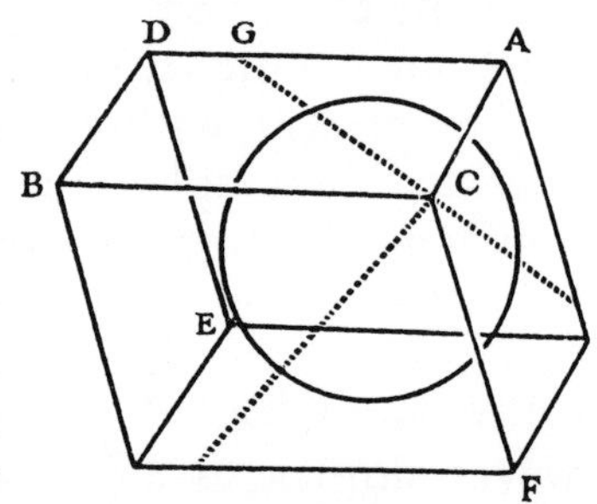

CA, and the third perpendicular to the face AF passing through the edge BC; I knew that the refractions of the incident rays belonging to these three planes were all similar. But there could be no position of the spheroid which would have the same relation to these three sections except that in which the axis was also the axis of the solid angle C. Consequently I saw that the axis of this angle, that is to say the straight line which traversed the crystal from the point C with equal inclination to the edges CF, CA, CB was the line which determined the position of the axis of all the spheroidal waves which onc imagined to originate from some point, taken within or on the surface of the crystal, since all these spheroids ought to be alike, and have their axes parallel to one another.

26. Considering after this the plane of one of these three sections, namely that through GCF, the angle of which is 109 degrees 3 minutes, since the angle F was shown above to be 70 degrees 57 minutes; and, imagining a spheroidal wave about the centre C, I knew, because I have just explained it, that its axis must be in the same plane, the half of which axis I have marked CS in the next figure: and seeking by calculation (which will be given with others at the end of this discourse) the value of the angle CGS, I found it 45 degrees 20 minutes.

27. To know from this the form of this spheroid, that is to say the proportion of the semi-diameters CS, CP, of its elliptical section, which are perpendicular to one another, I considered that the point M where the ellipse is touched by the straight line FH, parallel to CG, ought to be so situated that CM makes with the perpendicular CL an angle of 6 degrees 40 minutes; since, this being so, this ellipse satisfies what has been said about the refraction of the

the ray perpendicular to the surface CG, which is inclined to the perpendicular CL by the same angle. This, then, being thus disposed, and taking CM at 100,000 parts, I found by the calculation which will be given at the end, the semi-major diameter CP to be 105,032, and the semi-axis CS to be 93,410, the ratio of which numbers is very nearly 9 to 8; so that the spheroid was of the kind which resembles a compressed sphere, being generated by the revolution of an ellipse about its smaller diameter. I found also the value of CG the semi-diameter parallel to the tangent ML to be 98,779.

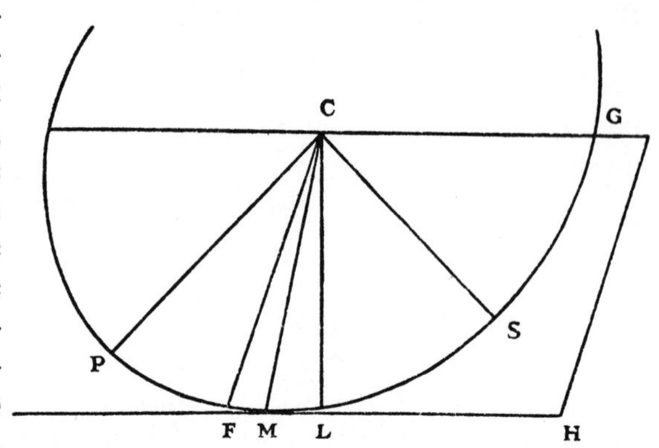

28. Now passing to the investigation of the refractions which obliquely incident rays must undergo, according to our hypothesis of spheroidal waves, I saw that these refractions depended on the ratio between the velocity of movement of the light outside the crystal in the ether, and that within the crystal. For supposing, for example, this proportion to be such that while the light in the crystal forms the spheroid GSP, as I have just said, it forms outside a sphere the semi-diameter of which is equal to the line N which will be determined hereafter, the following is the way of finding the refraction of the incident rays. Let there be such a ray RC falling upon the surface

surface CK. Make CO perpendicular to RC, and across the angle KCO adjust OK, equal to N and perpendicular to CO; then draw KI, which touches the Ellipse GSP, and from the point of contact I join IC, which will be the required refraction of the ray RC. The demonstration of this is, it will be seen, entirely similar to that of which we made use in explaining ordinary refraction. For the

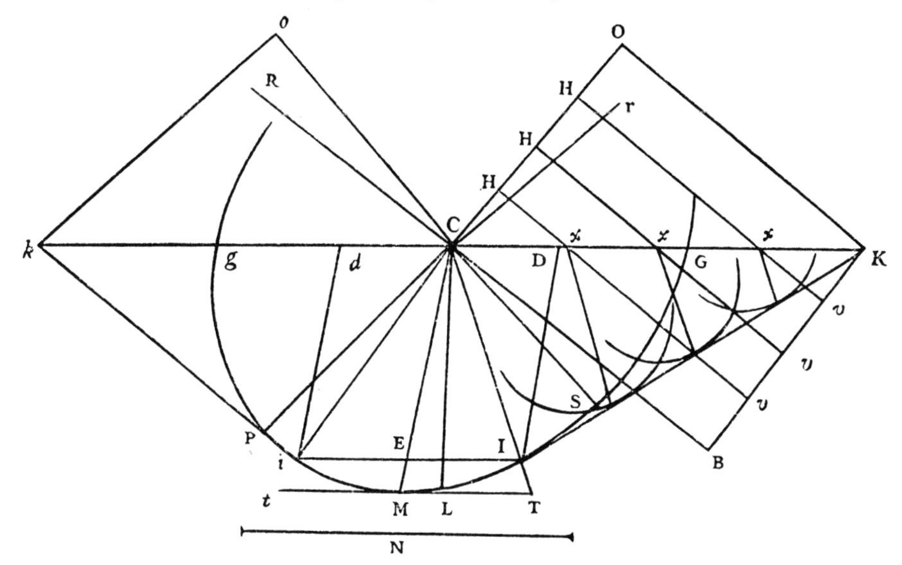

refraction of the ray RC is nothing else than the progression of the portion C of the wave CO, continued in the crystal. Now the portions H of this wave, during the time that O came to K, will have arrived at the surface CK along the straight lines H*x*, and will moreover have produced in the crystal around the centres *x* some hemi-spheroidal partial waves similar to the hemi-spheroidal GSP*g*, and similarly disposed, and of which the major and

and minor diameters will bear the same proportions to the lines *xv* (the continuations of the lines H*x* up to KB parallel to CO) that the diameters of the spheroid GSP*g* bear to the line CB, or N. And it is quite easy to see that the common tangent of all these spheroids, which are here represented by Ellipses, will be the straight line IK, which consequently will be the propagation of the wave CO; and the point I will be that of the point C, conformably with that which has been demonstrated in ordinary refraction.

Now as to finding the point of contact I, it is known that one must find CD a third proportional to the lines CK, CG, and draw DI parallel to CM, previously determined, which is the conjugate diameter to CG; for then, by drawing KI it touches the Ellipse at I.

29. Now as we have found CI the refraction of the ray RC, similarly one will find C*i* the refraction of the ray *r*C, which comes from the opposite side, by making C*o* perpendicular to *r*C and following out the rest of the construction as before. Whence one sees that if the ray *r*C is inclined equally with RC, the line C*d* will necessarily be equal to CD, because C*k* is equal to CK, and C*g* to CG. And in consequence I*i* will be cut at E into equal parts by the line CM, to which DI and *di* are parallel. And because CM is the conjugate diameter to CG, it follows that *i*I will be parallel to *g*G. Therefore if one prolongs the refracted rays CI, C*i*, until they meet the tangent ML at T and *t*, the distances MT, M*t*, will also be equal. And so, by our hypothesis, we explain perfectly the phenomenon mentioned above; to wit, that when there are two rays equally inclined, but coming from opposite sides, as here the rays RC, *rc*, their refractions diverge equally from the line followed

followed by the refraction of the ray perpendicular to the surface, by considering these divergences in the direction parallel to the surface of the crystal.

30. To find the length of the line N, in proportion to CP, CS, CG, it must be determined by observations of the irregular refraction which occurs in this section of the crystal; and I find thus that the ratio of N to GC is just a little less than 8 to 5. And having regard to some other observations and phenomena of which I shall speak afterwards, I put N at 156,962 parts, of which the semi-diameter CG is found to contain 98,779, making this ratio 8 to $5\frac{1}{29}$. Now this proportion, which there is between the line N and CG, may be called the Proportion of the Refraction; similarly as in glass that of 3 to 2, as will be manifest when I shall have explained a short process in the preceding way to find the irregular refractions.

31. Supposing then, in the next figure, as previously, the surface of the crystal *g*G, the Ellipse GP*g*, and the line N; and CM the refraction of the perpendicular ray FC, from which it diverges by 6 degrees 40 minutes. Now let there be some other ray RC, the refraction of which must be found.

About the centre C, with semi-diameter CG, let the circumference *g*RG be described, cutting the ray RC at R; and let RV be the perpendicular on CG. Then as the line N is to CG let CV be to CD, and let DI be drawn parallel to CM, cutting the Ellipse *g*MG at I; then joining CI, this will be the required refraction of the ray RC. Which is demonstrated thus.

Let CO be perpendicular to CR, and across the angle OCG let OK be adjusted, equal to N and perpendicular to CO, and let there be drawn the straight line KI, which if it is is demonstrated to be a tangent to the Ellipse at I, it will be evident by the things heretofore explained that CI is the refraction of the ray RC. Now since the angle RCO is a right angle, it is easy to see that the right-angled triangles RCV, KCO, are similar. As then, CK is to KO, so also

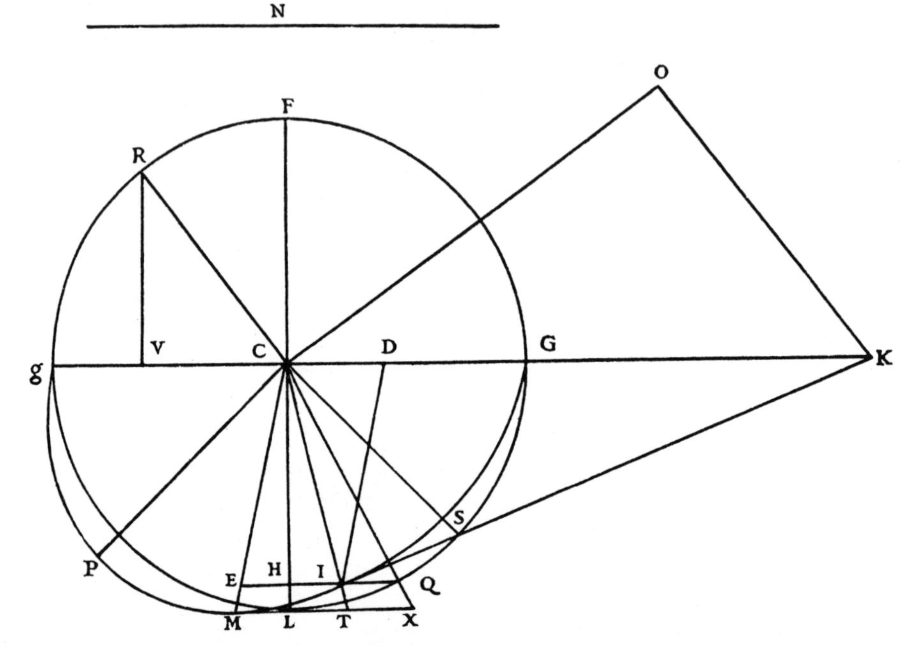

is RC to CV. But KO is equal to N, and RC to CG: then as CK is to N so will CG be to CV. But as N is to CG, so, by construction, is CV to CD. Then as CK is to CG so is CG to CD. And because DI is parallel to CM, the conjugate diameter to CG, it follows that KI touches the Ellipse at I; which remained to be shown.

32. One sees then that as there is in the refraction of ordinary

ordinary media a certain constant proportion between the sines of the angles which the incident ray and the refracted ray make with the perpendicular, so here there is such a proportion between CV and CD or IE; that is to say between the Sine of the angle which the incident ray makes with the perpendicular, and the horizontal intercept, in the Ellipse, between the refraction of this ray and the diameter CM. For the ratio of CV to CD is, as has been said, the same as that of N to the semi-diameter CG.

33. I will add here, before passing away, that in comparing together the regular and irregular refraction of this crystal, there is this remarkable fact, that if ABPS be the spheroid by which light spreads in the Crystal in a certain space of time (which spreading, as has been said, serves for the irregular refraction), then the inscribed sphere BVST is the extension in the same space of time of the light which serves for the regular refraction.

For we have stated before this, that the line N being the radius of a spherical wave of light in air, while in the crystal it spread through the spheroid ABPS, the ratio of N to CS will be 156,962 to 93,410. But it has also been stated that the proportion of the regular refraction was 5 to 3; that is to say, that N being the radius of a spherical wave of light in air, its extension in the crystal would, in the same space of time, form a sphere the radius of which would be to N as 3 to 5. Now 156,962 is to 93,410 as 5 to 3 less $\frac{1}{41}$. So that it is sufficiently nearly, and may be

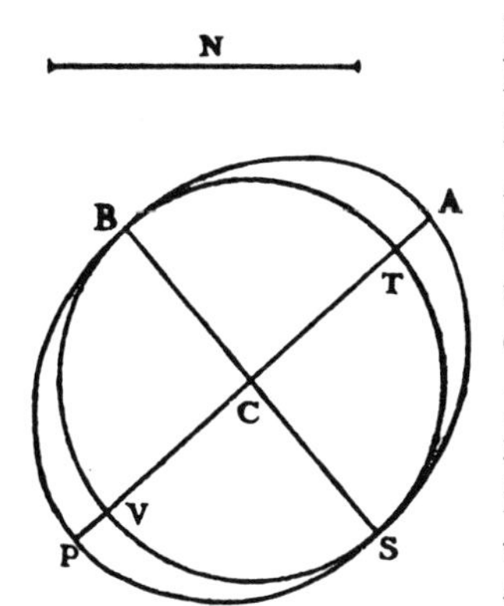

be exactly, the sphere BVST, which the light describes for the regular refraction in the crystal, while it describes the spheroid BPSA for the irregular refraction, and while it describes the sphere of radius N in air outside the crystal.

Although then there are, according to what we have supposed, two different propagations of light within the crystal, it appears that it is only in directions perpendicular to the axis BS of the spheroid that one of these propagations occurs more rapidly than the other; but that they have an equal velocity in the other direction, namely, in that parallel to the same axis BS, which is also the axis of the obtuse angle of the crystal.

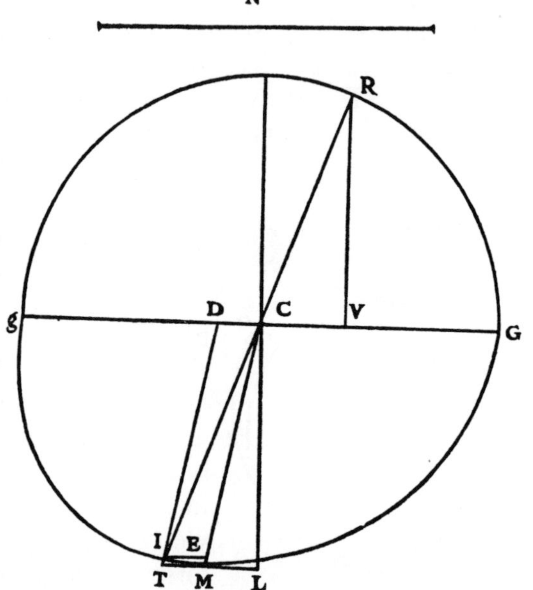

34. The proportion of the refraction being what we have just seen, I will now show that there necessarily follows thence that notable property of the ray which falling obliquely on the surface of the crystal enters it without suffering refraction. For supposing the same things as before, and that the ray RC makes with the same surface *g*G the angle RCG of 73 degrees

73 degrees 20 minutes, inclining to the same side as the crystal (of which ray mention has been made above); if one investigates, by the process above explained, the refraction CI, one will find that it makes exactly a straight line with RC, and that thus this ray is not deviated at all, conformably with experiment. This is proved as follows by calculation.

CG or CR being, as precedently, 98,779; CM being 100,000; and the angle RCV 73 degrees 20 minutes, CV will be 28,330. But because CI is the refraction of the ray RC, the proportion of CV to CD is 156,962 to 98,779, namely, that of N to CG; then CD is 17,828.

Now the rectangle *g*DC is to the square of DI as the square of CG is to the square of CM; hence DI or CE will be 98,353. But as CE is to EI, so will CM be to MT, which will then be 18,127. And being added to ML, which is 11,609 (namely the sine of the angle LCM, which is 6 degrees 40 minutes, taking CM 100,000 as radius) we get LT 27,936; and this is to LC 99,324 as CV to VR, that is to say, as 29,938, the tangent of the complement of the angle RCV, which is 73 degrees 20 minutes, is to the radius of the Tables. Whence it appears that RCIT is a straight line; which was to be proved.

35. Further it will be seen that the ray CI in emerging through the opposite surface of the crystal, ought to pass out quite straight, according to the following demonstration, which proves that the reciprocal relation of refraction obtains in this crystal the same as in other transparent bodies; that is to say, that if a ray RC in meeting the surface of the crystal CG is refracted as CI, the ray CI emerging through the opposite parallel surface of the crystal

crystal, which I suppose to be IB, will have its refraction IA parallel to the ray RC.

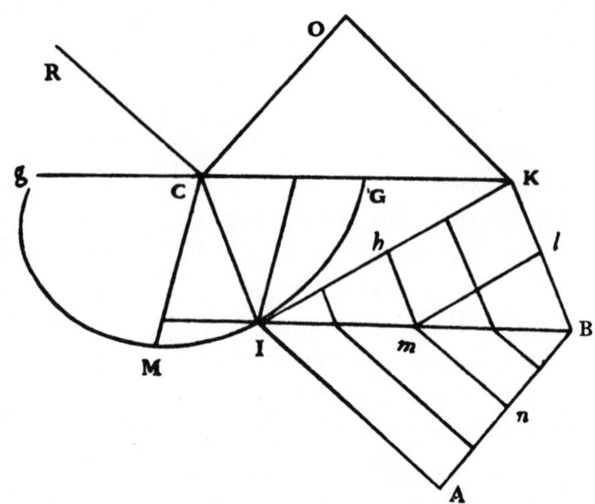

Let the same things be supposed as before; that is to say, let CO, perpendicular to CR, represent a portion of a wave the continuation of which in the crystal is IK, so that the piece C will be continued on along the straight line CI, while O comes to K. Now if one takes a second period of time equal to the first, the piece K of the wave IK will, in this second period, have advanced along the straight line KB, equal and parallel to CI, because every piece of the wave CO, on arriving at the surface CK, ought to go on in the crystal the same as the piece C; and in this same time there will be formed in the air from the point I a partial spherical wave having a semi-diameter IA equal to KO, since KO has been traversed in an equal time. Similarly, if one considers some other point of the wave IK, such as *h*, it will go along *hm*, parallel to CI, to meet the surface IB, while the point K traverses K*l* equal to *hm*; and while this accomplishes the remainder *l*B, there will start from the point *m* a partial wave the semi-diameter of which, *mn*, will have the same ratio to *l*B as IA to KB.

KB. Whence it is evident that this wave of semi-diameter *mn*, and the other of semi-diameter IA will have the same tangent BA. And similarly for all the partial spherical waves which will be formed outside the crystal by the impact of all the points of the wave IK against the surface of the Ether IB. It is then precisely the tangent BA which will be the continuation of the wave IK, outside the crystal, when the piece K has reached B. And in consequence IA, which is perpendicular to BA, will be the refraction of the ray CI on emerging from the crystal. Now it is clear that IA is parallel to the incident ray RC, since IB is equal to CK, and IA equal to KO, and the angles A and O are right angles.

It is seen then that, according to our hypothesis, the reciprocal relation of refraction holds good in this crystal as well as in ordinary transparent bodies; as is thus in fact found by observation.

* * * * * * *

Before finishing the treatise on this Crystal, I will add one more marvellous phenomenon which I discovered after having written all the foregoing. For though I have not been able till now to find its cause, I do not for that reason wish to desist from describing it, in order to give opportunity to others to investigate it. It seems that it will be necessary to make still further suppositions besides those which I have made; but these will not for all that cease to keep their probability after having been confirmed by so many tests.

The phenomenon is, that by taking two pieces of this crystal and applying them one over the other, or rather holding them with a space between the two, if all the sides of one are parallel to those of the other, then a ray of light, such as AB, is divided into two in the first piece, namely into BD and BC, following the two refractions, regular

regular and irregular. On penetrating thence into the other piece each ray will pass there without further divid-

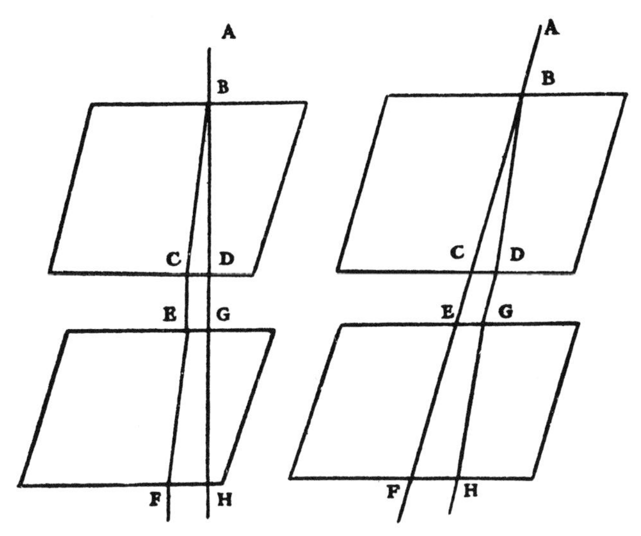

ing itself in two; but that one which underwent the regular refraction, as here DG, will undergo again only a regular refraction at GH; and the other, CE, an irregular refraction at EF. And the same thing occurs not only in this disposition, but also in all those cases in which the principal section of each of the pieces is situated in one and the same plane, without it being needful for the two neighbouring surfaces to be parallel. Now it is marvellous why the rays CE and DG, incident from the air on the lower crystal, do not divide themselves the same as the first ray AB. One would say that it must be that the ray DG in passing through the upper piece has lost something which is necessary to move the matter which serves for the irregular refraction; and that likewise CE has lost that which was

was necessary to move the matter which serves for regular refraction: but there is yet another thing which upsets this reasoning. It is that when one disposes the two crystals in such a way that the planes which constitute the principal sections intersect one another at right angles, whether the neighbouring surfaces are parallel or not, then the ray which has come by the regular refraction, as DG, undergoes only an irregular refraction in the lower piece; and on the contrary the ray which has come by the irregular refraction, as CE, undergoes only a regular refraction.

But in all the infinite other positions, besides those which I have just stated, the rays DG, CE, divide themselves anew each one into two, by refraction in the lower crystal, so that from the single ray AB there are four, sometimes of equal brightness, sometimes some much less bright than others, according to the varying agreement in the positions of the crystals: but they do not appear to have all together more light than the single ray AB.

When one considers here how, while the rays CE, DG, remain the same, it depends on the position that one gives to the lower piece, whether it divides them both in two, or whether it does not divide them, and yet how the ray AB above is always divided, it seems that one is obliged to conclude that the waves of light, after having passed through the first crystal, acquire a certain form or disposition in virtue of which, when meeting the texture of the second crystal, in certain positions, they can move the two different kinds of matter which serve for the two species of refraction; and when meeting the second crystal in another position are able to move only one of these kinds of matter. But to tell how this occurs, I have hitherto found nothing which satisfies me.

3

Reprinted from *Opticks,* 4th ed., London, 1730, pp. 354–361

OPTICKS

ISAAC NEWTON

Qu. 25. Are there not other original Properties of the Rays of Light, besides those already described? An instance of another original Property we have in the Refraction of Island Crystal, described first by *Erasmus Bartholine*, and afterwards more exactly by *Hugenius*, in his Book *De la Lumiere*. This Crystal is a pellucid fissile Stone, clear as Water or Crystal of the Rock, and without Colour; enduring a red Heat without losing its transparency, and in a very strong Heat calcining without Fusion. Steep'd a Day or two in Water, it loses its natural Polish. Being rubb'd on Cloth, it attracts pieces of Straws and other light things, like Ambar or Glass; and with *Aqua fortis* it makes an Ebullition. It seems to be a sort of Talk, and is found in form of an oblique Parallelopiped, with six parallelogram Sides and eight solid Angles.

The obtuse Angles of the Parallelograms are each of them 101 Degrees and 52 Minutes; the acute ones 78 Degrees and 8 Minutes. Two of the solid Angles opposite to one another, as C and E, are compassed each of them with three of these obtuse Angles, and each of the other six with one obtuse and two acute ones. It cleaves easily in planes parallel to any of its Sides, and not in any other Planes. It cleaves with a glossy polite Surface not perfectly plane, but with some little unevenness. It is easily scratch'd, and by reason of its softness it takes a Polish very difficultly. It polishes better upon polish'd Looking-glass than upon Metal, and perhaps better upon Pitch, Leather or Parchment. Afterwards it must be rubb'd with a little Oil or white of an Egg, to fill up its Scratches; whereby it will become very transparent and polite. But for several Experiments, it is not necessary to polish it. If a piece of this crystalline Stone be laid upon a Book, every Letter of the Book seen through it will appear double, by means of a double Refraction. And if any beam of Light falls either perpendicularly, or in any oblique Angle upon any Surface of this Crystal, it becomes divided into two beams by means of the same double Refraction. Which beams are of the same Colour with the incident beam of Light, and seem equal to one another in the quantity of their Light, or very nearly equal. One of these Refractions is perform'd by the usual Rule of Opticks, the Sine of Incidence out of Air into this Crystal being to the Sine of Refraction, as five to three. The

See the following Scheme, p. 356.

other Refraction, which may be called the unusual Refraction, is perform'd by the following Rule.

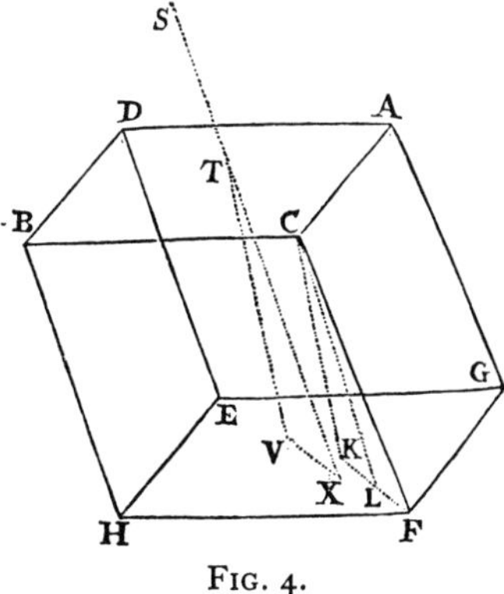

FIG. 4.

Let ADBC represent the refracting Surface of the Crystal, C the biggest solid Angle at that Surface, GEHF the opposite Surface, and CK a perpendicular on that Surface. This perpendicular makes with the edge of the Crystal CF, an Angle of 19 Degr. 3′. Join KF, and in it take KL, so that the Angle KCL be 6 Degr. 40′. and the Angle LCF 12 Degr. 23′. And if ST represent any beam of Light incident at T in any Angle upon the refracting Surface ADBC, let TV be the refracted beam determin'd by the given Portion of the Sines 5 to 3, according to the usual Rule of Opticks. Draw VX parallel and equal to KL. Draw it the same way from V in which L lieth from K; and joining TX, this line TX shall be the other refracted beam carried from T to X, by the unusual Refraction.

If therefore the incident beam ST be perpendicular to the refracting Surface, the two beams TV and TX, into which it shall become divided, shall be parallel to the lines CK and CL; one of those beams going through the Crystal perpendicularly, as it ought to do by the usual Laws of Opticks, and the other TX by an unusual Refraction diverging from the perpendicular, and making with it an Angle VTX of about $6\frac{2}{3}$ Degrees, as is found by Experience. And hence, the Plane VTX, and such like Planes which are parallel to the Plane CFK, may be called the Planes of perpendicular Refraction. And the Coast towards which the lines KL and VX are drawn, may be call'd the Coast of unusual Refraction.

In like manner Crystal of the Rock has a double Refraction: But the difference of the two Refractions is not so great and manifest as in Island Crystal.

When the beam ST incident on Island Crystal is divided into two beams TV and TX, and these two beams arrive at the farther Surface of the Glass; the beam TV, which was refracted at the first Surface after the usual manner, shall be again refracted entirely after the usual manner at the second Surface; and the beam TX, which was refracted after the unusual manner in the first Surface, shall be again refracted entirely after the unusual manner in the second Surface; so that both these beams shall emerge out of the second Surface in lines parallel to the first incident beam ST.

And if two pieces of Island Crystal be placed one after another, in such manner that all the Surfaces of the latter be parallel to all the corresponding Surfaces of the former: The Rays which are refracted after the usual manner in the first Surface of the first Crystal, shall be refracted after the usual manner in all the following Surfaces; and the Rays which are refracted after the unusual manner in the first Surface, shall be refracted after the unusual manner in all the following Surfaces. And the same thing happens, though the Surfaces of the Crystals be any ways inclined to one another, provided that their Planes of perpendicular Refraction be parallel to one another.

And therefore there is an original difference in the Rays of Light, by means of which some Rays are in this Experiment constantly refracted after the usual manner, and others constantly after the unusual manner: For if the difference be not original, but arises from new Modifications impress'd on the Rays at their first Refraction, it would be alter'd by new Modifications in the three following Refractions; whereas it suffers no alteration, but is constant, and has the same effect upon the Rays in all the Refractions. The unusual Refraction is therefore perform'd by an original property of the Rays. And it remains to be enquired, whether the Rays have not more original Properties than are yet discover'd.

Qu. 26. Have not the Rays of Light several sides, endued with several original Properties? For if the Planes of perpendicular Refraction of the second Crystal be at right Angles with the Planes of perpendicular Refraction of the first Crystal, the Rays which are refracted after the usual manner in passing through the first Crystal, will be all of them refracted after the unusual manner in passing through the second Crystal; and the Rays which are refracted after the unusual manner in passing through the first Crystal, will be all of them refracted after the usual manner in passing through the second Crystal. And therefore there are not two sorts of Rays differing in their nature from one another, one of which is constantly and in all Positions refracted after the usual manner, and the other constantly and in all Positions after the unusual manner. The difference between the two sorts of Rays in the Experiment mention'd in the 25th Question, was only in the Positions of the Sides of the Rays to the Planes of perpendicular Refraction. For one and the same Ray is here refracted sometimes after the usual, and sometimes after the unusual manner, according to the Position which its Sides have to the Crystals. If the Sides of the Ray are posited the same way to both Crystals, it is refracted after the same manner in them both: But if that side of the Ray which looks towards the Coast of the unusual Refraction of the first Crystal, be 90 Degrees from that side of the same Ray which looks toward the Coast of the unusual Refraction of the second Crystal, (which may be effected by varying the Position of the second Crystal to the first, and by consequence to the Rays of Light,) the Ray shall be refracted after several manners in the several Crystals. There is nothing more required to determine whether the Rays of Light which fall upon the second Crystal shall be refracted after the usual or after the unusual manner, but to turn about this Crystal, so that the Coast of this Crystal's unusual Refraction may be on this or on that side of the Ray. And therefore every Ray may be consider'd as having four Sides or Quarters, two of which opposite to one another incline the Ray to be refracted after the unusual manner, as often as either of them are turn'd towards the Coast of unusual Refraction; and the other two, whenever either of them are turn'd towards the Coast of unusual Refraction, do not incline it to be otherwise refracted than after the usual manner. The two first may therefore be call'd the Sides of unusual Refraction. And since these Dispositions were in the Rays before their Incidence on the second, third, and fourth Surfaces of the two Crystals, and suffered no alteration (so far as appears,) by the Refraction of the Rays in their passage through those Surfaces, and the Rays were refracted by the same Laws in all the four Surfaces; it appears that those Dispositions were in the Rays originally, and suffer'd no alteration by the first Refraction, and that by means of those Dispositions the Rays were refracted at their Incidence on the first Surface of the first Crystal, some of them after the usual, and some of them after the unusual manner, accordingly as their Sides of unusual Refraction were then turn'd towards the Coast of the unusual Refraction of that Crystal, or sideways from it.

Every Ray of Light has therefore two opposite Sides, originally endued with a Property on which the unusual Refraction depends, and the other two opposite Sides not endued with that Property. And it remains to be enquired, whether there are not more Properties of Light by which the Sides of the Rays differ, and are distinguished from one another.

In explaining the difference of the Sides of the Rays above mention'd, I have supposed that the Rays fall perpendicularly on the first Crystal. But if they fall obliquely on it, the Success is the same. Those Rays which are refracted after the usual manner in the first Crystal, will be refracted after the unusual manner in the second Crystal, supposing the Planes of perpendicular Refraction to be at right Angles with one another, as above; and on the contrary.

If the Planes of the perpendicular Refraction of the two Crystals be neither parallel nor perpendicular to one another, but contain an acute Angle: The two beams of Light which emerge out of the first Crystal, will be each of them divided into two more at their Incidence on the second Crystal. For in this case the Rays in each of the two beams will some of them have their Sides of unusual Refraction, and some of them their other Sides turn'd towards the Coast of the unusual Refraction of the second Crystal.

4

Reprinted from *Mém. Phys. Chim. Soc. D'Arcueil,* **2,** 143–158 (1809)

*Sur une propriété de la lumière réfléchie**

PAR M. MALUS.

LORSQU'UN rayon solaire est réfléchi ou réfracté, il conserve en général ses propriétés physiques; et, soumis à de nouvelles épreuves, il se comporte comme s'il émanoit directement du corps lumineux : le prisme, en dispersant les rayons colorés, ne fait que changer leur direction respective, sans altérer leur nature. Il y a cependant des circonstances où l'influence de certains corps imprime aux rayons qu'ils réfléchissent, ou qu'ils réfractent, des caractères et des propriétés qu'ils transportent avec eux, et qui les distinguent essentiellement de la lumière directe.

La propriété de la lumière que je vais décrire est une modification de ce genre. Elle avoit déja été apperçue dans une circonstance particulière de la duplication des images, offerte par le spath calcaire; mais le phénomène qui en résultoit étant attribué aux propriétés de ce cristal,

*Editor's Note: An English translation follows this article.

on ne soupconnoit pas qu'il pût être produit non-seulement par tous les corps cristallisés qui donnent une double réfraction, mais encore par toutes les autres substances diaphanes solides ou liquides et même par les corps opaques.

Si on reçoit un rayon de lumière perpendiculairement à la face d'un rhomboïde de spath calcaire, ce rayon se divise en deux faisceaux, l'un qui se prolonge dans la direction du rayon incident, et l'autre qui fait avec celui-ci un angle de quelques degrés. Le plan qui passe par ces deux rayons jouit de plusieurs propriétés particulières, et on le nomme plan de la section principale. Il est toujours parallèle à l'axe des molécules intégrantes du cristal et perpendiculaire à la face réfringente naturelle ou artificielle. Lorsque le rayon incident est incliné à la face réfringente il se divise également en deux faisceaux, l'un qui est réfracté suivant la loi ordinaire, et l'autre suivant une loi extraordinaire qui dépend des angles que le rayon incident forme avec la surface réfringente et la section principale. Lorsque la face d'émergence est parallèle à la face d'incidence, les deux rayons émergens sont parallèles au rayon incident, parce que chaque rayon éprouve aux deux faces opposées le même genre de réfraction.

Si actuellement on reçoit sur un second rhomboïde dont la section principale soit parallèle à celle du premier, les deux rayons qui ont déja traversé celui-ci, ils ne seront plus divisés en deux faisceaux comme l'eussent été des rayons de lumière directe. Le faisceau provenant de la réfraction ordinaire du premier cristal sera réfracté par le second suivant la loi de la réfraction ordinaire, comme si celui-ci avoit perdu la faculté de doubler les images. De même le faisceau provenant de la réfraction extraordinaire du premier cristal sera réfracté par le second suivant la loi de la réfraction extraordinaire.

Si le premier cristal restant immobile on fait tourner le second de manière que la face d'incidence reste parallèle à elle-même, chacun des deux rayons provenant de la réfraction du premier cristal commence à se diviser en deux faisceaux; en sorte, par exemple, qu'une partie du rayon provenant de la réfraction ordinaire, commence à se réfracter extraordinairement, ce qui produit quatre images. Enfin, après un quart de révolution le faisceau provenant de la réfraction ordinaire du premier cristal est en entier réfracté extraordinairement par le second; et réciproquement, le faisceau provenant de la réfraction extraordinaire du premier cristal est en entier réfracté par le second suivant la loi ordinaire, ce qui réduit de nouveau à deux le nombre des images. Ce phénomène est indépendant des angles d'incidence, puisque dans le mouvement du second cristal les faces réfringentes des deux rhomboïdes conservent entre elles la même inclinaison.

Ainsi le caractère qui distingue la lumière directe de celle qui a été soumise à l'action d'un premier cristal, c'est que l'une a constamment la faculté d'être divisée en deux faisceaux, tandis que dans l'autre cette faculté dépend de l'angle compris entre le plan d'incidence et celui de la section principale.

Cette faculté de changer le caractère de la lumière et de lui imprimer une nouvelle propriété qu'elle transporte avec elle, n'est pas particulière au spath d'Islande; je l'ai retrouvée dans toutes les substances connues qui doublent les images, et ce qu'il y a de remarquable dans ce phénomène c'est qu'il n'est pas nécessaire pour le produire d'employer deux cristaux d'une même espèce. Ainsi le second cristal, par exemple, pourroit être de carbonate de plomb ou de sulfate de barite : le premier pourroit être un cristal de soufre et le second un cristal de roche Toutes ces substances se comportent entre elles de la même manière que deux rhomboïdes de spath calcaire. En général cette disposition de la lumière à se réfracter en deux faisceaux ou en un seul, ne dépend que de la position respective de l'axe des molécules intégrantes des cristaux qu'on emploie, quels que soient d'ailleurs leurs principes chimiques et les faces naturelles ou artificielles sur lesquelles s'opère la réfraction. Ce résultat prouve que la modification que la lumière reçoit de ces différens corps est parfaitement identique.

Pour rendre plus sensibles les phénomènes que je viens de décrire, on peut regarder la flamme d'une bougie à travers deux prismes de matières différentes donnant la double réfraction et posés l'un sur l'autre. On aura en général quatre images de la flamme ; mais si on fait tourner lentement un des prismes, autour du rayon visuel comme axe, les quatre images se réduiront à deux, toutes les fois que les sections principales des faces contigues seront parallèles ou rectangulaires. Les deux images qui disparoissent ne se confondent pas avec les deux autres, on les voit s'éteindre peu-à-peu tandis que les autres augmentent d'intensité. Lorsque les deux sections principales sont parallèles, une des images est formée par des rayons réfractés ordinairement par les deux prismes, et la seconde par des rayons réfractés extraordinairement. Lorsque les deux sections principales sont rectangulaires, une des images est formée par des rayons réfractés ordinairement par le premier cristal, et extraordinairement par le second, et l'autre image par des rayons réfractés extraordinairement par le premier cristal et ordinairement par le second.

Non-seulement tous les cristaux qui doublent les images peuvent donner à la lumière cette faculté d'être réfractée en deux faisceaux ou en un seul, suivant la position du cristal réfringent, mais tous les corps diaphanes solides ou liquides, et les corps opaques eux-mêmes, peuvent imprimer aux molécules lumineuses cette singulière disposition qui sembloit être un des effets de la double réfraction.

Lorsqu'un faisceau de lumière traverse une substance diaphane, une partie des rayons est réfléchie par la surface réfringente, et une autre partie par la surface d'émergence. La cause de cette réflexion partielle qui a jusqu'ici échappé aux recherches des physiciens, semble avoir, dans plusieurs circonstances, quelque analogie avec les forces qui produisent la double réfraction.

Par exemple, la lumière réfléchie par la surface de l'eau sous un angle de 52°. 45′ avec la verticale, a tous les caractères d'un des faisceaux produits par la double réfraction d'un cristal de spath calcaire dont la section principale seroit parallèle ou perpendiculaire au plan qui passe par le rayon incident et le rayon réfléchi que nous nommerons plan de réflexion.

Si on reçoit ce rayon réfléchi sur un cristal quelconque, ayant la propriété de doubler les images, et dont la section principale soit parallèle au plan de réflexion, il ne sera pas divisé en deux faisceaux comme l'eût été un rayon de lumière directe, mais il sera réfracté tout entier suivant la loi ordinaire, comme si ce cristal avoit perdu la faculté de doubler les images. Si, au contraire, la section principale du cristal est perpendiculaire au plan de réflexion, le rayon réfléchi sera réfracté tout entier suivant la loi extraordinaire. Dans les positions intermédiaires il sera divisé en deux faisceaux suivant la même loi et dans la même proportion que s'il avoit acquis son nouveau caractère par l'influence de la double réfraction. Le rayon réfléchi par la surface du liquide a donc, dans cette circonstance, tous les caractères d'un rayon ordinaire formé par un cristal dont la section principale seroit perpendiculaire au plan de réflexion.

Pour analyser complètement ce phénomène j'ai disposé verticalement la section principale d'un cristal, et après avoir divisé un rayon lumineux, à l'aide de la double réfraction, j'ai reçu les deux faisceaux qui en provenoient sur la surface de l'eau et sous l'angle de 52°. 45'. Le rayon ordinaire, en se réfractant, a abandonné à la réflexion partielle une partie de ses molécules comme l'eût fait un faisceau de lumière directe, mais le rayon extraordinaire a pénétré en entier le liquide; aucune de ses molécules n'a échappé à la réfraction. Au contraire, quand la section principale du cristal étoit perpendiculaire au plan d'incidence, le rayon extraordinaire produisoit seul une réflexion partielle, et le rayon ordinaire étoit réfracté en entier.

L'angle sous lequel la lumière éprouve cette

modification en se réfléchissant à la surface des corps diaphanes, est variable pour chacun d'eux, il est, en général, plus grand pour les corps qui réfractent davantage la lumière. Au-delà et en deça de cet angle une partie du rayon est plus ou moins modifiée, et d'une manière analogue à ce qui se passe entre deux cristaux dont les sections principales cessent d'être parallèles ou rectangulaires.

Lorsqu'on veut simplement prendre connoissance de ce phénomène sans le mesurer avec exactitude, il faut placer en avant d'une bougie ou le corps diaphane ou le vase contenant le liquide qu'on veut soumettre à l'expérience. On examine à travers un prisme de cristal l'image de la flamme réfléchie à la surface du corps ou du liquide, on voit généralement deux images; mais en tournant le cristal autour du rayon visuel comme axe, on s'apperçoit qu'une des images s'affoiblit à mesure que l'autre augmente d'intensité. Au-delà d'une certaine limite, l'image qui s'étoit affoiblie recommence à augmenter d'intensité aux dépens de la seconde. Il faut saisir à-peu-près le point où l'intensité de lumière est au *minimum*, et rapprocher ou éloigner de la bougie le corps réfléchissant, jusqu'à ce que l'angle d'incidence soit tel qu'une des deux images disparoisse totalement ; cette distance déterminée, si on continue à faire tourner lentement le cristal, on s'appercevra qu'une des deux images s'éteindra alternativement à chaque quart de révolution.

Le phénomène que nous avons remarqué dans les rayons qui se réfléchissent sous un certain angle à la surface d'un corps diaphane, a lieu aussi sous un autre angle dans les faisceaux réfléchis intérieurement par la surface d'émergence, et le sinus du premier angle est au sinus du second, dans le même rapport que les sinus d'incidence et de réfraction ; ainsi, en supposant la face d'incidence et la face d'émergence parallèles, et l'angle d'incidence tel que le rayon réfléchi à la première surface présente le phénomène que nous avons décrit, le rayon réfléchi à la seconde surface sera modifié de la même manière. Si le rayon incident est tel que toutes ses molécules échappent à la réflexion partielle en traversant la face d'entrée, elles y échapperont de même en traversant la face de sortie. Cette nouvelle propriété de la lumière offre un moyen de mesurer d'une manière précise la quantité de rayons absorbés à la surface des corps diaphanes, problême que la réflexion partielle rendoit presque impossible à résoudre.

Lorsqu'un corps, qui donne la double réfraction, réfléchit la lumière à sa première surface, il se comporte comme une substance diaphane ordinaire. La lumière réfléchie sous un certain angle d'incidence acquiert la propriété que j'ai décrite ; et cet angle est indépendant de la position de la section principale qui n'influe que sur la double réfraction, ou sur les réflexions qui ont lieu dans l'intérieur du cristal.

En effet, les rayons qui se réfléchissent intérieurement à la seconde surface, présentent des phénomènes particuliers qui dépendent à la fois des forces réfringentes, et des propriétés de la lumière réfléchie que j'ai déja exposées.

Lorsqu'un faisceau lumineux a été divisé en deux rayons à la première surface d'un rhomboïde de spath calcaire ; ces deux rayons sortent par la seconde face en deux faisceaux parallèles au rayon incident, parce que chacun d'eux éprouve, à cette face, le même genre de réfraction qu'à la première. Il n'en est pas de même de la lumière réfléchie. Quoique le rayon réfracté ordinairement à la première face, soit réfracté ordinairement à la seconde, il est néanmoins réfléchi à cette surface en deux faisceaux, l'un ordinaire, l'autre extraordinaire. De même le rayon réfracté extraordinairement se réfléchit en deux autres ; en sorte qu'il y a quatre rayons réfléchis, tandis qu'il n'y en a que deux émergens. Ces quatre rayons revenant à la première face du cristal en sortent par quatre faisceaux parallèles, qui font, avec cette surface, mais en sens contraire, le même angle que le rayon incident, et qui sont parallèles au plan d'incidence. Pour lier ce genre de réflexion à celui de la double réfraction, il faut concevoir, par les deux points d'émergence de la seconde face, deux rayons incidens, faisant, avec cette surface, mais en sens contraire, le même angle que les rayons émergens. Ces deux rayons, par leur réfraction à travers le cristal, produiront quatre faisceaux qui suivront exactement la route des rayons réfléchis. Ainsi, la loi de la double réfraction étant connue, celle de la double réflexion peut s'en déduire facilement.

Nous allons passer actuellement au genre de phénomène qui fait l'objet de ce Mémoire, et qui est relatif non à la loi suivant laquelle se dirigent les rayons, mais à la quantité et aux propriétés de la lumière qu'ils contiennent.

Supposons l'angle d'incidence constant et le cristal posé horisontalement. Si on fait tourner le rhomboïde autour de la verticale de manière à rapprocher sa section principale du rayon incident, on voit diminuer peu-à-peu l'intensité du rayon ordinaire réfléchi extraordinairement, et du rayon extraordinaire réfléchi ordinairement. Enfin, lorsque le plan de la section principale passe par le rayon incident, ces deux rayons réfléchis disparoissent totalement, et il ne reste que le rayon ordinaire réfléchi ordinairement, et le rayon extraordinaire réfléchi extraordinairement. Ce dernier a néanmoins une intensité beaucoup moindre que le premier.

Si actuellement le rayon incident continuant à être compris dans la section principale, on augmente ou on diminue l'angle d'incidence, jusqu'à ce qu'il soit égal à 56°.30′, alors le dernier rayon réfléchi disparoît totalement, et il ne reste que celui qui a été réfracté ordinairement et réfléchi ordinairement. Au-delà et en deça de cet angle, le rayon extraordinaire réfléchi extraordinairement, reparoît avec d'autant plus d'intensité qu'on s'éloigne davantage de cette limite L'angle d'incidence dont je viens de parler est celui sous lequel un rayon réfléchi à la première surface, auroit acquis la propriété de se diviser en deux faisceaux ou en un seul, comme cela à lieu à la surface de tout autre corps diaphane. Le phénomène précédent se lie facilement à l'expérience dans laquelle nous avons pris l'eau pour exemple ; car si on fait tomber sur la surface du rhomboïde et sous l'angle d'environ 56°.30′, un rayon disposé à ne se réfracter qu'en un seul faisceau extraordinaire, ce rayon ne produit pas de réflexion partielle à la première surface, ce qui semble expliquer pourquoi il n'en produit pas à la seconde.

Cependant, il n'en est pas de même lorsque le plan d'incidence fait un angle sensible avec la section principale. Si on fait tomber dans ce plan et sous l'angle d'environ 56°.30′, le rayon dont nous venons de parler, il se comporte à la première surface comme dans le cas précédent ; il la traverse sans se réfléchir, mais à la seconde surface il est réfléchi en deux faisceaux qui parviennent à leur *maximum* d'intensité lorsque le plan d'incidence est perpendiculaire à la section principale.

On sent que la lumière réfléchie à la seconde face ne se comporte pas ici comme dans le cas précédent, parce que dans la première expérience le rayon incident réfracté et réfléchi est toujours dans un même plan, au lieu que dans le dernier cas la force répulsive qui produit la réfraction extraordinaire, détourne la lumière du plan d'incidence, en sorte qu'elle cesse d'être dans les mêmes circonstances par rapport aux forces qui agissent sur elle.

Si on examine la lumière qui provient de la réflexion partielle des corps opaques, tels que le marbre noir, le bois d'ébène, etc., on trouve également un angle pour lequel cette lumière jouit des propriétés de celle qui a traversé un cristal de spath d'Islande. Les substances métalliques polies sont les seules qui ne semblent pas susceptibles de fournir ce phénomène, mais si elles n'impriment pas aux rayons lumineux cette disposition particulière, elles ne l'altèrent pas lorsque la lumière l'a déja acquise par l'influence d'un autre corps.

Cette propriété se conserve aussi dans les faisceaux qui traversent les corps qui réfractent simplement la lumière.

J'exposerai dans la seconde partie de ce Mémoire les circonstances, où à l'aide de la réflexion, sur les miroirs métalliques, on peut changer la disposition mutuelle des molécules d'un même rayon ordinaire ou extraordinaire, de manière que les unes se réfractent toujours ordinairement, tandis que les autres se réfractent extraordinairement. L'examen de ces diverses circonstances nous conduira à la loi de ces phénomènes, qui dépend d'une propriété générale des forces répulsives qui agissent sur la lumière.

4

Concerning a Property of Reflected Light

E. LOUIS MALUS

Part of this article
(taken from pages 148 and 149)
was translated expressly for this
Benchmark volume by Willliam Swindell,
University of Arizona, from "Sur une
propriété de la lumière réflechie," in
Mém. Phys. Chim. Soc. D'Arcueil, **2,**
142–158 (1809)

It is not only the crystals which give double images that are able to give to light the property of refracting into one or two beams (depending on the position of the crystal), but also all transparent solids or liquids and opaque objects are able to impress on the molecules of light that strange property that appears to be one of the effects of double refraction.

When a beam of light passes through a transparent substance, part of the rays are reflected by the refringent (entrance) surface and another part by the exit surface. The cause of this reflection, which has so far escaped the researches of physicists, seems to possess in many ways several analogies with the forces that produce double refraction.

For example, the light reflected from the surface of water at an angle of 52°45′ to the vertical has all the characteristics of a beam produced by double refraction in a crystal of Iceland spar, whose principal section is parallel or perpendicular to the plane which passes through the incident ray and the reflected ray, which we call the plane of reflection.

5

Reprinted from *Oeuvres Complètes de Fresnel,* Vol. 1, H. de Senarmont, É. Verdet, and L. Fresnel, 1866, pp. 629–630 (Johnson Reprint Corporation, New York, 1965)

Considérations mécaniques sur la polarisation de la lumière*

AUGUSTIN FRESNEL

10. Lorsque je m'occupais de la rédaction de mon premier Mémoire sur la coloration des lames cristallisées (en septembre 1816), je remarquai que les ondes lumineuses polarisées agissaient les unes sur les autres comme des forces perpendiculaires aux rayons qui seraient dirigées dans leurs plans de polarisation, puisqu'elles ne s'affaiblissent ni ne se fortifient mutuellement quand ces plans sont rectangulaires, et que deux systèmes d'ondes présentent une opposition de signe indépendante de la différence des chemins parcourus, lorsque leurs plans de polarisation, d'abord réunis, se séparent et rentrent ensuite dans un plan commun, en se plaçant sur le prolongement l'un de l'autre. M. Ampère, à qui j'avais communiqué ces résultats de l'expérience, fit la même réflexion relativement à l'opposition de signe résultant de la marche des plans de polarisation. Nous sentîmes l'un et l'autre que ces phénomènes s'expliqueraient avec la plus grande

Editor's Note: An English translation follows this article.

N° XXII. simplicité, si les mouvements oscillatoires des ondes polarisées n'avaient lieu que dans le plan même de ces ondes. Mais que devenaient les oscillations longitudinales suivant les rayons? Comment se trouvaient-elles détruites par l'acte de la polarisation, et comment ne reparaissaient-elles pas lorsque la lumière polarisée était réfléchie ou réfractée obliquement par une plaque de verre?

Ces difficultés me semblaient si embarrassantes que je négligeai notre première idée, et continuai de supposer des oscillations longitudinales dans les rayons polarisés, en y admettant en même temps des mouvements transversaux, sans lesquels il m'a toujours paru impossible de concevoir la polarisation et la non-influence mutuelle des rayons polarisés à angle droit. Ce n'est que depuis quelques mois qu'en méditant avec plus d'attention sur ce sujet, j'ai reconnu qu'il était très probable que les mouvements oscillatoires des ondes lumineuses s'exécutaient uniquement suivant le plan de ces ondes, pour la lumière directe comme pour la lumière polarisée. Je ne puis pas entrer ici dans le détail des calculs sur les diverses combinaisons de mouvements longitudinaux et transversaux qui m'ont conduit à cette conséquence. Je m'attacherai seulement à faire voir que l'hypothèse que je présente n'a rien de physiquement impossible, et qu'elle peut déjà servir à l'explication des principales propriétés de la lumière polarisée au moyen de considérations mécaniques très-simples.

5

Mechanical Considerations on the Polarization of Light

AUGUSTIN FRESNEL

This article was translated expressly for this Benchmark volume by Jean-Paul Deiss, University of Strasbourg, from "Considérations mécaniques sur la polarisation de la lumière," in Oeuvres Complètes de Fresnel, *Vol. 1, H. de Senarmont, É. Verdet, and L. Fresnel, 1866, pp. 629–630*

As I was occupied with the writing of my first paper on the coloration of crystalline plates (in September 1846), I remarked that the polarized light waves were acting each on the other as if through forces perpendicular to the beams, which should be located in their polarization planes, because the waves don't weaken nor mutually enhance when these planes are perpendicular, and because two wave systems have a sign (phase) opposition independent of the path difference when their polarization planes, first together, are separated and then come together in a common plane, placing themselves in prolongation each to the other.* Mr. Ampère, to whom I spoke about the results of this experience, made the same remark about the (phase) sign opposition resulting from the path of the polarization planes. We both felt that these facts would be explained very simply, if the vibrations (oscillatory movements) of the polarized waves took place in the plane itself of these waves. But, what became of the longitudinal oscillations along the light beams? How were these oscillations destroyed by the polarization phenomenon and why didn't they reappear when the polarized light was reflected or refracted obliquely on a glass plate?

These difficulties seemed so puzzling to me, that I neglected our first idea and continued to suppose longitudinal oscillations in the polarized beams, admitting however at the same time the presence of transverse movements, without which it was impossible for me to conceive the polarization and the mutual non-influence of the

*This direct translation leaves the meaning somewhat unclear in the latter part of this sentence. Apparently, in a rather long-winded way, Fresnel is saying that two beams whose linear polarizations are at right angles show no interference effects, whereas beams with parallel linear polarizations do show interference effects.

beams polarized at right angles. It is only during the last months, in thinking more deeply about the subject, that I acknowledged that it was very probable that the vibrations of the light waves take place only along the plane of these waves, for the direct light as well as for the polarized light. I cannot here go into more detail of the calculations on the diverse combinations of longitudinal and transverse movements which led me to this consequence. I will only try to indicate that the hypothesis I present here is not physically impossible, and that it can be used to explain the main properties of the polarized light on the basis of very simple mechanical considerations.

6

Reprinted from *Oeuvres Complètes de Fresnel,* Vol. 1, H. de Senarmont, É. Verdet, and L. Fresnel, 1866, pp. 767–775 (Johnson Reprint Corporation, New York, 1965)

Mémoire sur la réflexion de la lumière polarisée

AUGUSTIN FRESNEL

N° XXX.

MÉMOIRE

SUR LA LOI DES MODIFICATIONS

QUE LA RÉFLEXION IMPRIME A LA LUMIÈRE POLARISÉE [a].

1. L'hypothèse que j'ai adoptée sur la nature des vibrations lumineuses m'a conduit à deux formules générales de l'intensité de la lumière réfléchie par les corps transparents, pour toutes les inclinaisons des rayons incidents; l'une de ces formules est relative aux rayons polarisés suivant le plan d'incidence, et l'autre à ceux qui l'ont été dans un plan perpendiculaire. On conçoit qu'elles devaient être différentes, puisque la lumière polarisée suivant le plan d'incidence éprouve une réflexion dont l'intensité croît toujours à mesure que l'obliquité des rayons augmente; tandis que, pour la lumière polarisée perpendiculairement au plan d'incidence, il existe, entre les directions perpendi-

---

[a] Les éditeurs des Annales ont accompagné la publication de ce Mémoire de la note suivante :

« Ce Mémoire, qu'on croyait égaré, vient d'être retrouvé dans les papiers de M. Fourier [*]; comme il n'est connu que par des extraits tout à fait insuffisants (voyez *Ann.* t. XXIX, p. 175), nous nous empressons d'en enrichir les Annales. »

On peut voir, comme introduction à ce Mémoire, les nos XVI, XVII, XXI et XXIX.

[*] Mort dans les premiers mois de 1830.

N° XXX. culaires et parallèles à la surface, un certain degré d'obliquité, qui rend la réflexion nulle, comme Malus l'a reconnu le premier. Ces formules ont été publiées dans les Annales de chimie et de physique, t. XVII, cahier de juillet 1821. J'ai fait voir comment j'étais arrivé à la première, mais je n'ai pas indiqué le chemin qui m'avait conduit à la seconde. Je vais exposer ici le principe ou la supposition mécanique qu'il faut ajouter à l'hypothèse fondamentale sur la nature des vibrations lumineuses pour arriver à ces deux formules, en considérant toujours, comme je l'ai fait jusqu'à présent, le cas où les deux milieux contigus ont la même élasticité et ne diffèrent que par leur densité.

2. Il faut se rappeler d'abord que cette hypothèse fondamentale consiste en ce que les vibrations lumineuses s'exécutent dans le sens même de la surface de l'onde perpendiculairement au rayon; d'où il résulte qu'un faisceau de lumière polarisée est celui dont les mouvements vibratoires conservent une direction unique et constante, et que son plan de polarisation est le plan perpendiculaire à cette direction constante des petites oscillations des molécules éthérées. Ainsi, quand le faisceau est polarisé suivant le plan d'incidence, les vibrations sont perpendiculaires à ce plan, et par conséquent sont toujours parallèles à la surface réfringente, quelle que soit l'inclinaison des rayons. Il n'en est plus de même pour ceux qui ont été polarisés perpendiculairement au plan d'incidence, parce que leurs vibrations, s'exécutant alors dans ce plan, ne sont parallèles à la surface réfringente que dans le cas de l'incidence perpendiculaire, puis forment avec elle des angles d'autant plus grands que les rayons s'inclinent davantage, et lui deviennent enfin perpendiculaires quand les rayons lui sont parallèles; c'est ce qui rend le problème de la réflexion plus difficile à résoudre dans ce second cas que dans le premier. Dans celui-ci, les mouvements oscillatoires s'exécutant uniquement suivant les directions parallèles à la surface pour les ondes réfléchies et réfractées, comme pour l'onde incidente, on peut admettre que les amplitudes de ces oscillations, ou que les vitesses absolues des molécules dans un élément quelconque de l'onde réfléchie ou de l'onde réfractée ne changent pas, tandis

qu'elles s'éloignent de la surface [1]; du moins il me semble que ce principe ne serait pas difficile à démontrer rigoureusement. J'adopte aussi la même supposition pour le cas de la lumière polarisée perpendiculairement au plan d'incidence, c'est-à-dire celui où les vibrations s'exécutent dans ce plan; bien entendu qu'il ne s'agit plus alors que des composantes des vitesses absolues parallèles à la surface réfléchissante; ainsi je suppose que ces composantes ont la même intensité lorsque l'ébranlement réfléchi ou réfracté touche encore à la surface, et lorsqu'il s'en est éloigné.

3. Cela posé, d'après la nature de l'élasticité que je considère, qui est celle qui s'oppose au glissement d'une tranche d'un même milieu sur la tranche suivante, ou au déplacement relatif des tranches en contact de deux milieux différents, les tranches contiguës des deux milieux doivent exécuter parallèlement à la surface qui les sépare des oscillations de même amplitude, sans quoi l'une de ces tranches aurait glissé sur l'autre d'une quantité d'un ordre bien supérieur aux déplacements relatifs des tranches contiguës de chaque milieu considéré séparément, d'où naîtrait une résistance beaucoup plus grande, qui s'opposerait à ce déplacement. Ainsi l'on peut admettre, comme une conséquence évidente de notre hypothèse fondamentale sur la nature de l'élasticité mise en jeu par les vibrations lumineuses, que les vitesses absolues des molécules voisines de la surface réfringente parallèlement à cette surface doivent être égales dans les deux milieux : or ces mouvements dans le premier milieu se composent à la fois de l'ébranlement apporté par l'onde incidente et de celui de l'onde réfléchie, c'est-à-dire que la composante parallèle à la surface réfringente du mouvement imprimé à chaque molécule du premier milieu par l'onde incidente et l'onde réfléchie doit être égale à la composante parallèle de la vitesse absolue des molécules dans le second milieu; ou, en d'autres termes, et supposant la surface réfringente horizontale pour simplifier les ex-

[1] Je suppose ici, bien entendu, que le centre de l'onde incidente est infiniment éloigné, en sorte qu'elle est plane, ainsi que les ondes réfléchie et réfractée, et que leurs intensités ne sont point affaiblies par leur propagation.

 pressions, la composante horizontale de la vitesse absolue apportée par l'onde incidente, ajoutée à la composante horizontale de la vitesse absolue imprimée par l'onde réfléchie (prise avec le signe qui lui convient), doit être égale à la composante horizontale de la vitesse absolue des molécules du second milieu dans l'onde transmise. Il est clair que cette égalité doit avoir lieu près de la surface de contact, et la supposition que nous avons énoncée d'abord, et dont nous allons nous servir, consiste seulement à admettre que ces composantes horizontales restent constantes pendant que les éléments successifs des ondes réfléchies et réfractées s'éloignent de la surface, et que par conséquent l'équation dont il s'agit a lieu à toutes distances. Avant de donner les raisons sur lesquelles je fonde cette conservation des composantes horizontales, j'attendrai que je puisse traiter la question plus à fond, et présenter en même temps la solution du problème pour le cas où les deux élasticités sont différentes. Je ne me propose actuellement que de déduire de cette hypothèse subsidiaire et du principe de la conservation des forces vives les formules que j'avais publiées en 1821, et dont nous tirerons les lois qui font l'objet de ce Mémoire.

4. Pour appliquer ici le principe de la conservation des forces vives, il faut pouvoir comparer les masses ébranlées dans les deux milieux, ce qui devient facile au moyen de la loi connue de la réfraction.

Soit EF la surface réfringente, AB l'onde incidente, *ab* la même onde réfractée; si du point A on abaisse sur *ab* le rayon perpendiculaire A*a*, et que par le point *b* on conçoive pareillement un rayon B*b* perpendiculaire à l'onde incidente, il est clair que AB et *ab* seront des étendues correspondantes des deux ondes dans les deux milieux, c'est-à-dire que la partie AB de l'onde incidente occupera dans le second milieu l'étendue *ab*; quant aux es-

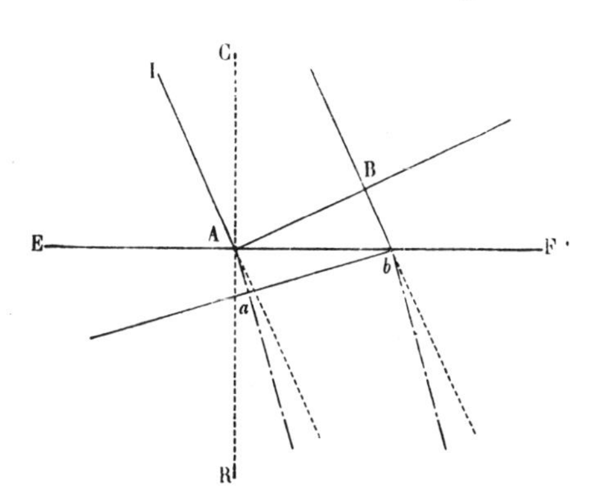

paces relatifs qu'elles occupent dans le sens perpendiculaire, suivant la direction des rayons IA et Aa, ce sont précisément les longueurs d'ondulation dans les deux milieux, dont le rapport est celui du sinus de l'angle d'incidence IAC au sinus de l'angle de réfraction RAa. Si donc nous appelons i le premier angle et i' le second, les dimensions relatives des ondes dans le sens des rayons pourront être représentées par $\sin i$ et $\sin i'$; et conséquemment les volumes des deux portions correspondantes que nous considérons dans les ondes incidentes et réfractées seront entre eux comme $\mathrm{AB}\sin i$ est à $ab\sin i$. Mais en prenant Ab pour rayon, AB et ab sont les cosinus respectifs des angles BAb et Aba, ou des angles i et i', auxquels ceux-ci sont égaux; les deux volumes sont donc entre eux comme $\sin i \cos i$ est à $\sin i' \cos i'$. Il nous reste à les multiplier par les densités pour avoir le rapport des masses. Or, comme les deux milieux sont supposés avoir la même élasticité et différer seulement en densité, les vitesses de propagation dans ces deux milieux sont en raison inverse des racines carrées de leurs densités; ainsi l'on a :

$$\sin i : \sin i' :: \frac{1}{\sqrt{d}} : \frac{1}{\sqrt{d'}},$$

ou

$$d : d' :: \frac{1}{\sin^2 i} : \frac{1}{\sin^2 i'};$$

multipliant ce rapport par celui des volumes, nous aurons pour celui des masses :

$$\frac{\sin i \cos i}{\sin^2 i} : \frac{\sin i' \cos i'}{\sin^2 i'},$$

ou

$$\frac{\cos i}{\sin i} : \frac{\cos i'}{\sin i'}.$$

Si donc on prend $\frac{\cos i'}{\sin i'}$ pour représenter la masse ébranlée dans l'onde réfractée, $\frac{\cos i}{\sin i}$ sera la masse ébranlée dans l'onde incidente, et en même temps la masse de la partie correspondante de l'onde réfléchie, puisque les parties correspondantes des ondes incidentes et réfléchies ont le même volume, et que d'ailleurs elles sont dans le même milieu.

N° XXX. Cela posé, je prends pour unité le coefficient commun de toutes les vitesses absolues des molécules dans l'onde incidente, et je représente par v celui des vitesses absolues dans l'onde réfléchie et par u celui des mêmes vitesses dans l'onde réfractée : en divisant par la pensée l'onde incidente en une série d'une infinité d'ébranlements successifs, et les ondes réfléchies et réfractées en un même nombre d'éléments pareils, il est évident que le rapport entre les vitesses absolues de deux éléments correspondants de l'onde incidente et de l'onde réfractée, par exemple, sera constant pour toutes les parties de ces deux ondes, puisqu'il doit être indépendant de l'intensité plus ou moins grande des vitesses absolues dans les divers éléments de l'onde incidente. Si donc on prend pour unité l'intensité du mouvement vibratoire dans l'onde incidente, v et u seront les coefficients par lesquels il faut multiplier chacune des vitesses absolues des éléments de l'onde incidente pour avoir les vitesses absolues des éléments correspondants de l'onde réfractée et de l'onde réfléchie, et indiqueront ainsi le degré d'intensité des vitesses absolues dans ces deux ondes. Par conséquent, la masse de l'onde réfractée multipliée par u^2, plus la masse de l'onde réfléchie multipliée par v^2, doivent donner une somme égale à la masse de l'onde incidente multipliée par 1, pour que la somme des forces vives reste constante; on a donc :

$$\frac{\cos i}{\sin i} 1 = \frac{\cos i'}{\sin i'} u^2 + \frac{\cos i}{\sin i} v^2,$$

ou

$$\frac{\cos i}{\sin i}(1 - v^2) = \frac{\cos i'}{\sin i'} u^2,$$

ou

$$\sin i' \cos i (1 - v^2) = \sin i \cos i' u^2 \; . \; . \; . \; . \; . \; (A).$$

Telle est l'équation qui résulte du principe de la conservation des forces vives et qui doit être satisfaite dans tous les cas, soit que le rayon incident ait été polarisé parallèlement ou perpendiculairement au plan d'incidence.

5. Nous avons admis que dans ces deux cas les mouvements parallèles à la surface réfringente devaient être égaux de chaque côté de cette

surface, c'est-à-dire que les vitesses horizontales de l'onde incidente ajoutées aux vitesses horizontales de l'onde réfléchie prises avec leur signe devaient être égales aux vitesses horizontales de l'onde transmise, et cela non-seulement contre la surface, où le principe est évident, mais encore à des distances contenant un grand nombre de fois la longueur d'ondulation. Lorsque l'onde incidente est polarisée suivant le plan d'incidence, c'est-à-dire que ses vibrations s'exécutent perpendiculairement à ce plan, elles sont toujours horizontales ainsi que celles des ondes réfléchie et transmise, et par conséquent les coefficients des vitesses horizontales sont 1, v et u pour les ondes incidente, réfléchie et réfractée, et l'on doit avoir, d'après notre hypothèse subsidiaire.

$$1+v=u \quad \text{ou} \quad (1+v)^2=u^2.$$

Divisant par cette équation celle que nous venons d'obtenir au moyen du principe de la conservation des forces vives, on a :

$$\sin i' \cos i\left(\frac{1-v}{1+v}\right)=\sin i \cos i'.$$

ou

$$\sin i' \cos i\,(1-v)=\sin i \cos i'\,(1+v):$$

d'où l'on tire

$$v=-\frac{\sin i \cos i'-\sin i' \cos i}{\sin i \cos i'+\sin i' \cos i},$$

ou

$$v=-\frac{\sin(i-i')}{\sin(i+i')}\cdot\cdot\cdot\cdot\cdot(1).$$

6. Dans le second cas, c'est-à-dire celui où la lumière est polarisée perpendiculairement au plan d'incidence, les vibrations s'exécutent alors parallèlement à ce plan et toujours perpendiculairement aux rayons incidents, réfléchis et réfractés, les composantes horizontales des vitesses absolues 1, v et u, sont $\cos i$, $v\cos i$ et $u\cos i'$; on doit donc avoir, d'après l'hypothèse subsidiaire,

$$\cos i+v\cos i=u\cos i', \quad \text{ou} \quad (1+v)\cos i=u\cos i'.$$

ou élevant au carré.

$$(1+v)^2\cos^2 i=u^2\cos^2 i'.$$

N° XXX. Divisant l'équation (A), qui résulte du principe de la conservation des forces vives, par cette dernière équation, l'on a :

$$\left(\frac{1-v}{1+v}\right)\frac{1}{\sin i \cos i}=\frac{1}{\sin i' \cos i'},$$

ou

$$(1-v)\sin i'\cos i'=(1+v)\sin i \cos i;$$

d'où l'on tire

$$v=-\frac{\sin i\cos i-\sin i'\cos i'}{\sin i\cos i+\sin i'\cos i'}\cdots\cdots(2).$$

Telle est l'expression de la vitesse absolue dans l'onde réfléchie, quand le plan de réflexion est perpendiculaire au plan de polarisation de la lumière incidente. On voit que cette expression devient nulle pour une certaine obliquité des rayons, lorsqu'on a $\sin i\cos i=\sin i'\cos i'$, ou $\sin 2i=\sin 2i'$, c'est-à-dire quand $2i=180°-2i'$, ou $i=90°-i'$, c'est-à-dire enfin quand l'angle de réfraction est le complément de l'angle d'incidence, ou, ce qui revient au même, lorsque le rayon réfracté est perpendiculaire au rayon réfléchi, conformément à la loi de Brewster. Il n'en est pas de même pour la formule (1); elle ne pourrait devenir nulle que dans le cas particulier où i' serait égal à i, c'est-à-dire où les ondes lumineuses auraient la même longueur dans les deux milieux en contact. Mais d'ailleurs les deux formules donnent la même vitesse réfléchie pour l'incidence perpendiculaire, et pour l'autre limite $i=90°$, et, dans le second cas, elles indiquent l'une et l'autre que la totalité de la lumière est réfléchie; ce qu'on trouverait sans doute aussi par l'expérience, si l'on pouvait atteindre à cette limite. Dans le cas de l'incidence perpendiculaire, les deux formules donnent :

$$v=-\frac{\sin i-\sin i'}{\sin i+\sin i'}=-\frac{\frac{\sin i}{\sin i'}-1}{\frac{\sin i}{\sin i'}+1},\quad \text{ou}\quad v=-\frac{r-1}{r+1},$$

en appelant r le rapport constant du sinus d'incidence au sinus de réfraction. C'est précisément la formule que M. Young a donnée le premier, et à laquelle M. Poisson est arrivé ensuite par une analyse plus savante et plus rigoureuse; mais en ne considérant l'un et l'autre que le genre d'élasticité auquel les géomètres ont attribué uniquement jusqu'à ce jour la propagation des ondes sonores, je veux dire la résistance des milieux vibrants à la compression.

N° XXX.

7

Reprinted from *Phil. Trans.*, **105,** 125–130, 158–159 (1815)

IX. *On the laws which regulate the polarisation of light by reflexion from transparent bodies. By* David Brewster, *LL.D. F.R.S. Edin. and F.S.A. Edin. In a letter addressed to the Right Hon. Sir* Joseph Banks, *Bart. K. B. P. R. S.*

Read March 16, 1815.

Dear Sir,

The discovery of the polarisation of light by reflexion, constitutes a memorable epoch in the history of optics; and the name of Malus, who first made known this remarkable property of bodies, will be for ever associated with a branch of science which he had the sole merit of creating. By a few brilliant and comprehensive experiments he established the general fact, that light acquired the same property as one of the pencils formed by double refraction, when it was reflected at a particular angle from the surfaces of all transparent bodies: he found that the angle of incidence at which this property was communicated, was greater in bodies of a high refractive power, and he measured, with considerable accuracy, the polarising angles for glass and water. In order to discover the law which regulated the phenomena, he compared these angles with the refractive and dispersive powers of glass and water, and finding that there was no relation between these properties of transparent bodies, he draws the following general conclusion. "The polarising angle neither "follows the order of the refractive powers, nor that of the

Editor's Note: A row of asterisks indicates that material has been omitted from the original article.

" dispersive forces. It is a property of bodies independent " of the other modes of action which they exercise upon " light."

This premature generalisation of a few imperfectly ascertained facts, is perhaps equalled only by the mistake of Sir ISAAC NEWTON, who pronounced the construction of an achromatic telescope to be incompatible with the known principles of optics. Like NEWTON, too, MALUS himself abandoned the enquiry; and even his learned associates in the Institute, to whom he bequeathed the prosecution of his views, have sought for fame in the investigation of other properties of polarised light.

In the summer of 1811, when my attention was first turned to this subject, I repeated the experiments of MALUS, and measured the polarising angles of a great number of transparent bodies. I endeavoured, in vain, to connect these results by some general principle: the measures for *water* and the *precious stones* afforded a surprising coincidence between the indices of refraction and the tangents of the polarising angles; but the results for glass formed an exception, and resisted every method of classification. Disappointed in my expectations, I abandoned the enquiry for more than twelve months, but having occasion to measure the polarising angle of *topaz*, I was astonished at its coincidence with the preceding law, and again attempted to reduce the results obtained from *glass* under the same principle. The piece which I used had two surfaces excellently polished. The polarising angle of one of these surfaces almost exactly accorded with the law of the tangents, but with the other surface there was a deviation of no less than two degrees. Upon examining the cause of this anomalous result, I found that one of the surfaces had suffered some chemical change, and reflected less light than any other part of the glass. This artificial substance acquires an incrustation, or experiences a decomposition by exposure to the air, which alters its polarising angle without altering its general refractive power. The perplexing anomalies which BOUGUER observed in the reflective power of plate glass, were owing to the same cause, and so liable is this substance to these changes, that by the aid of heat alone, I have produced a variation of 9° on the polarising angle of flint glass, and given it the power of acting upon light like the coloured oxides of steel.

Having thus ascertained the cause of the anomalies presented by glass, I compared the various angles which I had measured, and found that they were all represented by the following simple law.

The index of refraction is the tangent of the angle of polarisation.

In the course of last summer, when I had the pleasure of seeing M. ARAGO, I mentioned to him the relation which I had discovered between the refractive powers, and the tangents of the polarising angles. He informed me, that he had found the polarising angle of air to be 45° or 47°, which being at the very extremity of the scale would afford a good test of the accuracy of the law. Now, if we take the refractive power of air at 1.00031 the polarising angle will be 45° 0′ 32″, a result which agrees most strikingly with the observed angle.

In the following table I have given the polarising angles of eighteen transparent bodies, as determined by experiment, and as deduced from the law of the tangents. I have added in the *fourth* column the differences between the calculated and observed angles, and in the *fifth* column the calculated angles of polarisation for the second surfaces of the bodies subjected to experiment.

Table containing the calculated and observed polarising angles for various bodies.

Names of the Bodies..	Calculated polarising angles for the *first* surface.	Observed polarising angles for the *first* surface.	Difference between the calculated and observed angles.	Calculated polarising angles for the *second* surface.
	° ′ ″	° °		° ′ ″
Air - - - - - -	45 0 32	45 or 47		44 59 28
Water - - - - -	53 11 0	52° 45′	0° 26′ —	36° 49′
Fluor spar - - -	55 9 0	54 50	0 19 —	34 51
Obsidian - - -	56 6 0	56 3	0 3 —	33 54
Birdlime - - - -	56 40 0	56 46	0 6 +	33 20
Sulphate of lime -	56 45 0	56 28	0 17 —	33 15
Rock crystal - -	56 58 0	57 22	0 24 +	33 2
Opal coloured glass	58 33 0	58 1	0 32 —	31 27
Topaz - - - - -	58 34 0	58 40	0 6 +	31 26
Mother of pearl -	58 50 0	58 47	0 3 —	31 10
Iceland spar - -	58 51 0	58 23	0 28 —	31 9
Orange coloured glass	59 28 0	59 12	0 16 —	30 32
Spinelle ruby - -	60 25 0	60 16	0 9 —	29 35
Zircon - - - - -	63 0 0	63 8	0 8 +	27 0
Glass of antimony	64 30 0	64 45	0 15 +	25 30
Sulphur - - - - -	63 45 0	64 10	0 25 +	26 15
Diamond - - -	68 1 0	68 2	0 1 +	21 59
Chromate of Lead	68 3 0	67 42	0 21 —	21 56

The coincidence between the calculated and observed angles, as shown in the preceding Table, must appear very remarkable to those who are aware of the difficulty of measuring correctly the index of refraction for the *mean* refrangible ray, and the still greater difficulty of determining the angle at which the intensity of the evanescent pencil is a minimum. The total amount of the errors in seventeen observations is 259 minutes, which gives an average error of 15′ for each observation. In general the observed angles are less than the calculated angles, the number of negative being to the number of positive differences as 174′ is to 85′.

This circumstance arises from two separate causes, which ought to be carefully kept in view in all experiments on the polarising angles of bodies.

1. In order to illustrate the first of these causes, let us take the case of *Zircon*, in which the intensity of the evanescent pencil is a minimum at 63° of incidence. At 64° the intensity of the pencil which vanishes at 63° is much greater than that of the pencil at 62° on account of its falling more obliquely upon the reflecting surface, and consequently the intensity of that pencil varies more rapidly in passing from 64° to 63° than from 62° to 63°. Hence, in determining the point of minimum intensity, it is more likely, from the way in which the observation must be made, that the observed angle will fall *below* than above the real polarising angle.

2. As the differently coloured rays have different angles of polarisation, and as the most luminous rays of the spectrum have less refractive power than the mean refrangible rays, the observed polarising ought always to be less than the polarising angle for the mean rays. Hence, all the observed angles in the preceding Table ought to be increased by a certain quantity, or, what is the same thing, the index of refraction for the most luminous rays ought to be employed instead

of the mean index of refraction in computing the first column.

The law of the polarisation of light by reflexion being thus experimentally established, we shall now proceed to point out its geometrical consequences, and to arrange, under separate propositions, the new truths to which it leads, as well as those which I have obtained from direct experiment. It will thus be seen, that the subject assumes a scientific form, and that we can calculate *a priori*, the result of every experiment, whether the light is incident upon the first or second surface of transparent bodies, or upon the separating surface of different media, or whether it undergoes a series of successive reflexions in the same plane, or in planes at right angles to each other.

* * * * * * *

In these enquiries I have made use of no hypothetical assumptions. In imitation of MALUS, the language of theory has been occasionally employed, but the terms thus introduced are merely expressive of experimental results, and enable us to avoid frequent and perplexing circumlocutions. The science of physical optics is not yet in such a state as to authorise the construction of a new nomenclature. When discovery shall have accumulated a greater number of facts, and connected them together by general laws, we may then safely begin to impose better names, and to speculate respecting the

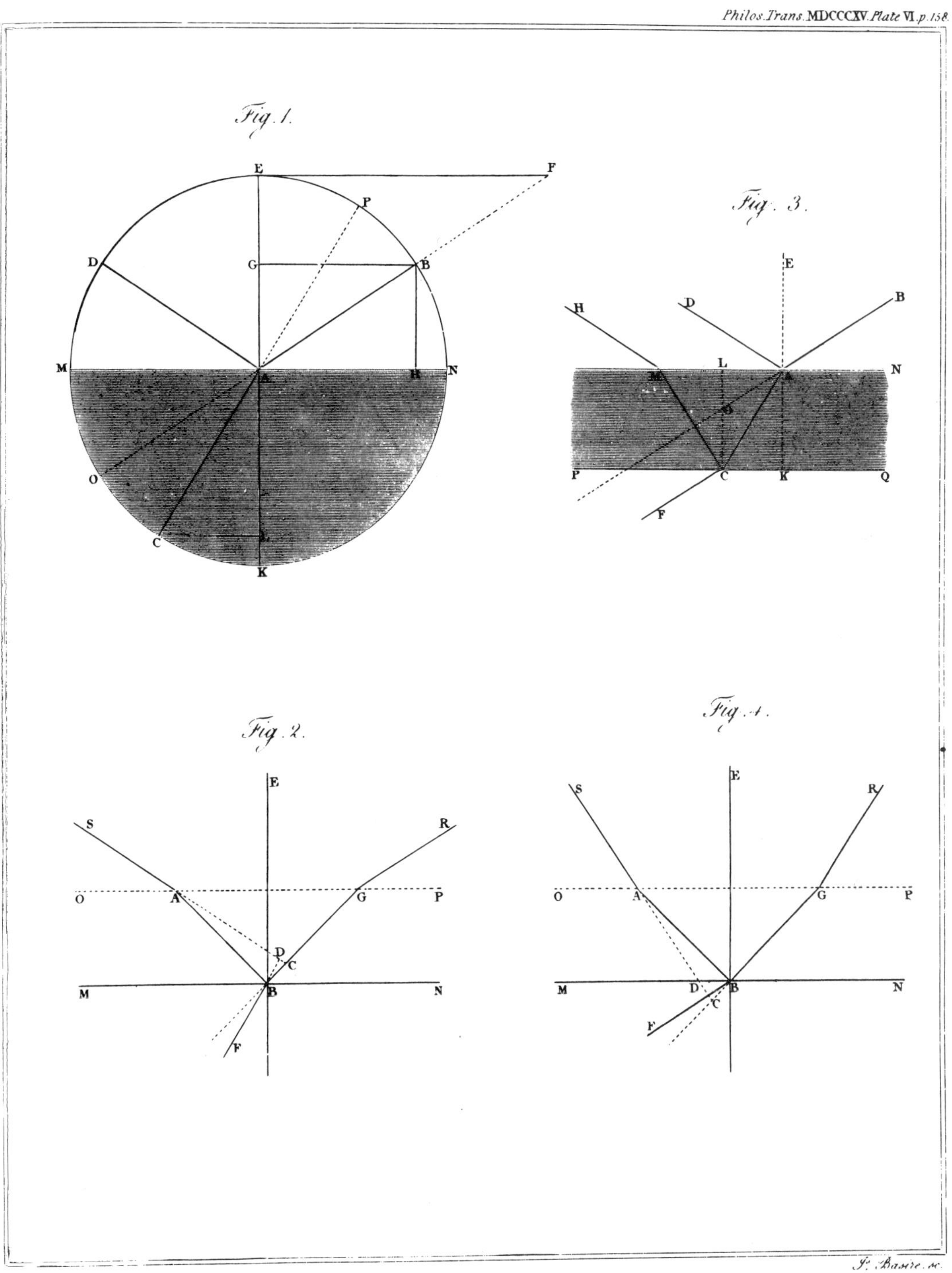

Fig. 1.
E
F
P
D
G
B
M
A
H
N
O
C
L
K
Fig. 3.
E
H
D
B
L
N
M
A
O
P
C
K
Q
F
Fig. 2.
E
S
R
O
A
G
P
D
C
M
B
N
F
Fig. 4.
E
S
R
O
A
G
P
M
D
B
N
C
F

Js. Basire sc.

cause of those wonderful phenomena which light exhibits under all its various modifications.

In the preceding pages, I have more than once had occasion to establish conclusions opposite to those which MALUS had deduced from less numerous experiments; and indeed the whole of this paper is founded on relations which he believed to have no existence. In differing, however, from this eminent philosopher, I trust I have always done it with that respect which it is impossible not to feel for his character and labours. It has fallen to the lot of few to enrich science with so many new and striking discoveries, and if he has failed in pursuing them through all their consequences, we must ascribe it to the limited interval which he was allowed to devote to science, and to the influence of that cruel disease which terminated so prematurely his short but brilliant career. Those, who without repeating his experiments endeavoured during his life to depreciate his labours, are alone capable of wounding his memory. Those who, like him, have pursued science under the oppression of bodily suffering;—who have been instructed and delighted with his discoveries, and who have patiently followed him in the path of research, will feel it their truest pride to do justice to his memory, and will never be able to review his labours without mingling their sorrow with their admiration.

I have the honour to be, &c.

DAVID BREWSTER.

Edinburgh, February 11, 1815.

8

Reprinted from *Edinburgh J. Phil.*, **6**, 83–84 (1828)

On a Method of so far increasing the Divergency of the two Rays in Calcareous-spar, that only one Image may be seen at a time. By WILLIAM NICOL, Esq. Lecturer on Natural Philosophy. Communicated by the Author.

THE following simple method of constructing a prism of calcareous-spar, so that only one image may be seen at a time, will, perhaps, prove interesting to those who are in the habit of examining the optical properties of crystallised bodies by polarised light.

Let a rhomboid of calcareous-spar one inch long be reduced in breadth and thickness to three-tenths of an inch ; let the obliquity of its terminal planes be increased about three degrees ; or, in other words, let the angles formed by the terminal planes, and the adjoining obtuse lateral edges, be made equal to 68°, by operating on the terminal planes : these planes may now be polished. The rhomboid is then to be divided into two equal portions, by a plane passing through the acute lateral edges, and nearly touching the two obtuse solid angles. The sectional plane of each of the two halves must now be made to form exactly an angle of 90° with the terminal plane, and then carefully polished. The two portions are now to be firmly cemented together by means of Canada balsam, so as to form a rhomboid similar to what it was before its division.

If a ray of common light fall on the end of such a rhomboid in a direction parallel to the lateral edges, the two rays into which it is divided, in passing through the spar, will deviate so far from each other, that only the ordinary image will be seen. That image, too, will appear exactly in its true position, and free from colour. The range of the ordinary ray will be found considerably greater than the whole field of vision, as may easily be seen by making the rhomboid revolve on an axis parallel to the longer diagonal of the terminal planes. There is a tinge of blue where the ordinary ray vanishes on one side, and a tinge of orange, accompanied by a number of extremely minute obscurely coloured fringes, where it terminates on the other. If the rhomboid revolve beyond the fringes, the ordinary image will disappear, and the extraordinary image come into view. The latter, however, from the great obliquity of the incident light, occupies a smaller range, and is less distinct than the other. The ordinary ray passing out of the rhomboid in a direction parallel to its lateral edges, is therefore the best adapted for analytical purposes; and as calcareous-spar, when pure, and free from flaws, is not only transparent, but perfectly colourless, a rhomboid of that substance, of the above construction, developes the coloured rings of crystallized bodies with a degree of brilliancy not to be equalled by a plate of tourmaline, or perhaps by any other substance.

With the view of rendering the structure of the analyzing rhomboid more easily understood, I have supposed a piece of the spar to be divided into two equal portions. Such a division, however, would be a difficult task; but if two similar pieces of spar be taken, it will be found a very easy matter to remove one-half of each of them, either by grinding, or by the action of a file. The pieces of spar should not be much less than an inch long, and they need not be longer than 1.4 inch. If the latter dimension be adopted, the breadth and thickness will require to be about .48 of an inch.

In cementing the two pieces together, it will be proper to let the pointed end of the one project a little over the terminal plane of the other. By so doing, a more firm contact is obtained at the edges; and when the cement is sufficiently indurated, the whole of the projecting parts may easily be removed, according to their cleavages. The lateral planes should be left quite rough, to prevent the reflection of extraneous light.

II
Foundations of Modern Methods

Editor's Comments on Papers 9 Through 12

9 **Faraday:** *Experimental Researches in Electricity*

10 **Stokes:** *On the Composition and Resolution of Streams of Polarized Light from Different Sources*

11 **Poincaré:** *Polarisation rotatoire: Théorie de M. Mallard*

12 **Wiener:** *Generalized Harmonic Analysis*

Around the middle of the nineteenth century there was a revolution in the development of understanding the nature of light that culminated in Maxwell's electromagnetic theory of light. This was presented in final form in 1873 ("Treatise on Electricity and Magnetism," Oxford, 1873). Maxwell was guided by the earlier experimental work of Faraday, Oersted, and Henry. Paper 9, by Michael Faraday, describes research in which for the first time a direct connection among electricity, magnetism, and light was established. Undoubtedly this work was an essential part of Maxwell's guidance. Here is reproduced a part of Faraday's book that was published in 1855, ten years after the important discovery was first announced.

Today, the practical applications of the Faraday effect include the remote, contactless sensing of electric currents and magnetic fields, the investigation of magnetization processes in transparent and semitransparent materials, and use in optical computer memories.

In Paper 10, by George Gabriel Stokes (1819–1903), we see the foundations of the modern mathematical theories used to describe unpolarized and partially polarized light. The most important result of his work is that any beam of light can be completely and uniquely characterized (at least as far as the polarization properties are concerned) by four parameters, now known universally as the Stokes parameters.

In the early part of the work, he provides a discussion of the synthesis of polarized light from other polarized light beams and introduces the concept of "opposite" (orthogonal) polarization. The Stokes parameters are introduced in section 9; and using this description he shows later, in section 19, that partially polarized light may always be regarded as a unique mixture of unpolarized and elliptically polarized light beams. Stokes is also known for his work in fluorescence, viscosity, and mathematics.

A significant article regarding theoretical descriptions of polarized light. Paper 11, appeared in H. Poincaré's book *Théorie mathématique de la lumière* in 1892. Poincaré showed that all possible states of polarization could be represented by points on the surface of a sphere, the latitude and longitude of each point defining the eccentricity and major axis azimuth of the polarization ellipse, respectively.

Poincaré's development starts with a two-dimensional complex-plane representation of polarized light forms and then defines the sphere as the generating sphere for that plane using a stereographic projection.

The usual way to think of the sphere nowadays is as the three-dimensional surface generated when the Stokes parameters are used as orthogonal cartesian coordinates. Apparently, Poincaré was not aware of this connection.

When fully understood, the Poincaré sphere becomes a marvelous tool for analyzing and synthesizing situations involving polarizers, retarders, and rotators. A later section of the article (not reproduced here) shows an application of this type. A modern description of the uses of the Poincaré sphere may be found in *Polarized Light* by W. A. Shurcliff (Harvard University Press, Cambridge, Mass., 1962).

Paper 12, by Norbert Wiener, is perhaps the most important contribution ever made to the development of the theory of polarized light. It is the third of a series of three articles in which Wiener develops a theory of generalized harmonic analysis to give a rigorous interpretation of the stochastic processes, of which white light is a special case. He introduces the concept of the coherency matrix that forms the link between the unobservable electric and magnetic field of Maxwell's theory and the physically observable quantity, irradiance. Wiener also showed the fundamental relationship that the coherency matrix is a linear combination of the Pauli spin matrices with the Stokes parameters as expansion coefficients (although he was apparently unaware of Stokes's earlier work), and he gives details of an experiment by which the Stokes parameters may be measured.

The previous papers in the series are "Harmonic Analysis and the Quantum Theory" [*J. Franklin Inst.,* **207,** 525–534 (1929)] and "Coherency Matrices and Quantum Theory" [J. Math. Phys., **7,** 109–125 (1927–1928)]. The article presented here is the relevant portion of his classical paper "Generalized Harmonic Analysis."

9

Reprinted from *Philosophical Transactions of 1846–1852,* Bernard Quaritch, 1855, pp. 1–26

I. *Experimental Researches in Electricity.—Nineteenth Series.*

By Michael Faraday, *Esq., D.C.L. F.R.S., Fullerian Prof. Chem. Royal Institution, Foreign Associate of the Acad. Sciences, Paris, Cor. Memb. Royal and Imp. Acadd. of Sciences, Petersburgh, Florence, Copenhagen, Berlin, Göttingen, Modena, Stockholm, &c. &c.*

Received November 6,—Read November 20, 1845.

§ 26 *On the magnetization of light and the illumination of magnetic lines of force*.*

¶ i. *Action of magnets on light.* ¶ ii. *Action of electric currents on light.* ¶ iii. *General considerations.*

¶ i. *Action of magnets on light.*

2146. I HAVE long held an opinion, almost amounting to conviction, in common I believe with many other lovers of natural knowledge, that the various forms under which the forces of matter are made manifest have one common origin ; or, in other

* The title of this paper has, I understand, led many to a misapprehension of its contents, and I therefore take the liberty of appending this explanatory note. Neither accepting nor rejecting the hypothesis of an ether, or the corpuscular, or any other view that may be entertained of the nature of light; and, as far as I can see, nothing being really known of a ray of light more than of a line of magnetic or electric force, or even of a line of gravitating force, except as it and they are manifest in and by substances; I believe that, in the experiments I describe in the paper, light has been magnetically affected, *i. e.* that that which is magnetic in the forces of matter has been affected, and in turn has affected that which is truly magnetic in the force of light: by the term magnetic I include here either of the peculiar exertions of the power of a magnet, whether it be that which is manifest in the magnetic or the diamagnetic class of bodies. The phrase " illumination of the lines of magnetic force" has been understood to imply that I had rendered them luminous. This was not within my thought. I intended to express that the line of magnetic force was illuminated as the earth is illuminated by the sun, or the spider's web illuminated by the astronomer's lamp. Employing a ray of light, we can tell, *by the eye*, the direction of the magnetic lines through a body; and by the alteration of the ray and its optical effect on the eye, can see the course of the lines just as we can see the course of a thread of glass, or any other transparent substance, rendered visible by the light: and this was what I meant by *illumination*, as the paper fully explains.—December 15, 1845. M. F.

words, are so directly related and mutually dependent, that they are convertible, as it were, one into another, and possess equivalents of power in their action*. In modern times the proofs of their convertibility have been accumulated to a very considerable extent, and a commencement made of the determination of their equivalent forces.

2147. This strong persuasion extended to the powers of light, and led, on a former occasion, to many exertions, having for their object the discovery of the direct relation of light and electricity, and their mutual action in bodies subject jointly to their power†; but the results were negative and were afterwards confirmed, in that respect, by WARTMANN‡.

2148. These ineffectual exertions, and many others which were never published, could not remove my strong persuasion derived from philosophical considerations; and, therefore, I recently resumed the inquiry by experiment in a most strict and searching manner, and have at last succeeded in *magnetizing and electrifying a ray of light, and in illuminating a magnetic line of force.* These results, without entering into the detail of many unproductive experiments, I will describe as briefly and clearly as I can.

2149. But before I proceed to them, I will define the meaning I connect with certain terms which I shall have occasion to use:—thus, by *line of magnetic force*, or *magnetic line of force*, or *magnetic curve*, I mean that exercise of magnetic force which is exerted in the lines usually called magnetic curves, and which equally exist as passing from or to magnetic poles, or forming concentric circles round an electric current. By *line of electric force*, I mean the force exerted in the lines joining two bodies, acting on each other according to the principles of static electric induction (1161, &c.), which may also be either in curved or straight lines. By a *diamagnetic*, I mean a body through which lines of magnetic force are passing, and which does not by their action assume the usual magnetic state of iron or loadstone.

2150. A ray of light issuing from an Argand lamp, was polarized in a horizontal plane by reflection from a surface of glass, and the polarized ray passed through a NICHOL's eye-piece revolving on a horizontal axis, so as to be easily examined by the latter. Between the polarizing mirror and the eye-piece, two powerful electro-magnetic poles were arranged, being either the poles of a horse-shoe magnet, or the contrary poles of two cylinder magnets; they were separated from each other about two inches in the direction of the line of the ray, and so placed, that, if on the same side of the polarized ray, it might pass near them; or, if on contrary sides, it might go between them, its direction being always parallel, or nearly so, to the magnetic lines of force (2149.). After that, any transparent substance placed between the two poles, would have passing through it, both the polarized ray and the magnetic lines of force at the same time and in the same direction.

* Experimental Researches, 57, 366, 376, 877, 961, 2071.

† Philosophical Transactions, 1834. Experimental Researches, 951–955.

‡ Archives de l'Electricité, ii. pp. 596–600.

2151. Sixteen years ago I published certain experiments made upon optical glass*, and described the formation and general characters of one variety of heavy glass, which, from its materials, was called silicated borate of lead. It was this glass which first gave me the discovery of the relation between light and magnetism, and it has power to illustrate it in a degree beyond that of any other body; for the sake of perspicuity I will first describe the phenomena as presented by this substance.

2152. A piece of this glass, about two inches square and 0·5 of an inch thick, having flat and polished edges, was placed as a *diamagnetic* (2149.) between the poles (not as yet magnetized by the electric current), so that the polarized ray should pass through its length; the glass acted as air, water, or any other indifferent substance would do; and if the eye-piece were previously turned into such a position that the polarized ray was extinguished, or rather the image produced by it rendered invisible, then the introduction of this glass made no alteration in that respect. In this state of circumstances the force of the electro-magnet was developed, by sending an electric current through its coils, and immediately the image of the lamp-flame became visible, and continued so as long as the arrangement continued magnetic. On stopping the electric current, and so causing the magnetic force to cease, the light instantly disappeared; these phenomena could be renewed at pleasure, at any instant of time, and upon any occasion, showing a perfect dependence of cause and effect.

2153. The voltaic current which I used upon this occasion, was that of five pair of Grove's construction, and the electro-magnets were of such power that the poles would singly sustain a weight of from twenty-eight to fifty-six, or more, pounds. A person looking for the phenomenon for the first time would not be able to see it with a weak magnet.

2154. The character of the force thus impressed upon the diamagnetic is that of *rotation*; for when the image of the lamp-flame has thus been rendered visible, revolution of the eye-piece to the right or left, more or less, will cause its extinction; and the further motion of the eye-piece to the one side or other of this position will produce the reappearance of the light, and that with complementary tints, according as this further motion is to the right- or left-hand.

2155. When the pole nearest to the observer was a marked pole, *i. e.* the same as the north end of a magnetic needle, and the further pole was unmarked, the rotation of the ray was right-handed; for the eye-piece had to be turned to the right-hand, or clock fashion, to overtake the ray and restore the image to its first condition. When the poles were reversed, which was instantly done by changing the direction of the electric current, the rotation was changed also and became left-handed, the altera-

* Philosophical Transactions, 1830, p. 1. I cannot resist the occasion which is thus offered to me of mentioning the name of Mr. Anderson, who came to me as an assistant in the glass experiments, and has remained ever since in the Laboratory of the Royal Institution. He has assisted me in all the researches into which I have entered since that time, and to his care, steadiness, exactitude, and faithfulness in the performance of all that has been committed to his charge, I am much indebted.—M. F.

tion being to an equal degree in extent as before. The direction was always the same for the same *line of magnetic force* (2149.).

2156. When the diamagnetic was placed in the numerous other positions, which can easily be conceived, about the magnetic poles, results were obtained more or less marked in extent, and very definite in character, but of which the phenomena just described may be considered as the chief example: they will be referred to, as far as is necessary, hereafter.

2157. The same phenomena were produced in the silicated borate of lead (2151.) by the action of a good ordinary steel horse-shoe magnet, no electric current being now used. The results were feeble, but still sufficient to show the perfect identity of action between electro-magnets and common magnets in this their power over light.

2158. Two magnetic poles were employed end-ways, *i. e.* the cores of the electro-magnets were hollow iron cylinders, and the ray of polarized light passed along their axes and through the diamagnetic placed between them: the effect was the same.

2159. One magnetic pole only was used, that being one end of a powerful cylinder electro-magnet. When the heavy glass was beyond the magnet, being close to it but between the magnet and the polarizing reflector, the rotation was in one direction, dependent on the nature of the pole; when the diamagnetic was on the near side, being close to it but between it and the eye, the rotation for the same pole was in the contrary direction to what it was before; and when the magnetic pole was changed, both these directions were changed with it. When the heavy glass was placed in a corresponding position to the pole, but above or below it, so that the *magnetic curves* were no longer passing through the glass parallel to the ray of polarized light, but rather perpendicular to it, then no effect was produced. These particularities may be understood by reference to fig. 1, where *a* and *b* represent the first positions of the diamagnetic, and *c* and *d* the latter positions, the course of the ray being marked by the dotted line. If also the glass were placed directly at the end of the magnet, then no effect was produced on a ray passing in the direction here described, though it is evident, from what has been already said (2155.), that a ray passing *parallel* to the magnetic lines through the glass so placed, would have been affected by it.

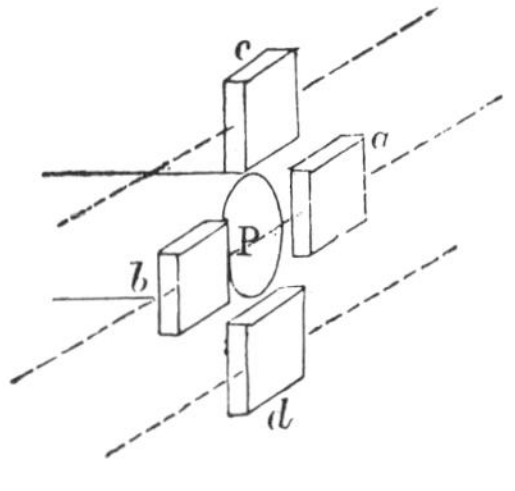

2160. Magnetic lines, then, in passing through silicated borate of lead, and a great number of other substances (2173.), cause these bodies to act upon a polarized ray of light when the lines are parallel to the ray, or in proportion as they are parallel to it: if they are perpendicular to the ray, they have no action upon it. They give the diamagnetic the power of rotating the ray; and the *law* of this action on light is, that if a magnetic line of force be *going from* a north pole, or *coming* from a south pole, along the path of a polarized ray coming to the observer, it will rotate that ray to the right-hand; or, that if such a line of force be coming from a north pole, or going from a south pole, it will rotate such a ray to the left-hand.

2161. If a cork or a cylinder of glass, representing the diamagnetic, be marked at its ends with the letters N and S, to represent the poles of a magnet, the line joining these letters may be considered as a magnetic line of force; and further, if a line be traced round the cylinder with arrow heads on it to represent direction, as in the figure, such a simple model, held up before the eye, will express the whole of the law, and give every position and consequence of direction resulting from it. If a watch be considered as the diamagnetic, the north pole of a magnet being imagined against the face, and a south pole against the back, then the motion of the hands will indicate the direction of rotation which a ray of light undergoes by magnetization.

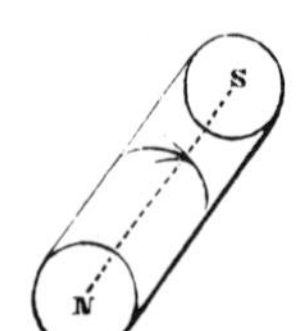

2162. I will now proceed to the different circumstances which affect, limit, and define the extent and nature of this new power of action on light.

2163. In the first place, the rotation appears to be in proportion to the extent of the diamagnetic through which the ray and the magnetic lines pass. I preserved the strength of the magnet and the interval between its poles constant, and then interposed different pieces of the same heavy glass (2151.) between the poles. The greater the extent of the diamagnetic in the line of the ray, whether in one, two, or three pieces, the greater was the rotation of the ray; and, as far as I could judge by these first experiments, the amount of rotation was exactly proportionate to the extent of diamagnetic through which the ray passed. No addition or diminution of the heavy glass on the *side* of the course of the ray made any difference in the effect of that part through which the ray passed.

2164. The power of rotating the ray of light *increased* with the intensity of the magnetic lines of force. This general effect is very easily ascertained by the use of electro-magnets; and within such range of power as I have employed, it appears to be directly proportionate to the intensity of the magnetic force.

2165. Other bodies, besides the heavy glass, possess the same power of becoming, under the influence of magnetic force, active on light (2173.). When these bodies possess a rotative power of their own, as is the case with oil of turpentine, sugar, tartaric acid, tartrates, &c., the effect of the magnetic force is to add to, or subtract from, their specific force, according as the natural rotation and that induced by the magnetism is right or left-handed (2231.).

2166. I could not perceive that this power was affected by any degree of motion which I was able to communicate to the diamagnetic, whilst jointly subject to the action of the magnetism and the light.

2167. The interposition of copper, lead, tin, silver, and other ordinary non-magnetic bodies in the course of the magnetic curves, either between the pole and the diamagnetic, or in other positions, produced no effect either in kind or degree upon the phenomena.

2168. Iron frequently affected the results in a very considerable degree; but it always appeared to be, either by altering the direction of the magnetic lines, or disposing within itself of their force. Thus, when the two contrary poles were on one side of the polarized ray (2150.), and the heavy glass in its best position between them and in the ray (2152.), the bringing of a large piece of iron near to the glass on the other side of the ray, caused the power of the diamagnetic to fall. This was because certain lines of magnetic force, which at first passed through the glass parallel to the ray, now crossed the glass and the ray; the iron giving two contrary poles opposite the poles of the magnet, and thus determining a new course for a certain portion of the magnetic power, and that across the polarized ray.

2169. Or, if the iron, instead of being applied on the opposite side of the glass, were applied on the same side with the magnet, either near it or in contact with it, then, again, the power of the diamagnetic fell, simply because the power of the magnet was diverted from it into a new direction. These effects depend much of course on the intensity and power of the magnet, and on the size and softness of the iron.

2170. The electro-helices (2190.) without the iron cores were very feeble in power, and indeed hardly sensible in their effect. With the iron cores they were powerful, though no more electricity was then passing through the coils than before (1071.). This shows, in a very simple manner, that the phenomena exhibited by light under these circumstances, is directly connected with the magnetic form of force supplied by the arrangement. Another effect which occurred illustrated the same point. When the contact at the voltaic battery is made, and the current sent round the electro-magnet, the image produced by the rotation of the polarized ray does not rise up to its full lustre immediately, but increases for a couple of seconds, gradually acquiring its greatest intensity; on breaking the contact, it sinks instantly and disappears apparently at once. The gradual rise in brightness is due to the *time* which the iron core of the magnet requires to evolve all that magnetic power which the electric current can develope in it; and as the magnetism rises in intensity, so does its effect on the light increase in power; hence the progressive condition of the rotation.

2171. I cannot as yet find that the heavy glass (2151.), when in this state, *i. e.* with magnetic lines of force passing through it, exhibits any increased degree, or has any specific magneto-inductive action of the recognized kind. I have placed it in large quantities, and in different positions, between magnets and magnetic needles, having at the time very delicate means of appreciating any difference between it and air, but could find none.

2172. Using water, alcohol, mercury, and other fluids contained in very large delicate thermometer-shaped vessels, I could not discover that any difference in volume occurred when the magnetic curves passed through them.

2173. It is time that I should pass to a consideration of this power of magnetism over light as exercised, not only in the silicated borate of lead (2151.), but in many other substances; and here we perceive, in the first place, that if all transparent bodies possess the power of exhibiting the action, they have it in very different degrees, and that up to this time there are some that have not shown it at all.

2174. Next, we may observe, that bodies that are exceedingly different to each other in chemical, physical, and mechanical properties, develope this effect; for solids and liquids, acids, alkalies, oils, water, alcohol, ether, all possess the power.

2175. And lastly, we may observe, that in all of them, though the degree of action may differ, still it is always the same in kind, being a rotative power over the ray of light; and further, the direction of the rotation is, in every case, independent of the nature or state of the substance, and dependent upon the direction of the magnetic line of force, according to the law before laid down (2160.).

2176. Amongst the substances in which this power of action is found, I have already distinguished the *silico-borate of lead* (2151.) as eminently fitted for the purpose of exhibiting the phenomena. I regret that it should be the best, since it is not likely to be in the possession of many, and few will be induced to take the trouble of preparing it. If made, it should be well annealed, for otherwise the pieces will have considerable power of depolarizing light, and then the particular phenomena under consideration are much less strikingly observed. The *borate of lead*, however, is a substance much more fusible, softening at the heat of boiling oil, and therefore far more easily prepared in the form of glass plates and annealed; and it possesses as much magneto-rotative power over light as the silico-borate itself. *Flint-glass* exhibits the property, but in a less degree than the substances above. Crown-glass shows it, but in a still smaller degree.

2177. Whilst employing crystalline bodies as diamagnetics, I generally gave them that position in which they did not affect the polarized ray, and then induced the magnetic curves through them. As a class, they seemed to resist the assumption of the rotating state. *Rock-salt* and *fluor-spar* gave evidence of the power in a slight degree; and I think that a crystal of alum did the same, but its ray length in the transparent part was so small that I could not ascertain the fact decisively. Two specimens of transparent fluor, lent me by Mr. TENNANT, gave the effect.

2178. Rock-crystal, four inches across, gave no indications of action on the ray, neither did smaller crystals, nor cubes about three-fourths of an inch in the side, which were so cut as to have two of their faces perpendicular to the axis of the crystal (1692, 1693.), though they were examined in every direction.

2179. *Iceland spar* exhibited no signs of effect, either in the form of rhomboids, or of cubes like those just described (1695.).

2180. *Sulphate of baryta, sulphate of lime,* and *carbonate of soda,* were also without action on the light.

2181. A piece of fine clear *ice* gave me no effect. I cannot however say there is none, for the effect of water in the same mass would be very small, and the irregularity of the flattened surface from the fusion of the ice and flow of water, made the observation very difficult.

2182. With some degree of curiosity and hope, I put gold-leaf into the magnetic lines, but could perceive no effect. Considering the extremely small dimensions of the length of the path of the polarized ray in it, any positive result was hardly to be expected.

2183. In experiments with liquids, a very good method of observing the effect, is to inclose them in bottles from $1\frac{1}{2}$ to 3 or 4 inches in diameter, placing these in succession between the magnetic poles (2150.), and bringing the analysing eye-piece so near to the bottle, that, by adjustment of the latter, its cylindrical form may cause a diffuse but useful image of the lamp-flame to be seen through it: the light of this image is easily distinguished from that which passes by irregular refraction through the striæ and deformations of the glass, and the phenomena being looked for in this light are easily seen.

2184. Water, alcohol, and ether, all show the effect; water most, alcohol less, and ether the least. All the fixed oils which I have tried, including almond, castor, olive, poppy, linseed, sperm, elaine from hog's lard, and distilled resin oil, produce it. The essential oils of turpentine, bitter almonds, spike lavender, lavender, jessamine, cloves, and laurel, produce it. Also naphtha of various kinds, melted spermaceti, fused sulphur, chloride of sulphur, chloride of arsenic, and every other liquid substance which I had at hand and could submit in sufficient bulk to experiment.

2185. Of aqueous solutions I tried 150 or more, including the soluble acids, alkalies and salts, with sugar, gum, &c., the list of which would be too long to give here, since the great conclusion was, that the exceeding diversity of substance caused no exception to the general result, for all the bodies showed the property. It is indeed more than probable, that in all these cases the water and not the other substance present was the ruling matter. The same general result was obtained with alcoholic solutions.

2186. Proceeding from liquids to air and gaseous bodies, I have here to state that, as yet, I have not been able to detect the exercise of this power in any one of the substances in this class. I have tried the experiment with bottles 4 inches in diameter, and the following gases: oxygen, nitrogen, hydrogen, nitrous oxide, olefiant gas, sulphurous acid, muriatic acid, carbonic acid, carbonic oxide, ammonia, sulphuretted hydrogen, and bromine vapour, at ordinary temperatures; but they all gave negative results. With air, the trial has been carried, by another form of apparatus, to a much higher degree, but still ineffectually (2212.).

2187. Before dismissing the consideration of the substances which exhibited this power, and in reference to those in which it was superinduced upon bodies possessing, naturally, rotative force (2165. 2231.), I may record, that the following are the substances submitted to experiment: castor oil, resin oil, oil of spike lavender, of laurel, Canada balsam, alcoholic solution of camphor, alcoholic solution of camphor and corrosive sublimate, aqueous solutions of sugar, tartaric acid, tartrate of soda, tartrate of potassa and antimony, tartaric and boracic acid, and sulphate of nickel, which rotated to the right-hand; copaiba balsam, which rotated the ray to the left-hand; and two specimens of camphine or oil of turpentine, in one of which the rotation was to the right-hand, and in the other to the left. In all these cases, as already said (2165.), the superinduced magnetic rotation was according to the general law (2160.), and without reference to the previous power of the body.

2188. Camphor being melted in a tube about an inch in diameter, exhibited high natural rotative force, but I could not discover that the magnetic curves induced additional force in it. It may be, however, that the shortness of the ray length and the quantity of coloured light left, even when the eye-piece was adjusted to the most favourable position for darkening the image produced by the naturally rotated ray, rendered the small magneto-power of the camphor insensible.

¶ ii. *Action of electric currents on light.*

2189. From a consideration of the nature and position of the lines of magnetic and electric force, and the relation of a magnet to a current of electricity, it appeared almost certain that an electric current would give the same result of action on light as a magnet; and, in the helix, would supply a form of apparatus in which great lengths of diamagnetics, and especially of such bodies as appeared to be but little affected between the poles of the magnet, might be submitted to examination and their effect exalted: this expectation was, by experiment, realized.

2190. Helices of copper wire were employed, three of which I will refer to. The first, or *long helix,* was 0·4 of an inch internal diameter; the wire was 0·03 of an inch in diameter, and having gone round the axis from one end of the helix to the other, then returned in the same manner, forming a coil sixty-five inches long, double in its whole extent, and containing 1240 feet of wire.

2191. The second, or *medium helix,* is nineteen inches long, 1·87 inch internal diameter, and three inches external diameter. The wire is 0·2 of an inch in diameter, and eighty feet in length, being disposed in the coil as two concentric spirals. The electric current, in passing through it, is not divided, but traverses the whole length of the wire.

2192. The third, or *Woolwich helix,* was made under my instruction for the use of Lieut.-Colonel Sabine's establishment at Woolwich. It is 26·5 inches long, 2·5 inches internal diameter, and 4·75 inches external diameter. The wire is 0·17 of an inch in diameter, and 501 feet in length. It is disposed in the coil in four concentric spirals connected end to end, so that the whole of the electric current employed passes through all the wire.

2193. The long helix (2190.) acted very feebly on a magnetic needle placed at a little distance from it; the medium helix (2191.) acted more powerfully, and the Woolwich helix (2192.) very strongly; the same battery of ten pairs of Grove's plate being employed in all cases.

2194. Solid bodies were easily subjected to the action of these electro-helices, being for that purpose merely cut into the form of bars or prisms with flat and polished ends, and then introduced as cores into the helices. For the purpose of submitting liquid bodies to the same action, tubes of glass were provided, furnished at the ends with caps; the cylindrical part of the cap was brass, and had a tubular aperture for the introduction of the liquids, but the end was a flat glass plate. When the tube was intended to contain aqueous fluids, the plates were attached to the caps, and the caps to the tube by Canada balsam; when the tube had to contain alcohol, ether or essential oils, a thick mixture of powdered gum with a little water was employed as the cement.

2195. The general effect produced by this form of apparatus may be stated as follows:—The tube within the long helix (2190.) was filled with distilled water and placed in the line of the polarized ray, so that by examination through the eye-piece (2150.), the image of the lamp-flame produced by the ray could be seen through it. Then the eye-piece was turned until the image of the flame disappeared, and, afterwards, the current of ten pairs of plates sent through the helix; instantly the image of the flame reappeared, and continued as long as the electric current was passing through the helix; on stopping the current the image disappeared. The light did not rise up gradually, as in the case of electro-magnets (2170.), but instantly. These results could be produced at pleasure. In this experiment we may, I think, justly say that a ray of light is electrified and the electric forces illuminated.

2196. The phenomena may be made more striking, by the adjustment of a lens of long focus between the tube and the polarizing mirror, or one of short focus between the tube and the eye; and where the helix, or the battery, or the substance experimented with, is feeble in power, such means offer assistance in working out the effects: but, after a little experience, they are easily dispensed with, and are only useful as accessories in doubtful cases.

2197. In cases where the effect is feeble, it is more easily perceived if the Nichol eye-piece be adjusted, not to the perfect extinction of the ray, but a little short of or beyond that position; so that the image of the flame may be but just visible. Then, on the exertion of the power of the electric current, the light is either increased in intensity, or else diminished, or extinguished, or even re-illuminated on the other side of the dark condition; and this change is more easily perceived than if the eye began to observe from a state of utter darkness. Such a mode of observing also assists in demonstrating the rotatory character of the action on light; for, if the light be made visible beforehand by the motion of the eye-piece in one direction, and the power of the current be to *increase* that light, an instant only suffices, after stopping the current, to move the eye-piece in the other direction until the light is apparent as at first, and then the power of the current will be to *diminish* it; the tints of the lights being affected also at the same time.

2198. When the current was sent round the helix in one direction, the rotation induced upon the ray of light was one way; and when the current was changed to the contrary direcfion, the rotation was the other way. In order to express the direction, I will assume, as is usually done, that the current passes from the zinc through the acid to the platinum in the same cell (663. 667. 1627.): if such a current pass under the ray towards the right, upwards on its right side, and over the ray towards the left, it will give left-handed rotation to it; or, if the current pass over the ray to the right, down on the right side, and under it towards the left, it will induce it to rotate to the right-hand.

2199. The LAW, therefore, by which an electric current acts on a ray of light is easily expressed. When an electric current passes round a ray of polarized light in a plane perpendicular to the ray, it causes the ray to revolve on its axis, as long as it is under the influence of the current, in the *same direction* as that in which the current is passing.

2200. The simplicity of this law, and its identity with that given before, as expressing the action of magnetism on light (2160.), is very beautiful. A model is not wanted to assist the memory; but if that already described (2161.) be looked at, the line round it will express at the same time the direction both of the current and the rotation. It will indeed do much more; for if the cylinder be considered as a piece of iron, and not a piece of glass or other diamagnetic, placed between the two poles N and S, then the line round it will represent the direction of the currents, which, according to AMPERE's theory, are moving round its particles; or if it be considered as a core of iron (in place of a core of water), having an electric current running round it in the direction of the line, it will also represent such a magnet as would be formed if it were placed between the poles whose marks are affixed to its ends.

2201. I will now notice certain points respecting the degree of this action under different circumstances. By using a tube of water (2194.) as long as the helix, but placing it so that more or less of the tube projected at either end of the helix, I was able, in some degree, to ascertain the effect of length of the diamagnetic, the force of the helix and current remaining the same. The greater the column of water subjected to the action of the helix, the greater was the rotation of the polarized ray; and the amount of rotation seemed to be directly proportionate to the length of fluid round which the electric current passed.

2202. A short tube of water, or a piece of heavy glass, being placed in the axis of the Woolwich helix (2192.), seemed to produce equal effect on the ray of light, whether it were in the middle of the helix or at either end; provided it was always within the helix and in the line of the axis. From this it would appear that every part of the helix has the same effect; and, that by using long helices, substances may be submitted to this kind of examination which could not be placed in sufficient length between the poles of magnets (2150.).

2203. A tube of water as long as the Woolwich helix (2192.), but only 0·4 of an inch in diameter, was placed in the helix parallel to the axis, but sometimes in the axis and sometimes near the side. No apparent difference was produced in these different situations; and I am inclined to believe (without being quite sure) that the action on the ray is the same, wherever the tube is placed, within the helix, in relation to the axis. The same result was obtained when a larger tube of water was looked through, whether the ray passed through the axis of the helix and tube, or near the side.

2204. If bodies be introduced into the helix possessing, naturally, rotating force, then the rotating power given by the electric current is superinduced upon them, exactly as in the cases already described of magnetic action (2165. 2187.).

2205. A helix, twenty inches long and 0·3 of an inch in diameter, was made of uncovered copper wire, 0·05 of an inch in diameter, in close spirals. This was placed in a large tube of water, so that the fluid, both in the inside and at the outside of the helix, could be examined by the polarized ray. When the current was sent *through* the helix, the water within it received rotating power; but no trace of such an action on the light was seen on the outside of the helix, even in the line most close to the uncovered wire.

2206. The water was inclosed in brass and copper tubes, but this alteration caused not the slightest change in the effect.

2207. The water in the brass tube was put into an *iron* tube, much longer than either the Woolwich helix or the brass tube, and quite one eighth of an inch thick in the side; yet when placed in the Woolwich helix (2192.), the water rotated the ray of light apparently as well as before.

2208. An iron bar, one inch square and longer than the helix, was put into the helix, and the small water tube (2203.) upon it. The water exerted as much action on the light as before.

2209. Three iron tubes, each twenty-seven inches long and one-eighth of an inch in thickness in the side, were selected of such diameters as to pass easily one into the other, and the whole into the Woolwich helix (2192.). The smaller one was supplied with glass ends and filled with water; and being placed in the axis of the Woolwich helix, had a certain amount of rotating power over the polarized ray. The second tube was then placed over this, so that there was now a thickness of iron equal to two-eighths of an inch between the water and the helix; the water had *more* power of rotation than before. On placing the third tube of iron over the two former, the power of the water *fell*, but was still very considerable. These results are complicated, being dependent on the new condition which the character of iron gives to its action on the forces. Up to a certain amount, by increasing the development of magnetic forces, the helix and core, *as a whole*, produce increased action on the water; but on the addition of more iron and the disposal of the forces through it, their action is removed in part from the water and the rotation is lessened.

2210. Pieces of heavy glass (2151.), placed in iron tubes in the helices, produced similar effects.

2211. The bodies which were submitted to the action of an electric current in a helix, in the manner already described, were as follows:—Heavy glass (2151. 2176.), water, solution of sulphate of soda, solution of tartaric acid, alcohol, ether, and oil of turpentine; all of which were affected, and acted on light exactly in the manner described in relation to magnetic action (2173.).

2212. I submitted *air* to the influence of these helices carefully and anxiously, but could not discover any trace of action on the polarized ray of light. I put the long helix (2190.) into the other two (2191. 2192.), and combined them all into one consistent series, so as to accumulate power, but could not observe any effect of them on light passing through air.

2213. In the use of helices, it is necessary to be aware of one effect, which might otherwise cause confusion and trouble. At first, the wire of the long helix (2190.) was wound directly upon the thin glass tube which served to contain the fluid. When the electric current passed through the helix it raised the temperature of the metal, and that gradually raised the temperature of the glass and the film of water in contact with it, and so the cylinder of water, warmer at its surface than its axis, acted as a lens, gathering and sending rays of light to the eye, and continuing to act for a time after the current was stopped. By separating the tube of water from the helix, and by other precautions, this source of confusion is easily avoided.

2214. Another point of which the experimenter should be aware, is the difficulty, and almost impossibility, of obtaining a piece of glass which, especially after it is cut, does not depolarize light. When it does depolarize, difference of position makes an immense difference in the appearance. By always referring to the parts that do not depolarize, as the black cross, for instance, and by bringing the eye as near as may be to the glass, this difficulty is more or less overcome.

2215. For the sake of supplying a general indication of the amount of this induced rotating force in two or three bodies, and without any pretence of offering correct numbers, I will give, generally, the result of a few attempts to measure the force, and compare it with the natural power of a specimen of oil of turpentine. A very powerful electro-magnet was employed, with a *constant* distance between its poles of $2\frac{1}{2}$ inches. In this space was placed different substances; the amount of rotation of the eye-piece observed several times and the average taken, as expressing the rotation for the ray length of substance used. But as the substances were of different dimensions, the ray lengths were, by calculation, corrected to one standard length, upon the assumption that the power was proportionate to this length (2163.). The oil of turpentine was of course observed in its natural state, i. e. without magnetic action. Making water 1, the numbers were as follows:—

Oil of turpentine	11·8
Heavy glass (2151.) . . .	6·0
Flint-glass	2·8
Rock salt	2·2

Water	1·0
Alcohol	less than water.
Ether	less than alcohol.

2216. In relation to the action of magnetic and electric forces on light, I consider, that to know the conditions under which there is no apparent action, is to add to our knowledge of their mutual relations; and will, therefore, very briefly state how I have lately combined these forces, obtaining no apparent result (955.).

2217. Heavy glass, flint-glass, rock crystal, Iceland spar, oil of turpentine, and air, had a polarized ray passed through them; and, at the same time, lines of electro-static tension (2149.) were, by means of coatings, the Leyden jar, and the electric machine, directed across the bodies, parallel to the polarized ray, and perpendicular to it, both in and across the plane of polarization; but without any visible effect. The tension of a rapidly recurring, induced secondary current, was also directed upon the same bodies and upon water (as an electrolyte), but with the same negative result.

2218. A polarized ray, powerful magnetic lines of force, and the electric lines of force (2149.) just described, were combined in various directions in their action on heavy glass (2151. 2176.), but with no other result than that due to the mutual action of the magnetic lines of light, already described in this paper.

2219. A polarized ray and electric currents were combined in every possible way in electrolytes (951–954). The substances used were distilled water, solution of sugar, dilute sulphuric acid, solution of sulphate of soda, using platinum electrodes; and solution of sulphate of copper, using copper electrodes: the current was sent along the ray, and perpendicular to it in two directions at right angles with each other; the ray was made to rotate, by altering the position of the polarizing mirror, that the plane of polarization might be varied; the current was used as a continuous current, as a rapidly intermitting current, and as a rapidly alternating double current of induction; but in no case was any trace of action perceived.

2220. Lastly, a ray of polarized light, electric currents, and magnetic lines of force, were directed in every possible way through dilute sulphuric acid and solution of sulphate of soda, but still with negative results, except in those positions where the phenomena already described were produced. In one arrangement, the current passed in the direction of radii from a central to a circumferential electrode, the contrary magnetic poles being placed above and below; and the arrangements were so good, that when the electric current was passing, the fluid rapidly rotated; but a polarized ray sent horizontally across this arrangement was not at all affected. Also, when the ray was sent vertically through it, and the eye-piece moved to correspond to the rotation impressed upon the ray in this position by the magnetic curves alone, the superinduction of the passage of the electric current made not the least difference in the effect upon the ray.

¶ iii. *General considerations.*

2221. Thus is established, I think for the first time*, a true, direct relation and dependence between light and the magnetic and electric forces; and thus a great addition made to the facts and considerations which tend to prove that all natural forces are tied together, and have one common origin (2146.). It is, no doubt, difficult in the present state of our knowledge to express our expectation in exact terms; and, though I have said that another of the powers of nature is, in these experiments, directly related to the rest, I ought, perhaps, rather to say that another form of the great power is distinctly and directly related to the other forms; or, that the great power manifested by particular phenomena in particular forms, is here further identified and recognised, by the direct relation of its form of light to its forms of electricity and magnetism.

2222. The relation existing between *polarized* light and magnetism and electricity, is even more interesting than if it had been shown to exist with common light only. It cannot but extend to common light; and, as it belongs to light made, in a certain respect, more precise in its character and properties by polarization, it collates and connects it with these powers, in that duality of character which they possess, and yields an opening, which before was wanting to us, for the appliance of these powers to the investigation of the nature of this and other radiant agencies.

2223. Referring to the conventional distinction before made (2149.), it may be again stated, that it is the magnetic lines of force *only* which are effectual on the rays of light, and they *only* (in appearance) when parallel to the ray of light, or as they tend to parallelism with it. As, in reference to matter not magnetic after the manner of iron, the phenomena of electric induction and electrolyzation show a vast superiority in the energy with which electric forces can act as compared to magnetic forces, so here, in another direction and in the peculiar and correspondent effects which belong to magnetic forces, they are shown, in turn, to possess great superiority, and to have their full equivalent of action on the same kind of matter.

2224. The magnetic forces do not act on the ray of light directly and without the intervention of matter, but through the mediation of the substance in which they and the ray have a simultaneous existence; the substances and the forces giving to and receiving from each other the power of acting on the light. This is shown by the

* I say, for the first time, because I do not think that the experiments of Morrichini on the production of magnetism by the rays at the violet end of the spectrum prove any such relation. When in Rome with Sir H. Davy in the month of May 1814, I spent several hours at the house of Morrichini, working with his apparatus and under his directions, but could not succeed in magnetising a needle. I have no confidence in the effect as a *direct* result of the action of the sun's rays; but think, that when it has occurred it has been secondary, incidental, and perhaps even accidental; a result that might well happen with a needle that was preserved during the whole experiment in a north and south position.

January 2, 1846.—I should not have written "for the first time" as above, if I had remembered Mr. Christie's experiments and papers on the Influence of the Solar Rays on Magnets, communicated in the Philosophical Transactions for 1826, p. 219, and 1828, p. 379.—M. F.

non-action of a vacuum, of air or gases; and it is also further shown by the special degree in which different matters possess the property. That magnetic force acts upon the ray of light always with the same character of manner and in the same direction, independent of the different varieties of substance, or their states of solid or liquid, or their specific rotative force (2232.), shows that the magnetic force and the light have a direct relation: but that substances are necessary, and that these act in different degrees, shows that the magnetism and the light act on each other through the intervention of the matter.

2225. Recognizing or perceiving *matter* only by its powers, and knowing nothing of any imaginary nucleus, abstract from the idea of these powers, the phenomena described in this paper much strengthen my inclination to trust in the views I have on a former occasion advanced in reference to its nature*.

2226. It cannot be doubted that the magnetic forces act upon and affect the internal constitution of the diamagnetic, just as freely in the dark as when a ray of light is passing through it; though the phenomena produced by light seem, as yet, to present the only means of observing this constitution and the change. Further, any such change as this must belong to opake bodies, such as wood, stone, and metal; for as diamagnetics, there is no distinction between them and those which are transparent. The degree of transparency can at the utmost, in this respect, only make a distinction between the individuals of a class.

2227. If the magnetic forces had made these bodies magnets, we could, by light, have examined a transparent magnet; and that would have been a great help to our investigation of the forces of matter. But it does not make them magnets (2171.), and therefore the molecular condition of these bodies, when in the state described, must be specifically distinct from that of magnetized iron, or other such matter, and must be *a new magnetic condition*; and as the condition is a state of tension (manifested by its instantaneous return to the normal state when the magnetic induction is removed), so the *force* which the matter in this state possesses and its mode of action, must be to us a *new magnetic force* or *mode of action* of matter.

2228. For it is impossible, I think, to observe and see the action of magnetic forces, rising in intensity, upon a piece of heavy glass or a tube of water, without also perceiving that the latter acquire properties which are not only *new* to the substance, but are also in subjection to very definite and precise laws (2160. 2199.), and are equivalent in proportion to the magnetic forces producing them.

2229. Perhaps this state is a state of *electric tension tending to a current*; as in magnets, according to AMPÈRE's theory, the state is a state of *current*. When a core of iron is put into a helix, every thing leads us to believe that currents of electricity are produced within it, which rotate or move in a plane perpendicular to the axis of the helix. If a diamagnetic be placed in the same position, it acquires power to make light rotate in the same plane. The state it has received is a state of tension, but it

* A speculation, &c. Philosophical Magazine, 1844, vol. xxiv. p. 136.

has not passed on into currents, though the acting force and every other circumstance and condition are the same as those which do produce currents in iron, nickel, cobalt, and such other matters as are fitted to receive them. Hence the idea that there exists in diamagnetics, under such circumstances, a tendency to currents, is consistent with all the phenomena as yet described, and is further strengthened by the fact, that, leaving the loadstone or the electric current, which by inductive action is rendering a piece of iron, nickel, or cobalt magnetic, perfectly unchanged, a mere change of temperature will take from these bodies their extra power, and make them pass into the common class of diamagnetics.

2230. The present is, I believe, the first time that the molecular condition of a body, required to produce the circular polarization of light, has been artificially given; and it is therefore very interesting to consider this known state and condition of the body, comparing it with the relatively unknown state of those which possess the power naturally: especially as some of the latter rotate to the right-hand and others to the left; and, as in the cases of quartz and oil of turpentine, the same body chemically speaking, being in the latter instance a liquid with particles free to move, presents different specimens, some rotating one way and some the other.

2231. At first one would be inclined to conclude that the natural state and the state conferred by magnetic and electric forces must be the same, since the effect is the same; but on further consideration it seems very difficult to come to such a conclusion. Oil of turpentine will rotate a ray of light, the power depending upon its particles and not upon the arrangement of the mass. Whichever way a ray of polarized light passes through this fluid, it is rotated in the same manner; and rays passing in every possible direction through it *simultaneously* are all rotated with equal force and according to one common law of direction; *i. e.* either all right-handed or else all to the left. Not so with the rotation superinduced on the *same* oil of turpentine by the magnetic or electric forces: it exists only in one direction, *i. e.* in a plane perpendicular to the magnetic line; and being limited to this plane, it can be changed in direction by a reversal of the direction of the inducing force. The direction of the rotation produced by the natural state is connected invariably with the direction of the ray of light; but the power to produce it appears to be possessed in every direction and at all times by the particles of the fluid: the direction of the rotation produced by the induced condition is connected invariably with the direction of the magnetic line or the electric current, and the condition is possessed by the particles of matter, but strictly limited by the line or the current, changing and disappearing with it.

2232. Let *m*, in fig. 3, represent a glass cell filled with oil of turpentine, possessing naturally the power of producing right-hand rotation, and *a b* a polarized ray of light. If the ray proceed from *a* to *b*, and the eye be placed at *b*, the rotation will

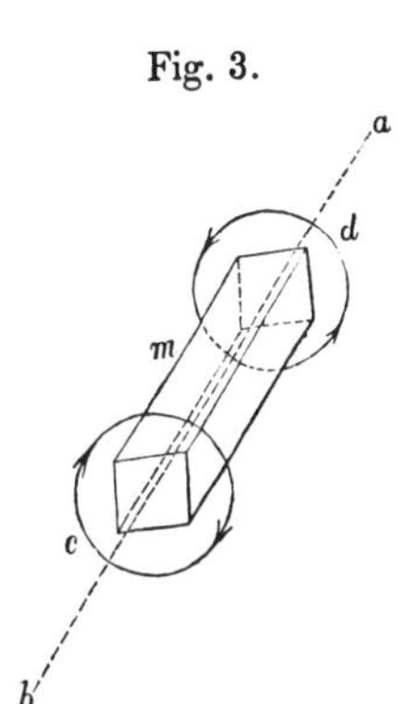

be right-handed, or according to the direction expressed by the arrow heads on the circle *c*; if the ray proceed from *b* to *a*, and the eye be placed at *a*, the rotation will still be right-handed *to the observer, i. e.* according to the direction indicated on the circle *d*. Let now an electric current pass round the oil of turpentine in the direction indicated on the circle *c*, or magnetic poles be placed so as to produce the same effect (2155.); the particles will acquire a further rotative force (which no motion amongst themselves will disturb), and a ray coming from *a* to *b* will be seen by an eye placed at *b* to rotate to the right-hand more than before, or in the direction on the circle *c*; but pass a ray from *b* to *a*, and observe with the eye at *a*, and the phenomenon is no longer the same as before; for instead of the new rotation being according to the direction indicated on the circle *d*, it will be in the contrary direction, or to the observer's left-hand (2199.). In fact the induced rotation will be added to the natural rotation as respects a ray passing from *a* to *b*, but it will be subtracted from the natural rotation as regards the ray passing from *b* to *a*. Hence the particles of this fluid which rotate by virtue of their natural force, and those which rotate by virtue of the induced force, cannot be in the same condition.

2233. As respects the power of the oil of turpentine to rotate a ray in whatever direction it is passing through the liquid, it may well be, that though all the particles possess the power of rotating the light, only those whose planes of rotation are more or less perpendicular to the ray affect it; and that it is the resultant or sum of forces in any one direction which is active in producing rotation. But even then a striking difference remains, because the resultant in the same plane is not absolute in direction, but relative to the course of the ray, being in the one case as the circle *c*, and in the other as the circle *d*, fig. 3; whereas the resultant of the magnetic or electric induction is absolute, and not changing with the course of the ray, being always either as expressed by *c* or else as indicated by *d*.

2234. All these differences, however, will doubtless disappear or come into harmony as these investigations are extended; and their very existence opens so many paths, by which we may pursue our inquiries, more and more deeply, into the powers and constitution of matter.

2235. Bodies having rotating power of themselves, do not seem by that to have a greater or a less tendency to assume a further degree of the same force under the influence of magnetic or electric power.

2236. Were it not for these and other differences, we might see an analogy between these bodies, which possess at all times the rotating power, as a specimen of quartz which rotates only in one plane, and those to which the power is given by the induction of other forces, as a prism of heavy glass in a helix, on the one hand; and, on the other, a natural magnet and a helix through which the current is passing.

The natural condition of the magnet and quartz, and the constrained condition of the helix and heavy glass, form the link of the analogy in one direction; whilst the supposition of currents existing in the magnet and helix, and only a tendency or tension to currents existing in the quartz and heavy glass, supplies the link in the transverse direction.

2237. As to those bodies which seem as yet to give no indication of the power over light, and therefore none of the assumption of the new magnetic conditions; these may be divided into two classes, the one including air, gases and vapours, and the other rock crystal, Iceland spar, and certain other crystalline bodies. As regards the latter class, I shall give, in the next series of these researches, proofs drawn from phenomena of an entirely different kind, that they do acquire the new magnetic condition; and these being so disposed of for the moment, I am inclined to believe that even air and gases have the power to assume the peculiar state, and even to affect light, but in a degree so small that as yet it has not been made sensible. Still the gaseous state is such a remarkable condition of matter, that we ought not too hastily to assume that the substances which, in the solid and liquid state, possess properties even general in character, always carry these into their gaseous condition.

2238. Rock-salt, fluor-spar, and, I think, alum, affect the ray of light; the other crystals experimented with did not; these are equiaxed and singly refracting, the others are unequiaxed and doubly refracting. Perhaps these instances, with that of the rotation of quartz, may even now indicate a relation between magnetism, electricity, and the crystallizing forces of matter.

2239. All bodies are affected by helices as by magnets, and according to laws which show that the causes of the action are identical as well as the effects. This result supplies another fine proof in favour of the identity of helices and magnets, according to the views of AMPÈRE.

2240. The theory of static induction which I formerly ventured to set forth (1161, &c.), and which depends upon the action of the contiguous particles of the dielectric intervening between the inductric and the inducteous bodies, led me to expect that the same kind of dependence upon the intervening particles would be found to exist in magnetic action; and I published certain experiments and considerations on this point seven years ago (1709—1736.). I could not then discover any peculiar condition of the intervening substance or diamagnetic; but now that I have been able to make out such a state, which is not only a state of tension (2227.), but dependent entirely upon the magnetic lines which pass through the substance, I am more than ever encouraged to believe that the view then advanced is correct.

2241. Although the magnetic and electric forces *appear* to exert no power on the ordinary or on the depolarized ray of light, we can hardly doubt but that they have some special influence, which probably will soon be made apparent by experiment. Neither can it be supposed otherwise than that the same kind of action should take place on the other forms of radiant agents as heat and chemical force.

2242. This mode of magnetic and electric action, and the phenomena presented by it, will, I hope, greatly assist hereafter in the investigation of the nature of transparent bodies, of light, of magnets, and their action one on another, or on magnetic substances. I am at this time engaged in investigating the new magnetic condition, and shall shortly send a further account of it to the Royal Society. What the possible effect of the force may be in the earth as a whole, or in magnets, or in relation to the sun, and what may be the best means of causing light to evolve electricity and magnetism, are thoughts continually pressing upon the mind; but it will be better to occupy both time and thought, aided by experiment, in the investigation and development of real truth, than to use them in the invention of suppositions which may or may not be founded on, or consistent with fact.

Royal Institution,
Oct. 29, 1845.

10

Reprinted from *Trans. Cambridge Phil. Soc.*, **9**, 399–416 (1852)

XVI. *On the Composition and Resolution of Streams of Polarized Light from different Sources. By* G. G. STOKES, M.A., *Fellow of Pembroke College, and Lucasian Professor of Mathematics in the University of Cambridge.*

[Read *Feb.* 16 and *March* 15, 1852.]

WHEN a stream of polarized light is decomposed into two streams which, after having been modified in a slightly different manner, are reunited, the mixture is found to have acquired properties which are quite distinct from those of the original stream, and give rise to a number of curious and apparently complicated phenomena. These phenomena have now, however, through the labours of Young and Fresnel, been completely reduced to law, and embraced in a theory, the wonderful simplicity of which is such as to bear with it the stamp of truth. But when two polarized streams from different sources mix together, the mixture possesses properties intermediate between those of the original streams, and none of the curious phenomena depending upon the interference of polarized light are manifested. The properties of such mixtures form but an uninviting subject of investigation; and accordingly, though to a certain extent they are obvious, and must have forced themselves upon the attention of all who have paid any special attention to the physical theory of light, they do not seem hitherto to have been studied in detail.

Were the only object of such a study to enable us to calculate with greater facility the results obtained by means of certain complicated combinations, the subject might deservedly be deemed of small importance. For the object of the philosopher is not to complicate, but to simplify and analyze, so as to reduce phenomena to laws, which in their turn may be made the stepping-stones for ascending to a general theory which shall embrace them all; and when such a theory has been arrived at, and thoroughly verified, the task of deducing from it the results which ought to be observed under a combination of circumstances which has nothing to recommend it for consideration but its complexity, may well be abandoned for new and more fertile fields of research. But in the present case certain difficulties seem to have arisen respecting the connexion between common and elliptically polarized light which it needed only a more detailed study of the laws of combination of polarized light to overcome; and accordingly the subject may be deemed not wholly devoid of importance.

The early part of the following paper is devoted to a demonstration of various properties of elliptically polarized light, and of oppositely polarized streams. When two streams of light are called oppositely polarized, it is meant that, so far as relates to its state of polarization, one stream is what the other becomes when it is turned in azimuth through 90^0, and has its nature reversed as regards right-handed and left-handed. Most, if not all, of these properties have doubtless already occurred to persons studying the subject, but I am not aware of any formal demonstrations of them which have been published; and indeed some artifices were required in order to avoid being encumbered in the demonstrations with long analytical expressions. The combination of several independent polarized streams is next considered, and with respect to this subject a proposition is proved which may be regarded as the capital theorem of the paper. It is as follows.

When any number of independent polarized streams, of given refrangibility, are mixed together, the nature of the mixture is completely determined by the values of four constants, which are certain functions of the intensities of the streams, and of the azimuths and eccentricities of the ellipses by which they are respectively characterized; so that any two groups of polarized streams which furnish the same values for each of these four constants are optically equivalent.

It is a simple consequence of this theorem, that any group of polarized streams is equivalent to a stream of common light combined with a stream of elliptically polarized light from a different source. The intensities of these two streams, as well as the azimuth and eccentricity of the ellipse which characterizes the latter, are determined by certain formulæ, which will be found in their place.

The general principles established in this paper bear on two questions of physical interest. Strong reasons are adduced in favour of the universality of the law, that the two polarized pencils which a doubly refracting medium of any nature is capable of propagating independently in a given direction are polarized oppositely. In strictness, we ought to speak of two series of waves rather than two pencils; for it is the fronts of the waves, not the rays, which are supposed to have a common direction. The other point alluded to relates to the distinction between common, and elliptically polarized light. It is shewn that the changes which are continually taking place in the mode of vibration may be of any nature, and that there is no occasion, in the case of common light, to suppose the transition from a series of vibrations of one kind to a series of another kind to be abrupt.

At the end of the paper the general formulæ are applied to the case of some actual experiments, but these applications are not of sufficient importance to deserve separate mention.

1. Consider a stream of light polarized in the most general way, that is, elliptically polarized, and propagated through the free ether. Let the medium be referred to the rectangular axes of x, y, z, the axis of z being measured in the direction of propagation. Let α and $\alpha + 90^0$ be the azimuths of the principal planes, that is, the planes of maximum and minimum polarization, azimuths being measured about the axis of z from x towards y. Let the rectangular components of the displacements of the ether be represented by lines drawn in the planes of polarization of the plane-polarized streams which these components, taken separately, would constitute. I make this assumption to avoid entering into the question whether the vibrations of plane-polarized light are parallel or perpendicular to the plane of polarization. If we adopt the former theory, the actual lines in the figures which we are to suppose drawn will represent in magnitude and direction the ethereal displacements; if we adopt the latter, the same will still be the case if we first suppose all our figures turned round the axis of z, in a given direction, through 90^0.

Let the co-ordinates x', y' be measured in the principal planes whose azimuths are α, $\alpha + 90^0$; let β be the angle whose tangent is equal to the ratio of the axis of the ellipse described, the numerical value of β being supposed not to lie beyond the limits 0 and 90^0; let v be the velocity of propagation, t the time, λ the length of a wave, and put for shortness,

$$\frac{2\pi}{\lambda}(vt - x) = \phi. \qquad (1).$$

Then about the time t, and at no very great distance from a given point, suppose the origin, we may represent the diplacements belonging to the given stream of elliptically polarized light by

$$x' = c\cos\beta\sin(\phi + \epsilon), \quad y' = c\sin\beta\cos(\phi + \epsilon). \qquad (2).$$

If the light be convergent or divergent, c will depend upon ϕ, but for our present object any variation of c arising from this cause will not enter into account. The value of a, which determines the direction in which x' is measured, as well as that of β, is given by the nature of the polarization. The polarization is right-handed or left-handed according to the sign of β. As to c and ϵ, the phenomena of optics oblige us to suppose that they are constant, or sensibly constant, for a great number of consecutive undulations, but that they change in an irregular manner a great number of times in the course of one second. The known rapidity of the luminous vibrations allows abundant scope for such a supposition, since c and ϵ may be constant for millions of consecutive undulations, and yet change millions of times in a second. This series of changes, rapid with respect to the duration of impressions on the retina, but slow compared with the periodic changes in the motion of the ethereal particles, is exactly what we might have expected beforehand from a consideration of the circumstances under which light is produced, so far at least as its sources are accessible to us; and thus in this point, as in so many others, the theory of undulations commends itself for its simplicity.

If c were constant c^2 would be a measure of the intensity, so long as we were only comparing different streams having the same refrangibility. But since c is liable to the changes just mentioned, if we wish to express ourselves exactly, avoiding conventional abbreviations, we must say that the intensity is measured, not by c^2, but by the mean value of c^2, which may conveniently be represented by $\mathfrak{m}(c^2)$.

2. Let us examine now whether it be always possible to resolve the given disturbance into two which, taken separately, would correspond to two elliptically polarized streams of given nature. For the sake of clear ideas, it may be supposed that the azimuths and eccentricities of the ellipses belonging to these two streams are given and invariable, while the azimuth and eccentricity of the ellipse belonging to the first stream are given for that stream, but vary from one to another of a set of streams which we wish to consider in succession.

Let x_1, c_1, &c. be for the first, and x_2, c_2, &c. be for the second stream of the pair, what x', c, &c. were for the original stream; and resolve all the displacements along the principal axes of the latter stream. Then, in order that the original disturbance may be equivalent to the pair, we must have, independently of ϕ,

$$\left.\begin{aligned} x_1\cos(a_1 - a) - y_1\sin(a_1 - a) + x_2\cos(a_2 - a) - y_2\sin(a_2 - a) &= x'; \\ x_1\sin(a_1 - a) + y_1\cos(a_1 - a) + x_2\sin(a_2 - a) + y_2\cos(a_2 - a) &= y'. \end{aligned}\right\} \quad (3)$$

Conceive x_1, y_1, x_2, y_2, x', and y' expressed in terms of ϕ by the formulæ (2) and the similar formulæ whereby x_1, y_1, &c. are expressed, and then let the sines and cosines of $\phi + \epsilon$, $\phi + \epsilon_1$, and $\phi + \epsilon_2$ be developed. In order that equations (3) may be satisfied independently of ϕ, the coefficients of $\sin\phi$ and $\cos\phi$ must separately be equal to zero, so that each of these equations will split into two. We shall thus have four equations to determine the four unknown quantities c_1, c_2, ϵ_1, and ϵ_2. For the sake of shortness, let

$$c \cos\epsilon = g, \qquad c \sin\epsilon = h,$$

whether the letters be or be not affected with suffixes; and further put

$$a_1 - a = \gamma_1, \qquad a_2 - a = \gamma_2;$$

then our four equations become

$$\left.\begin{aligned}
\cos\beta_1 \cos\gamma_1 . g_1 + \sin\beta_1 \sin\gamma_1 . h_1 + \cos\beta_2 \cos\gamma_2 . g_2 + \sin\beta_2 \sin\gamma_2 . h_2 &= \cos\beta . g,\\
\cos\beta_1 \cos\gamma_1 . h_1 - \sin\beta_1 \sin\gamma_1 . g_1 + \cos\beta_2 \cos\gamma_2 . h_2 - \sin\beta_2 \sin\gamma_2 . g_2 &= \cos\beta . h,\\
\cos\beta_1 \sin\gamma_1 . g_1 - \sin\beta_1 \cos\gamma_1 . h_1 + \cos\beta_2 \sin\gamma_2 . g_2 - \sin\beta_2 \cos\gamma_2 . h_2 &= -\sin\beta . h,\\
\cos\beta_1 \sin\gamma_1 . h_1 + \sin\beta_1 \cos\gamma_1 . g_1 + \cos\beta_2 \sin\gamma_2 . h_2 + \sin\beta_2 \cos\gamma_2 . g_2 &= \sin\beta . g.
\end{aligned}\right\} \quad (4)$$

Multiplying the first and second of these equations by 1, $\sqrt{-1}$, and adding, then multiplying the third and fourth by $-\sqrt{-1}$, 1, and adding, and putting generally

$$g + \sqrt{-1}\, h = G, \ldots\ldots\ldots\ldots\ldots\ldots\ldots (5)$$

we have

$$\left.\begin{aligned}
(\cos\beta_1 \cos\gamma_1 - \sqrt{-1} \sin\beta_1 \sin\gamma_1)\, G_1 + (\cos\beta_2 \cos\gamma_2 - \sqrt{-1} \sin\beta_2 \sin\gamma_2)\, G_2 &= \cos\beta . G,\\
(\sin\beta_1 \cos\gamma_1 - \sqrt{-1} \cos\beta_1 \sin\gamma_1)\, G_1 + (\sin\beta_2 \cos\gamma_2 - \sqrt{-1} \cos\beta_2 \sin\gamma_2)\, G_2 &= \sin\beta . G;
\end{aligned}\right\} \quad (6)$$

which two equations are equivalent to the four (4).

Putting for shortness p_1, p_2, q_1, q_2 for the coefficients in the left-hand members of equations (6), we have

$$\frac{G_1}{q_2 \cos\beta - p_2 \sin\beta} = \frac{G_2}{p_1 \sin\beta - q_1 \cos\beta} = \frac{G}{p_1 q_2 - p_2 q_1}. \quad (7)$$

On substituting for p_1, q_2, &c. their values, we find

$$p_1 q_2 - p_2 q_1 = \cos(\gamma_1 - \gamma_2) \sin(\beta_2 - \beta_1) + \sqrt{-1} \sin(\gamma_1 - \gamma_2) \cos(\beta_1 + \beta_2).$$

Now the equations (6) cannot be incompatible or identical unless the above quantity vanish. But this can only take place when

$$\sin(\beta_2 - \beta_1) = 0 \text{ and } \sin(\gamma_1 - \gamma_2) = 0,$$

or else

$$\cos(\beta_1 + \beta_2) = 0 \text{ and } \cos(\gamma_1 - \gamma_2) = 0,$$

or lastly

$$\sin(\beta_2 - \beta_1) = 0 \text{ and } \cos(\beta_1 + \beta_2) = 0.$$

The first case gives $\beta_2 = \beta_1$, $\gamma_1 - \gamma_2 = a_1 - a_2 = 0$, or $\pm 180^0$, so that the two streams into which it was proposed to resolve the first are of the same nature. The second case gives $\beta_2 = 90^0 - \beta_1$, $\gamma_1 - \gamma_2 = a_1 - a_2 = \pm 90^0$, which shews that the two streams are identical in their nature, only the first and second principal planes of the first of these streams are accounted respectively the second and first of the second stream. The third case gives $\beta_2 = \beta_1 = \pm 45^0$, so that the two streams are circularly polarized and of the same kind, which is a particular instance of the first case.

Hence, universally, a stream of elliptically polarized light may be resolved into two streams of elliptically polarized light in which the polarizations are of any kind that we please, but different from one another.

Substituting for q_2, &c. their values in (7), and replacing γ_1, γ_2 by $\alpha_1 - \alpha$, $\alpha_2 - \alpha$, for which they had been temporarily written, we find

$$\left.\begin{aligned} &\{\sin(\beta_2 - \beta)\cos(\alpha_2 - \alpha) - \sqrt{-1}\cos(\beta_2 + \beta)\sin(\alpha_2 - \alpha)\}^{-1} G_1 \\ = &\{\sin(\beta - \beta_1)\cos(\alpha_1 - \alpha) - \sqrt{-1}\cos(\beta + \beta_1)\sin(\alpha_1 - \alpha)\}^{-1} G_2 \\ = &\{\sin(\beta_2 - \beta_1)\cos(\alpha_2 - \alpha_1) - \sqrt{-1}\cos(\beta_2 + \beta_1)\sin(\alpha_2 - \alpha_1)\}^{-1} G \end{aligned}\right\} . \quad (8)$$

3. Among these various modes of resolution there is one which possesses several peculiar properties, any one of which might serve to define it. Let us in the first place examine under what circumstances the intensity of the stream made up of the two components is independent of any retardation which the phase of vibration of one component may have undergone relatively to the phase of vibration of the other previously to the recomposition.

For this purpose there is evidently no occasion to consider the manner in which a_1, b_1, ϵ_1, a_2, b_2, ϵ_2 are made up of a, b, ϵ, but we may start with the components. Let ρ_1, ρ_2 be the retardations of phase which take place before recomposition, and resolve the disturbances along the axes of x, y. We shall have for the resolved parts

$$x = S\{a_1 \cos\alpha_1 \sin(\phi + \epsilon_1 - \rho_1) - b_1 \sin\alpha_1 \cos(\phi + \epsilon_1 - \rho_1)\}$$
$$y = S\{a_1 \sin\alpha_1 \sin(\phi + \epsilon_1 - \rho_1) + b_1 \cos\alpha_1 \cos(\phi + \epsilon_1 - \rho_1)\},$$

where S denotes the sum of the expression written down and that formed from it by replacing the suffix 1 by 2. To form the expression for the intensity, or rather what would be the intensity if the quantities c and ϵ were absolutely constant, not merely, constant for a great number of successive undulations, we must develope the expressions for x and y so as to contain the sine and cosine of $\phi + \kappa$, and take the sum of the squares of the coefficients. κ is here a constant quantity which may be chosen at pleasure, and which it will be convenient to take equal to $\epsilon_1 - \rho_1$. If I be the intensity, in the sense above explained, or as it may be called the *temporary intensity*, we find, putting δ for $\epsilon_2 - \rho_2 - \epsilon_1 + \rho_1$,

$$\begin{aligned} I = &\{a_1\cos\alpha_1 + a_2\cos\alpha_2\cos\delta + b_2\sin\alpha_2\sin\delta\}^2 + \{-b_1\sin\alpha_1 + a_2\cos\alpha_2\sin\delta - b_2\sin\alpha_2\cos\delta\}^2 \\ &+ \{a_1\sin\alpha_1 + a_2\sin\alpha_2\cos\delta - b_2\cos\alpha_2\sin\delta\}^2 + \{b_1\cos\alpha_1 + a_2\sin\alpha_2\sin\delta + b_2\cos\alpha_2\cos\delta\}^2 \\ &= a_1^2 + b_1^2 + a_2^2 + b_2^2 + 2(a_1a_2 + b_1b_2)\cos(\alpha_2 - \alpha_1)\cos\delta + 2(a_1b_2 + a_2b_1)\sin(\alpha_2 - \alpha_1)\sin\delta. \end{aligned}$$

On putting for a, b their values $c\cos\beta$, $c\sin\beta$, this expression becomes

$$I = c_1^2 + c_2^2 + 2c_1c_2\{\cos(\alpha_2 - \alpha_2)\cos(\beta_2 - \beta_1)\cos\delta + \sin(\alpha_2 - \alpha_1)\sin(\beta_2 + \beta_1)\sin\delta\} \quad . \ (9).$$

In order that I may be independent of the difference of phase $\rho_2 - \rho_1$, and therefore of δ, we must have either

$$\cos(\alpha_2 - \alpha_1) = 0, \qquad \sin(\beta_2 + \beta_1) = 0, \quad . \ . \ . \ . \quad (10),$$
$$\text{or } \sin(\alpha_2 - \alpha_1) = 0, \qquad \cos(\beta_2 - \beta_1) = 0, \quad . \ . \ . \ . \quad (11).$$

The equations (10) give $\alpha_2 - \alpha_1 = \pm 90^0$, $\beta_2 = -\beta_1$, so that the ellipses described in the case of the two streams are similar, their major axes perpendicular to each other, and the direction of revolution in the one stream contrary to that in the other. It will be easily seen that the equations (11) differ from (10) only in this, that what are regarded as the first and second principal planes of the second stream when equations (10) are satisfied, are accounted respectively the second and first when (11) are satisfied.

Two streams thus related may be said to be *oppositely polarized.* Two streams of plane-polarized light in which the planes of polarization are at right angles to each other, and two streams of circularly polarized light, one right-handed and the other left-handed, are particular cases of streams oppositely polarized.

In the reasoning of this article, nothing depends upon the precise relation between the two polarized streams and the original stream. All that it is necessary to suppose is, that the two polarized streams came originally from the same polarized source, so that the changes in epoch and intensity, that is, the changes in the quantities ϵ, c, are the same for the two streams. Nothing depends upon the precise nature of these changes, which may be either abrupt or continuous, but must be sufficiently infrequent if abrupt, or sufficiently gradual if continuous, to allow of our regarding c and ϵ as constant for a great number of successive undulations. Our results will apply just as well to the disturbance produced by the union of two neighbouring streams coming originally from the same polarized source, but having had their polarizations modified, as to that produced by the union, after recomposition, of the components of a single polarized stream. Since the resulting intensity is independent of δ, it follows that two oppositely polarized streams coming originally from the same polarized source are incapable of interfering, but two streams polarized otherwise than oppositely necessarily interfere, to a greater or less degree, when the difference in their retardation of phase is sufficiently small. Of course the interference here spoken of means only that which is exhibited without analyzation.

4. Two interfering streams may be said to interfere perfectly when the fluctuations of intensity are the greatest that the difference in the intensities of the interfering streams admits of, so that in case of equality the minima are absolutely equal to zero. Referring to (9), we see that in order that this may be the case the maximum value of the coefficient of $2c_1c_2$ must be equal to 1. Now the maximum value of $A \cos \delta + \beta \sin \delta$ is $\sqrt{(A^2 + B^2)}$, and therefore we must have

$$\cos^2(\alpha_2 - \alpha_1)\cos^2(\beta_2 - \beta_1) + \sin^2(\alpha_2 - \alpha_1)\sin^2(\beta_2 + \beta_1) = 1 = \cos^2(\alpha_2 - \alpha_1) + \sin^2(\alpha_2 - \alpha_1),$$

whence,

$$\cos^2(\alpha_2 - \alpha_1)\sin^2(\beta_2 - \beta_1) + \sin^2(\alpha_2 - \alpha_1)\cos^2(\beta_2 + \beta_1) = 0,$$

which lead to the very same conditions that have been already discussed in Art. 2. Hence two polarized streams coming from the same polarized source are capable of interfering perfectly if the polarizations are the same, not at all if the polarizations are opposite, and in intermediate cases of course in intermediate degrees.

5. When a stream of polarized light is resolved into two oppositely polarized streams, which are again compounded after their phases have been differently altered, we have from (9), taking account of (10) or (11),

$$I = c_1^2 + c_2^2, \quad . \quad . \quad . \quad . \quad . \quad . \quad . \quad . \quad . \quad . \quad (12)$$

so that the intensity of the resultant is equal to the sum of the intensities of the components, and is therefore constant, that is, independent of $\rho_1 - \rho_2$, and is accordingly equal to what it was at first, when ρ_1 and ρ_2 were each equal to zero, that is, equal to the intensity of the original stream.

It may be readily proved from the formulæ (8) that it is only in the case in which the polarizations of the two components of the original polarized stream are opposite that the intensity of the original stream, whatever be the nature of its polarization, is equal to the sum of the intensities of the component streams. For, changing the sign of $\sqrt{-1}$ in these formulæ, multiplying the resulting equations, member for member, by the equations (8), and observing that if G' be what G becomes, $GG' = g^2 + h^2 = c^2$, we find

$$\left.\begin{aligned} &\{\sin^2(\beta_2 - \beta)\cos^2(\alpha_2 - \alpha) + \cos^2(\beta_2 + \beta)\sin^2(\alpha_2 - \alpha)\}^{-1}c_1^2 \\ = &\{\sin^2(\beta - \beta_1)\cos^2(\alpha_1 - \alpha) + \cos^2(\beta + \beta_1)\sin^2(\alpha_1 - \alpha)\}^{-1}c_2^2 \\ = &\{\sin^2(\beta_2 - \beta_1)\cos^2(\alpha_2 - \alpha_1) + \cos^2(\beta_2 + \beta_1)\sin^2(\alpha_2 - \alpha_1)\}^{-1}c^2. \end{aligned}\right\} \quad \dots \quad (13)$$

In order that $\mathfrak{m}(c^2)$ may be equal to $\mathfrak{m}(c_1^2) + \mathfrak{m}(c_2^2)$, it is necessary that c^2 be equal to $c_1^2 + c_2^2$, because, whatever fluctuations c_1 and c_2 may undergo in a moderate time, such as the tenth part of a second, c_1 and c_2 are always proportional to c. Hence the sum of the quantities whose reciprocals are the coefficients of c_1^2 and c_2^2 must be equal to that whose reciprocal is the coefficient of c^2. Since this has to be true independently of β, let the quantities $\sin^2(\beta_2 - \beta)$, &c. be replaced by sines and cosines of multiple arcs, and let our equation be put under the form

$$A + B\cos 2\beta + C\sin 2\beta = 0.$$

Then A, B, C must be separately equal to zero, or

$$\left.\begin{aligned} \sin^2(\beta_2 - \beta_1)\cos^2(\alpha_2 - \alpha_1) + \cos^2(\beta_2 + \beta_1)\sin^2(\alpha_2 - \alpha_1) &= 1; \\ \cos 2\beta_2\cos 2(\alpha_2 - \alpha) + \cos 2\beta_1\cos 2(\alpha_1 - \alpha) &= 0; \\ \sin 2\beta_2 + \sin 2\beta_1 &= 0. \end{aligned}\right\} \quad \dots \quad (14)$$

Replacing unity in the right-hand member of the first of these equations by

$$\cos^2(\alpha_2 - \alpha_1) + \sin^2(\alpha_2 - \alpha_1),$$

we find

$$\cos^2(\beta_2 - \beta_1)\cos^2(\alpha_2 - \alpha_1) + \sin^2(\beta_2 + \beta_1)\sin^2(\alpha_2 - \alpha_1) = 0;$$

whence $\beta_2 = -\beta_1$, $\alpha_2 = \alpha_1 + 90^0$, or else β_2 and β_1 differ by 90^0, and $\alpha_2 = \alpha_1$, except in the particular case in which $\beta_1 = \pm 45^0$, when $\beta_2 = \mp 45^0$ satisfies the equation independently of α_2. Hence the streams must be polarized oppositely, a condition which may always be expressed by

$$\beta_2 = -\beta_1, \quad \alpha_2 = \alpha_1 + 90^0,$$

which equations satisfy the second and third of equations (14) independently of α, as it might have been foreseen that they would, since it has been already shewn that the condition (12) is satisfied in the case of oppositely polarized streams. It now appears that it is only in the case of such streams that this is satisfied.

6. The properties of oppositely polarized pencils which have been proved, render it in a high degree probable that it is a general law that in a doubly refracting medium the two polarized pencils transmitted in a given direction are oppositely polarized. Were this not

52—2

the case, the two pencils, polarized otherwise than oppositely, into which a polarized pencil is resolved on entering into the medium, would at emergence compound a pencil of which the intensity would depend upon the retardations of phase of one pencil relatively to the other, so that such a medium, when examined with polarized light, ought to exhibit rings or colours without the employment of an analyzer. It is here supposed that the light enters the medium at the first surface, and leaves it at the second, in a direction normal to the surface, or very nearly so, and likewise that the refracting power of the medium for the two pencils is not very different, so that the effect of reflexion at the two surfaces may be disregarded, the only sensible effect being to diminish the intensity of each of the two pencils in the same proportion, without affecting its state of polarization. These views would lead us to scrutinize very carefully any experimental evidence brought forward which would lead to the conclusion, that the two polarized pencils which a doubly refracting medium was capable of propagating in a given direction were polarized otherwise than oppositely.

7. In ordinary doubly refracting crystals, whether uniaxal or biaxal, and in doubly refracting liquids, such as syrop of sugar, it is generally admitted that the two pencils transmitted in a given direction are oppositely polarized. In the former case the two pencils are polarized in rectangular planes, in the latter they are circularly polarized, one being right-handed and the other left-handed. The same is the case with quartz for pencils transmitted in the direction of the axis; but in following out the researches in which he so successfully connected the phenomena, identical with those of a doubly refracting liquid, which quartz exhibits in the direction of the axis, with the phenomena which it exhibits as a uniaxal crystal, Mr Airy met with an experimental result which seemed to shew that while the ellipses which characterize the two streams transmitted in a given direction, oblique to the axis, have their major axes situated, one in a principal plane, and the other perpendicular to the principal plane, and the directions of revolution are opposite, the eccentricities of the ellipses, though nearly, are not quite equal*. This conclusion depended upon the result of certain experiments made by means of a Fresnel's rhomb. The nature of the experiments seemed to eliminate the effect of an error in the rhomb, that is, a deviation from 90^0 in the retardation of phase which it produced in a pencil polarized in the plane of reflexion relatively to a pencil polarized in a plane perpendicular to the former. The effect of a possible index error in the plane of reflexion seemed also to be eliminated; and the quantities on which the results depended, though small, seemed to be beyond mere errors of pointing. Being impressed however with a strong conviction that the result depended in some way on the mode of observation, I was led to scrutinize the different steps of the process, and it occurred to me that the apparent inequality of eccentricities was probably due to defects of annealing in the rhomb. It is next to impossible to procure a piece of glass of such a size free from defects of that nature, for which reason I believe that even a good Fresnel's rhomb is not to be trusted for minute quantities, except in the case of merely differential observations.

* *Cambridge Philosophical Transactions*, Vol. IV. p. 204.

The same views lead to the conclusion that the two pencils transmitted through magnetized glass in a direction oblique to the lines of magnetic force are oppositely polarized. Some theoretical investigations in which I was engaged some time ago led me to the result that these polarized streams are circularly polarized, as well as those transmitted along the line of magnetic force, and that the difference between the wave-velocities varies as the cosine of the inclination of the wave-normal to the line of magnetic force. I hope at some future time to bring these researches before the notice of this Society.

8. After these preliminary investigations respecting the nature of opposite polarization, which indeed contain, it is probable, but little that has not already occurred to persons who have studied the subject, it is time to come to the more immediate object of this paper, which relates to the combination of independent streams. But first it will be convenient to state explicitly a principle which is generally recognized.

When any number of polarized streams from different sources mix together, after having been variously modified by reflexion, refraction, transmission through doubly refracting media, tourmalines, &c., the intensity of the mixture is equal to the sum of the intensities due to the separate streams.

The reason of this law may be easily seen. The components whereby the disturbance due to any one stream is originally expressed have to be resolved, their components resolved again, and so on; and of these partial disturbances the phases of vibration have to be altered by quantities independent of the time, and the coefficients in some cases diminished in given ratios, and in some cases suppressed altogether. Each stream has to be treated in a similar way. The final disturbance being resolved in any two rectangular directions, each component must be put under the form $U \cos \phi + V \sin \phi$, and the sum of the squares of U and V must be taken to form the expression for the temporary intensity. All the quantities such as U and V will evidently be linear functions of $c \cos \epsilon$, $c \sin \epsilon$, $c' \cos \epsilon'$, $c' \sin \epsilon'$, &c., where c, c'... and ϵ, ϵ'... refer to the different streams, so that U for instance will be of the form

$$Ac \cos \epsilon + Bc \sin \epsilon + A'c' \cos \epsilon' + B'c' \sin \epsilon' + \ldots$$

where A, B, A', B'... are independent of the time. The temporary intensity will involve U^2, but the actual intensity will involve $\mathfrak{m}(U^2)$, or

$$\mathfrak{m}\Sigma (Ac \cos \epsilon + Bc \sin \epsilon)^2 + 2\mathfrak{m}\Sigma \{(Ac \cos \epsilon + Bc \sin \epsilon)(A'c' \cos \epsilon' + B'c' \sin \epsilon')\}.$$

Now the products such as $\cos \epsilon \cos \epsilon'$, $\cos \epsilon \sin \epsilon'$, &c. will have a mean value zero, since the changes in ϵ and those in ϵ' have no relation to each other, and therefore the expression for $\mathfrak{m}(U^2)$ becomes

$$\mathfrak{m}\Sigma(Ac \cos \epsilon + Bc \sin \epsilon)^2, \text{ or } \Sigma\mathfrak{m}(Ac \cos \epsilon + Bc \sin \epsilon)^2,$$

that is, the sum of the quantities by which it would be expressed were the different streams taken separately.

Two streams which come from different sources, or which, though in strictness they come from the same source, are such that the changes of epoch and intensity in the one have no relation to the changes of epoch and intensity in the other, may be called *independent*.

9. Suppose that there are any number of independent polarized streams mixing together; let the mixture be resolved in any manner into two oppositely polarized streams, and let us examine the intensity of each.

Let us take one stream first. The intensities of its components are given by the formulæ (13), which become somewhat simpler in the case of opposite polarizations, since $\beta_2 = -\beta_1$, and $\alpha_2 = 90_0 + \alpha_1$. Hence

$$c_1^2 = \{\sin^2(\beta_1+\beta)\sin^2(\alpha_1-\alpha) + \cos^2(\beta_1-\beta)\cos^2(\alpha_1-\alpha)\}c^2;$$

whence we find

$$\mathfrak{m}(c_1^2) = \tfrac{1}{2}\{1+\sin 2\beta_1 \sin 2\beta + \cos 2\alpha_1 \cos 2\alpha \cos 2\beta_1 \cos 2\beta + \sin 2\alpha_1 \sin 2\alpha \cos 2\beta_1 \cos 2\beta\}\mathfrak{m}(c^2). \quad (15)$$

There will be no occasion to write down the value of $\mathfrak{m}(c_2^2)$, since c_1 may be taken to refer to either component.

Let us pass now to the consideration of a group consisting of any number of independent polarized streams, let

$$\left.\begin{array}{lll} \Sigma\mathfrak{m}(c^2) = A, & \Sigma\sin 2\beta\mathfrak{m}(c^2) = B, & \Sigma\cos 2\alpha\cos 2\beta\mathfrak{m}(c^2) = C, \\ \Sigma\sin 2\alpha\cos 2\beta\mathfrak{m}(c^2) = D, & & \end{array}\right\} \quad (16)$$

and let c_1 now refer to one of the components of the whole group; then

$$2\mathfrak{m}(c_1^2) = A + B\sin 2\beta_1 + C\cos 2\alpha_1\cos 2\beta_1 + D\sin 2\alpha_1\cos 2\beta_1. \quad (17)$$

It follows that if there are two groups of independent polarized streams which are such as to give the same values to each of the four quantities A, B, C, D defined by (16), if the groups be resolved in any manner whatsoever, which is the same for both, into two oppositely polarized streams, the intensities of the components of the one group will be respectively equal to the intensities of the components of the other group. Conversely, if two groups of oppositely polarized streams are such that when they are resolved in any manner, the same for both, into two oppositely polarized streams, the intensities of the components of the one group are respectively equal to the intensities of the components of the other group, the quantities A, B, C, D must be the same for the two groups. For, if we take accented letters to refer to the second group, the second member of equation (17), and the expression thence derived by accenting A, B, C, D, must be equal independently of α_1 and β_1, which requires that A', B', C', D' be respectively equal to A, B, C, D.

DEFINITION. Two such groups will be said to be *equivalent*.

10. The theoretical definition of equivalence which has just been given agrees completely with the experimental tests of equivalence. One of the most ready as well as delicate modes of detecting minute traces of polarization, and at the same time determining qualitatively the nature of the polarization, consists in viewing the light to be examined through a plate of calcareous spar, or other crystal, cut for shewing rings, followed by a Nicol's prism. The plane-polarized pencils respectively stopped by and transmitted through the Nicol's prism consisted, on entering the crystal of pencils elliptically polarized in opposite ways; and the nature of this elliptic polarization changes in every possible manner from one point to another of the field of view. If these two streams of light be equivalent according to the definition

given in the preceding article, they will present exactly the same appearance on being viewed through a crystal followed by a Nicol's prism or other analyzer.

11. THEOREM. Let a polarized stream be resolved into two oppositely polarized streams; let the phase of vibration of one of the streams be altered by a given quantity relatively to that of the other, and let the streams be then compounded. If the polarization of the original stream be now changed to its opposite, the polarization of the final stream will also be changed to its opposite.

The straightforward mode of demonstrating this theorem, by making use of the general expressions, would lead to laborious analytical processes, which are wholly unnecessary. For the formulæ which determine the components of a given stream are expressed by simple equations, so that the results are unique, and accordingly whenever we can foresee what the result will be, it is sufficient to shew that the formulæ themselves, or the geometrical conditions of which the formulæ are merely the expressions, are satisfied.

For shortness' sake call the original stream X, and its components O, E. Let ρ be the given quantity, positive or negative, by which the phase of vibration of O is retarded relatively to that of E. Let o, e denote the streams O, E after the changes of phase, and Y the stream resulting from their reunion. Conceive now all the vibrations with which we are concerned to be turned in azimuth through 90^0. This will not affect the geometrical relations connecting components and resultants. Let X', O', E', o', e', Y' be the streams which X, O, E, o, e, Y thus become. The streams O', E' are evidently polarized in the same manner respectively as E, O, except that *right-handed* is changed into *left-handed,* and *vice versa*; and in passing from O' to o' the phase is retarded by ρ. Now conceive the direction of motion of a given particle reversed, the motions of all other particles being derived from that of the first according to the general law of wave propagation. The relations between components and resultants will evidently not be violated; and if X'', O'', E'', o'', e'', Y'' denote the streams into which X', O', E', o', e', Y' are thus changed, it is evident that the polarizations of X'', O'', E'', o'', e'', Y'' are respectively opposite to those of X, O, E, o, e, Y. But on account of the reversion in direction of motion it is plain that there is reversion as regards acceleration and retardation of phase, so that in passing from the pair O'', E'' to the pair o'', e'' the phase of O'' is *accelerated* by ρ relatively to the phase of E''.

Hence the stream X'', polarized oppositely to X, is resolved into two E'', O'', polarized in the same manner respectively as O, E, which are recompounded after the phase of the one polarized in the same manner as O has been retarded by ρ relatively to the phase of the other, and the result is Y'', a stream polarized in a manner opposite to Y, which proves the theorem.

12. This theorem may be applied to the case of light transmitted through a slice of a doubly refracting crystal, and shews that two streams going in oppositely polarized come out oppositely polarized. Also, since nothing in the demonstration depends upon the order in which the decompositions and recompositions take place, it is immaterial whether X, X'' denote a pair of oppositely polarized incident streams, which give rise to the emergent streams

Y, Y'', or X, X'' denote a pair of oppositely polarized emergent streams, which came from the incident streams Y, Y''. A particular case of this theorem was assumed in the preceding article, when it was stated that the pencils, polarized in perpendicular planes, which on coming out of a crystal of calcareous spar are respectively stopped by and transmitted through a Nicol's prism, went into the crystal oppositely polarized.

The theorem will evidently be true of a train consisting of any number of crystalline plates, each possessing the property of resolving the incident light into two oppositely polarized streams, which are propagated within the medium with different velocities. For two oppositely polarized streams incident on the first plate give rise to two emergent streams which fall oppositely polarized on the second plate, and so on. Since the number of plates may be supposed to increase and their thickness to decrease indefinitely, while at the same time the steps by which the doubly refracting nature of the plates alters from one to another become separately insensible, the theorem will be true if the whole train, or part of it, consist of a substance of which the doubly refracting nature alters continuously, as for example a piece of strained or unannealed glass. If, however, the train contain a member which performs a partial analysis of the light, as for example a plate of smoky quartz, or a plate of glass inclined to the incident light at a considerable angle, it will no longer be true that two pencils going in oppositely polarized will come out oppositely polarized.

13. THEOREM. If two equivalent groups of polarized streams be resolved in any manner, which is the same for both, into two oppositely polarized groups, and these be recombined after the phase of one of the components has been retarded by a given quantity relatively to that of the other, the two groups of resultant streams will be equivalent.

Let the groups of resultants be each resolved in any manner into two oppositely polarized streams, and call these O, E. By Art. 11, if O', E' be the streams which furnish O, E respectively, O', E' are oppositely polarized. Now by Art. 3, the intensities of O, E are the same respectively as those of O', E'; but these are the same for the one group as for the other, by the definition of equivalence. Therefore the intensities of O, E are the same for the one group as for the other; but O, E are any two oppositely polarized components of the resultant groups; therefore these groups are equivalent.

Hence, if two equivalent groups be transmitted through a crystalline plate, the emergent groups will be equivalent; and by the same reasoning as in Art. 12 the theorem may be extended to an optical train consisting of any number of crystalline plates, pieces of unannealed glass, &c.

14. THEOREM. If two equivalent groups be resolved in any manner, the same for both, into two polarized streams, the intensities of the components of the one group will be respectively equal to the intensities of the components of the other group.

The proof of this theorem is very easy. It is sufficient to treat the general expressions (13) exactly as in Art. 9, only that no relations are to be introduced between α_2, β_2, and α_1, β_1. Since the coefficient of c^2 in (13) is constant, if we consider as variable such quantities only as may change in passing from the one group to the other, it will be easily seen that the intensity of either component depends on the same four constants A, B, C, D as before; and since these have the same values for equivalent groups, it follows that in the most general mode of resolution the components of any two such groups have the same intensities respectively.

15. THEOREM. If two equivalent groups be each resolved in the same manner into two oppositely polarized streams, and these be recombined after their vibrations have been diminished in two different given ratios, the resultant groups will be equivalent.

Let each of the resultant groups be resolved into two oppositely polarized streams O, E, and let O', E' denote the streams which furnished O, E respectively. It is easily seen that the streams O', E' are both polarized, though not in general oppositely. Were there any occasion to determine the nature of the polarizations, it might easily be done by following a process the reverse of that by which the modified are deduced from the original groups. Thus, if it were required to determine the polarization of O', we should resolve O into streams oppositely polarized in the manner originally given, augment the vibrations in ratios the inverse of those in which they are actually diminished, and recombine the streams so obtained. For our present purpose, however, it is sufficient to observe that the polarizations of O', E' depend only on the mode of resolution, and not on the nature of the original group, and that therefore they are the same for the one group as for the other. The intensity of O' is not the same as that of O, as in the case considered in Art. 13, but bears to it a ratio depending only on the mode of resolution, and therefore the same for each of the two original groups. The same is true of the intensity of E' compared with that of E. Now by Art. 14 the intensities of O', E' are the same for the two original groups, and the intensities of O, E bear to those of O', E' respectively, ratios the same for the two groups: therefore the intensities of O, E are the same for the two final groups; but O, E are any two oppositely polarized components of these groups; therefore these groups are equivalent.

16. It follows from this theorem that no partial analysis of light, such, for example, as would be produced by reflexion from the surface of glass or metal, or by transmission through a doubly absorbing medium, can from equivalent groups produce groups which are not equivalent to each other; and we have seen already that this cannot be done by means of the alteration of phase accompanying double refraction. It follows, therefore, that equivalent groups are optically undistinguishable.

In proving this property of equivalent groups, it has been supposed that the polarizations of the two streams into which any group was resolved were opposite, such being the case in nature. But did a medium exist such that the two streams of light which it transmitted independently were polarized otherwise than oppositely, it would still not enable us to distinguish between equivalent groups.

17. The experimental definition of common light is, light which is incapable of exhibiting rings of any kind when examined by a crystal of Iceland spar and an analyzer, or by some equivalent combination. Consequently, a group of independent polarized streams will together be equivalent to common light when, on being resolved in any manner into two

oppositely polarized pencils, the intensities of the two are the same, and of course equal to half that of the original group. Accordingly, in order that the group should be equivalent to common light, it is necessary and sufficient that the constants B, C, D should vanish.

18. Let us now see under what circumstances two independent streams of polarized light can together be equivalent to common light.

Let α, β refer to the first, and α', β' to the second stream, and let the intensities of the two streams be as 1 to n; then we get from the formulæ (16)

$$\sin 2\beta + n \sin 2\beta' = 0;$$
$$\cos 2\alpha \cos 2\beta + n \cos 2\alpha' \cos 2\beta' = 0;$$
$$\sin 2\alpha \cos 2\beta + n \sin 2\alpha' \cos 2\beta' = 0.$$

Transposing, squaring, and adding, we find $n^2 = 1$, and therefore $n = 1$, since n is essentially positive. Since β and β' are supposed not to lie beyond the limits -90^0 and $+90^0$, we get from the first equation $\beta' = -\beta$, or $\beta' = \pm 90^0 - \beta$, + or − according as β is positive or negative. Now it is plain that any one solution must be expressed analytically in two ways, in which the values of β' are complementary, and the values of α' differ by 90^0, since either principal axis of the ellipse belonging to the second stream may be that whose azimuth is α'. Accordingly, we may reject the second solution as being nothing more than the first expressed in a different way, and may therefore suppose $\beta' = -\beta$. The second and third equations then give $\cos 2\alpha' = -\cos 2\alpha$, $\sin 2\alpha' = -\sin 2\alpha$, and therefore α and α' differ by 90^0. The equations are indeed satisfied by $\beta = -\beta' = \pm 45^0$, but this solution is only a particular case of the former.

It follows therefore that common light is equivalent to any two independent oppositely polarized streams of half the intensity; and no two independent polarized streams can together be equivalent to common light, unless they be polarized oppositely, and have their intensities equal.

19. We have seen that the nature of the mixture of a given group of independent polarized streams is determined by the values of the four constants A, B, C, D. Consider now the mixture of a stream of common light having an intensity J, and a stream, independent of the former, consisting of elliptically polarized light having an intensity J', and having α' for the azimuth of its plane of maximum polarization, and $\tan \beta'$ for the ratio of the axes of the ellipse which characterizes it.

By the preceding article, the stream of common light is equivalent to two independent streams, plane-polarized in azimuths 0^0 and 90^0, having each an intensity equal to $\frac{1}{2}J$. Hence, applying the formulæ (16) and (17) to the mixture, we have

$$2\mathfrak{m}\,(c_1)^2 = J + J' + J' \sin 2\beta' \sin 2\beta_1 + J' \cos 2\alpha' \cos 2\beta' \cos 2\alpha_1 \cos 2\beta_1$$
$$+ J' \sin 2\alpha' \cos 2\beta' \sin 2\alpha_1 \cos 2\beta_1;$$

and this mixture will be equivalent to the original group of polarized streams, provided

$$\left.\begin{aligned} J + J' &= A; & J' \sin 2\beta' &= B; \\ J' \cos 2\alpha' \cos 2\beta' &= C; & J' \sin 2\alpha' \cos 2\beta' &= D. \end{aligned}\right\} \quad . \quad . \quad . \quad . \quad . \qquad (18)$$

These equations give

$$\left.\begin{aligned} J' &= \sqrt{(B^2 + C^2 + D^2)}\,; & J &= A - \sqrt{(B^2 + C^2 + D^2)}\,; \\ \sin 2\beta' &= \frac{B}{\sqrt{(B^2 + C^2 + D^2)}}\,; & \tan 2\alpha' &= \frac{D}{C}\,. \end{aligned}\right\} \quad . \quad . \quad . \quad . \quad . \quad . \quad (19)$$

These formulæ can always be satisfied, and therefore it is always possible to represent the given group by a stream of common light combined with a stream of elliptically polarized light independent of the former. Moreover, there is only one way in which the group can be so represented. For, though the third of equations (19) gives two values for β' complementary to each other, these values, as before explained, lead only to two ways of expressing the same result. If we choose that value of β which is numerically the smaller, then among the different values of α', differing by 90^0, which satisfy the fourth equation, we must choose one which gives to $\cos 2\alpha'$ the same sign as C.

20. Let us now apply the principles and formulæ which have just been established to a few examples. And first let us take one of the fundamental experiments by which MM. Arago and Fresnel established the laws of interference of polarized light, or rather an analogous experiment mentioned by Sir John Herschel. The experiment selected is the following.

Two neighbouring pencils of common light from the same source are made to form fringes of interference. A tourmaline, carefully worked to a uniform thickness, is cut in two, and its halves interposed in the way of the two streams respectively. It is found that when the planes of polarization of the two tourmalines are parallel the fringes are formed perfectly; but as one of the tourmalines is turned round in azimuth the fringes become fainter, and at last, when the planes of polarization become perpendicular to each other, the fringes disappear.

Let the planes of polarization of the tourmalines be inclined at an angle α, and let it be required to investigate an expression for the intensity of the fringes. Since common light is equivalent to two independent streams, of equal intensity, polarized in opposite ways, let the original light be represented by two independent streams, having each an intensity equal to unity, polarized in planes respectively parallel and perpendicular to the plane of polarization of the first tourmaline. If c, c' be the coefficients of vibration in the two streams respectively, $c\cos\alpha$, $c'\sin\alpha$ will be the coefficients of the resolved parts in the direction of the vibrations transmitted by the second tourmaline. Hence we shall have, mixing together, two independent streams, in one of which the temporary intensity fluctuates between the limits $(c \pm c\cos^2\alpha)^2 + (c\cos\alpha\sin\alpha)^2$ as the interval of retardation changes in passing from one point to another in the field of view. The temporary intensity of the other stream being $(c'\sin\alpha)^2$, the intensity at different points will fluctuate between the limits

$$\mathfrak{m}\,(c^2 \pm 2c^2\cos^2\alpha + c^2\cos^2\alpha) + \mathfrak{m}\,(c'\sin\alpha)^2.$$

It is needless to take account of the absorption which takes place even on the pencils which the tourmalines do transmit, because it affects both pencils equally. Since $\mathfrak{m}\,(c^2) = \mathfrak{m}\,(c'^2) = 1$, we have for the limits of fluctuation of the intensity $2 \pm 2\cos^2\alpha$. When $\alpha = 0$ these limits become 4 and 0, and the interference is perfect. When $\alpha = 90^0$ the limits coalesce, becoming each equal to 2, and there are no fringes. As α increases from 0 to 90^0, the superior

53—2

limit continually decreases, and the inferior increases, and consequently the fringes become fainter and fainter.

21. It is a well-known law of interference that if two rays of common light from the same source be polarized in rectangular planes, and afterwards be brought to the same plane of polarization, or in other words, analyzed so as to retain only light polarized in a particular plane, they will not interfere, but if the light be primitively polarized in one plane they will interfere. This law seems to have presented a difficulty to some, because, it would be argued, the most general kind of vibrations are elliptic, so that we must suppose the vibrations of the ether in the case of common light to be of this kind; and yet the phenomena of interference are exhibited perfectly well if the light be at first elliptically polarized instead of plane-polarized. For my own part, I never could see the difficulty, but on the other hand it seems to me that it would be an immense difficulty were the law anything else than what it is. For, if we consider the rectangular components of the vibrations which make up common light, these components being measured along any two rectangular axes perpendicular to the ray, we must suppose them to be independent of each other, or at least to have no fixed relation to each other so far as regards the changes in the mode of vibration, which we must suppose to be taking place continually, though slowly, it may be, in comparison with the time of a luminous vibration. To suppose otherwise would be contrary to the idea of common light, in which it is implied that on the average whatever we can say of one plane passing through the ray, we can say of another: whatever we can say of the direction one way round we can say of the other way round.

At the end of his excellent Tract on the Undulatory Theory, Mr Airy has shewn how the simple supposition of the existence, in common light, of successive series of undulations, in which the vibrations of one series have no relation to those of another, would account at the same time for the interference of common light and the non-interference of the pencils, polarized in rectangular planes, into which common light may be conceived to be decomposed. But he has, I think, introduced a gratuitous difficulty into the subject, by asserting that it is necessary to suppose the transition from one series into another to be abrupt, and that a gradual change in the nature of the vibrations is inadmissible. This assertion, which seems to have led others to conceive that there was here a difficulty with which the undulatory theory had to contend, seems to have resulted from an investigation from which it appeared that common light could not be represented by an indefinite series of elliptic vibrations, in which the major axis of the ellipse was supposed to revolve uniformly, rapidly, with regard to the duration of impressions on the retina, though slowly with regard to the time of a luminous vibration. I have elsewhere pointed out on what grounds I conceive that the instance of the revolving ellipse is not a case in point, namely, that it is not a fair representation of common light, because it gives a preponderance on the average to one direction of revolution over the contrary, which is contrary to the idea of common light*. Let us now apply the general formulæ (16) and (17) to this case

Let $c \cos \beta$, $c \sin \beta$ be the semi-axes of the ellipse, α the azimuth of the first axis at a given

* Report of the meeting of the British Association at Swansea in 1848. Transactions of the Sections, p. 5.

time. So far as relates to the stream for which the azimuth of the first axis lies between a and $a + da$, we have $\mathfrak{m}(c^2) = (2\pi)^{-1} c^2 da$; and in the application of the formulæ (16) the summation, to which Σ refers, will of course pass into an integration. We have therefore

$$A = \frac{c^2}{2\pi}\int_0^{2\pi} da = c^2; \quad B = \frac{c^2}{2\pi}\sin 2\beta \int_0^{2\pi} da = c^2 \sin 2\beta; \quad C = 0; \quad D = 0;$$

whence we get from the formulæ (19), supposing β positive,

$$J' = c^2 \sin 2\beta; \quad J = c^2(1 - \sin 2\beta); \quad \beta' = 45^0;$$

while a' remains indeterminate. Hence the mixture is equivalent, not to common light alone, but to a stream of common light having an intensity equal to $c^2(1 - \sin 2\beta)$, combined with a stream of circularly polarized light, independent of the former, having an intensity equal to $c^2 \sin 2\beta$, and being of the same character as regards right-handed or left-handed as the original stream would be were the ellipse stationary. The result of supposing β negative is here assumed as obvious.

22. Suppose that a polarizing prism and a mica plate, which produce elliptic polarization, are made to revolve together with great rapidity. The stream of light thus produced will be equivalent to the former. The only difference is that in the former case c was supposed constant, whereas in the case of actual experiment it will be subject to the fluctuations mentioned at the beginning of this paper; but the mean values represented by $\mathfrak{m}$ will not be affected when these fluctuations are taken into account, and therefore the same formulæ will continue to apply. Hence, if the polarization be circular the rotation will make no difference; if it be plane, the light will appear completely depolarized; in intermediate cases the result will be intermediate, and the light will be equivalent to a mixture of common light and circularly polarized light. The reader may compare these conclusions of theory with some experiments by Professor Dove*.

23. As a last example, let light polarized by transmission through a Nicol's prism be transmitted through a second Nicol's prism, which is made to revolve uniformly and rapidly while the first remains fixed.

Let a be the azimuth of the plane of polarization of the second Nicol's prism, measured from that of the first, c the coefficient of vibration in the stream transmitted through the first prism. The stream passing through the second prism is made up of an infinite number of independent streams such as that whose intensity is $(2\pi)^{-1}\cos^2 a da$ multiplied by the mean value of c^2. Hence we have from the formulæ (16)

$$A = \tfrac{1}{2}\mathfrak{m}(c^2); \quad B = 0; \quad C = \tfrac{1}{4}\mathfrak{m}(c^2); \quad D = 0;$$

whence, taking the intensity of the original stream as unity, we have

$$J' = \tfrac{1}{4}; \quad J = \tfrac{1}{4}; \quad \beta' = 0; \quad a' = 0;$$

or the light is equivalent to a mixture of common light having an intensity $\frac{1}{4}$, and light

* See *Philosophical Magazine*, Vol. XXX. (1847) p. 465.

independent of the former, of the same intensity, polarized at an azimuth zero. This result may be compared with one of Professor Dove's experiments.

If a rotating crystalline plate, cut from a doubly refracting crystal in a direction not nearly coinciding with the optic axis or one of the optic axes, and not sufficiently thin to exhibit colours in polarized light, be substituted for the rotating Nicol's prism, since the plate is too thick to allow of the exhibition of any phenomena depending on the interference of the oppositely polarized pencils, the effect will be just the same as in the case of the Nicol's prism, only the intensity of each steam will be doubled.

In applying the formulæ of this paper to experiments in which one part of an optical train is made to revolve rapidly, it must be understood that the other parts of the train are at rest, or at least do not revolve with a velocity nearly equal to the former. Otherwise, particular phenomena will be exhibited depending on the simultaneous movements of two or more parts of the train, as appeared in Professor Dove's experiments; and in the calculation of these phenomena it will not be allowable to substitute for the stream of light emerging from the polarizer, or first revolving piece whatever it be, the streams of common and elliptically polarized light to which, for general purposes, it is equivalent.

G. G. STOKES.

11

Reprinted from *Théorie mathématique de la lumière,* Georges Carré, ed., Paris, 1892, pp. 275–285

POLARISATION ROTATOIRE.—THÉORIE DE M. MALLARD

155. La théorie que Fresnel a donnée du pouvoir rotatoire n'est pas, à proprement parler, une théorie physique, mais seulement la constatation d'une identité cinématique.

De même les théories d'Airy se réduisent à l'adjonction, aux equations des termes nécessaires pour faire concorder leurs conséquences avec les résultats observés.

Les expériences de Reusch sur les piles de mica ont donné l'idée d'autres théories, en particulier, de celle que nous allons exposer due à M. Mallard.

L'exposé de cette théorie comporte des calculs assez longs, que nous chercherons à simplifier, à l'aide de quelques considérations préliminaires.

Rappelons d'abord en quoi consistent les expériences de Reusch. Reusch empilait des lames de mica, très minces, à faces parallèles, identiques entre elles, de manière que la section principale d'une lame fit avec la section principale de la suivante un angle de 45° ou de 60°. Cet angle pourrait être quelconque : seulement ces deux valeurs particulières avaient été choisies parce qu'elles correspondaient à la symétrie quaternaire et à la symétrie ternaire. Le système ainsi obtenu possède le pouvoir rotatoire comme une lame de quartz perpendiculaire à l'axe.

Si la vibration incidente est rectiligne, la vibration émergente est en général elliptique : mais l'ellipse est fort allongée, d'autant plus allongée que les lames sont plus minces. Pour des lames très minces la polarisation elliptique est très faible, et les phénomènes se rapprochent d'autant plus de ceux du quartz.

Supposons que nous ayons ainsi une série de lames planes, à faces parallèles, empilées les unes sur les autres : nous prendrons comme plan des xy un plan parallèle à ceux des lames. Une onde lumineuse, parallèle à ce plan, tombant sur le système de lames ne subira pas de déviation et restera constamment parallèle à cette direction.

Soit une vibration située dans le plan de l'onde :

$$\xi = Ae^{\sqrt{-1}pt}$$
$$\eta = Be^{\sqrt{-1}pt}.$$

(sur ces expressions imaginaires, voir chap. préc.). A sera en général une quantité imaginaire.

$$A = A_0 e^{\sqrt{-1}\varphi}$$

de même :

$$B = B_0 e^{\sqrt{-1}\psi}$$

A_0 et B_0 sont les amplitudes des vibrations ; φ, ψ leurs phases. Alors

$$\xi = A_0 \cos(pt + \varphi)$$
$$\eta = B_0 \cos(pt + \psi).$$

Le rapport $\frac{B}{A}$ sera aussi imaginaire en général, nous poserons donc :

$$\frac{B}{A} = u + \sqrt{-1}\, v$$

La valeur de ce rapport nous fait connaître la forme de l'ellipse et son orientation.

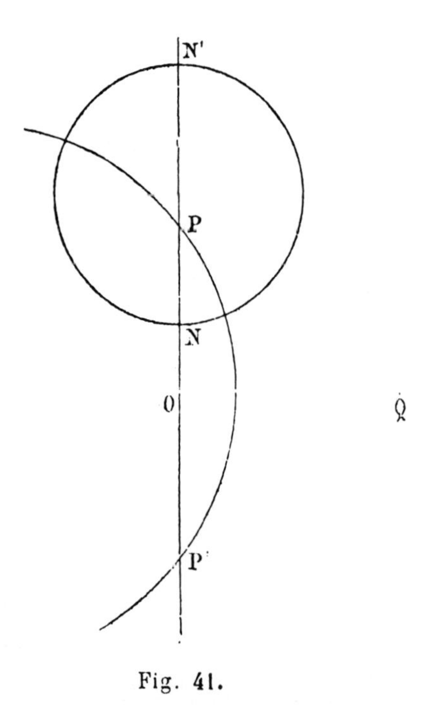

Fig. 41.

156. Mode de représentation des vibrations. — Nous representerons les variations de cette ellipse par les déplacements de l'affixe de l'imaginaire $u + \sqrt{-1}v$, c'est-à-dire du point ayant u pour abscisse et v pour ordonnée (*fig.* 41).

Si $\frac{B}{A}$ est réel, $v = o$, le point représentatif Q est sur l'axe des u : la vibration est rectiligne puisque $\frac{\eta}{\xi}$ est réel, l'angle de sa direction avec Ox étant θ, on a :

$$\frac{B}{A} = \text{tang}\,\theta = u$$

Donc :

$$OQ = \text{tang}\,\theta.$$

Si $\frac{B}{A}$ est purement imaginaire, les deux composantes présentent une différence de phase de $\frac{\pi}{2}$; la vibration est elliptique et ses axes sont dirigés suivant les axes de coordonnées. Le point représentatif N est sur l'axe des v. ON est le rapport des axes $\frac{B}{A}$.

Si on a $OP = 1$, le point P représente une vibration circulaire $\frac{B}{A} = \sqrt{-1}$, de même le point P' tel que $OP' = 1$ ou $\frac{B}{A} = -\sqrt{-1}$. La première vibration est droite, la deuxième est gauche.

D'une façon générale d'ailleurs, les points situés au-dessus de Ou ($v < o$) représentent des vibrations droites : les points situés au dessous ($v < o$), des vibrations gauches.

Supposons que le rayon vienne à traverser une lame cristallisée, dont les sections principales sont orientées suivant les axes de coordonnées : ξ et η se propagent avec des vitesses inégales ; leurs phases varient de quantités inégales, les modules A_0 et B_0 ne changent pas, mais $\frac{B}{A}$ change et devient par exemple

$$\frac{B}{A} e^{\sqrt{-1}\omega}.$$

Le point M′, qui représente la nouvelle vibration, sera donc tel que :

$$OM' = OM \qquad MOM' = \omega$$

Tout se passe comme si le plan avait tourné d'un angle ω autour du point O, ω étant la différence de phase introduite par la lame cristalline. Pour des rayons de même couleur, cette différence est proportionnelle à la différence des temps employés par les deux vibrations pour traverser la lame, par conséquent proportionnelle à l'épaisseur de la lame et à son pouvoir biréfringent $n - n'$.

Pour des rayons de couleurs différentes traversant une même lame, $n - n'$ est indépendant de la longueur d'onde si nous négligeons la dispersion ; la différence de temps est donc la même pour tous les rayons. Or ω est égal à 2π quand cette différence est égale à une période entière : ω est donc en raison inverse de la période, ou en raison inverse de la longueur d'onde.

157. Proposons-nous de chercher quel sera, dans le mode de représentation que nous avons adopté, le lieu des points correspondants à des vibrations dont on donne soit l'orientation des axes, soit le rapport des axes.

Pour trouver ces lieux, nous nous appuierons sur le théorème suivant :

Théorème. — Soit une quantité complexe

$$w = u + \sqrt{-1}\, v = \frac{a + bt}{c + dt}$$

a, b, c, d étant des constantes imaginaires, t une variable réelle — Quand t varie de $-\infty$ à $+\infty$, le point (u, v) décrit un cercle.

Désignons en effet par a_0, b_0, c_0, d_0 les imaginaires conjuguées de a, b, c, d :

$$w_0 = u - \sqrt{-1}\, v = \frac{a_0 + b_0 t}{c_0 + d_0 t}.$$

D'où :

$$u^2 + v^2 = w w_0 = \frac{P_1}{(c + dt)(c_0 + d_0 t)} = \frac{P_1}{P_4}$$

$$u = \frac{w + w_0}{2} = \frac{P_2}{P_4}$$

$$v = \frac{w - w_0}{2\sqrt{-1}} = \frac{P_3}{P_4}$$

$$1 = \frac{P_4}{P_4}$$

p_1, p_2, p_3, p_4 étant des polynômes du second degré en t. Ces polynômes ne peuvent donc être indépendants et nous aurons une relation de la forme :

$$C_1 (u^2 + v^2) + C_2 u + C_3 v + C_4 = 0,$$

équation qui représente un cercle.

Cela posé, considérons les points qui représentent les ellipses ayant leurs axes dirigés suivant Ox', Oy', faisant un angle θ avec les axes de coordonnées.

Les projections des vibrations sur Ox' et Oy' sont :

$$\xi' = \xi \cos\theta + \eta \sin\theta = (A \cos\theta + B \sin\theta)\, e^{\sqrt{-1}pt}$$

$$\eta' = -\xi \sin\theta + \eta \cos\theta = (-A \sin\theta + B \cos\theta)\, e^{\sqrt{-1}pt}$$

Soit τ le rapport des axes :

$$\frac{-A\sin\theta + B\cos\theta}{A\cos\theta + B\sin\theta} = \sqrt{-1}\,\tau$$

ou comme :

$$\frac{B}{A} = u + \sqrt{-1}\,v$$

$$\frac{u + \sqrt{-1}\,v - \operatorname{tang}\theta}{1 + (u + \sqrt{-1}\,v)\operatorname{tang}\theta} = \sqrt{-1}\,\tau.$$

Si nous cherchons le lieu des points tels que $\theta = \text{const}$, τ sera une variable réelle, liée à $u + \sqrt{-1}\,v$ par une relation homographique. Nous venons de voir que, quand τ varie de $-\infty$ à $+\infty$, le point (u, v) décrit un cercle. Ce cercle passera par les points P et P′, précédemment définis (*fig.* 41). En effet, ces points représentent des vibrations circulaires, dont les axes ont une direction indéterminée.

Si nous laissons $\tau = \text{const}$, c'est-à-dire si nous nous donnons la forme de l'ellipse et que nous fassions varier θ, c'est-à-dire l'orientation de cette ellipse, $\operatorname{tang}\theta$ est encore une variable réelle liée à $u + \sqrt{-1}\,v$ par une relation homographique. Le point (u, v) décrit encore un cercle.

Si nous prenons $ON = \tau$, $ON' = \frac{1}{\tau}$, le cercle passera par ces points N et N′, ces points correspondent à des ellipses ayant leurs axes dirigés suivant les axes de coordonnées : l'ellipse N′ est égale à l'ellipse N, mais elle a tourné de 90°. Par raison de symétrie, NN′ doit être un diamètre. Comme d'autre part

$$1 = ON \,.\, ON' = \overline{OP}^2 = \overline{OP'}^2$$

les deux cercles NN′ et PP′ se coupent orthogonalement.

158. Il nous sera commode dans la suite de remplacer cette représentation plane par une représentation analogue sur la sphère; nous effectuerons cette transformation par une projection stéréographique, le plan des (u, v) étant le plan du tableau et l'origine O des coordonnées ; uv, le point de contact de la sphère avec ce plan. Cette projection, comme on le sait, conserve les angles, et les cercles se projettent sur des cercles.

Au point M du plan correspond le point m où la droite VM rencontre la sphère. Nous conviendrons de représenter l'ellipse M par le point m (*fig.* 42).

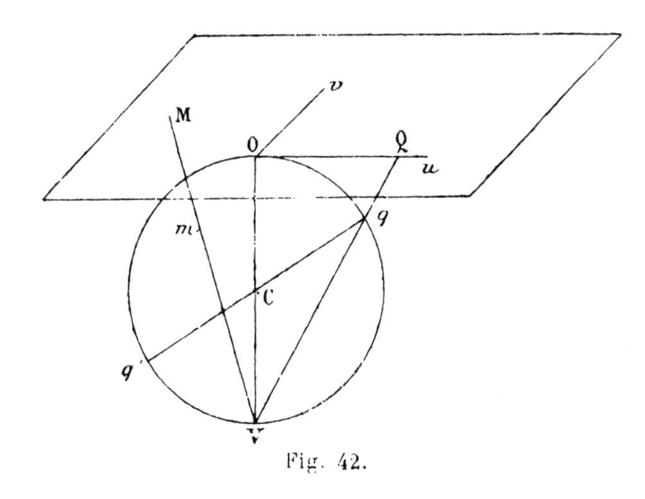

Fig. 42.

L'axe des u se projettera suivant un grand cercle passant par O et que nous appellerons équateur. Les points de l'équateur représenteront par conséquent les vibrations rectilignes.

Soient Q un point de l'axe des u, q sa projection sur la sphère,

$$OQ = \operatorname{tang} \theta$$

si nous prenons le diamètre de la sphère égal à 1, donc :

$$\theta = \widehat{OVq}$$

$$2\theta = \widehat{OCq}$$

l'angle O des axes de l'ellipse avec les axes de coordonnées sera donc égal à la demi-longitude du point q.

Deux points diamétralement opposés ont des longitudes qui diffèrent de π. Elles représentent donc des vibrations rectilignes rectangulaires entre elles.

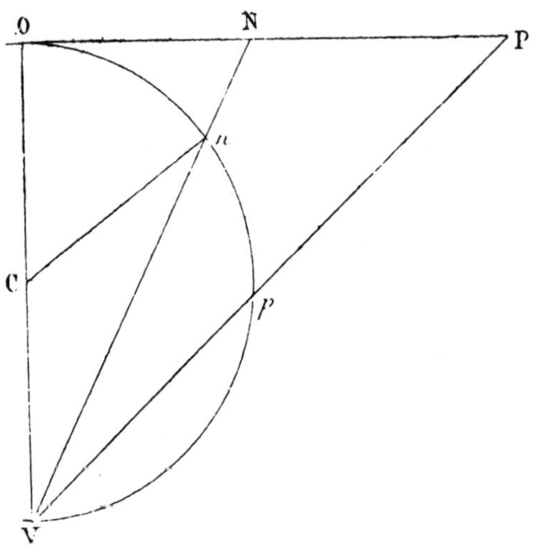

Fig. 43

Les points de l'axe des v se projettent sur un grand cercle perpendiculaire à l'équateur, que nous appellerons premier méridien. Un point N se projette en n (*fig.* 43)

$$\widehat{OCn} = l = \text{latitude}$$

$$\widehat{OVn} = \frac{l}{2}$$

$$ON = \text{tang}\,\frac{l}{2}$$

Le rapport des axes de l'ellipse représentée par le point n est donc égal à la tangente de la demi-latitude de ce point n.

Pour le point P, OP = OV et par suite $\widehat{OVP} = 45^\circ$, et

$OCp = 90°$. Le point P se projette donc au pôle p de l'équateur. Les deux pôles de l'équateur correspondent aux vibrations circulaires, les divers points du premier méridien, aux ellipses dont les axes sont divisés suivant les axes de coordonnées. Ces ellipses sont droites dans l'hémisphère nord, gauches dans l'hémisphère sud.

L'orientation des axes ne dépend que de la longitude. Les lieux des points tels que $\theta =$ const sont des cercles passant par p et p', c'est-à-dire des méridiens.

La forme de l'ellipse ne dépend que de la latitude ; les lieux des points correspondant à une forme donnée sont des parallèles.

Supposons que le rayon lumineux traverse une lame biréfringente, mais dénuée de pouvoir rotatoire. Si les sections principales de la lame sont dirigées suivant les axes, tout se passe comme si le plan autour de O ou la sphère autour de OV tournaient d'un angle ω ; les points O et V correspondent aux sections principales ; l'azimut des axes de l'ellipse est proportionnel à la longitude. Mais les points O et V ne jouent aucun rôle particulier ; quand on fait tourner les axes dans le plan des xy, cela revient simplement à changer l'origine des longitudes.

Considérons une lame dont les sections principales aient une direction quelconque, l'une correspondant par exemple au point q, l'autre au point q' diamétralement opposé, sur l'équateur ; l'axe qq' jouera le même rôle que OV précédemment. Le résultat sera le même que si la sphère tournait de ω autour de qq'. Le passage à travers une lame correspond ainsi à une rotation autour d'un axe situé dans le plan de l'équateur.

Si la lame possède le pouvoir rotatoire, sans avoir le pouvoir biréfringent, comme une lame de quartz perpendiculaire à l'axe, l'ellipse conserve sa forme, mais son orientation change ; tout se passera comme si la sphère avait tourné d'un certain angle autour de pp' perpendiculaire au plan de l'équateur.

Si les deux effets de la biréfringence et du pouvoir rotatoire se superposent, les deux rotations de la sphère se composeront ; on peut alors les remplacer par une rotation autour d'un axe quelconque.

12

Reprinted from *Acta Math.*, **55**, 182–195 (1930)

GENERALIZED HARMONIC ANALYSIS.[1]

BY

NORBERT WIENER

of CAMBRIDGE, MASS., U. S. A.

Page

9. Coherency matrices.

The spectrum theory of our earlier sections is a theory of the spectrum of an individual function. There are, however, many phenomena intimately connected with harmonic analysis which refer to several functions considered simultaneously. Chief among these are the phenomena of coherency and incoherency, of interference, and of polarization.

It is known to every beginner in physics that two rays of light from the same source may interfere: that is, they may be superimposed to form a darkness, or else a light more intense than is ordinarily formed by two rays of light of their respective intensities. On the other hand, two rays of light from independent sources or from different parts of the same source never exhibit this phenomenon. The former rays are said to be *coherent*, and the latter to be *incoherent*. Although it is mathematically impossible for two truly sinusoidal oscillations to be incoherent, even the most purely monochromatic light which we can sensibly produce never coheres with similar light from another source.

The physicist's explanation of incoherency is the following: the interference pattern produced by two sources of light depends on their relative displacement in phase. Now, the relative phase of two sensibly monochromatic sources of light is able to assume all possible values, and since light probably consists in a series of approximately sinusoidal trains of oscillations each lasting but a small portion of a millionth of a second, this relative phase assumes in any sensible interval all possible values with a uniform distribution which averages out light and dark bands into a sensibly uniform illumination.

This explanation of incoherency is unquestionable adequate to account for the phenomenon which it was invented to explain. Nevertheless, it is desirable to have a theory of coherency and incoherency which does not postulate a hypothetical set of constituent harmonic trains of oscillations, which at any rate must become merged in the general electromagnetic oscillation constituting the light. The present section is devoted to the development of a theory of coherency which is as direct as the theory of this paper concerning the harmonic analysis of a single function, and indeed forms a natural extension of the latter.

In interference, the components of the electromagnetic field of the constituent light rays combine additively. Accordingly, the theory of coherency and interference must be a theory of the harmonic analysis of all functions which can be obtained from a given set by linear combination. Let us see what the outlines of this theory are.

We start from a class of functions, $f_k(t)$, in general complex, and defined for all real arguments between $-\infty$ and ∞. For the present we shall assume this class of functions to be finite, although this restriction is not essential. Let

$$f(t)=a_1 f_1(t)+a_2 f_2(t)+\cdots+a_n f_n(t) \tag{9.01}$$

be the general linear combination of functions of the set. We shall have

$$\varphi(\tau)=\lim_{T\to\infty}\frac{1}{2T}\int_{-T}^{T} f(t+\tau)\bar{f}(t)\,dt=\sum_{j,k=1}^{n}\sum a_j\bar{a}_k\lim_{T\to\infty}\frac{1}{2T}\int_{-T}^{T} f_j(t+\tau)\bar{f}_k(t)\,dt \tag{9.02}$$

in case the latter limits exist. The necessary and sufficient condition for this to exist for all linear combinations $f(t)$ of functions of the set is that

$$\varphi_{jk}(\tau)=\lim_{T\to\infty}\frac{1}{2T}\int_{-T}^{T} f_j(t+\tau)\bar{f}_k(t)\,dt \tag{9.03}$$

should exist for every j, k, and τ. Then

$$\varphi(\tau)=\sum_j\sum_k a_j\bar{a}_k\varphi_{jk}(\tau). \tag{9.04}$$

Again, we shall have formally

$$S(u)=\frac{1}{2\pi}\int_{-\infty}^{\infty}\varphi(\tau)\frac{e^{iu\tau}-1}{i\tau}\,d\tau=\sum_j\sum_k a_j\bar{a}_k\frac{1}{2\pi}\int_{-\infty}^{\infty}\varphi_{jk}(\tau)\frac{e^{iu\tau}-1}{i\tau}\,d\tau, \tag{9.05}$$

where we may write

$$S_{jk}(u)=\frac{1}{2\pi}\int_{-\infty}^{\infty}\varphi_{jk}(\tau)\frac{e^{iu\tau}-1}{i\tau}\,d\tau. \tag{9.06}$$

If $\varphi_{jk}(\tau)$ exists for every j, k, and τ, we may readily show that each $S_{jk}(u)$ exists. Clearly

$$\varphi_{kj}(\tau)=\lim_{T\to\infty}\frac{1}{2T}\int_{-T}^{T} f_k(t+\tau)\bar{f}_j(t)\,dt=\lim_{T\to\infty}\frac{1}{2T}\int_{-T}^{T}\bar{f}_j(t-\tau)f_k(t)\,dt=\bar{\varphi}_{jk}(-\tau) \tag{9.07}$$

and

$$S_{kj}(u) = \frac{1}{2\pi} \int_{-\infty}^{\infty} \bar{\varphi}_{jk}(-\tau) \frac{e^{iu\tau} - 1}{i\tau} d\tau = \frac{1}{2\pi} \int_{-\infty}^{\infty} \bar{\varphi}_{jk}(\tau) \frac{e^{-iu\tau} - 1}{-i\tau} d\tau = \bar{S}_{jk}(u), \tag{9.08}$$

so that the matrix

$$\| S_{jk}(u) \|$$

is Hermitian. This matrix determines the spectra of all possible linear combinations of $f_1(t), \ldots, f_n(t)$. Since it determines the precise coherency relations of the functions in question, we shall call it the coherency matrix.

Let us subject the functions $f_k(t)$ to the linear transformation

$$g_j(t) = \sum_k a_{jk} f_k(t). \tag{9.09}$$

Then

$$\psi_{jk}(\tau) = \lim_{T \to \infty} \frac{1}{2T} \int_{-T}^{T} g_j(t+\tau) \bar{g}_k(t)\, dt = \sum_l \sum_m a_{jl} \bar{a}_{km} \varphi_{lm}(\tau) \tag{9.10}$$

and

$$T_{jk}(u) = \frac{1}{2\pi} \int_{-\infty}^{\infty} \psi_{jk}(\tau) \frac{e^{iu\tau} - 1}{i\tau} d\tau = \sum_l \sum_m a_{jl} \bar{a}_{km} S_{lm}(u). \tag{9.11}$$

Thus the new coherency matrix $\| T_{jk}(u) \|$ may be written

$$M_1 \cdot \| S_{jk}(u) \| \cdot M_2, \tag{9.12}$$

where

$$M_1 = \| a_{jk} \|, \tag{9.13}$$

and

$$M_2 = \| \bar{a}_{kj} \|. \tag{9.14}$$

In case the transformation with matrix M_1 has the property

$$M_1 \cdot M_2 = M_2 \cdot M_1 = 1, \tag{9.15}$$

it is said to be *unitary*. For such transformations,

$$\| T_{jk} \| = M_1 \cdot \| S_{jk} \| \cdot M_1^{-1}. \tag{9.16}$$

A matrix $\| a_{jk} \|$ is said to be in diagonal form if all the terms a_{jk} for which $j \neq k$ are identically zero. By a theorem of Weyl[5], every Hermitian matrix may be transformed into a diagonal matrix by a unitary transformation. Since we may regard a diagonal matrix as representing a set of completely incoherent phenomena, this transformation is of fundamental importance in the characterization of the state of coherency of the functions determining the matrix. Together with the numerical values of the diagonal elements of the diagonal matrix, it indeed consitutes a complete characterization of the state of coherency of the original function. In the case the values of the diagonal elements are distinct, this characterization is indeed to be carried through in but a single way.

The production of incoherent functions is a simple matter, when we have once settled the existence theory of functions with given types of spectra. Let $f_1(t)$ be any bounded function such that $\varphi(\tau)$ and consequently $S(u)$ exists. Then if

$$f_2(t) = f_1(t)\, e^{i\sqrt{|t|}}, \tag{9. 17}$$

we have

$$\lim_{T\to\infty} \frac{1}{2T}\int_{-T}^{T} f_2(t+\tau)\bar{f}_2(t)\,dt = \lim_{T\to\infty} \frac{1}{2T}\int_{-T}^{T} f_1(t+\tau)\bar{f}_1(t)\, e^{i(\sqrt{|t+\tau|}-\sqrt{|t|})}\,dt$$

$$= \lim_{T\to\infty} \frac{1}{2T}\int_{-T}^{T} f_1(t+\tau)\bar{f}_1(t) \exp\left(i\,\frac{|t+\tau|-|t|}{\sqrt{|t+\tau|}+\sqrt{|t|}}\right)dt, \tag{9. 18}$$

and hence, since $\lim\limits_{t\to\pm\infty} \exp\left(i\,\frac{|t+\tau|-|t|}{\sqrt{|t+\tau|}+\sqrt{|t|}}\right) = 1$,

$$\varphi_{22}(\tau) = \varphi_{11}(\tau) \tag{9. 19}$$

and

$$S_{22}(u) = S_{11}(u). \tag{9. 20}$$

On the other hand, if for example $f(t) = e^{i\lambda t}$,

$$\left.\begin{aligned} \varphi_{12}(\tau) &= \lim_{T\to\infty} \frac{1}{2T}\int_{-T}^{T} f_1(t+\tau)\bar{f}_1(t)\, e^{-i\sqrt{|t|}}\,dt \equiv 0; \\ S_{12}(u) &= S_{21}(u) = 0. \end{aligned}\right\} \tag{9. 21}$$

Thus the coherency matrix of $f_1(t)$ and $f_2(t)$ is

24 — 29764. *Acta mathematica.* 55. Imprimé le 14 avril 1930.

$$S_{11}(u)\begin{Vmatrix} 1 & 0 \\ 0 & 1 \end{Vmatrix}. \tag{9. 22}$$

The coherency matrix of $\sqrt{2}\, f_1(t)$ and 0 is

$$S_{11}(u)\begin{Vmatrix} 2 & 0 \\ 0 & 0 \end{Vmatrix}; \tag{9. 23}$$

that of $f_1(t)$ and $f_1(t)$ is

$$S_{11}(u)\begin{Vmatrix} 1 & 1 \\ 1 & 1 \end{Vmatrix}; \tag{9. 24}$$

that of $f_1(t)$ and $if_1(t)$ is

$$S_{11}(u)\begin{Vmatrix} 1 & -i \\ i & 1 \end{Vmatrix}. \tag{9. 25}$$

Let it be noted that the coherency matrices of real functions are in general complex. Thus if

$$f_2(t) = f_1(t+\lambda), \tag{9. 26}$$

we have

$$\left.\begin{aligned} \varphi_{22}(\tau) &= \varphi_{11}(\tau); \\ \varphi_{12}(\tau) &= \lim_{T\to\infty} \frac{1}{2T}\int_{-T}^{T} f_1(t+\lambda+\tau) f_1(t)\, dt \\ &= \varphi_{11}(\tau+\lambda); \end{aligned}\right\} \tag{9. 27}$$

and hence

$$\left.\begin{aligned} S_{22}(u) &= S_{11}(u); \\ S_{12}(u) &= \int_{-\infty}^{\infty} \varphi_{11}(\tau+\lambda)\frac{e^{iu\tau}-1}{i\tau}\, d\tau = \int_{-\infty}^{u} e^{-iv\lambda}\, dS_{11}(v) + S_{12}(-\infty) \end{aligned}\right\} \tag{9. 28}$$

giving the coherency matrix with derivative

$$S'_{11}(u)\begin{Vmatrix} 1 & e^{-iv\lambda} \\ e^{iv\lambda} & 1 \end{Vmatrix}. \tag{9. 29}$$

In optics, coherency is generally considered for light of one particular frequency. From that standpoint, the coherency of a set of functions $f_1(t), \ldots, f_n(t)$

with continuous differentiable spectra may be regarded as determined for frequency u by the matrix

$$\left\|\begin{matrix} S_{11}^{1}(u), & \ldots, & S_{1n}^{1}(u) \\ \cdot & \cdot & \cdot \\ \cdot & \cdot & \cdot \\ S_{n1}^{1}(u), & \ldots, & S_{nn}^{1}(u) \end{matrix}\right\|; \tag{9. 30}$$

or in the case of functions with line spectra, by

$$\left\|\begin{matrix} S_{11}(u+0)-S_{11}(u-0), & \ldots, & S_{1n}(u+0)-S_{1n}(u-0) \\ \cdot & \cdot & \cdot \\ \cdot & \cdot & \cdot \\ S_{n1}(u+0)-S_{n1}(u-0), & \ldots, & S_{nn}(u+0)-S_{nn}(u-0) \end{matrix}\right\|. \tag{9. 31}$$

We may regard these matrices in a secondary sense as coherency matrices.

Coherency matrices of two functions are of particular interest in connection with the characterization of the state of polarization of light. As everyone knows, this characterization is identically the characterization of the state of coherency between two components of the electric vector at right angles to one another. With this interpretation, matrix (9. 22) represents unpolarized light, matrix (9. 23) light polarized completely in one plane, while

$$\left\|\begin{matrix} 0 & 0 \\ 0 & 2 \end{matrix}\right\| \tag{9. 32}$$

represents light completely polarized in a plane perpendicular to the first. Matrix (9. 24) and matrix

$$\left\|\begin{matrix} 1 & -1 \\ -1 & 1 \end{matrix}\right\| \tag{9. 33}$$

represent light polarized respectively at 45° and at 135° to the first direction. Matrix (9. 25) and matrix

$$\left\|\begin{matrix} 1 & i \\ -i & 1 \end{matrix}\right\| \tag{9. 34}$$

represent respectively light polarized circularly in a counter-clockwise and a clockwise direction.

When the matrix of completely polarized light, whether linearly, elliptically, or circularly polarized, is brought into diagonal form by a linear unitary transformation, the resulting diagonal matrix will have only one element distinct from 0. On the other hand, completely unpolarized light has the diagonal terms equal. This suggests as a measure of the amount of polarization of the diagonal matrix

$$\left\| \begin{matrix} a & 0 \\ 0 & b \end{matrix} \right\|, \tag{9. 35}$$

or of any other matrix equivalent to it under a unitary transformation, the quantity

$$a - b. \tag{9. 36}$$

If we subtract from our original diagonal matrix the completely incoherent matrix

$$\left\| \begin{matrix} \frac{a+b}{2} & 0 \\ 0 & \frac{a+b}{2} \end{matrix} \right\|, \tag{9. 37}$$

which is invariant under every unitary transformation, we get the matrix

$$\left\| \begin{matrix} \frac{a-b}{2} & 0 \\ 0 & \frac{b-a}{2} \end{matrix} \right\|, \tag{9. 38}$$

which may be regarded as a representative of the quantity $a-b$. This suggests that given any coherency matrix

$$\left\| \begin{matrix} A & B+Ci \\ B+Ci & D \end{matrix} \right\|, \tag{9. 39}$$

we may take $A+D$ to represent the intensity of the corresponding light, and the matrix

$$\left\| \begin{matrix} \frac{A-D}{2} & B-Ci \\ B+Ci & \frac{D-A}{2} \end{matrix} \right\| \tag{9. 40}$$

as its polarization. Thus horizontal polarization is represented by the matrix

$$\begin{Vmatrix} 1 & 0 \\ 0 & -1 \end{Vmatrix}; \tag{9.41}$$

polarization at 45° by the matrix

$$\begin{Vmatrix} 0 & 1 \\ 1 & 0 \end{Vmatrix}; \tag{9.42}$$

and circular polarization by the matrix

$$\begin{Vmatrix} 0 & i \\ -i & 0 \end{Vmatrix}. \tag{9.43}$$

These are the same matrices which Jordan, Dirac and Weyl have employed to such advantage in the theory of quanta.

Since the most general Hermitian matrix of the second order may be written

$$\begin{Vmatrix} \alpha+\beta & \gamma+\delta i \\ \gamma-\delta i & \alpha-\beta \end{Vmatrix} = \alpha \begin{Vmatrix} 1 & 0 \\ 0 & 1 \end{Vmatrix} + \beta \begin{Vmatrix} 1 & 0 \\ 0 & -1 \end{Vmatrix} + \gamma \begin{Vmatrix} 0 & 1 \\ 1 & 0 \end{Vmatrix} + \delta \begin{Vmatrix} 0 & i \\ -i & 0 \end{Vmatrix}, \tag{9.44}$$

it appears that all light may characterized as to its state of polarization by given the total amount of light it contains, the excess of the amount polarized at 0° over that polarized at 90°, the excess of the amount polarized at 45° over that polarized at 135° and the excess of that polarized circularly to the right over that polarized circularly to the left. This characterization is complete and univocal. The total intensity of the light may be read off any sort of a photometer. The excess of light polarized horizontally over that polarized vertically may be determined by a doubly refracting crystal in one orientation, and the excess of light polarized at 45° over that polarized at 135° by the same crystal in another orientation. It is possible furthermore to devise an instrument which will read off the amount of circular polarization in the light in question. The three latter instruments possess some very remarkable group properties with respect to one another. Either portion of the light emerging from any one of the instruments will behave towards the other two exactly like unpolarized light. Rotation of the plane of polarization of the light through 45° will change the reading of the first of the last three instruments into that of the second, and the reading of the second into minus that of the first, leaving the reading of the third unchanged. There are precisely analogous unitary transformations interchanging any other pair of the three instruments, leaving the reading of the remaining instrument untouched. These transformations together with their powers and the identical transformation form a group.

A fact concerning polarized light which is so apparently obvious that it is generally regarded as not needing any proof is that all light is a case or limiting case of partially elliptically polarized light. It is nevertheless desirable to prove this statement. Completely elliptically polarized light with the coordinate axes as principal axes has a coherency matrix of the form

$$\begin{Vmatrix} A^2 & -ABi \\ ABi & B^2 \end{Vmatrix}; \tag{9.45}$$

and hence partially polarized light with the same principal axes has a coherency matrix of the form

$$P = \begin{Vmatrix} A^2+D^2 & -ABi \\ ABi & B^2+D^2 \end{Vmatrix}. \tag{9.46}$$

We wish to show that the general coherency matrix

$$M = \begin{Vmatrix} \alpha & \gamma-\delta i \\ \gamma+\delta i & \beta \end{Vmatrix} \tag{9.47}$$

may be transformed into this form by a real unitary transformation in such a way that

$$T \,.\, M \,.\, T^{-1} = P.$$

To do this, we need only put

$$T = \begin{Vmatrix} \cos\varphi & \sin\varphi \\ \sin\varphi & -\cos\varphi \end{Vmatrix}; \text{ where } \tan 2\varphi = \frac{2\gamma}{\alpha-\beta}. \tag{9.48}$$

Thus the axes of polarization of M are 1 and 2 directions when we replace $f_1(t)$ and $f_2(t)$ by

$$\left.\begin{aligned} g_1(t) &= \quad f_1(t)\cos\varphi + f_2(t)\sin\varphi; \\ g_2(t) &= -f_2(t)\cos\varphi + f_1(t)\sin\varphi; \end{aligned}\right\} \tag{9.49}$$

the »lengths» of these axes are respectively

$$\left.\begin{aligned} A &= \left\{\frac{1}{2}\left[(\alpha-\beta)^2+4(\gamma^2+\delta^2)\right]^{1/2}+\frac{1}{2}\left[(\alpha-\beta)^2+4\gamma^2\right]^{1/2}\right\}^{1/2}; \\ B &= \left\{\frac{1}{2}\left[(\alpha-\beta)^2+4(\gamma^2+\delta^2)\right]^{1/2}-\frac{1}{2}\left[(\alpha-\beta)^2+4\gamma^2\right]^{1/2}\right\}^{1/2}; \end{aligned}\right\} \tag{9. 50}$$

and the percentage of polarization

$$100\left(1-\frac{4(\gamma^2+\delta^2)}{(\alpha+\beta)^2}\right)^{1/2}. \tag{9. 51}$$

The connection between coherency matrices and optical instruments, which we have already mentioned in the case of polarized light, is of far more general applicability. An optical instrument is a method, linear in electric and magnetic field vectors, of transforming a light input into a light output. In general, this transformation, in the language of Volterra[5], belongs to the group of the closed cycle with respect to the time, in the sense that it is independent of the position of our initial instant in time. Such a transformation leaves a simple harmonic input still simply harmonic in the time, although in general with a shift in phase.

An example of an optical instrument is a microscope. This may be regarded as a means of making an electromagnetic disturbance in the image plane depend linearly on a given electromagnetic disturbance in the object plane. Telescopes, spectroscopes, Nicol prisms, etc., serve as further examples of optical instruments in the sense in question. Among these, a particularly interesting ideal type is the conservative optical instrument, in which the power of the input and the power of the output are identical. This power depends quadratically on the electric and magnetic vectors, so that a conservative optical instrument has a quadratic invariant for the corresponding transformations. When only terms of the one frequency of e^{iut} are considered, this quadratic positive invariant becomes Hermitian, and has essentially the same properties as the expression

$$x_1\bar{x}_1 + x_2\bar{x}_2 + \cdots + x_n\bar{x}_n \tag{9. 52}$$

which is invariant under all unitary transformations of $x_1, \ldots, x_n$. Thus the theory of the group of unitary transformations is physically applicable, not only in quantum mechanics, where Weyl has already employed it so successfully, but even in classical optics. It is the conviction of the author that this analogy is not merely an accident, but is due to a deep-lying connection between the two theories.

In quantum mechanics, while all the terms of a matrix enter in an essential way into its transformation theory, only diagonal terms are given an immediate physical significance. This is also in precise accord with the optical situation. Every optical observation ends with the measurement of an energy or power, either by direct bolometric or thermometric means, or by the observations of a visual intensity or the blackening of a photographic plate. Every such observation means the more or less complete determination of some diagonal term. The non-diagonal terms of a coherency matrix of light only have signifiance in so far as they enable us to predict the energies or intensities which the light will show after having been subjected to some linear transformation or optical instrument. This fact that new diagonal terms after a transformation cannot be read off from the old diagonal terms before a transformation, without the intervention of non-diagonal terms, is the optical analogue for the principle of indetermination in quantum theory, according to which observations on the momentum of an electron alone cannot yield a single value if its position is known, and vice versa. The statement that every observation of an electron affects its properties has the following analogy: if two optical instruments are arranged in series, the taking of a reading from the first will involve the interposition of a ground-glass screen or photographic plate between the two, and such a plate will destroy the phase relations of the coherency matrix of the emitted light, replacing it by the diagonal matrix with the same diagonal terms. Thus the observation of the output of the first instrument alters the output of the second. In this case, the possibility of taking part of the output of the first instrument for reading by a thinly silvered mirror warns us not to try to push the analogy with quantum theory too far.

Coherency matrices form a close analogue to the correlation matrices long familiar in statistical theory. If we have a set of n observations $x^{(1)}, x^{(2)}, \ldots, x^{(n)}$ all made together, and this set of observations is repeated on the occasions $1, 2, \ldots, m$ thus yielding sets $x_1^{(1)}, \ldots, x_1^{(n)}$; $x_2^{(1)}, \ldots, x_2^{(n)}$; $\ldots$; $\ldots$; $x_m^{(1)}, \ldots, x_m^{(n)}$, the correlation matrix of these observations is

$$\left\| \begin{array}{cccc} \sum_1^m (x_k^{(1)})^2, & \sum_1^m x_k^{(1)} x_k^{(2)}, & \cdots\cdots, & \sum_1^m x_k^{(1)} x_k^{(n)} \\ \sum_1^m x_k^{(2)} x_k^{(1)}, & \sum_1^m (x_k^{(2)})^2, & \cdots\cdots, & \sum_1^m x_k^{(2)} x_k^{(n)} \\ \cdots & \cdots & \cdots & \cdots \\ \cdots & \cdots & \cdots & \cdots \\ \sum_1^m x_k^{(n)} x_k^{(1)}, & \sum_1^m x_k^{(n)} x_k^{(2)}, & \cdots\cdots, & \sum_1^m (x_k^{(n)})^2 \end{array} \right\|. \tag{9.53}$$

This symmetrical matrix represents the entire amount and kind of linear relationship to be observed between the different observations in question. The further analysis of the information yielded by a correlation matrix depends on the nature of the data to be analysed. Thus if the two observations of each set are the x and y coordinates of the position of a shot on a target, the rotations of the x and y axes have a concrete geometrical meaning, and the question of reducing the matrix to diagonal form by a rotation of axes is the significant question of determining the ellipse which best represents the distribution of holes in the target. On the other hand, if the quantities whose correlation we are investigating are the price of wheat x in dollars per bushel and the marriage rate y, rotations have no significance, as there is no common scale, while on the other hand, the significant information yielded by the matrix must be invariant under the transformations

$$\left.\begin{array}{l} x_1 = kx; \\ y_1 = ly. \end{array}\right\} \tag{9.54}$$

The so-called coefficients of correlation and of partial correlation and the lines of regression of x on y and of y on x have this type of invariance.

Correlation matrices and their derived quantities are the tool for the statistical analysis of what is known as frequency series, series of data where no such variable as the time enters as a parameter. In the study of meteorology, of business cycles, and of many other phenomena of interest to the statistician, on the other hand, we must discuss time series, where the relations of the data in time are essential. The proper analysis of these has long been a moot point among statisticians and economists. As far as it is linear relationships which we are seeking for, it is only reasonable to suppose that coherency matrices

25—29764. Acta mathematica. 55. Imprimé le 14 avril 1930.

must play the same rôle for time series which correlation matrices play for frequency series. In statistical work, the group of transformations which will most frequently be permissible is as before

$$\begin{aligned} g_1(t) &= A f_1(t); \\ g_2(t) &= B f_2(t). \end{aligned} \qquad [A \text{ and } B \text{ real}] \tag{9.55}$$

Under this group, the significant invariants of the Hermitian matrix

$$\begin{Vmatrix} S^1_{11}(u) & S^1_{12}(u) \\ S^1_{21}(u) & S^1_{22}(u) \end{Vmatrix} \tag{9.56}$$

are

$$r(u) = S^1_{12}(u)\,[S^1_{11}(u)\,S^1_{22}(u)]^{-1/2}; \tag{9.57}$$

which we may call the coefficient of coherency of f_1 and f_2 for frequency u, and

$$\sigma_1(u) = \frac{S^1_{12}(u)\sqrt{S^1_{11}(u)}}{S^1_{22}(u)} \text{ and } \sigma_2(u) = \frac{S^1_{21}(u)\sqrt{S^1_{22}(u)}}{S^1_{11}(u)}, \tag{9.58}$$

the coefficients of regression respectively of f_1 on f_2 and of f_2 on f_1. The modulus of the coefficient of coherency represents the amount of linear coherency between $f_1(t)$ and $f_2(t)$ and the argument the phase-lag of this coherency. The coefficients of regression determine in addition the relative scale for equivalent changes of f_1 and f_2.

The computation of coefficients of coherency and of regression is to be done in the steps indicated in their definition. In the case where only a finite set of real data are at our disposal, distributed at equal unit intervals from 0 to n, say $x_0, x_1, \ldots, x_n$ and $y_0, y_1, \ldots, y_n$, the steps of our computation are:

$$\left.\begin{aligned} (\varphi_k)_{11} &= \frac{1}{n}\sum_0^{n-k} x_j\, x_{j+k}; \\ (\varphi_k)_{12} &= \frac{1}{n}\sum_0^{n-k} x_j\, y_{j+k}; \\ (\varphi_k)_{21} &= \frac{1}{n}\sum_0^{n-k} y_j\, x_{j+k}; \\ (\varphi_k)_{22} &= \frac{1}{n}\sum_0^{n-k} y_i\, y_{j+k}; \end{aligned}\right\} \qquad [0 \le k \le n] \tag{9.59}$$

$$r(u)=\frac{\sum_{0}^{n}\{[(\varphi_k)_{21}+(\varphi_k)_{12}]\cos ku+i\,[(\varphi_k)_{21}-(\varphi_k)_{12}]\sin ku\}-(\varphi_0)_{12}/2}{2\left[\sum_{0}^{n}(\varphi_k)_{11}\cos ku-(\varphi_0)_{11}/2\right]^{1/2}\left[\sum_{0}^{n}(\varphi_k)_{22}\cos ku-(\varphi_0)_{22}/2\right]^{1/2}}; \qquad (9.60)$$

$$\left.\begin{aligned}\sigma_1(u)&=r(u)\frac{\left[\sum_{0}^{n}(\varphi_k)_{11}\cos ku-(\varphi_0)_{11}/2\right]^{1/2}}{\left[\sum_{0}^{n}(\varphi_k)_{22}\cos ku-(\varphi_0)_{22}/2\right]^{1/2}};\\ \sigma_2(u)&=\bar{r}(u)\frac{\left[\sum_{0}^{n}(\varphi_k)_{22}\cos ku-(\varphi_0)_{22}/2\right]^{1/2}}{\left[\sum_{0}^{n}(\varphi_k)_{11}\cos ku-(\varphi_0)_{11}/2\right]^{1/2}}.\end{aligned}\right\} \qquad (9.61)$$

In case we have at our disposal methods for performing the Fourier transformation, we may compute these coefficients directly from graphs. Several devices for this purpose are now being developed in the laboratory of Professor V. Bush in the Department of Electrical Engineering of the Massachusetts Institut of Technology.

III
Unpolarized Light

Editor's Comments on Papers 13 Through 15

13 **Birge:** *On the Nature of Unpolarized Light*

14 **DuBridge:** Comments on the Paper by R. T. Birge *"On the Nature of Unpolarized Light"*

15 **Hurwitz:** *The Statistical Properties of Unpolarized Light*

In Paper 13 Raymond T. Birge argues that no experiment can provide information as to the precise nature of the form of the vibration. This paper is one of the first discussions specifically on the nature of unpolarized light. An earlier paper by Stokes (Paper 10 in this volume) provides some insight to the problem.

Since 1955 Birge has been Emeritus Professor in the Department of Physics, University of California, Berkeley. He is well known for his work on the most likely value of physical constants and was chairman of the committee on physical constants of the National Research Council from 1930 until 1937. He is also well known for his work in spectroscopy and the history of physics.

The corroborating remarks in Paper 14 by L. A. DuBridge were made on R. T. Birge's article "On the Nature of Unpolarized Light." They follow immediately after Birge's article in the same journal.

Lee A. DuBridge has held many distinguished positions during his career. These have included President, California Institute of Technology, 1946–1969; Member of the President's Air Quality Advisory Board, 1968–1969; and Science Advisor to the President, 1969–1970.

Paper 15 is important because it provides the first detailed analysis of the statistical behavior of the electric vector in an unpolarized light beam. It discusses only the case of almost monochromatic light such as is emitted by a spectrum line.

Since 1957 Hurwitz has been at General Electric Corporate Research and Development in Schenectady. He was one of the 1961 recipients of the E. O. Lawrence Award for important contributions to the theory and design of nuclear reactors, and he has authored many papers in the fields of mathematics, theoretical physics, and nuclear energy.

13

Reprinted from *J. Opt. Soc. Am.*, **25**, 179–182 (June 1935)

On the Nature of Unpolarized Light

Raymond T. Birge, *University of California, Berkeley*

(Received February 20, 1935)

All that can be affirmed regarding the nature of unpolarized light is that it shows no preferential polarization. When, for instance, unpolarized light enters a quartz crystal in a direction parallel to the axis, the light is decomposed into right- and left-circularly polarized beams of equal intensity. This was demonstrated experimentally by Fresnel. By the use of right- and left-handed quartz crystals in the two paths of an interferometer, interference effects occur whose nature can be definitely predicted on the assumption of the existence of such circularly polarized components. An experiment of this nature has been performed by Langsdorf and DuBridge, but the experiment, in itself, furnishes no information as to the nature of unpolarized light.

LANGSDORF and DuBridge[1] have described an interesting experiment in a paper entitled *Optical Rotation of Unpolarized Light.* In a note at the end of their paper R. W. Wood mentions an earlier experiment by Stefan[2] which is similar in principle to that of L. and D. These experiments are also described by Wood in his well-known text.[3]

In L. and D.'s experiment, which alone is discussed here, unpolarized monochromatic light was broken into two beams of equal intensity by an interferometer (in their apparatus, by a Fresnel biprism) so that a set of interference fringes was formed in the region where the two beams were reunited. Into the path of each of the two beams there was then inserted an optically active substance, such as quartz,[4] with its axis parallel to the direction of propagation of the light. One quartz was right-handed and the other left-handed, and each was of such thickness as to rotate the plane of polarization of incident plane-polarized light through 45°. By using unpolarized light under these conditions the fringes completely disappeared.

L. and D. remark "whether any effect would be expected depends on the view we adopt as to the nature of an unpolarized light beam. Let us regard it as made up of a large number of elementary components (photons?), each emanating from an individual atom or molecule in the source and each being plane-polarized, the polarization planes for the various components having a random distribution about the axis of the main beam." With this assumption L. and D. show that the interference fringes should disappear. Wood[3] writes "A. Langsdorf and L. DuBridge have described an interesting experiment which may be regarded as showing the presence of the variously oriented polarized components of ordinary unpolarized light." It appears to me, however, that this experiment gives no information whatsoever on the nature of unpolarized light. The fringes should disappear, regardless of the assumptions made. The experiment merely confirms the well-known action of quartz on light traveling parallel to the optic axis. L. and D. also state "nor have we found any statement in the standard treatises on optics which would enable us to predict with certainty what the result of the experiment would be." Because of this last statement, with which the writer concurs, it seems desirable to publish the following rather detailed discussion.

Fresnel assumed that any sort of light, polarized or unpolarized, when incident normally on a quartz surface cut perpendicular to the optic axis, is broken into two beams, one of right-circular polarization, and the other of left-circular, which travel with different velocities. Fresnel made a direct experimental test of these assumptions. He used a train of alternate right- and left-handed quartz prisms to separate in angle the two beams, and then tested the polarization of each. The only positive statement that we can make regarding the nature of unpolarized light is that it exhibits no preferential polarization, and this, as a matter of fact, is

[1] A. Langsdorf and L. A. DuBridge, J. Opt. Soc. Am. **24**, 1 (1934). To be referred to as "L. and D."

[2] Stefan, Sitz. Ber. Wien. Akad. **50**, 380 (1864).

[3] R. W. Wood, *Physical Optics* 3rd ed., Macmillan Co., 1934, pp. 362, 829.

[4] Sugar solutions were actually used by L. and D., because of lack of suitable quartz plates.

merely a working definition of the term "unpolarized light." If such light is tested with a Nicol, the intensity of the transmitted plane-polarized light is found to be independent of the orientation of the Nicol. If it is decomposed in any way into right- and left-circularly polarized beams, the intensity of the two beams is the same. Hence in the experiment under discussion we can feel confident that each light beam, on entering the inserted quartz crystal, breaks up into two beams of *equal* intensity, one right-circularly polarized and the other left-circularly polarized. What effects should now be produced, when all of the beams are reunited?

Let us consider first the situation before the quartz plates are inserted. For some point on the screen the optical path lengths for the two interfering beams will be equal. We take this as our fiducial point O. At this point there occurs the center of a bright fringe. If we now insert into each beam, in place of quartz, an ordinary uniaxial crystal (i.e., one not optically active) with its axis parallel to the direction of propagation, our fiducial point O will be displaced to the right or left, unless the two pieces of crystal are of precisely the same thickness. Such a shift is immaterial to the argument and we therefore assume, for simplicity, that the thicknesses *are* identical and that there is no resulting shift of O. We now replace the two crystals by right- and left-handed quartz, respectively, of such thickness that in one the right-circular beam gets one-quarter of a wavelength ahead of the left-circular beam, and in the other one-quarter of a wavelength behind. The proper thickness to use is determined experimentally by the fact that it is sufficient to rotate incident plane-polarized light through a 45° angle.

We now have four beams of light, two right-circularly polarized, and two left-circularly polarized, and all four of equal intensity. The two right-circular beams are coherent and are capable of interference, when reunited. However, the *optical* path for one in the quartz is longer than that for the other. In fact the speed of the *right*-circularly polarized light in *left*-handed quartz is just that of *left*-circularly polarized light in *right*-handed quartz. Since we have chosen the thickness of the right-handed quartz such that the right-circular beam gains $\lambda/4$ over the left-circular, this means that it also gains $\lambda/4$ over the *right*-circular beam that has passed through the *left*-handed quartz. Hence at our fiducial point O the two right-circular beams arrive 90° out of phase, and the *resulting fringe system formed by right-circularly polarized light* has its fiducial point O' shifted to one side—let us say to the right—by one-quarter the distance between fringes. The two left-circular beams similarly arrive at O 90° out of phase, but the velocities are interchanged, so that the fiducial (zero path difference) point O'' of the fringe system consisting of *left*-circularly polarized light is shifted to the *left* by one-quarter of the fringe separations. Our two sets of fringes, one due to the interference of right-circular beams, and the other of left-circular beams, are thus displaced relative to one another by just half the fringe separation. The two fringe systems are necessarily of equal intensity, since they are formed of the equal amounts of right- and left-circularly polarized light into which the original unpolarized beam was decomposed by the quartz. The final resultant intensity is therefore uniform.

Although the *intensity* is uniform at all points on the screen, the state of polarization is not. At one set of points, centered on O', the light is wholly right-circularly polarized, since at these points the intensity of the left-circular light is zero. At another set of points, centered on O'' and lying just intermediate between the first set, the light is wholly left-circularly polarized. At all other points there is a mixture of right- and left-circularly polarized light. By using a quarter-wave plate and properly oriented analyzing Nicol, one can cut out completely the right-circularly polarized set of fringes, leaving only the left-polarized set. By then turning the Nicol through 90°, the right-polarized set of fringes is alone transmitted, the illumination on the screen in the two cases being just complementary. This experiment was performed by L. and D., with exactly these expected results.

It is evident, from the foregoing discussion, that the displacement of one set of fringes, relative to the other, is directly proportional to the path difference between the right- and left-circular beams produced by the quartz. Hence it is only for $\lambda/4$ path difference (or $n\lambda/2+\lambda/4$ in general) that the resulting intensity is uniform.

If the path difference is $n\lambda/2$ (n=integer), there will be no resultant blurring of the fringes, when the quartz plates are inserted. Now the optical path difference of right- and left-circularly polarized light in quartz of a given thickness varies with the frequency of the light used. Hence the fringes may completely disappear for one frequency, and be quite distinct for another. This was shown by Stefan,[2] in the original experiment on this matter.

For the purpose of this paper the important characteristic of the foregoing argument is that it includes no assumption regarding the nature of unpolarized light save that such light is decomposed by quartz into equal amounts of right- and left-circularly polarized light, traveling with different velocities. But in view of Fresnel's work this is more nearly an experimental fact than an assumption. Hence the L. and D. experiment gives no further information on the nature of unpolarized light.

It is important to notice also that only classical physics is needed in order to predict the experimental results. In the classical physics two elementary coherent light beams, i.e., two beams capable of interference, are thought of as produced originally by a single real source, without any specification as to whether this source is an atom, a molecule, or an aggregate of molecules. The maximum path difference over which interference is observed to occur may be interpreted as the length of such an elementary wave train. In the case of quantum phenomena, as is well known, we are forced to postulate the existence of corpuscular radiant energy ($h\nu$) and momentum ($h\nu/c$), to which the term "photon" is applied. This term is sometimes applied also to any elementary coherent light train of classical physics. One way in which classical interference phenomena, and the superposition principle on which they are based, can be incorporated into the newer quantum mechanics has been indicated by Dirac.[5] Such a discussion, although very illuminating, is quite irrelevant to the subject matter in question here.

If we wish, on the basis of classical physics, to make any assumption as to the nature of the polarization of *each* elementary wave train, our best experimental evidence comes from the Zeeman effect. This effect shows us that if all the atoms in an assembly are oriented with respect to some preferred direction (the direction of the magnetic field) the emitted light is *elliptically* polarized. The ellipses are of all possible shapes, depending on the direction in which the light is viewed, and in certain special directions the ellipse takes on the special form of a straight line or of a circle. But the amount of such linear or circularly polarized light is infinitesimal, compared to the amount of elliptically polarized light, when all directions of emission are considered. Now the integrated light emitted in *all* directions by an assembly of atoms oriented with respect to some *one* preferred direction, should be similar in character to the light emitted in *one* direction by an assembly of atoms oriented (at random) in *all* directions. Hence it seems natural to consider unpolarized light as a mixture of *elliptically* polarized components, the ellipses being of every possible shape. I can see no scientific reason for the assumption, so commonly made in texts, that unpolarized light consists of *plane*-polarized components, oriented in all azimuths.

This is, however, merely a matter of mathematics. An elliptic vibration can be considered as the superposition of two linear vibrations, at right angles, and of different amplitudes and phases. But it can equally well be considered as the superposition of two circular motions, one right-circular and the other left-circular, of different amplitudes and phases. If elliptic polarized light enters a quartz crystal so as to travel inside the crystal at right angles to the optic axis, we know that it breaks up into two plane-polarized components, traveling with different velocities. If this same light enters the crystal parallel to the optic axis, it breaks up into right- and left-circularly polarized components, traveling with different velocities. It is thus evident that any particular set of components cannot be said to be present in the original light. On the contrary, such a set is *manufactured* by the quartz,[6] just as the series of monochromatic

[5] P. A. M. Dirac, *Principles of Quantum Mechanics*, 1930, Chapter I.

[6] The action of quartz on unpolarized incident light is a good illustration of the arbitrary character of the manufactured components. In general the light splits into two

beams reflected from a grating, when white light is incident upon it, is *manufactured* by the grating. If we are given any number of coherent beams of light, of specified polarizations, intensities, and phase relations, we can by means of the principle of superposition calculate uniquely the character of the resulting light produced by interference. But given *only* the nature of the resultant light, we can deduce nothing as to the specific character of its components, for there are an infinite number of different sets of components that may be superposed to produce exactly the same resultant light beam. This is just the situation in the L. and D. experiment. From a knowledge of the presence of right- and left-circularly polarized beams inside the quartz, we can deduce nothing as to the polarization of the individual light trains that produced them.

Let us return, however, to our assumption that each individual wave train consists of elliptically polarized light. This is split into two parts by the interferometer, and then each part is again split into two parts by the quartz. The right- and left-circular components will be of equal intensity only in the special case where the original wave train was plane-polarized. We now assume that *each* elementary wave train may produce its own set of interference fringes, and when it is broken into right- and left-circular components, as in the L. and D. experiment, each such train produces not one, but two such sets of fringes, one of right- and the other of left-circularly polarized light. The two sets of fringes are exactly out of step. They are, however, not in general of equal intensity, so far as one original wave train is concerned. But the position on the screen is the same for all sets of right-circular polarized fringes,—similarly for left-circular polarized. If the original light is unpolarized there must be, in the aggregate, equal amounts of the two sorts of polarization. Hence the sum of the intensities of all the sets of right-circularly polarized fringes must equal the sum of all left-polarized fringes, and the final resultant intensity is uniform.

It seems safe to conclude that no experiment can give us any information on the nature of unpolarized light, further than that which we now have; namely, that *if* such light is split into components, in any given apparatus, no preferential polarization will be found. If the splitting is into plane-polarized components, all azimuths will be equally intense. If it is split into circularly polarized components, the right- and left-circular members will be equally intense, etc. I am confident that many persons, in the past, must have themselves gone through the foregoing train of reasoning. As already noted, however, such a discussion does not appear in any of the best known texts on physical optics (in particular, texts written in English), and it has therefore seemed desirable to publish these details.

---

elliptic vibrations, one right handed and the other left handed. The shape and size of the two ellipses are the same, but the major axes are at right angles. For the special direction parallel to the optic axis the two ellipses become circles. For the special set of directions at right angles to the axis the ellipses degenerate into two linear vibrations, at right angles to each other. In all cases the two components are of equal intensity. With incident *polarized* light of any character whatsoever the only difference in the result is that the two components produced by the quartz are in general of *unequal* intensity.

14

Reprinted from *J. Opt. Soc. Am.*, **25**, 182–183 (June 1935)

Comments on the Paper by R. T. Birge "On the Nature of Unpolarized Light"

L. A. DuBridge, *University of Rochester, Rochester, N. Y.*

PROFESSOR BIRGE has greatly clarified, by his discussion in the above paper, certain questions raised in our paper on *The Optical Rotation of Unpolarized Light*. I am quite in agreement with his conclusion that any particular method of resolving unpolarized light into components is quite arbitrary. As far as the light beam in the optically active substance is concerned, it is immaterial whether we deal with linearly, elliptically or circularly polarized components since they all lead to the same result. Our suggestion of a resolution into plane-polarized components was a simple and obvious one, although, as Professor Birge has shown, it is not unique.

However, the chief purpose of our experiment (as indicated by the title of the paper) was to determine whether optically active substances produced *any observable* effect on unpolarized light, and such an effect was indeed found. So far as we were able to learn at the time, this question had not been answered by previous experiments of this kind. Professor Wood called attention to the paper by Stefan, and quite recently my attention has been called to a paper published in 1932 by S. Wawilow entitled *On Some Cases of Interference of Natural Bundles* (in Russian).[1] In this paper several experiments bearing on this subject are described, one of which is quite similar to ours, except that a Rayleigh interferometer was used in place of the Fresnel biprism. These experiments show that by means of optically active materials it is possible to produce two beams of "natural" (unpolarized) light which originated in the same source but which do not produce an interference pattern of the usual type. Wawilow and Brumberg[2] have described another experiment which shows that a similar result may be obtained without the use of optically active materials. This experiment consists essentially of selecting two beams which emerge in antiparallel directions from a small source and then reflecting each through 90° by mirrors to bring them into parallelism. In this case also it is found that the interference field is uniformly illuminated, a fact which is readily understood if the beams are resolved into plane or circularly polarized components.

Now normally when we find two beams of unpolarized light which combine to produce uniform illumination we say that the beams are "non-coherent." In our experiment, however, a closer examination of the field with a quarter wave plate and Nicol prism shows that there are really two overlapping sets of fringes, one produced by left- and the other by right-circularly polarized light. In other words the corresponding components of the beam are still coherent. It is this peculiar type of coherence to which we wished to call attention. It is not a new phenomenon and is, of course, readily understood in terms of present knowledge concerning the behavior of light, but it seems to be of sufficient interest to merit more attention than it has been given in usual discussions of the subject. The type of coherence possessed by two unpolarized beams is not revealed by ordinary polarization experiments.

In describing and understanding interference experiments of this kind it is most convenient—possibly necessary for most of us—to resolve (mentally) the interfering beams into some combination of polarized components. We find indeed that these components produce separate sets of interference fringes and may, as in Fresnel's experiments, actually become separated in angle. We do not of course think of these components as having an independent existence in the original beam any more than we think of the arbitrary components into which we might resolve a vector as having "existence" in the vector itself.

In conclusion I may say that the purpose of our paper was to call attention to these phenomena and not to interpret them as giving any new information about the nature of unpolarized light. We hoped that its publication would result in a further discussion of the subject and Professor Birge's paper now justifies this hope.[3]

[1] S. I. Wawilow, Bull. Acad. Sci. U. S. S. R., p. 1451 (1932). (I am indebted to Professor J. K. Kostko of Washington University for his translation of this paper.)

[2] S. I. Wawilow and E. M. Brumberg, Physik. Zeits. d. Sowjetunion **3**, 103 (1933).

[3] I have discussed these matters in some detail by correspondence with Professor Birge and we are now in substantial agreement as to the interpretation of these experiments.

15

Reprinted from *J. Opt. Soc. Am.*, **35**, 525–531 (Aug. 1945)

The Statistical Properties of Unpolarized Light

HENRY HURWITZ, JR.
Cornell University, Ithaca, New York
(Received November 1, 1944)

In a beam of monochromatic unpolarized light the electric field vector at a point traces out an ellipse whose size, eccentricity, and orientation are slowly varying functions of time. The statistical properties of the parameters of this ellipse are investigated. It is shown that the quantity S which is defined as twice the product of the principle axes of the ellipse divided by the sum of the squares is uniformly distributed between zero and one. It therefore has median value $\frac{1}{2}$ which corresponds to a ratio of minor to major axis equal to .268. Hence fairly thin ellipses predominate. The square root of the sum of the squares of the semi-major and semi-minor axes, R, is statistically independent of S and has the distribution function $(r^3/2p^4) \exp(-r^2/2p^2)$ where $2p^2$ is the average value of R^2.

INTRODUCTION

MANY textbooks and teachers of optics are rather vague in their descriptions of unpolarized light. They pass over the subject with a cryptic statement of the sort "unpolarized light is a mixture of plane-polarized light in all directions" or "unpolarized light is a random mixture of light with all kinds of polarizations." Students find difficulty in obtaining from these brief remarks a clear picture of how the electric field vector at a point varies as a function of time. In this note we shall perform some simple statistical calculations whose results might serve as a pedagogical aid in making the situation clear to students.

By natural, or unpolarized light we shall mean radiation which is emitted by molecules which have been randomly excited (e.g. thermally, or by an electrical discharge), and which are not in a region of space where large external fields provide a preferred direction. We shall assume that the beam of light under consideration is monochromatic in the sense that its line breadth is small compared to the central frequency ν_0. It is immaterial whether the monochromatic character arises because the light has been passed through a narrow filter, or because it consists of a single spectral line. In order to avoid the necessity of considering quantum effects we shall limit the discussion to fairly intense beams. This means effectively that the number of photons reaching the detector in a time interval equal to the reciprocal line breadth must be large compared to one.

As will be seen from the formulas given below, a beam of unpolarized monochromatic light would, if observed over a time interval short compared to the reciprocal of the line breadth, appear to be elliptically polarized. But the type of polarization changes continually so that in any experiment which requires a longer period of observation one measures only the average effect of a large variety of different polarizations. In other words, in a beam of unpolarized monochromatic light the electric vector at a point traces out an ellipse whose size, eccentricity, and orientation gradually change. After a time somewhat larger than the reciprocal of the line breadth the new ellipse is completely unrelated to the original one. The purpose of our calculation is to find the statistical properties of the various parameters of the "instantaneous ellipses" traced out by the electric vector.

STATISTICAL FOUNDATIONS

Our statistical calculations will be based on the following mathematical representation for a monochromatic beam of unpolarized light:[1]

$$\begin{aligned}
E_x &= X(t)\cos\omega_0 t + U(t)\sin\omega_0 t \\
&= A(t)\cos(\omega_0 t - \delta_x(t)), \\
E_y &= Y(t)\cos\omega_0 t + V(t)\sin\omega_0 t \\
&= B(t)\cos(\omega_0 t - \delta_y(t)), \\
A(t) &= [X^2(t)+U^2(t)]^{\frac{1}{2}}; \quad B(t) = [Y^2(t)+V^2(t)]^{\frac{1}{2}}; \\
\delta_x(t) &= \tan^{-1}[U(t)/X(t)]; \quad \delta_y(t) = \tan^{-1}[V(t)/Y(t)]; \quad \omega_0 = 2\pi\nu_0.
\end{aligned} \tag{1}$$

[1] We assume that the electric field is constant over the part of the cross section of the beam under consideration.

Here E_x and E_y are the x and y components of the electric field at a point P in the beam. The quantities $X(t)$, $Y(t)$, $U(t)$, and $V(t)$ are statistically independent Gaussianly distributed random functions. The probability that a particular one of them has a value between x and $x+dx$ is

$$P_X(x)dx = \frac{1}{p} P_G\left(\frac{x}{p}\right)dx = \frac{1}{(2\pi)^{\frac{1}{2}}} \exp\left(-\frac{x^2}{2p^2}\right)\frac{dx}{p}, \quad (2)$$

where p^2 is the average of the square of each one of these quantities. In accordance with Eq. (2) and the independence of $X(t)$, $Y(t)$, $U(t)$, and $V(t)$, the quantities $A(t)$, $B(t)$, $\delta_x(t)$, and $\delta_y(t)$ are all independent. Furthermore $\delta_x(t)$ and $\delta_y(t)$ are uniformly distributed between 0 and 2π, and $A(t)$ and $B(t)$ are distributed according to the law

$$P_A(x)dx = P_B(x)dx = \frac{x}{p^2} \exp\left(-\frac{x^2}{2p^2}\right)dx. \quad (3)$$

(In this notation $P_A(x)dx$ is the probability that $A(t)$ lies between x and $x+dx$, etc.) The eight functions $X(t)$, $Y(t)$, $U(t)$, $V(t)$, $A(t)$, $B(t)$, $\delta_x(t)$, and $\delta_y(t)$ vary slowly compared to $\cos \omega_0 t$ and $\sin \omega_0 t$. Thus the values of one of these functions at two different times are not statistically independent unless the time difference is large compared to the reciprocal line breadth. For smaller time differences the correlation depends on the shape of the line.

We shall indicate two different methods of obtaining the representation given in Eqs. (1) to (3). The first is based on a simple classical model of light emission while the second makes use of a natural independence assumption.

In the classical model which we shall employ in the first derivation the molecules are assumed to emit radiation in the form of wave trains which contain only Fourier components with frequencies close to ν_0 and which last for a time interval of order of magnitude of the reciprocal line breadth. (In accordance with quantum theory the pulses may be imagined to contain an amount of energy equal to $h\nu_0$.) The nature of the polarization of the individual pulses is unimportant so long as it is sufficiently random. For example, if the pulses are plane polarized, then all planes of polarization must be equally probable, while if the pulses are circularly polarized, right- and left-handed polarizations must be equally likely. A typical field component in a pulse will be equal to $f(t-t_i)$ where t_i is the time when the pulse begins. Expanding $f(t)$ in a Fourier series in the long time interval 0 to T we may write

$$f(t) = \sum_{k=0}^{\infty} (\alpha_k \cos \omega_k t + \beta_k \sin \omega_k t), \quad \omega_k = 2\pi k/T. \quad (4)$$

The quantity $\alpha_k^2+\beta_k^2$ is small if ω_k differs from ω_0 by an amount large compared to the line breadth. The problem of finding the resultant of a large number of pulses of the sort described above has been studied in great detail in other connections.[2] Therefore without going into details about the mathematical calculation we shall simply quote the result. From the fact that the times t_i when the individual pulses are emitted are randomly distributed it can be shown that the x and y components of the electric field at a point in the beam may be written in the following form:

$$E_x = \sum_{k=0}^{\infty} p_k(X_k \cos \omega_k t + U_k \sin \omega_k t),$$
$$E_y = \sum_{k=0}^{\infty} p_k(Y_k \cos \omega_k t + V_k \sin \omega_k t). \quad (5)$$

[2] H. Hurwitz and M. Kac, Ann. Math. Stat. **15**, 173 (1944); S. O. Rice, Bell Sys. Tech. J. **23**, 282 (1944), **24**, 46 (1945). Some insight as to why the quantities in Eq. (5) are Gaussianly distributed can be had from the following considerations. Let

$$F(t) = \sum_1^N f(t-t_i) = \sum_{k=0}^N (a_k \cos \omega_k t + b_k \sin \omega_k t).$$

From Eq. (4)

$$a_k = \sum_1^N (\alpha_k^2+\beta_k^2)^{\frac{1}{2}} \cos(\omega_k t_i - \varphi_k),$$
$$b_k = \sum_1^N (\alpha_k^2+\beta_k^2)^{\frac{1}{2}} \sin(\omega_k t_i - \varphi_k), \quad \varphi_k = \tan^{-1}\frac{\beta_k}{\alpha_k}.$$

If the t_i's are randomly distributed the quantities a_k and b_k can be regarded as the two components of a vector which is the sum of N two-dimensional vectors of constant length but random direction. Hence from the familiar results of the problem of "random walk" we can immediately conclude that a_k and b_k are Gaussianly distributed in the limit that N becomes infinite.

The quantities p_k^2 are proportional to $\alpha_k^2+\beta_k^2$ and are equal to $2\pi/c$ times the radiant-energy current in the beam per unit frequency range at frequency ω_k. The quantities X_k, Y_k, U_k, and V_k are independently distributed Gaussian random variables such that

$$\langle X_k\rangle_{\text{Av}}=\langle Y_k\rangle_{\text{Av}}=\langle U_k\rangle_{\text{Av}}=\langle V_k\rangle_{\text{Av}}=0,$$
$$\langle X_k^2\rangle_{\text{Av}}=\langle Y_k^2\rangle_{\text{Av}}=\langle U_k^2\rangle_{\text{Av}}=\langle V_k^2\rangle_{\text{Av}}=1. \tag{6}$$

Thus the probability that a particular one of these variables has a value between x and $x+dx$ is

$$P_G(x)dx=[1/(2\pi)^{\frac{1}{2}}]\exp(-x^2/2)dx. \tag{7}$$

The representation given in Eqs. (5) and (7) is strictly correct only in the limit that the number of pulses per unit time becomes infinite. However it is a sufficiently good approximation if the number of pulses emitted in a time equal to the reciprocal line breadth is large compared to one.

Equation (5) can be immediately written in the form of Eq. (1) by using the definitions

$$\begin{aligned}
X(t)&=\sum p_k[X_k\cos(\omega_k-\omega_0)t\\
&\qquad+U_k\sin(\omega_k-\omega_0)t],\\
U(t)&=\sum p_k[-X_k\sin(\omega_k-\omega_0)t\\
&\qquad+U_k\cos(\omega_k-\omega_0)t],\\
Y(t)&=\sum p_k[Y_k\cos(\omega_k-\omega_0)t\\
&\qquad+V_k\sin(\omega_k-\omega_0)t],\\
V(t)&=\sum p_k[-Y_k\sin(\omega_k-\omega_0)t\\
&\qquad+V_k\cos(\omega_k-\omega_0)t].
\end{aligned} \tag{8}$$

The quantities $X(t)$, $Y(t)$, $U(t)$, and $V(t)$, being linear combinations of Gaussianly distributed variables, are themselves Gaussianly distributed. Furthermore they are independently distributed because of the orthogonality property of the coefficients in their expansions in terms of X_k, Y_k, U_k, and V_k. The quantity p appearing in Eq. (2) is given by

$$p^2=\sum_{k=0}^{\infty}p_k^2. \tag{9}$$

Thus p^2 is $2\pi/c$ times the average power flow per unit area in the beam. The only remaining property of the functions $X(t)$, $Y(t)$, $U(t)$, and $V(t)$ to be proved is that they are slowly varying compared to $\cos\omega_0 t$ and $\sin\omega_0 t$. But this follows immediately from Eq. (8) and the fact that p_k is appreciable only if ω_k is close to ω_0.

Our second method of obtaining the fundamental representation given in Eqs. (1) to (3) is similar to the derivation of the Maxwellian distribution of velocities in a gas from the assumption that the Cartesian components of the velocities are independently distributed. The analogous assumption which we shall make is that E_x and E_y are independently distributed. More specifically we shall assume that the functions $X(t)$ and $U(t)$ in Eq. (1) are distributed independently of the functions $Y(t)$ and $V(t)$.[3] This independence property followed directly from the model assumed in the above derivation. However it is more general than the model there assumed and so may logically be taken as the starting point in our second derivation. In this derivation we begin by arbitrarily writing the field components in the form given in Eq. (1). Since the electric field is assumed to have appreciably large Fourier components only for angular frequencies close to ω_0 it follows immediately that the functions $X(t)$, $Y(t)$, $U(t)$, and $V(t)$ are slowly varying compared to $\cos\omega_0 t$ and $\sin\omega_0 t$. It remains to be proved that these four functions are Gaussianly distributed and that $X(t)$ is independent of $U(t)$ and $Y(t)$ is independent of $V(t)$. In addition to the fundamental assumption that $X(t)$ and $U(t)$ are independent of $Y(t)$ and $V(t)$ we shall assume that the statistical properties of the field components are independent of the orientation of the x and y axes. This means, for example, that for any value of θ the quantity $X(t)\cos\theta+Y(t)\sin\theta$ must have the same distribution function as the quantity $X(t)$. The fact that $X(t)$, $Y(t)$, $U(t)$, and $V(t)$ are Gaussianly distributed is then an immediate consequence of the following theorem:

[3] The assumption that E_x and E_y are statistically independent can be justified in terms of the somewhat more physical assumption that the statistical properties of an intense beam of unpolarized light cannot be changed by passing the light through a wave plate. By the use of a wave plate of suitable thickness, one field component can be retarded with respect to the other by any desired amount. It is evident that no possible type of correlation between E_x and E_y could remain invariant under this general class of transformations.

If A and B are independently distributed random variables and for every θ the variable $A\cos\theta+B\sin\theta$ has the same distribution as A (or B), then A and B are Gaussianly distributed. This theorem was proved in utmost generality by M. Kac and H. Steinhaus.[4] Their proof proceeds by obtaining a simple functional equation for the characteristic function of the distribution. Since A and $A\cos\theta+B\sin\theta$ have the same distribution it follows that for every real ξ the average value of $e^{i\xi A}$ satisfies the relation

$$f(\xi)=\langle e^{i\xi A}\rangle_{\text{Av}}=\langle e^{i\xi(A\cos\theta+B\sin\theta)}\rangle_{\text{Av}}. \tag{10}$$

But since A and B are independently distributed Eq. (10) can be rewritten

$$f(\xi)=\langle e^{i\xi A\cos\theta}\rangle_{\text{Av}}\langle e^{i\xi B\sin\theta}\rangle_{\text{Av}} = f(\xi\cos\theta)f(\xi\sin\theta); \tag{11}$$

or

$$f((x^2+y^2)^{\frac{1}{2}})=f(x)f(y), \tag{12}$$

where

$$x=\xi\cos\theta;\quad y=\xi\sin\theta. \tag{13}$$

It is easy to show that the only continuous solution of this well known equation is of the form

$$f(\xi)=\exp(\lambda\xi^2). \tag{14}$$

Furthermore λ is negative since

$$|f(\xi)|=|\langle e^{i\xi A}\rangle_{\text{Av}}|\leqslant 1. \tag{15}$$

Thus, since $f(\xi)$ is the characteristic function of a Gaussian distribution the theorem is proved. In our application of this theorem the random variable A corresponds to $X(t)$ (or $U(t)$) and the random variable B corresponds to $Y(t)$ (or $V(t)$).

The proof that $X(t)$ is independent of $U(t)$ and $Y(t)$ is independent of $V(t)$ is based on the fact that $X(t)$, $Y(t)$, $U(t)$, and $V(t)$ are Gaussianly distributed, and the additional assumption that $A(t)$ is independent of $\delta_x(t)$ and $B(t)$ is independent of $\delta_y(t)$. This latter assumption is clearly justified since $\delta_x(t)$ and $\delta_y(t)$ depend on the origin of time which is arbitrary. Let $P_A(a)da$ be the probability that $A(t)$ lies between a and $a+da$. Then, since $A(t)$ and $\delta_x(t)$ are independent, and in view of the relations given in Eq. (1), the probability that $X(t)$ lies between x and $x+dx$, and simultaneously $U(t)$ lies between u and $u+du$ is

$$P_{X,U}(x,u)dxdu=\frac{P_A((x^2+u^2)^{\frac{1}{2}})}{2\pi(x^2+u^2)^{\frac{1}{2}}}dxdu = g(x^2+u^2)dxdu. \tag{16}$$

Hence

$$P_X(x)=\frac{1}{p(2\pi)^{\frac{1}{2}}}\exp\left(-\frac{x^2}{2p^2}\right) = \int_{-\infty}^{+\infty}P_{X,U}(x,u)du = \int_{-\infty}^{+\infty}g(x^2+u^2)du. \tag{17}$$

Replacing x^2 by x^2+z^2 we have

$$\frac{1}{p(2\pi)^{\frac{1}{2}}}\exp\left(-\frac{(x^2+z^2)}{2p^2}\right) = \int_{-\infty}^{+\infty}g(x^2+z^2+u^2)du; \tag{18}$$

so that

$$\frac{1}{p(2\pi)^{\frac{1}{2}}}\exp\left(-\frac{x^2}{2p^2}\right)\int_{-\infty}^{+\infty}\exp\left(-\frac{z^2}{2p^2}\right)dz = \int_{-\infty}^{+\infty}\int_{-\infty}^{+\infty}g(x^2+z^2+u^2)dudz, \tag{19}$$

or

$$\exp\left(-\frac{x^2}{2p^2}\right)=\pi\int_{x^2}^{\infty}g(y)dy. \tag{20}$$

By differentiating both sides of the above equation with respect to x^2 we find

$$\frac{1}{2p^2}\exp\left(-\frac{x^2}{2p^2}\right)=\pi g(x^2), \tag{21}$$

so that

$$P_{X,U}(x,u)dxdu = \frac{1}{2\pi p^2}\exp\left(-\frac{(x^2+u^2)}{2p^2}\right)dxdu. \tag{22}$$

Since the joint distribution function of $X(t)$ and $U(t)$ is the product of the individual distribution functions the variables $X(t)$ and $U(t)$ are independent. This completes our second derivation of Eqs. (1) to (3).

[4] M. Kac and H. Steinhaus, Studia Mathematica 6, 89 (1936).

The difference between polarized and unpolarized light can be easily understood in terms of our fundamental representation as given in Eqs. (1) to (3). For light which is elliptically polarized in the usual sense the quantities $A(t)$, $B(t)$, $\delta_x(t)$, and $\delta_y(t)$ still are functions of time, but they vary in such a way that $A(t)/B(t)$ and $\delta_x(t)-\delta_y(t)$ remain constant. Hence though the amplitude and phase of the ellipse traced by the electric vector vary with time, the eccentricity and orientation remain constant. This can be accomplished in the laboratory by first removing one of the components of the electric field by a Nicol prism or Polaroid disk and then passing the resultant plane polarized beam through a wave plate. Clearly variations in the phase and amplitude of the plane polarized beam cannot cause variations in phase differences and amplitude ratios of components of the field after the wave plate.

An interesting paradox is encountered if one considers the fluctuations in the energy flowing past a point in the beam in a time interval τ. Let

$$W_\tau(t)=\int_t^{t+\tau}(A^2(t')+B^2(t'))dt'. \qquad (23)$$

The statistical behavior of the quantity $W_\tau(t)$ has been carefully investigated in another connection by M. Kac. His calculations are based on an equation corresponding to Eq. (5). It is easy to show that the quantity

$$\Delta=\frac{\langle W_\tau^2(t)\rangle_{Av}-\langle W_\tau(t)\rangle_{Av}^2}{\langle W_\tau(t)\rangle_{Av}^2}$$

depends only on τ and not on $\langle W_\tau(t)\rangle_{Av}$. Its limiting value for small τ is $\frac{1}{2}$ while for large τ it approaches zero. For values of τ of order of magnitude of the reciprocal line breadth the value of Δ depends on the line shape. According to quantum theory the quantity $W_\tau(t)$ is proportional to the number of photons incident in the time interval τ. One would expect this number n to have a Poisson distribution which would mean that Δ should be inversely proportional to n. As was pointed out by V. Weisskopf, the explanation of this paradox lies in the fact that the photons obey Einstein-Bose statistics. The photons therefore have a greater tendency for bunching than they would have if they were independent.[5] Hence the fluctuations in $W_\tau(t)$ are increased so that if the intensity is sufficiently large Δ is independent of $\langle W_\tau(t)\rangle_{Av}$ instead of being inversely proportional to it.

STATISTICAL PROPERTIES OF THE INSTANTANEOUS ELLIPSE

The distribution function of the parameters of the instantaneous ellipse can be calculated quite easily from Eqs. (1) to (3). Let $M(t)$ and $N(t)$ be respectively the semi-major and semi-minor axes of the instantaneous ellipse. Then[6]

$$M^2(t)+N^2(t)=A^2(t)+B^2(t)=R^2(t),$$
$$S(t)=\frac{2M(t)N(t)}{R^2(t)}=\frac{2A(t)B(t)}{R^2(t)}|\sin\delta(t)|, \qquad (24)$$
$$\delta(t)=\delta_x(t)-\delta_y(t).$$

Let

$$\Phi(t)=\tan^{-1}\frac{B(t)}{A(t)} \quad 0\leqslant\Phi(t)\leqslant\frac{\pi}{2}. \qquad (25)$$

Then the probability that $R(t)$ lies between r and $r+dr$ and simultaneously $\Phi(t)$ lies between φ and $\varphi+d\varphi$ is

$$P_{R,\Phi}(r,\varphi)drd\varphi=P_A(r\cos\varphi)P_B(r\sin\varphi)rdrd\varphi$$
$$=\frac{r^3}{2p^4}\sin 2\varphi\exp\left(-\frac{r^2}{2p^2}\right)drd\varphi. \qquad (26)$$

The quantity $rdrd\varphi$ represents the element of area in the A-B plane, and we have used Eq. (3) for $P_A(x)$ and $P_B(x)$. From Eq. (24)

$$S(t)=\sin 2\Phi(t)|\sin\delta(t)|. \qquad (27)$$

Since $\delta(t)$ is uniformly distributed between zero and 2π the probability that $|\sin\delta(t)|$ lies between σ and $\sigma+d\sigma$ is

$$P_{|\sin\delta|}(\sigma)d\sigma=\frac{2}{\pi}d\sin^{-1}\sigma=\frac{2}{\pi}\frac{d\sigma}{(1-\sigma^2)^{\frac{1}{2}}}. \qquad (28)$$

If the values of R and Φ are fixed, the probability that S lies between s and $s+ds$, which is obtained by substituting $ds/\sin 2\Phi$ for $d\sigma$ and $s/\sin 2\Phi$ for

[5] L. Brillouin, *Die Quantenstatistik* (Julius Springer, Berlin, 1931), p. 111, 177.
[6] M. Born, *Optik* (Julius Springer, Berlin, 1933), p. 23.

σ in Eq. (28), is

$$P_{|\sin\delta|}\left(\frac{s}{\sin 2\Phi}\right)\frac{ds}{\sin 2\Phi}$$
$$=\frac{2}{\pi}\frac{ds}{\sin 2\Phi\left(1-\frac{s^2}{\sin^2 2\Phi}\right)^{\frac{1}{2}}}. \quad (29)$$

Hence the probability that R lies between r and $r+dr$ while S lies between s and $s+ds$ is

$$P_{S,R}(s,r)dsdr$$
$$=dsdr\int\frac{2}{\pi}\frac{P_{R,\Phi}(r,\varphi)d\varphi}{\sin^2\varphi\left(1-\frac{s^2}{\sin^2 2\varphi}\right)^{\frac{1}{2}}}. \quad (30)$$

The integral is to be extended over that part of the first quadrant for which $\sin 2\varphi$ is greater than s. From Eq. (26)

$$P_{S,R}(s,r)dsdr=dsdr\frac{r^3}{\pi p^2}\exp\left(-\frac{r^2}{2p^2}\right)$$
$$\times\int\frac{d\varphi}{\left(1-\frac{s^2}{\sin^2 2\varphi}\right)^{\frac{1}{2}}}. \quad (31)$$

But

$$\int\frac{d\varphi}{\left(1-\frac{s^2}{\sin^2 2\varphi}\right)^{\frac{1}{2}}}=-\frac{1}{2}\int\frac{dy}{(1-s^2-y^2)^{\frac{1}{2}}}=\frac{\pi}{2} \quad (32)$$
$$y=\cos 2\varphi$$

since the range of integration is clearly from $y=(1-s^2)^{\frac{1}{2}}$ to $y=-(1-s^2)^{\frac{1}{2}}$. Thus

$$P_{S,R}(s,r)dsdr=\frac{r^3}{2p^4}\exp\left(-\frac{r^2}{2p^2}\right)dsdr. \quad (33)$$

Equation (33) shows that S is uniformly distributed between zero and one, and that R is distributed independently of S according to the law

$$P_R(r)dr=\frac{r^3}{2p^4}\exp(-r^2/2p^2)dr. \quad (34)$$

The distribution of the ratio

$$L(t)=N(t)/M(t) \quad (35)$$

can be found immediately from Eq. (33). We have

$$S=2L/(1+L^2), \quad (36)$$

so that

$$dS=2(1-L^2)dL/(1+L^2)^2. \quad (37)$$

Hence the joint distribution function of $L(t)$ and $R(t)$ is

$$P_{L,R}(l,r)dldr=\frac{(1-l^2)}{(1+l^2)^2}\frac{r^3}{p^4}$$
$$\times\exp\left(-\frac{r^2}{2p^2}\right)dldr, \quad (38)$$

and the distribution function of $L(t)$ is

$$P_L(l)dl=[2(1-l^2)/(1+l^2)^2]dl. \quad (39)$$

The function $P_L(l)$ has its maximum value for l equal to zero, and decreases monotonically to zero when l equals one. The median value of $L(t)$ is related to the median value of $S(t)$ by Eq. (36). Since the median value of $S(t)$ is $\frac{1}{2}$ we find for the median value of $L(t)$

$$L_m=.268. \quad (40)$$

We therefore have the rather surprising result that more than half the time, the ellipse being traced out by the electric vector has a major axis more than three and one half times as large as the minor axis. The quantities $L(t)$ and $M(t)$ are not independently distributed but have the joint distribution function

$$P_{L,M}(l,m)dldm$$
$$=(1-l^2)\frac{m^3}{p^4}\exp\left[-\frac{m^2(1+l^2)}{2p^2}\right]dldm. \quad (41)$$

The quantities $M(t)$ and $N(t)$ have the joint distribution function

$$P_{M,N}(m,n)dmdn$$
$$=\frac{(m^2-n^2)}{p^4}\exp\left[-\frac{(m^2+n^2)}{2p^2}\right]dmdn. \quad (42)$$

It is interesting to note that the distribution function $P_S(s)$ of $S(t)$ can be derived directly from the requirement that it be unchanged if the light is passed through a quarter wave plate. This condition leads to an integral equation for $P_S(s)$ which has a constant as its only solution. An alternate and elegant derivation of the same result using Poincaré's representation for polarized light has been suggested by H. Mueller and L. Tisza.[7]

CONCLUDING REMARKS

The statistical properties of unpolarized light which are discussed above cannot be measured with ordinary optical equipment such as wave plates and Nicol prisms. The standard experiments cannot be performed in sufficiently short times and so give only the average values of the field components. However the difficulty is only one of experimental technique for the parameters of the instantaneous ellipse are measurable in principle if the beam is sufficiently intense. In fact an actual experiment might be possible if radiation in the radio or microwave region were used. Although a spectral line in the microwave region could conceivably be employed, it would probably be more convenient to use a narrow spectral range of high temperature black body radiation. The experimental procedure would be to employ two narrow band receivers fed by antennas responsive to electric fields in perpendicular directions. The output of the receivers (possibly after beating with a standard frequency) would be connected to the two pairs of deflection plates of an oscilloscope. The pattern on the oscilloscope screen would then correspond to the pattern traced out by that part of the incident electric field in the narrow frequency range to which the receivers were responsive. (The fact that the narrow frequency range is singled out by filters in the receivers rather than by a filter acting on the radiation before it reaches the antennas does not alter the final results.)

An interesting feature of the experiment just described is that if the receiver noise were large enough to completely overshadow the signal due to the incoming radiation, the statistical behavior of the pattern on the oscilloscope screen would be unchanged. The reason is that the most important forms of receiver noise are fundamentally due to shot effect and therefore have the statistical properties corresponding to Eqs. (5) and (7). This suggests that a simple way to obtain a model of the behavior of the electric vector in natural monochromatic light is to feed two narrow band amplifiers with shot noise from two diodes and put their outputs on the deflection plates of an oscilloscope. The electron beam will then trace out an ellipse whose parameters fluctuate at a rate inversely proportional to the band width and have the statistical properties described in the preceding section.

[7] Private communication.

IV
The Jones Calculus

Editor's Comments on Papers 16 Through 24

16 **Jones:** *A New Calculus for the Treatment of Optical Systems: I. Description and Discussion of the Calculus*

17 **Hurwitz and Jones:** *A New Calculus for the Treatment of Optical Systems: II. Proof of Three General Equivalence Theorems*

18 **Jones:** *A New Calculus for the Treatment of Optical Systems: III. The Sohncke Theory of Optical Activity*

19 **Jones:** *A New Calculus for the Treatment of Optical Systems: IV*

20 **Jones:** *A New Calculus for the Treatment of Optical Systems: V. A More General Formulation, and Description of Another Calculus*

21 **Jones:** *A New Calculus for the Treatment of Optical Systems: VI. Experimental Determination of the Matrix.*

22 **Jones:** *A New Calculus for the Treatment of Optical Systems: VII. Properties of the N-Matrices*

23 **Jones:** *New Calculus for the Treatment of Optical Systems: VIII. Electromagnetic Theory*

24 **Jones:** *A Comedy of Errors*

Papers 16 through 23 form a group that must be considered as one of the most important contributions to the mathematics of polarized light. Paper 16 establishes the principle that an optical system may be represented by a 2×2 complex matrix, and the incident light by a vector. The nature of the outgoing light is obtained by multiplying the vector by the matrix. Papers 17, 18, and 19 expand this initial concept, providing additional theorems and examples.

The Jones algebra is limited to nonscattering, nondepolarizing systems. In Paper 20 he compares his algebra to the more powerful, although often less convenient, Mueller algebra, which does not suffer from these restrictions.

Paper 22 gives the theory of N matrices, which permit the electric field to be calculated at any point, even within a solid element of the system. The last paper in the series relates the N matrix to the dielectric and gyration tensors of the medium.

The great virtue of casting the theory in matrix form is that the whole armory of matrix algebra methods is available to help with the conceptual understanding of the operation of polarization systems in general, and with the analysis and synthesis of particular situations.

In Paper 23 of the Jones series there is an apparent error as pointed out by Pancharatnam [S. Pancharatnam, "Light Propagation in Absorbing Crystals Possessing Optical Activity—Electromagnetic Theory," *Proc. Indian Acad. Sci.,* **48,** Sec. A. (1958)]. It turned out that because of a double error, Jones was indeed correct and the editor (W.S.) is profoundly grateful to Dr. Jones for providing "A Comedy of Errors," published here for the first time.

Jones is also well known for his work in photographic theory, radiation detectors, and acoustics. He is recipient of the Adolf Lomb Medal in 1944 and the Frederick Ives Medal in 1972 (both awarded by the Optical Society of America).

16

Reprinted from *J. Opt. Soc. Am.*, **31**, 488–493 (July 1941)

A New Calculus for the Treatment of Optical Systems

I. Description and Discussion of the Calculus

R. CLARK JONES*
Research Laboratory, Polaroid Corporation, Cambridge, Massachusetts, and Research Laboratory of Physics, Harvard University, Cambridge, Massachusetts

(Received April 23, 1941)

The effect of a plate of anisotropic material, such as a crystal, on a collimated beam of polarized light may always be represented mathematically as a linear transformation of the components of the electric vector of the light. The effect of a retardation plate, of an anisotropic absorber (plate of tourmaline; Polaroid sheeting), or of a crystal or solution possessing optical activity, may therefore be represented as a matrix which operates on the electric vector of the incident light. Since a plane wave of light is characterized by the phases and amplitudes of the two transverse components of the electric vector, the matrices involved are two-by-two matrices, with matrix elements which are in general complex. A general theory of optical systems containing plates of the type mentioned is developed from this point of view.

INTRODUCTION

THE type of optical system with which this paper deals is not the more familiar type involving lenses, prisms, etc., but is rather the type which is composed of retardation plates, partial polarizers, and plates possessing the ability to rotate the plane of polarization. We shall therefore be concerned not with the directions of rays of light, but with the state of polarization and the intensity of the light as it passes through the optical system.

An example of a partial polarizer is a plate of tourmaline, or a sheet of light-polarizing film such as those sold under the trade-name "Polaroid." [1] For light incident perpendicularly on the plate, a partial polarizer may be characterized by two mutually perpendicular axes, with each of which is associated one of the two principal absorption coefficients. The retardation plate, or wave plate, is too well known to require description. In this paper we shall refer to a plate having the ability to rotate the plane of polarization as a *rotator*. A rotator is exemplified by a plate of quartz viewed along the optic axis, or by solutions of optically active molecules. We shall refer to the individual plates which make up the complete optical system as the elements of the system, or as the optical elements.

In the present paper it will be assumed that the light is always incident normally on the elements of the optical system.

The present paper is devoted to a description of the new calculus. When light passes through any of the three fundamental types of optical elements, the state of polarization and sometimes the intensity of the light will be changed. We shall find it possible to represent the effect of any optical element on the light as a linear operator acting upon the electric vector of the light wave. The operator is expressed in the convenient form of a two-by-two matrix, whose four matrix elements are in the general complex. From the associative property of matrices, it will then follow that the complete optical system may also be represented by a two-by-two complex matrix.

In Part II, the calculus is used to prove three theorems relating to optical systems of the type considered. The two theorems relating to systems containing partial polarizers are new, so far as the writer knows, and should be of some practical importance because of the recent development of a readily available and high quality polarizer in the form of Polaroid sheeting.

In Part III, the theory is applied to a rigorous treatment of the Sohncke theory of optical activity.

---

* National Scholar, Harvard University.

[1] Registered trade-mark of the Polaroid Corporation. For a description of Polaroid sheeting, see Martin Grabau, J. Opt. Soc. Am. **27**, 420 (1937).

Conventions and Notation

Consider a right-handed rectangular coordinate system, x, y and z. The elements of the optical system are considered to be arranged along the z axis, the z axis being perpendicular to the plane of the plates. The optical system will always be described as viewed from a point which is further out along the positive z axis than any of the elements; with this convention, the x and y axes have the relative position of the usual xy plane: the positive x axis may be superposed upon the positive y axis by rotating the x axis 90° counterclockwise.

Furthermore, let the two principal axes of the ith retardation plate or partial polarizer be indicated by x_i' and y_i'. We assume that the positive x_i' and y_i' axes have the same relative orientation as the positive x and y axes. We may now define the orientation of the ith element by stating the angle ω_i measured counterclockwise from the positive x axis to the positive x_i axis.

The light will be represented as plane waves progressing in either direction along the z axis. The state of polarization may be completely defined by stating the amplitudes and phases of the x and y components of the electric vector of the light wave. At any fixed point along the z axis, the components may be written in the usual complex form

$$E_x = A_x \exp\left[i(\epsilon_x + 2\pi\nu t)\right], \qquad E_y = A_y \exp\left[i(\epsilon_y + 2\pi\nu t)\right], \tag{1}$$

where A_x and A_y, ϵ_x and ϵ_y are real. If $\epsilon_x - \epsilon_y$ is an integral multiple of π, the light is plane polarized; otherwise, elliptically polarized.

Now let us consider the change in the character of a light wave as it passes through a retardation plate or a partial polarizer. As a matter of fact, in order to avoid giving a separate treatment of the two cases, we shall suppose that both the indices of refraction and the absorption coefficients are different along the two principal axes (x' and y') of the plate. If we know the x' and y' components ($E_{x'0}$ and $E_{y'0}$) of the electric vector as the light enters the plate, then the corresponding components of the light as it emerges from the other side of the plate are

$$\begin{aligned} E_{x'1} &= E_{x'0} \exp\left[-i(2\pi d/\lambda)(n_{x'} - ik_{x'})\right] \\ &= N_{x'}E_{x'0}, \\ E_{y'1} &= E_{y'0} \exp\left[-i(2\pi d/\lambda)(n_{y'} - ik_{y'})\right] \\ &= N_{y'}E_{y'0}. \end{aligned} \tag{2}$$

In these expressions, d is the thickness of the plate, λ is the wave-length of the light in vacuum, the n's are the principal indices of refraction, and the k's are the principal extinction coefficients. The extinction coefficient is a particular type of amplitude absorption coefficient. The N's are merely abbreviations for the exponential expressions. In the case of a retardation plate, the k's are the same and the n's different, whereas in the case of a partial polarizer, the n's are the same and the k's different. In practice, of course, it is difficult to secure a partial polarizer which is not also birefringent.

In general, however, we are interested in knowing the change in the x and y components of the light wave, rather than the change in the x' and y' components. If ω is the angle measured counterclockwise from the positive x axis to the positive x' axis, we have the relations

$$\begin{aligned} E_{x'} &= E_x \cos\omega + E_y \sin\omega, \\ E_{y'} &= -E_x \sin\omega + E_y \cos\omega. \end{aligned} \tag{3}$$

The elimination of the x' and y' components in (2) and (3) yields the result

$$\begin{aligned} E_{x1} &= (N_{x'}\cos^2\omega + N_{y'}\sin^2\omega)E_{x0} \\ &\quad + (N_{x'} - N_{y'})\sin\omega\cos\omega E_{y0}, \\ E_{y1} &= (N_{x'} - N_{y'})\sin\omega\cos\omega E_{x0} \\ &\quad + (N_{x'}\sin^2\omega + N_{y'}\cos^2\omega)E_{y0}. \end{aligned} \tag{4}$$

The equations (4) give the important relation between the x and y components of the entering and emergent light.

The Matrix Notation

The relations (4) may be put in a convenient and simple form by the use of a matrix notation.[2]

[2] In Parts I and II we shall need only the algebraic properties of matrices, as they are presented on pp. 348–352 of E. C. Kemble, *Fundamental Principles of Quantum Mechanics* (McGraw-Hill Book Company, Inc., New York, 1937). In Part III, we shall find it necessary to use also the transformation properties of matrices; see the reference just given, or V. Rojansky, *Introductory Quantum Mechanics* (Prentice-Hall, New York, 1938), pp. 285–340.

Let $\mathbf{M}$ be the two-by-two matrix*

$$\mathbf{M}\equiv\begin{pmatrix} m_1 & m_4 \\ m_3 & m_2 \end{pmatrix}, \tag{5}$$

where

$$\begin{aligned} m_1 &= N_{x'}\cos^2\omega + N_{y'}\sin^2\omega, \\ m_2 &= N_{x'}\sin^2\omega + N_{y'}\cos^2\omega, \\ m_3 &= m_4 = (N_{x'} - N_{y'})\sin\omega\cos\omega. \end{aligned} \tag{6}$$

Furthermore, let ε_0 and ε_1 be the one-column vectors

$$\varepsilon_0\equiv\begin{pmatrix} E_{x0} \\ E_{y0} \end{pmatrix};\quad \varepsilon_1\equiv\begin{pmatrix} E_{x1} \\ E_{y1} \end{pmatrix}. \tag{7}$$

The relations (4) are now equivalent to the vector equation

$$\varepsilon_1 = \mathbf{M}\varepsilon_0. \tag{8}$$

The matrix $\mathbf{M}$ may itself be written in a simple form. Let $\mathbf{S}(\omega)$ be the rotation matrix

$$\mathbf{S}(\omega)\equiv\begin{pmatrix} \cos\omega & -\sin\omega \\ \sin\omega & \cos\omega \end{pmatrix} \tag{9}$$

and let

$$\mathbf{N}\equiv\begin{pmatrix} N_{x'} & 0 \\ 0 & N_{y'} \end{pmatrix}. \tag{10}$$

We then have

$$\mathbf{M} = \mathbf{S}(\omega)\mathbf{N}\mathbf{S}(-\omega) \tag{11}$$

or

$$\varepsilon_1 = \mathbf{S}(\omega)\mathbf{N}\mathbf{S}(-\omega)\varepsilon_0. \tag{12}$$

We have thus been able to represent the effect of a retardation plate or a partial polarizer as a matrix operator which operates upon the vector describing the intensity and polarization of the entering light. The matrix operator has been factored into a product of two types of matrices, the first of which, $\mathbf{N}$, describes the optical element in a way independent of its orientation, and the second of which, $\mathbf{S}$, describes the orientation.

Multi-Element Systems

Suppose now that we have a series of n optical elements, represented by the n matrices $\mathbf{N}_1$, $\mathbf{N}_2$, $\cdots$, $\mathbf{N}_n$, with the respective orientations ω_1, ω_2, $\cdots$, ω_n, and suppose that the light passes through the elements in the order in which they are numbered. We then have

$$\begin{aligned} \varepsilon_1 &= \mathbf{S}(\omega_1)\mathbf{N}_1\mathbf{S}(-\omega_1)\varepsilon_0 = \mathbf{M}_1\varepsilon_0, \\ \varepsilon_0 &= \mathbf{M}_2\varepsilon_1, \\ &\cdots\cdots \\ \varepsilon_n &= \mathbf{M}_n\varepsilon_{n-1}, \end{aligned} \tag{13}$$

where ε_i is the light vector as it emerges from the ith element. By substituting each of the Eqs. (13) in the one following it, we have at once the relation between ε_n and ε_0

$$\begin{aligned} \varepsilon_n &= \mathbf{M}_n\mathbf{M}_{n-1}\cdots\mathbf{M}_2\mathbf{M}_1\varepsilon_0 \\ &= \mathbf{M}^{(n)}\varepsilon_0, \end{aligned} \tag{14}$$

where $\mathbf{M}^{(n)}$ is the two-by-two matrix

$$\mathbf{M}^{(n)}\equiv\mathbf{M}_n\mathbf{M}_{n-1}\cdots\mathbf{M}_2\mathbf{M}_1. \tag{15}$$

According to (6), each of the $\mathbf{M}_i$'s is a symmetrical matrix. In general, however, the product of two or more of the $\mathbf{M}$ matrices will be neither symmetrical nor anti-symmetrical. This criterion permits us to distinguish immediately between the matrices corresponding to a single element, and those corresponding to the superposition of two or more elements.

It is fruitful to examine the structure of $\mathbf{M}^{(n)}$ more closely. We have

$$\mathbf{M}^{(n)} = [\mathbf{S}(\omega_n)\mathbf{N}_n\mathbf{S}(-\omega_n)][\mathbf{S}(\omega_{n-1})\mathbf{N}_{n-1}\mathbf{S}(-\omega_{n-1})]\cdots[\mathbf{S}(\omega_2)\mathbf{N}_2\mathbf{S}(-\omega_2)][\mathbf{S}(\omega_1)\mathbf{N}_1\mathbf{S}(-\omega_1)]. \tag{16}$$

By the use of the easily proved relation:

$$\mathbf{S}(\omega_1+\omega_2)\equiv\mathbf{S}(\omega_1)\mathbf{S}(\omega_2), \tag{17}$$

the expression for $\mathbf{M}^{(n)}$ may be written in either of the following forms:

$$\mathbf{M}^{(n)} = \mathbf{S}(\omega_n)[\mathbf{N}_n\mathbf{S}(\omega_{n,n-1})\cdots\mathbf{N}_2\mathbf{S}(\omega_{2,1})\mathbf{N}_1\mathbf{S}(\omega_{1,n})]\mathbf{S}(-\omega_n), \tag{18}$$

$$= \mathbf{S}(\omega_1)[\mathbf{S}(\omega_{1,n})\mathbf{N}_n\mathbf{S}(\omega_{n,n-1})\cdots\mathbf{S}(\omega_{2,1})\mathbf{N}_1]\mathbf{S}(-\omega_1), \tag{19}$$

where

$$\omega_{i,j}\equiv\omega_j-\omega_i. \tag{20}$$

* We depart from the usual double-subscript notation for matrix elements solely for reasons of typographical convenience.

The matrices represented by the square brackets in Eqs. (18) and (19) depend only on the relative orientation of the elements of the optical system, and are independent of the orientation of the optical system as a whole. Thus, just as in the case of a single element, we have factored the operator representing the optical system into the product of two types of matrices, the first of which depends only on the nature of the optical elements and on their *relative* orientation, and the second of which describes only the orientation of the optical system as a whole with respect to the x, y axes.

So far we have excluded any consideration of rotators, which are plates possessing the ability to rotate the plane of polarization about the z axis. The formal change caused by the introduction of rotators into the system is very small, however. Suppose that a given rotator is able to rotate the plane of polarization through the angle $\bar{\omega}$ in the counterclockwise direction. The matrix operator corresponding to the rotator is then simply $\mathbf{S}(\bar{\omega})$. We shall not stop to prove this statement,* since its proof is elementary. If the rotator just considered is introduced between the ith and the $(i+1)$th element of the optical system considered in the last paragraph, the resulting optical system may still be represented in the form of Eqs. (18) and (19) if we redefine $\omega_{i+1,\,i}$ as

$$\omega_{i+1,\,i} \equiv \omega_i - \omega_{i+1} + \bar{\omega}. \tag{21}$$

The generalization to the case in which several rotators are placed in the optical system is obvious. If several rotators are placed in juxtaposition, they are equivalent to a single rotator, according to (17).

There is an important physical distinction between the behavior of retardation plates and partial polarizers on one hand, and the behavior of rotators on the other. With the coordinate system which we are using, it makes no difference in what direction the light passes through a retardation plate or partial polarizer. With the type of rotator represented by a crystal or solution possessing optical activity, however, the rotation changes sign if the light passes through the plate in the reverse direction. In determining the operator corresponding to a rotator, therefore, it is necessary to know in which direction the light passes through the plate.

* More precisely, the matrix of a rotator is $e^{i\phi}\mathbf{S}(\bar{\omega})$, where the phase factor takes into account the finite optical thickness of the plate. In this paper, however, rotators will be considered only in combination with retardation plates and partial polarizers, so that the phase factor may always be considered to be associated with the other elements of the optical system. For simplicity, we omit the phase factor.

Reversibility of the Optical System

It is, however, not necessary to make a recalculation if the light passes through the optical system in the opposite direction. Suppose that we have computed the matrix $\mathbf{M}^{(n)}$ for the case that the light passes through the system in the order in which the plates are numbered, and then suppose that the light is reversed so that it passes through the system in the opposite direction, passing first through the nth plate, and last through the first plate. In terms of the one-row vectors,

$$\bar{\varepsilon} \equiv (E_x\ E_y), \tag{22}$$

the relation between the vector $\bar{\varepsilon}_{\text{initial}}$ of the light entering the nth plate and the vector $\varepsilon_{\text{final}}$ of the light emerging from the first plate is

$$\bar{\varepsilon}_{\text{final}} = \bar{\varepsilon}_{\text{initial}} \mathbf{M}^{(n)}, \tag{23}$$

where $\mathbf{M}^{(n)}$ is exactly the same matrix as that previously defined. This statement may be proved easily, first by showing that a relation of the form (23) holds for each element of the system, and then by eliminating the intermediate $\bar{\varepsilon}$ vectors.

The statement made in the last paragraph holds only if all of the rotators in the system are of the type represented by ordinary optical activity, and not of the type represented by the Faraday effect, in which case the direction of the rotation depends on the direction of the magnetic field, and not on the direction in which the light passes through the material. If any of the rotators are of the type represented by the Faraday effect, then the $\mathbf{M}^{(n)}$ which appears in (23) is not the same as the $\mathbf{M}^{(n)}$ which appears in (14), the difference being that the $\bar{\omega}$'s appearing in the $\mathbf{S}$ matrices which represent the Faraday rotators must have opposite signs in the two $\mathbf{M}^{(n)}$'s.

Discussion of the Matrices

When the optical system does not contain any partial polarizers but consists entirely of rotators and retardation plates, *all of the matrices are unitary.*‡ This fact may be proved by either of two methods.

First proof

The **N** matrices are all unitary, since their characteristic values are all of absolute value unity; the latter is a necessary and sufficient condition that a matrix be unitary. Furthermore, the **S** matrices are easily shown to be unitary, since the conjugate transpose of $\mathbf{S}(\omega)$ is $\mathbf{S}(-\omega)$, so that $\mathbf{S}^\dagger\mathbf{S}=\mathbf{1}$. The general matrix $\mathbf{M}^{(n)}$ is the product of factors which are all unitary, and therefore $\mathbf{M}^{(n)}$ is unitary.

Second proof

This proof is not purely mathematical, but is based on the fact that an optical system composed only of rotators and retardation plates does not change the intensity of the light passing through it. The intensity of the light is proportional to the sum of the squares of the x and y components of the electric vector:

$$I \propto A_x^2 + A_y^2 = E^*_x E_x + E^*_y E_y = \bar{\varepsilon}^* \varepsilon = \bar{\varepsilon}\varepsilon^* \quad (24)$$

(cf. Eq. (1) for part of the notation). The intensity of the light is therefore proportional to the square of the length of the vector ε. Now a unitary matrix does not change the length of a vector, and conversely, if a matrix does not change the length of an arbitrary vector, then the matrix is unitary. Thus, since the optical system does not change the intensity of the light, the corresponding matrices must be unitary.

Conversely, if the first proof of the unitary character of the matrices is accepted, then the second may be regarded as a proof that an optical system composed only of retardation plates and rotators does not change the intensity of the light passing through it.

In nearly every case, we are not interested in the total phase change corresponding to the two axes of a retardation plate, but only in the phase difference of the two principal axes; we often refer to a retardation plate as a q wave plate, implying that the optical path difference for plane polarized light parallel to each of the two axes is q wave-lengths.

Since the two principal extinction coefficients of a retardation plate are both zero, we find by comparison of (2) with (10) that the matrix of a retardation plate whose axes are parallel to the x and y axes may always be written in the form $e^{i\phi}\mathbf{G}$, where

$$\mathbf{G} \equiv \begin{pmatrix} e^{i\gamma} & 0 \\ 0 & e^{-i\gamma} \end{pmatrix}. \quad (25)$$

If we are interested only in the phase difference, the phase factor $e^{i\phi}$ may be omitted, and we may consider **G** alone to be the diagonal representation of a wave plate.

If we choose to write the diagonal matrices of all the wave plates in the optical system in the form (25), then, since the determinant of an **S** matrix is unity, all of the matrices which occur as factors on the right-hand side of Eq. (16) will have determinants equal to unity, provided that the system contains only retardation plates and rotators. Then, since the determinant of a product of matrices is equal to the product of the determinants of the matrices, the determinant of $\mathbf{M}^{(n)}$ will have the value unity.

Our treatment of partial polarizers is similar. The diagonal form of the matrix corresponding to a partial polarizer may always be written in the form $e^{i\phi}\mathbf{P}$, where

$$\mathbf{P} \equiv \begin{pmatrix} p_1 & 0 \\ 0 & p_2 \end{pmatrix} \quad \begin{matrix} 0 \leqslant p_1 \leqslant 1 \\ 0 \leqslant p_2 \leqslant 1. \end{matrix} \quad (26)$$

Again, if we are not interested in the absolute phase of the two components of the light, but only in the relative phase, we may consider **P** alone to be the diagonal representation of a partial polarizer. If the plate is a perfect polarizer, that is to say, if the plate absorbs completely light parallel to one of its axes, then one of the two p's will be zero; in this case the determinant of **P** is zero. More generally, the determinant of **P** is real, with its value lying in the range zero to unity.

We thus find that if we use **G** to represent a retardation plate, and **P** to represent a partial polarizer, then the determinant of $\mathbf{M}^{(n)}$ for an optical system containing all three types of

‡ Let $\mathbf{M}^\dagger$ be the matrix obtained from **M** by first exchanging m_3 and m_4 and then replacing each of the matrix elements by its complex conjugate. The matrix **M** is unitary if $\mathbf{M}^\dagger\mathbf{M}=\mathbf{M}\mathbf{M}^\dagger=\mathbf{1}$. The matrix $\mathbf{M}^\dagger$ is said to be the Hermitian adjoint, or the conjugate transpose of **M**.

elements will be real and non-negative, with its value in the range zero to unity.

It will be shown in Part II that the most general unitary matrix may be represented by an optical system consisting of only two elements, one a rotator and the other a retardation plate. Similarly, it will be shown that the general two-by-two complex matrix may be represented by an optical system containing not more than four elements, the only restriction being that the matrix may not be one which would increase the intensity of the light.

Elliptical Polarization

For convenience in reference, we state here the relations giving the shape and orientation of the ellipse representing the state of polarization of the light. Let the amplitude of the x and y components be A_x and A_y, and let $\delta=\epsilon_y-\epsilon_x$, as in Eq. (1). Further, let $\tan\theta$ equal the ratio of axes of the ellipse, and let ψ be the orientation of the major axis of the ellipse, measured counter-clockwise from the x axis. The following relations determine θ and ψ:[3]

$$\tan 2\psi=\tan 2\alpha\cos\delta,\qquad \cos 2\theta=\sin 2\alpha|\sin\delta|, \tag{27}$$

where

$$\tan\alpha=A_y/A_x,\quad 0\leqslant\alpha\leqslant\tfrac{1}{2}\pi. \tag{28}$$

When $\sin\delta$ is positive, the ellipse is described in the clockwise direction, and *vice versa.*

[3] Max Born, *Optik* (Springer, Berlin, 1933), p. 23.

17

Reprinted from *J. Opt. Soc. Am.*, **31**, 493–499 (July 1941)

A New Calculus for the Treatment of Optical Systems

II. Proof of Three General Equivalence Theorems

HENRY HURWITZ, JR., *Research Laboratory of Physics, Harvard University, Cambridge, Massachusetts,*

AND

R. CLARK JONES,* *Research Laboratory, Polaroid Corporation, Cambridge, Massachusetts, and Research Laboratory of Physics, Harvard University, Cambridge, Massachusetts*

(Received April 23, 1941)

The general theory developed in Part I is used to prove three equivalence theorems about optical systems of the type under discussion. We prove that any optical system which contains only retardation plates and rotators is optically equivalent to a system containing only two plates—one a retardation plate, and the other a rotator. We then prove an exactly analogous theorem for systems containing only partial polarizers and rotators. Finally, it is proved that the most general optical system which contains any number of all three types of plates is optically equivalent to a system containing at most four plates—two retardation plates, one partial polarizer, and one rotator.

STATEMENT OF THE THEOREMS

THE power of the calculus described in Part I may be illustrated by using it to establish a few general theorems relating to optical systems of the type under consideration.

We shall prove the following three theorems:

I. For light of a given wave-length, an optical system containing any number of retardation plates and rotators is optically equivalent to a system containing only two elements—one a retardation plate, and the other a rotator.

II. For light of a given wave-length, an optical system containing any number of partial polarizers and rotators is optically equivalent to a system containing only two elements—one a partial polarizer and the other a rotator.

III. For light of a given wave-length, an optical system containing any number of re-

* National Scholar, Harvard University.

tardation plates, partial polarizers, and rotators is optically equivalent to a system containing four elements—two retardation plates, one partial polarizer, and one rotator. In a large and finite class of cases, the rotator is not necessary.

The method of proof of these theorems is essentially the same for each of them. We shall first determine the mathematical restrictions on the matrix representing the original optical system. Secondly, we shall write down the matrix corresponding to the substitute optical system. We shall then equate these two matrices. In each case, we shall find that the resulting equations are of such a form that they always admit a physically realizable solution for the parameters of the substitute optical system. The proof that the equations are always solvable will be obtained by actually solving the equations explicitly for the parameters of the substitute optical system. The proof will thus not only establish that the replacement is always possible, but it will also provide the explicit equations for computing the parameters of the substitute optical system corresponding to any given original optical system.

The first of the three theorems has been proved by H. Poincaré by an entirely different method (cf. below). So far as the writers have been able to learn, the remaining two theorems are new.

Proof of Theorem I

We saw in Part I that the matrix $\mathbf{M}^{(n)}$ representing any optical system composed of retardation plates and rotators was unitary. We shall demonstrate that any unitary matrix may be represented by a system consisting of one retardation plate and one rotator.

We shall write

$$\mathbf{M}^{(n)}=e^{i\alpha}\mathbf{U} \tag{1}$$

and we shall select the angle α so that the determinant of $\mathbf{U}$ is unity.

It is well known[1] that the general unitary matrix with two rows and columns with unit determinant may be written in the form:

$$\mathbf{U}\equiv\begin{pmatrix} e^{i\phi}\cos\theta & -e^{-i\psi}\sin\theta \\ e^{i\psi}\sin\theta & e^{-i\phi}\cos\theta \end{pmatrix}. \tag{2}$$

The real angles θ, ϕ and ψ may be determined from the matrix elements of $\mathbf{U}$ by the relations:

$$\begin{aligned} e^{2i\phi}&=u_1/u_2, & \sin\theta&=e^{-i\psi}u_3, \\ e^{2i\psi}&=u_3/u_4, & \cos\theta&=e^{-i\phi}u_4. \end{aligned} \tag{3}$$

For any given $\mathbf{M}^{(n)}$, it is thus possible to determine the values of α, θ, ϕ and ψ, although the determination is not unique; for example, if ϕ is a solution, then ϕ plus an integral multiple of π is also a solution.

The matrix of a retardation plate with the orientation ω is $\mathbf{S}(\omega)\mathbf{G}\mathbf{S}(-\omega)$, where $\mathbf{G}$ is defined by Eq. (25, I); the matrix of a rotator is $\mathbf{S}(\bar{\omega})$. Whether the rotator is the first or second element of the system, the matrix of the combination will be of the form

$$\mathbf{V}\equiv\mathbf{S}(A)\mathbf{G}\mathbf{S}(B), \tag{4}$$

where the angles A and B are in general different.

If now for any values of θ, ϕ and ψ, we are always able to choose A, B and γ so that $\mathbf{V}=\mathbf{U}$, our theorem is proved. We multiply out explicitly the expression on the right of (4), so that the matrix elements of $\mathbf{V}$ appear as explicit functions of A, B and γ. We then set the four elements of $\mathbf{U}$ equal, respectively, to the four elements of $\mathbf{V}$, and after a little algebraic reduction, we find

$$\begin{aligned} \cos\phi\cos\theta&=\cos(A+B)\cos\gamma, \\ \sin\phi\cos\theta&=\cos(A-B)\sin\gamma, \\ \cos\psi\sin\theta&=\sin(A+B)\cos\gamma, \\ \sin\psi\sin\theta&=\sin(A-B)\sin\gamma \end{aligned} \tag{5}$$

as the conditions which must be satisfied in order that $\mathbf{V}=\mathbf{U}$.

From (5), we find as the explicit solution for A, B and γ:

$$\tan(A+B)=\frac{\cos\psi}{\cos\phi}\tan\theta=\frac{u_3-u_4}{u_1+u_2}, \tag{6}$$

[1] B. L. van der Waerden, *Die Gruppentheoretische Methode in der Quantenmechanik* (Springer, Berlin, 1932), p. 58.

$$\tan (A-B)=\frac{\sin \psi}{\sin \phi} \tan \theta=\frac{u_3+u_4}{u_1-u_2}, \qquad (7)$$

$$\tan \gamma=\frac{\cos (A+B)}{\cos (A-B)} \tan \phi$$

$$=\frac{\sin (A+B)}{\sin (A-B)} \tan \psi. \qquad (8)$$

It is evident that these equations always have a solution. The theorem is thus established. If we set

$$\mathbf{N}\equiv e^{i\alpha}\mathbf{G}, \qquad (9)$$

then we have

$$\mathbf{M}^{(n)}=\mathbf{S}(A)\mathbf{N}\mathbf{S}(B). \qquad (10)$$

We may at once show also that any matrix in the form **V** may be expressed in the form **U**. From (5) we have as the explicit solution for θ, ϕ and ψ in terms of A, B and γ:

$$\tan \phi=\frac{\cos (A-B)}{\cos (A+B)} \tan \gamma, \qquad (11)$$

$$\tan \psi=\frac{\sin (A-B)}{\sin (A+B)} \tan \gamma, \qquad (12)$$

$$\tan \theta=\frac{\cos \phi}{\cos \psi} \tan (A+B)=\frac{\sin \phi}{\sin \psi} \tan (A-B). \qquad (13)$$

It is also evident that these equations always have a solution. We have thus shown that **U** and **V** are mathematically equivalent representations of the general unitary matrix with unit determinant. In the proof of Theorem III, we shall find it convenient to pass freely from one representation to the other.

Poincaré's Proof of Theorem I

Poincaré[2] has given a geometrical proof of Theorem I which appears at first sight to be quite different from ours. Actually, however, the two proofs are very closely related, as we shall now show.

Poincaré expresses the ratio of the two complex components of the light in the form

$$E_y/E_x=(A_y/A_x) \exp [i(\epsilon_y-\epsilon_x)]=\xi+i\eta. \qquad (14)$$

In this expression, ξ and η together define completely the shape and orientation of the ellipse representing the state of polarization of the light. Thus each point on the ξ, η plane corresponds to a state of polarization, and conversely, each state of polarization corresponds to a point on the ξ, η plane.

Each point on the plane is considered to be projected stereographically upon a sphere of unit diameter to which the plane is tangent at the origin of coordinates. There is thus established a one-to-one correspondence between points on the sphere and points on the plane.

If now the sphere is subjected to a rotation about any diameter, and if we consider that the points on the sphere rotate rigidly with the sphere, then it is clear that all of the corresponding points on the plane (except two) will suffer a change of position. Thus a rotation of the sphere always may be said to correspond to a change in the state of polarization, and conversely, any change in the state of polarization may be related to some rotation of the sphere. Poincaré shows that the rotation corresponding to the action of a wave plate with retardation $\Delta=2\gamma$ is a rotation of angle Δ about the diameter which is perpendicular to the plane; he also shows that the rotation which corresponds to a rotator of angle $\bar{\omega}$ is a rotation of angle $2\bar{\omega}$ about the diameter which is parallel to the ξ axis.

Once the facts stated in the last paragraph have been established, the proof of Theorem I follows immediately, since it is physically obvious that any rotation of the sphere may always be compounded from two successive rotations, one about the diameter normal to the plane, and the other about the diameter parallel to the ξ axis.

Now we learn from the group theory of three-dimensional rotations[3] that any unitary two-by-two matrix with unit determinant may be associated uniquely with a rotation of a sphere. The elements of this matrix are referred to as the Cayley-Klein parameters of the rotation. If we express the matrix of the optical system of Theorem I in the form **U**, then the Eulerian angles specifying the rotation of Poincaré's sphere are linearly related to the

[2] H. Poincaré, *Théorie Mathématique de la Lumière* (Paris, 1892), Vol. II, Chapter XII.

[3] Carl Eckart, Rev. Mod. Phys. **2**, 305 (1930), pp. 341, 346.

angles ϕ, θ and ψ. This is the relation between Poincaré's geometrical proof and our proof by the use of matrices.

From a purely formal point of view, the relation between the proofs consists in the fact that the group of unitary two-by-two matrices with unit determinant is isomorphic to the group of three-dimensional rotations.

Proof of Theorem II

We saw in Part I that all of the matrices representing partial polarizers and rotators could always be so written that the matrix elements were real and the determinants real and non-negative, except for phase factors occurring as factors of the matrix as a whole. For an optical system composed only of partial polarizers and rotators, we may therefore set

$$\mathbf{M}^{(n)}=e^{i\alpha}\mathbf{R}, \tag{15}$$

where $\mathbf{R}$ is a matrix whose matrix elements are real and whose determinant lies in the range zero to unity.

The matrix of a partial polarizer with the orientation ω is $\mathbf{S}(\omega)\mathbf{PS}(-\omega)$, where $\mathbf{P}$ is defined by Eq. (26, I). By the same argument that was used in the proof of Theorem I, the optical system composed of a rotator and a partial polarizer will always be represented by a matrix of the form

$$\mathbf{T}\equiv\mathbf{S}(A)\mathbf{PS}(B), \tag{16}$$

where A and B are related in an obvious manner to ω and $\bar{\omega}$.

In order to establish the theorem, we must show that any matrix $\mathbf{R}$ which satisfies the stated conditions may be written in the form $\mathbf{T}$. We multiply out explicitly the expression on the right of (16), so that the matrix elements of $\mathbf{T}$ appear as explicit functions of A, B, p_1, and p_2. We then set the four elements of $\mathbf{R}$ equal, respectively, to the four elements of $\mathbf{T}$, and after a little algebraic reduction, we find

$$\begin{aligned}(p_1+p_2)\sin(A+B)&=r_3-r_4,\\(p_1+p_2)\cos(A+B)&=r_1+r_2,\\(p_1-p_2)\sin(A-B)&=r_3+r_4,\\(p_1-p_2)\cos(A-B)&=r_1-r_2\end{aligned} \tag{17}$$

as the conditions which must be satisfied in order that $\mathbf{T}=\mathbf{R}$.

The quotient of the first two, and the quotient of the last two of the Eqs. (17) are

$$\tan(A+B)=(r_3-r_4)/(r_1+r_2), \tag{18a}$$

$$\tan(A-B)=(r_3+r_4)/(r_1=r_2). \tag{18b}$$

By the use of the trigonometric identity

$$\sin x\cos x\equiv\tan x/(1+\tan^2 x),$$

the product of the first two, and the product of the last two of the Eqs. (17) may be written

$$\begin{aligned}p_1p_2&=r_1r_2-r_3r_4\equiv\det\mathbf{R},\\p_1^2+p_2^2&=r_1^2+r_2^2+r_3^2+r_4^2.\end{aligned} \tag{19}$$

The Eqs. (18) always permit a solution for A and B, and, since the determinant of $\mathbf{R}$ is non-negative, the Eqs. (19) always permit a solution for p_1 and p_2 in which both p_1 and p_2 are positive. Although the Eqs. (19) always permit a solution for p_1 and p_2, they do not determine which is which; if $p_1=a$, $p_2=b$ is a solution of (19), then so is $p_1=b$, $p_2=a$. We must go back to the last two of the Eqs. (17) in order to determine which of the p's is the larger, and it is clear that the decision will depend on which of the two physically distinct solutions of (18b) has been chosen. Note also that only one of the two possible solutions of (18a) is admissible, since p_1+p_2 cannot be negative.

The theorem is thus established, provided only that we can show that neither of the p's is greater than unity. We proceed with this final part of the proof.

With every matrix $\mathbf{A}$ there may be associated a number $\Gamma(\mathbf{A})$ where Γ is the maximum value of the ratio $|\mathbf{A}\varepsilon|/|\varepsilon|$ with respect to any variation of the two components of ε. Such a maximum will always exist for a matrix $\mathbf{A}$ whose matrix elements are finite and do not depend on ε. For any two matrices $\mathbf{A}$ and $\mathbf{B}$, we obviously have the relation

$$\Gamma(\mathbf{AB})\leqslant\Gamma(\mathbf{A})\Gamma(\mathbf{B}). \tag{20}$$

Since a unitary matrix does not change the length of a vector, we have

$$\begin{aligned}\Gamma(\mathbf{U})&=1\\\Gamma(\mathbf{AU})&=\Gamma(\mathbf{UA})=\Gamma(\mathbf{A}),\end{aligned} \tag{21}$$

where **A** is any matrix and **U** is a unitary matrix.

The physical interpretation of Γ for a matrix which represents an element of an optical system is that Γ^2 is the maximum transmission factor of the given optical element. The value of Γ for such a matrix may therefore never be greater than unity. By repeated application of (20), we then find that the value of $\Gamma(\mathbf{M}^{(n)})$ may not be greater than unity. We clearly have $\Gamma(\mathbf{M}^{(n)})=\Gamma(\mathbf{R})$ and, since we have set $\mathbf{T}=\mathbf{R}$, we have also $\Gamma(\mathbf{M}^{(n)})=\Gamma(\mathbf{T})$. But by repeated application of (21), we have $\Gamma(\mathbf{T})=\Gamma(\mathbf{P})$, so that the value of Γ for **P** may not be greater than unity. The value of Γ for the matrix **P**, however, is clearly the larger of the two p's, so that neither of the p's may be greater than unity. The proof of Theorem II is now complete.

Proof of Theorem III

In this section we shall show that any optical system composed of retardation plates, partial polarizers and rotators may always be replaced by an optical system composed of four elements; the four-element system consists of a partial polarizer placed between two retardation plates, with the addition of a rotator inserted at any of the four possible positions in the system.

We showed in Part I that, except for a phase factor, the matrix $\mathbf{M}^{(n)}$ representing the optical system could always be written so that the value of its determinant was real and in the range zero to unity. We therefore set

$$\mathbf{M}^{(n)}=e^{i\alpha}\mathbf{W}, \tag{22}$$

where **W** has a determinant which is real and which lies in the range zero to unity.

The matrix representing a partial polarizer placed between two retardation plates is

$$[\mathbf{S}(\omega_3)\mathbf{G}_1\mathbf{S}(-\omega_3)][\mathbf{S}(\omega_2)\mathbf{P}\mathbf{S}(-\omega_2)]\times[\mathbf{S}(\omega_1)\mathbf{G}_2\mathbf{S}(-\omega_1)]. \tag{23}$$

This matrix may be written in the form

$$\mathbf{\Phi}\equiv\mathbf{S}(A_1)\mathbf{G}_1\mathbf{S}(B_1)\mathbf{P}\mathbf{S}(A_2)\mathbf{G}_2\mathbf{S}(\mathbf{B}_2), \tag{24}$$

where the A's and B's are subject to the condition

$$A_1+B_1+A_2+B_2=0. \tag{25}$$

If now we add a rotator to the three-element system just described it is evident that it is always possible to choose the three ω's and the angle $\tilde{\omega}$ of the rotator so that the A's and B's may all be chosen arbitrarily. That is to say, the addition of a rotator permits us to avoid the restriction (25).

We thus see that the four-element optical system just described may be represented by the matrix $\mathbf{\Phi}$, where the A's and B's are now all independent. The matrix $\mathbf{\Phi}$ is of the form $\mathbf{V}_1\mathbf{P}\mathbf{V}_2$, where **V** is the matrix defined by Eq. (4). We have shown that any matrix of the form **V** is mathematically equivalent to a matrix of the form **U**, where **U** is defined by Eq. (2). Any matrix of the form $\mathbf{\Phi}$ may therefore be written in the form $\mathbf{U}_1\mathbf{P}\mathbf{U}_2$, *and conversely*.

In order to establish the theorem, we must now show that any matrix **W** can be written in the form

$$\mathbf{\Phi}\equiv\mathbf{U}_1\mathbf{P}\mathbf{U}_2. \tag{26}$$

This fact will be proved in the following paragraphs. In anticipation of this result, we should like to make an important observation: The matrices **W** and $\mathbf{\Phi}$ are not completely general, because they are restricted by the fact that the imaginary parts of their determinants are zero. They therefore contain eight minus one, or seven *independent* variables. The matrix $\mathbf{\Phi}$, however, contains eight variables (three in each of the **U**'s and two in **P**), so that these variables cannot be uniquely determined. We shall find that the extra degree of freedom afforded by the eight variables will often, but by no means always, permit us to dispense with the rotator; whenever the extra degree of freedom permits us to satisfy the relation (25), the rotator is unnecessary.

We now multiply out explicitly the product on the right in Eq. (26). We find

$$\mathbf{\Phi}=\begin{bmatrix} \exp[i(\phi_1+\phi_2)]p_1\cos\theta_1\cos\theta_2 & -\exp[i(\phi_1-\psi_2)]p_1\cos\theta_1\sin\theta_2 \\ -\exp[-i(\psi_1-\psi_2)]p_2\sin\theta_1\sin\theta_2 & -\exp[-i(\phi_2+\psi_1)]p_2\sin\theta_1\cos\theta_2 \\ \exp[i(\phi_2+\psi_1)]p_1\sin\theta_1\cos\theta_2 & -\exp[i(\psi_1-\psi_2)]p_1\sin\theta_1\sin\theta_2 \\ +\exp[-i(\phi_1-\psi_2)]p_2\cos\theta_1\sin\theta_2 & +\exp[-i(\phi_1+\phi_2)]p_2\cos\theta_1\sin\theta_2 \end{bmatrix}. \tag{27}$$

By examination of this matrix, we see that the ϕ's and ψ's occur only in four different combinations

$$\chi_1 \equiv \phi_1+\phi_2, \quad \chi_3 \equiv \phi_2+\psi_1, \quad \chi_2 \equiv \psi_1-\psi_2, \quad \chi_4 \equiv \phi_1-\psi_2. \tag{28}$$

These four relations are not linearly independent, in view of the relation

$$\chi_1+\chi_2=\chi_3+\chi_4.$$

We therefore choose to replace the ϕ's and ψ's by three new *independent* variables η, ξ_1, and ξ_2:

$$\phi_1+\phi_2=\eta+\xi_1, \quad \phi_2+\psi_1=\eta+\xi_2, \quad \psi_1-\psi_2=\eta-\xi_1, \quad \phi_1-\psi_2=\eta-\xi_2. \tag{29}$$

We have the following definitions of η, ξ_1, and ξ_2:

$$\begin{aligned} 2\xi_1 &\equiv \phi_1+\phi_2-\psi_1+\psi_2, \\ 2\xi_2 &\equiv -\phi_1+\phi_2+\psi_1+\psi_2, \\ 2\eta &\equiv \phi_1+\phi_2+\psi_1-\psi_2, \\ 2\sigma &\equiv \phi_1-\phi_2+\psi_1+\psi_2, \end{aligned} \tag{30}$$

where we have added a fourth variable σ in order that we may also solve the equations for the ϕ's and ψ's:

$$\begin{aligned} 2\phi_1 &= \xi_1-\xi_2+\eta+\sigma, \quad 2\psi_1=-\xi_1+\xi_2+\eta+\sigma, \\ 2\phi_2 &= \xi_1+\xi_2+\eta-\sigma, \quad 2\psi_2=\xi_1+\xi_2-\eta+\sigma. \end{aligned} \tag{31}$$

We make the further changes of variable

$$q_1 \equiv e^{i\eta}p_1, \quad q_2 \equiv e^{-i\eta}p_2. \tag{32}$$

The q's are thus complex quantities.

With the changes of variable outlined in the last paragraph, the matrix $\mathbf{\Phi}$ may now be written in the simpler form

$$\mathbf{\Phi}=\begin{pmatrix} \exp(i\xi_1)(q_1\cos\theta_1\cos\theta_2-q_2\sin\theta_1\sin\theta_2) & -\exp(-i\xi_2)(q_1\cos\theta_1\sin\theta_2+q_2\sin\theta_1\cos\theta_2) \\ \exp(i\xi_2)(q_1\sin\theta_1\cos\theta_2+q_2\cos\theta_1\sin\theta_2) & -\exp(-i\xi_1)(q_1\sin\theta_1\sin\theta_2-q_2\cos\theta_1\cos\theta_2) \end{pmatrix}. \tag{33}$$

We now set the four matrix elements of $\mathbf{\Phi}$ equal to the four matrix elements of $\mathbf{W}$. The resulting four equations are equivalent to

$$\begin{aligned} (q_1+q_2)\sin(\theta_1+\theta_2) &= \exp(-i\xi_2)w_3-\exp(i\xi_2)w_4, \\ (q_1+q_2)\cos(\theta_1+\theta_2) &= \exp(-i\xi_1)w_1+\exp(i\xi_1)w_2, \\ (q_1-q_2)\sin(\theta_1-\theta_2) &= \exp(-i\xi_2)w_3+\exp(i\xi_2)w_4, \\ (q_1-q_2)\cos(\theta_1-\theta_2) &= \exp(-i\xi_1)w_1-\exp(i\xi_1)w_2, \end{aligned} \tag{34}$$

where the w's are the matrix elements of $\mathbf{W}$, and are in general complex.

We wish to show that these equations can always be solved by θ's and ξ's which are real, and by q's whose product is real and non-negative. The quotient of the first two, and the quotient of the last two of the above equations are

$$\begin{aligned} \tan(\theta_1+\theta_2) &= \frac{\exp(-i\xi_2)w_3-\exp(i\xi_2)w_4}{\exp(-i\xi_1)w_1+\exp(i\xi_1)w_2}, \\ \tan(\theta_1-\theta_2) &= \frac{\exp(-i\xi_2)w_3+\exp(i\xi_2)w_4}{\exp(-i\xi_1)w_1-\exp(i\xi_1)w_2}. \end{aligned} \tag{35}$$

In order that these equations be capable of solution by real values of the θ's, it is necessary that the imaginary parts of the expressions on the right side be zero. By rationalizing the denominators, we find that this condition is equivalent to

$$\begin{aligned} &\text{I.P.}\ \{w_1^*w_3\exp[i(\xi_1-\xi_2)]-w_2^*w_4\exp[-i(\xi_1-\xi_2)]\}=0, \\ &\text{I.P.}\ \{w_1^*w_4\exp[i(\xi_1+\xi_2)]-w_2^*w_3\exp[-i(\xi_1+\xi_2)]\}=0 \end{aligned}$$

or finally,

$$\xi_1-\xi_2=\arg(w_1w_3^*+w_2^*w_4),$$
$$\xi_1+\xi_2=\arg(w_1w_4^*+w_2^*w_3). \tag{36}$$

The last two equations serve to evaluate the ξ's, and with this choice of the ξ's, Eqs. (35) may always be solved for real values of the θ's.

We may now determine the q's from the Eqs. (34). In order, however, to show that the product q_1q_2 is real and non-negative, we form the product of the first two and the product of the last two of the Eqs. (34), and then subtract the two resulting equations. With the help of (35), we find the result reduces to

$$q_1q_2=w_1w_2-w_3w_4=\det \mathbf{W}. \tag{37}$$

The matrix $\mathbf{W}$ was obtained from $\mathbf{M}^{(n)}$ by requiring that the determinant of $\mathbf{W}$ be real and non-negative. After evaluating q_1 and q_2, we may determine the quantity η from the condition that p_1 and p_2 be individually real and non-negative. (Cf. Eq. (32).)

Up to this point we have shown that the equations obtained by setting $\mathbf{\Phi}$ equal to $\mathbf{W}$ can always be solved for the real quantities ξ_1, ξ_2, θ_1, θ_2, p_1, p_2 and η, in the order mentioned. If we assign an arbitrary real value to σ, the Eqs. (31) may always be solved for the ϕ's and ψ's. The proof that neither of the p's is greater than unity is exactly similar to the proof used in Theorem II. The present theorem is therefore established.

It remains to determine the condition that we be able to choose σ so that the equation

$$A_1+B_1+A_2+B_2=0 \tag{25}$$

is satisfied. If σ can be so chosen, the rotator is unnecessary. Now let us consider the equation

$$\tan(A_1+B_1)+\tan(A_2+B_2)=\text{constant}. \tag{38}$$

In order that we be able to choose the A's and B's so that they satisfy (25), it is necessary and sufficient that the constant in Eq. (38) be zero. The validity of the equation

$$\tan(A_1+B_1)+\tan(A_2+B_2)=0 \tag{39}$$

is therefore the necessary and sufficient condition that we be able to dispense with the rotator.

By the use of Eqs. (31) and Eq. (6), we find that (39) may be written:

$$\cos\psi_1\cos\phi_2\tan\theta_1+\cos\psi_2\cos\phi_1\tan\theta_2=0 \tag{40}$$

or

$$\cos\tfrac{1}{2}[-\xi_1+\xi_2+\eta+\sigma]\cos\tfrac{1}{2}[\xi_1+\xi_2+\eta-\sigma]\times\tan\theta_1+\cos\tfrac{1}{2}[\xi_1+\xi_2-\eta+\sigma]\cos\tfrac{1}{2}[\xi_1-\xi_2+\eta+\sigma]\tan\theta_2=0. \tag{41}$$

By trigonometric reduction we find that the last equation is equivalent to

$$k_1\cos\sigma+k_2\sin\sigma+k_3=0, \tag{42}$$

where

$$\begin{aligned}k_1&\equiv\cos\xi_1(\tan\theta_1+\tan\theta_2)\\ k_2&\equiv\sin\xi_1(\tan\theta_1-\tan\theta_2)\\ k_3&\equiv\cos(\eta+\xi_2)\tan\theta_1+\cos(\eta-\xi_2)\tan\theta_2.\end{aligned} \tag{43}$$

The condition which must be satisfied in order that Eq. (42) be solvable for σ is

$$k_3^2\leqslant k_1^2+k_2^2. \tag{44}$$

This inequality is the condition which must be satisfied in order that it be possible to dispense with the rotator. As it is written, the condition is purely a mathematical one. Although the condition can be simplified slightly, the authors have not been able to attach any simple physical significance to the inequality expressed by Eq. (44).

Reprinted from *J. Opt. Soc. Am.*, **31**, 500–503 (July 1941)

A New Calculus for the Treatment of Optical Systems

III. The Sohncke Theory of Optical Activity

R. Clark Jones*
Research Laboratory, Polaroid Corporation, Cambridge, Massachusetts, and Research Laboratory of Physics, Harvard University, Cambridge, Massachusetts

(Received April 23, 1941)

Reusch and Sohncke have examined the properties of a system containing a large number n of identical retardation plates, each of which is rotated with respect to the one preceding it through the angle ω. The product of n and ω must be equal to $\mu\pi$, where μ is an integer. Under certain conditions this system is optically equivalent to a simple rotator. This system, which would be very difficult to examine by ordinary methods, is given a treatment which is completely rigorous and which is more general than any given heretofore.

Introduction

Reusch[1] has examined experimentally the optical properties of a system composed of n identical retardation plates, the corresponding axes of which are distributed uniformly about a whole number of semicircles. If the number of semicircles is denoted by μ, we may describe the system by stating that the axis of each plate is rotated through the angle $-\pi\mu/n$ with respect to the corresponding axis of the plate immediately preceding it. Reusch found experimentally that when the phase retardation of each of the plates was small, the complete optical system behaved approximately like a simple rotator. A few years after Reusch published his results, Sohncke[2] provided a rigorous mathematical treatment of the system composed of three retardation plates with their axes distributed through 180°—that is to say, with the relative angle 60°. In his book[3] on crystal optics, Pockels derives an approximate expression for the angle of the rotation produced by the system of n retardation plates distributed through 180°:

$$\bar{\omega}=\tfrac{1}{8}n\Delta^2 \cot(\pi/n). \tag{1}$$

In this expression, Δ is the phase retardation of each of the plates, and is assumed to be small. Just how small Δ must be, will be indicated in the treatment to be given.

We present here a rigorous treatment of the optical system described at the beginning of the previous paragraph. The results will be exact for arbitrary integral values of n and μ. By use of the transformation properties of matrices, we shall find an exact expression for the matrix $\mathbf{M}^{(n)}$ of the entire optical system. Since the optical system under consideration contains only retardation plates, it satisfies the conditions of Theorem I of Part II, and may therefore be replaced by a system composed only of one rotator and one retardation plate. We shall find exact and reasonably simple expressions (cf. Eqs. (18) and (20)) for the angle $\bar{\omega}$ of the equivalent rotator and the phase retardation $2\bar{\gamma}$ of the equivalent retardation plate, and we shall derive the condition (cf. Eq. (21)) which must be satisfied in order that the optical system behave approximately as a rotator. Our results will confirm Pockels' approximate result, Eq. (1).

Determination of the Matrix $\mathbf{M}^{(n)}$

With respect to the fixed axes x and y, let the orientation of the plate which the light enters first be zero. Then in general the orientation of the kth plate will be

$$\omega_k=-(k-1)\pi\mu/n. \tag{2}$$

Let us represent the retardation plate in the form $\mathbf{G}$ (cf. Eq. (25, I)). The phase retardation of the plate is then $\Delta=2\gamma$. According to Eqs. (19, I) and

---

* National Scholar, Harvard University.

[1] E. Reusch, Berliner Monatsber. (Gesammtsitzung, July 8, 1869) 530 (1869); Pogg. Ann. **138**, 628 (1869).

[2] L. Sohncke, Math. Ann. **9**, 504 (1876). See also Pogg. Ann. Ergänzungsband **8**, 16 (1878), and Zeits. f. Krist. **13**, 214 (1888).

[3] F. Pockels, *Lehrbuch der Kristalloptik* (Teubner, Leipzig, 1906), pp. 289–290.

(20, I), the matrix $\mathbf{M}^{(n)}$ of the complete optical system reduces to

$$\begin{aligned}\mathbf{M}^{(n)}&=\mathbf{S}(\pi\mu)(\mathbf{S}(\omega)\mathbf{G})^n\\&=(-1)^\mu(\mathbf{S}(\omega)\mathbf{G})^n,\end{aligned}\tag{3}$$

where ω is here minus the relative orientation of the plates

$$\omega\equiv\pi\mu/n.\tag{4}$$

The determination of $\mathbf{M}^{(n)}$ thus reduces to the problem of finding the nth power of a matrix which in this case is unitary.

Now there exists a simple method of finding the nth power of any matrix which can be diagonalized. Let $\mathbf{A}$ be any matrix which can be diagonalized, and let $\mathbf{T}$ be the matrix of the transformation which accomplishes the diagonalization:

$$\begin{aligned}\mathbf{D}&=\mathbf{TAT}^{-1},\\\mathbf{A}&=\mathbf{T}^{-1}\mathbf{DT},\end{aligned}\tag{5}$$

where $\mathbf{D}$ is the diagonal form of $\mathbf{A}$, and where $\mathbf{T}^{-1}$ is the reciprocal of $\mathbf{T}$:

$$\mathbf{T}^{-1}\mathbf{T}=\mathbf{TT}^{-1}=1.\tag{6}$$

The nth power of $\mathbf{A}$ is then

$$\begin{aligned}\mathbf{A}^n&=\mathbf{T}^{-1}\mathbf{DTT}^{-1}\mathbf{DT}\cdots\mathbf{T}^{-1}\mathbf{DT}\\&=\mathbf{T}^{-1}\mathbf{D}^n\mathbf{T}.\end{aligned}\tag{7}$$

Since the nth power of a diagonal matrix can be written down at once, the determination of $\mathbf{A}^n$ reduces to the problem of determining the transformation $\mathbf{T}$.

The writer has found by detailed calculation the following result. Let a matrix $\mathbf{A}$ be diagonalized by the transformation $\mathbf{T}$, and let d_1 and d_2 be the diagonal elements of the matrix $\mathbf{D}$ so obtained. Let now d_1 and d_2 be replaced by new diagonal elements d_1' and d_2', and let the resulting matrix be transformed back by the reciprocal of the transformation $\mathbf{T}$. We denote the matrix so obtained by $\mathbf{A}'$. The matrix elements of $\mathbf{A}'$ are then related to those of $\mathbf{A}$ by the relations

$$\begin{aligned}a_1'&=\frac{d_1'-d_2'}{d_1-d_2}a_1+\frac{d_2'd_1-d_1'd_2}{d_1-d_2},\\a_2'&=\frac{d_1'-d_2'}{d_1-d_2}a_2+\frac{d_2'd_1-d_1'd_2}{d_1-d_2},\\a_3'&=\frac{d_1'-d_2'}{d_1-d_2}a_3.\\a_4'&=\frac{d_1'-d_2'}{d_1-d_2}a_4.\end{aligned}\tag{8}$$

These relations involve only the old and new diagonal elements, and do not involve the transformation $\mathbf{T}$.

In application to Eq. (3), the matrix $\mathbf{A}$ is

$$\mathbf{A}=\mathbf{S}(\omega)\mathbf{G}=\begin{pmatrix}e^{i\gamma}\cos\omega & -e^{-i\gamma}\sin\omega\\ e^{i\gamma}\sin\omega & e^{-i\gamma}\cos\omega\end{pmatrix}.\tag{9}$$

The diagonal elements, or characteristic values, of this matrix are the roots of

$$(e^{i\gamma}\cos\omega-d)(e^{-i\gamma}\cos\omega-d)+\sin^2\omega=0\tag{10}$$

or

$$d_1=e^{i\chi},\quad d_2=e^{-i\chi},\tag{11}$$

where

$$\cos\chi=\cos\omega\cos\gamma.\tag{12}$$

For the sake of definiteness, we assume that χ is the positive solution of this equation. The matrix $\mathbf{D}^n$ has the diagonal elements $e^{in\chi}$ and $e^{-in\chi}$, so that if we set

$$d_1'=e^{in\chi},\quad d_2'=e^{-in\chi},\tag{13}$$

then the matrix $\mathbf{A}'$ is equal to $\mathbf{A}^n$.

By the indicated procedure, we find

$$\mathbf{M}^{(n)}=(-1)^\mu\mathbf{A}^n=(-1)^\mu\mathbf{A}',\tag{14}$$

where

$$\begin{aligned}a_1'&=\frac{\sin n\chi}{\sin\chi}\cos\omega e^{i\gamma}-\frac{\sin(n-1)\chi}{\sin\chi},\\a_2'&=\alpha_1'^*,\\a_3'&=\frac{\sin n\chi}{\sin\chi}\sin\omega e^{i\gamma},\\a_4'&=-a_3'^*.\end{aligned}\tag{15}$$

We find that these expressions for the matrix elements of $\mathbf{A}'$ yield the following expressions for the elements of $\mathbf{M}^{(n)}$, which we denote simply by m_1, m_2, m_3 and m_4:

$$\begin{aligned} m_1 &= \cos n(\chi-\omega)+i[\sin n(\chi-\omega)]\frac{\cos\omega\sin\gamma}{\sin\chi}, \\ m_2 &= m_1^*, \\ m_3 &= [\sin n(\chi-\omega)]\frac{\sin\omega}{\sin\chi}e^{i\gamma}, \\ m_4 &= -m_3^*. \end{aligned} \tag{16}$$

We should like to stress that these expressions are exact, no approximations having been used in their derivation.

Resolution of the Matrix $\mathbf{M}^{(n)}$

It is evident that the matrix $\mathbf{M}^{(n)}$ does not represent a pure rotation, since the matrix elements of a rotator are real. We wish to know, however, under just what conditions and in just what sense we may say that the matrix $\mathbf{M}^{(n)}$ is an approximation to a rotator. The most direct and elegant way to determine these conditions is to express the matrix $\mathbf{M}^{(n)}$ as the combination of a rotator and a retardation plate. We know by Theorem I of Part II that this may always be done.

We shall express the matrix $\mathbf{M}^{(n)}$ as the combination of a rotator $\mathbf{S}(\bar{\omega})$ and a retardation plate $\bar{\mathbf{G}}$ with the orientation $\bar{\omega}$:

$$\mathbf{M}^{(n)}=\mathbf{S}(\bar{\omega})[\mathbf{S}(\bar{\omega})\bar{\mathbf{G}}\mathbf{S}(-\bar{\omega})]. \tag{17}$$

From Eqs. (6–8, II), (12) and (16), we find

$$\begin{aligned} \tan\bar{\omega} &= \tan(A+B)=\frac{m_3-m_4}{m_1+m_2} \\ &= [\tan n(\chi-\omega)]\frac{\tan\omega}{\tan\chi}, \end{aligned} \tag{18}$$

$$\begin{aligned} \tan(\bar{\omega}+2\bar{\omega}) &= \tan(A-B)=\frac{m_3+m_4}{m_1-m_2} \\ &= \tan\omega, \end{aligned} \tag{19}$$

$$\begin{aligned} \tan\bar{\gamma} &= \frac{\sin(A+B)}{\sin(A-B)}\tan\gamma \\ &= \sin\bar{\omega}\frac{\tan\gamma}{\sin\omega}. \end{aligned} \tag{20}$$

We have found that the complete optical system is equivalent to the combination of a rotator with angle $\bar{\omega}$ and a retardation plate with the relative phase retardation $2\bar{\gamma}$. In order that the matrix $\mathbf{M}^{(n)}$ represent a rotator, it is therefore reasonable to require that $\bar{\gamma}$ must be small compared with $\bar{\omega}$, and in order to insure this inequality, it is by Eq. (20) necessary and sufficient to require that

$$\left|\frac{\tan\gamma}{\sin\omega}\right|\ll 1. \tag{21}$$

This condition may be more stringent than the condition that γ be small compared with unity, because Eq. (21) is equivalent to the condition

$$\gamma\ll\omega=\pi\mu/n. \tag{22}$$

The following two relations may be derived from the Eq. (12) defining χ:

$$\sin(\chi-\omega)=\sin\omega\cos\omega\cos\gamma \times\left[\left(1+\frac{\tan^2\gamma}{\sin^2\omega}\right)^{\frac{1}{2}}-1\right], \tag{23}$$

$$\frac{\tan^2\chi}{\tan^2\omega}=1+\frac{\tan^2\gamma}{\sin^2\omega}. \tag{24}$$

When the condition (21) is satisfied, by Eqs. (18), (23) and (24) we then have for the angle $\bar{\omega}$:

$$\begin{aligned} \bar{\omega} &\cong n(\chi-\omega)\cong n\sin(\chi-\omega) \\ &\cong \tfrac{1}{2}n\cot\omega\sin\gamma\tan\gamma\cong\tfrac{1}{2}n\gamma^2\cot\omega \end{aligned} \tag{25}$$

or

$$\bar{\omega}=O(n^2\gamma^2/2\pi\mu). \tag{26}$$

The approximately equal sign in these equations should be understood to mean that the ratio of the expressions on the two sides approaches unity as $\tan\gamma/\sin\omega$ approaches zero.

Comparing (22) with (26), we find that the rotation $\bar{\omega}$ is subject to the condition

$$\bar{\omega}\ll\tfrac{1}{2}\pi\mu. \tag{27}$$

We see from this relation that in order to obtain a rotation which is not infinitesimal, it is necessary that $\mu\gg 1$—that is to say, that the axes of the plates be distributed uniformly through a large number of semicircles.

It should be noted that whereas the orientations of the plates advance in the clockwise direction, the angle of the equivalent rotator is in the counter clockwise direction if $\cot \omega$ is positive.

Rotator Formed by Three Retardation Plates

The considerations presented so far in this part had at one time considerable theoretical interest in connection with the theory of crystal structure.[4] If one wishes to form a rotator from retardation plates alone, however, it may be done in a much simpler manner than might be inferred from the foregoing. In fact, it is possible to produce an optical system which behaves rigorously as a rotator with any given $\bar{\omega}$, by the use of only three retardation plates.

We may produce a rotator by inserting a retardation plate between two crossed quarter-wave plates in such a position that its axes are at 45° with respect to the axes of the quarter-wave plates. Let us write the matrix defined by Eq. (25, I) as $\mathbf{G}(\gamma)$. The statement we have just made then follows directly from the identity

$$\mathbf{S}(ab\gamma) \equiv \mathbf{G}(a\pi/4)[\mathbf{S}(b\pi/4)\mathbf{G}(\gamma) \times \mathbf{S}(-b\pi/4)]\mathbf{G}(-a\pi/4), \quad (28)$$

where

$$a = \pm 1, \quad b = \pm 1.$$

Since the relative phase retardation of the middle plate is 2γ, we see that the angle of the rotation is one-half of the phase retardation of the inserted plate.

By the use of two achromatic quarter-wave plates, it is possible in this manner to use a Soleil compensator as a variable rotator.[5]

There exists another theorem which may be considered as the converse of the one just proved. We may produce a retardation plate of any given retardation from two quarter-wave plates and a rotator. The rotator should be placed between the two crossed quarter-wave plates. The optical system then behaves as a retardation plate with its axes at 45° with those of the quarter-wave plates. This statement follows directly from the identity

$$\mathbf{G}(ab\bar{\omega}) \equiv \mathbf{S}(a\pi/4)[\mathbf{G}(b\pi/4)\mathbf{S}(\bar{\omega}) \times \mathbf{G}(-b\pi/4)]\mathbf{S}(-a\pi/4), \quad (29)$$

where

$$a = \pm 1, \quad b = \pm 1.$$

We see that the phase retardation of the optical system is twice the angle of the rotator.

With two achromatic quarter-wave plates, it is thus possible to use a variable rotator as a Soleil compensator.

According to Billet,[6] the physical statements corresponding to the identities (28) and (29) were known to Fresnel.

Conclusion

Part IV of this series, which is now in preparation, will contain, in addition to emendations to the general theory, derivations of the matrices which represent (1) plates or solutions exhibiting circular birefringence or circular dichroism, and (2) plates cut from crystals of such low symmetry (monoclinic and triclinic) that the principal axes of absorption are not parallel to the principal axes of refraction.

The writer is indebted to Mr. Edwin H. Land, President and Research Director of the Polaroid Corporation, for suggesting the problem of developing an operational calculus of the type presented here. He wishes to thank Dr. Cutler D. West, also of the Polaroid Corporation, for providing all of the references to the older optical literature, and he is indebted to his co-author of Part II, Dr. Henry Hurwitz, Jr., for helpful suggestions about Parts I and III. The author wishes to express his appreciation to Professor M. S. Vallarta, Department of Physics, Massachusetts Institute of Technology, for reading the first draft of the manuscripts and for making helpful suggestions.

[4] L. Sohncke, *Entwickelung einer Theorie der Krystallstructur* (Teubner, Leipzig, 1879). See particularly pp. 238–246.

[5] T. M. Lowry, *Optical Rotatory Power* (Longmans Green, New York, 1935), p. 194, Fig. 79(a).

[6] M. F. Billet, *Traité D'optique Physique* (Paris, 1859), Vol. 2, Chapter 20, Art. 1, pp. 328–351.

19

Reprinted from *J. Opt. Soc. Am.*, **32**, 486–493 (Aug. 1942)

A New Calculus for the Treatment of Optical Systems. IV.

R. Clark Jones†

Research Laboratory, Polaroid Corporation, Cambridge, Massachusetts, and Research Laboratory of Physics, Harvard University, Cambridge, Massachusetts

(Received May 1, 1942)

Part IV is divided into two sections. The first is devoted to some additions to the general theory developed in Part I, and the second section to the derivation of the matrices representing two optical elements which were not treated in Parts II and III: (1) plates possessing circular dichroism, and (2) plates cut from crystals of such low symmetry that the principal axes of absorption and refraction are not parallel. In case (2), the discussion is limited to monoclinic and triclinic crystals which do not possess optical activity.

INTRODUCTION

In the first three parts of this series which appeared together in a recent issue[1] of this Journal, the writer presented the outlines of a new calculus for the treatment of the type of optical system which contains plane-parallel plates of crystalline or other anisotropic material. The usefulness and power of the calculus was demonstrated by the applications in Parts II and III.

The purpose of the present paper is to make some extensions of the general theory which were suggested by members of the Research Laboratory of the Polaroid Corporation, and also to present further applications of the theory: we here derive the matrices representing (1) plates possessing circular dichroism and (2) plates cut from monoclinic and triclinic crystals.

EMENDATIONS TO THE GENERAL THEORY

Eigenvectors of a Matrix

In the case of a partial polarizer or a retardation plate, we know that plane polarized light with its plane parallel to either of the principal axes is transmitted through the plate without change in the state of polarization, although there may be a change in the intensity of the light. Similarly, right-handed and left-handed circularly polarized light suffer no change in their state of polarization in passing through a rotator or through a crystal exhibiting circular dichroism. In terms of the general theory developed in Part I, we see that in each of these cases just mentioned, the vector $\mathcal{E}$ representing the emergent light is equal to the vector* representing the entering light multiplied by some (in general complex) constant. A vector which has this property with respect to a given optical system or with respect to the matrix representing it, will be termed an *eigenvector* of the system or matrix.

The questions now arise, does every optical system have just two eigenvectors, and how may the eigenvectors of a given optical system be determined? We proceed to answer these questions.

As in I, let $\mathbf{M}^{(n)}$ be the matrix of the optical system. Then the condition that $\mathcal{E}_{(i)}$ be an eigenvector of $\mathbf{M}^{(n)}$ is

$$\mathbf{M}^{(n)}\mathcal{E}_{(i)}=d_i\mathcal{E}_{(i)}, \tag{1}$$

where d_i is a constant, at present undetermined, which we shall term the *eigenvalue* corresponding to the eigenvector $\mathcal{E}_{(i)}$.

The condition that (1) possess a non-vanishing solution for the two components of $\mathcal{E}_{(i)}$ is

$$\det[\mathbf{M}^{(n)}-d_i]=0 \tag{2}$$

or

$$(m_1-d_i)(m_2-d_i)=m_3m_4. \tag{3}$$

Let the two solutions of (3) be d_1 and d_2. Then from (1) we find that the corresponding eigenvectors are

$$\mathcal{E}_{(1)}\equiv\begin{pmatrix}a_1\\ b_1\end{pmatrix}\doteq\begin{pmatrix}m_1-d_2\\ m_3\end{pmatrix}\doteq\begin{pmatrix}m_4\\ m_2-d_2\end{pmatrix}$$
$$\doteq\begin{pmatrix}d_1-m_2\\ m_3\end{pmatrix}\doteq\begin{pmatrix}m_4\\ d_1-m_1\end{pmatrix} \tag{4}$$

† Part of the work reported here was done while the writer held a National Scholarship at Harvard University; he is now a Member of the Technical Staff, Bell Telephone Laboratories, Inc., Murray Hill, New Jersey.

* The light vector $\mathcal{E}$ was mistakenly printed as a lower case epsilon in Parts I, II, and III. The error is here corrected.

and

$$\mathcal{E}_{(2)} \equiv \begin{pmatrix} a_2 \\ b_2 \end{pmatrix} \doteq \begin{pmatrix} m_1 - d_1 \\ m_3 \end{pmatrix} \doteq \begin{pmatrix} m_4 \\ m_2 - d_1 \end{pmatrix}$$
$$\doteq \begin{pmatrix} d_2 - m_2 \\ m_3 \end{pmatrix} \doteq \begin{pmatrix} m_4 \\ d_2 - m_1 \end{pmatrix}, \quad (5)$$

where the symbol $\doteq$ should be read "equal except for a constant factor." It is, of course, not possible to define the eigenvector uniquely, since if $\mathcal{E}_{(i)}$ is an eigenvector, then so is any multiple of $\mathcal{E}_{(i)}$.

Up to this point, we have found that every optical system has, at most, two eigenvectors, and we have found general expressions for these eigenvectors. We must now determine whether $\mathcal{E}_{(1)}$ and $\mathcal{E}_{(2)}$ are linearly independent, since if they are merely multiples of one another, they are essentially the same eigenvector.

We shall now show that the necessary and sufficient condition for $\mathbf{M}^{(n)}$ to possess two and only two** linearly independent eigenvectors, is that $\mathbf{M}^{(n)}$ be diagonalizable—that is to say, that a matrix $\mathbf{T}$ exists such that

$$\mathbf{T}^{-1}\mathbf{M}^{(n)}\mathbf{T} = \mathbf{D} \equiv \begin{pmatrix} d_1 & 0 \\ 0 & d_2 \end{pmatrix}. \quad (6)$$

We first point out the obvious fact that the linear independence of $\mathcal{E}_{(1)}$ and $\mathcal{E}_{(2)}$ is equivalent to the condition that

$$\Delta \equiv a_1 b_2 - a_2 b_1 \quad (7)$$

be different from zero. We now define the matrices

$$\mathbf{T} \equiv \begin{pmatrix} a_1 & a_2 \\ b_1 & b_2 \end{pmatrix}; \quad \mathbf{T}^{-1} \equiv \Delta^{-1} \begin{pmatrix} b_1 & -a_2 \\ -b_1 & a_1 \end{pmatrix} \quad (8)$$

and by actual substitution of (4), (5), and (8) in (6), we find that (6) is satisfied. We have thus found a matrix which diagonalizes $\mathbf{M}^{(n)}$. Although the condition (6) does not determine $\mathbf{T}$ uniquely, detailed examination confirms what is suggested by (8): when Δ is zero, no matrix $\mathbf{T}$ exists which satisfies (8). We have thus proved the assertion at the beginning of this paragraph.

** In making this statement we are ignoring the degenerate case in which every vector is an eigenvector of the matrix, as is the case with the matrix representing an isotropic crystal.

The Matrix with Given Eigenvectors and Given Eigenvalues

Suppose that a given optical system is found by experiment to have the eigenvectors

$$\mathcal{E}_{(1)} \equiv \begin{pmatrix} a_1 \\ b_1 \end{pmatrix}; \quad \mathcal{E}_{(2)} \equiv \begin{pmatrix} a_2 \\ b_2 \end{pmatrix} \quad (9)$$

with the corresponding eigenvalues d_1 and d_2. The matrix of the optical system is then uniquely determined, and is given by

$$\mathbf{M}^{(n)} = \mathbf{TDT}^{-1}$$
$$\equiv \Delta^{-1} \begin{pmatrix} d_1 a_1 b_2 - d_2 a_2 b_1 & -(d_1 - d_2) a_1 a_2 \\ (d_1 - d_2) b_1 b_2 & d_2 a_1 b_2 - d_1 a_2 b_1 \end{pmatrix}. \quad (10)$$

This statement may be confirmed by subjecting the matrix (10) to the analysis of the last section.

Other Representations of the Light Vector

In the development of the general theory in Part I, the state of polarization of the light was expressed by stating the amplitudes and phases of the two Cartesian components of the plane wave. That is to say, the light vector was expressed as a linear combination of the two *basis vectors*

$$\mathcal{E}_{(1)} \equiv \begin{pmatrix} 1 \\ 0 \end{pmatrix}; \quad \mathcal{E}_{(2)} \equiv \begin{pmatrix} 0 \\ 1 \end{pmatrix} \quad (11)$$

in the manner

$$\mathcal{E} \equiv \begin{pmatrix} E_x \\ E_y \end{pmatrix} \equiv E_x \mathcal{E}_{(1)} + E_y \mathcal{E}_{(2)}. \quad (12)$$

The writer is indebted to Dr. Martin Grabau for pointing out that it is sometimes expedient to use basis vectors which are not Cartesian—for example, in dealing with optical systems in which most of the elements are rotators or crystals exhibiting circular dichroism, it is clearly more convenient to express the light vector as the sum of its two circularly polarized components.

We shall now develop some of the relations which hold when the basis vectors are chosen arbitrarily. Let $\mathcal{E}_{(1)}$ and $\mathcal{E}_{(2)}$ be two arbitrary and linearly independent vectors, with Cartesian components as defined in (9). For the time being, the vectors $\mathcal{E}_{(1)}$ and $\mathcal{E}_{(2)}$ are *not* to be thought of as eigenvectors. Let us define a new light vector in the manner

$$\mathfrak{F} \equiv \begin{pmatrix} F_1 \\ F_2 \end{pmatrix} \equiv F_1 \mathcal{E}_{(1)} + F_2 \mathcal{E}_{(2)}. \qquad (13)$$

Evidently, the vector $\mathfrak{F}$ is defined only with respect to some given pair of basis vectors.

From (11), (12), and (13), we find that the relations between $\mathcal{E}$ and $\mathfrak{F}$ are

$$\mathcal{E} = \mathbf{T}\mathfrak{F}; \quad \mathfrak{F} = \mathbf{T}^{-1}\mathcal{E}, \qquad (14)$$

where $\mathbf{T}$ is defined by (8). If $\mathbf{M}^{(n)}$ is the matrix of an optical system in the $\mathcal{E}$-representation:

$$\mathcal{E}_n = \mathbf{M}^{(n)} \mathcal{E}_0, \qquad (15)$$

then $\mathbf{T}^{-1}\mathbf{M}^{(n)}\mathbf{T}$ is the matrix in the $\mathfrak{F}$-representation:

$$\mathfrak{F}_n = \mathbf{T}^{-1}\mathbf{M}^{(n)}\mathbf{T}\mathfrak{F}_0. \qquad (16)$$

Conversely, if $\mathbf{M}^{(n)}$ is the matrix in the $\mathfrak{F}$-representation, then $\mathbf{T}\mathbf{M}^{(n)}\mathbf{T}^{-1}$ is the matrix in the $\mathcal{E}$-representation.

If now we use as basis vectors the eigenvectors of $\mathbf{M}^{(n)}$, where $\mathbf{M}^{(n)}$ is the matrix of an optical system in the $\mathcal{E}$-representation, then the matrix of the optical system in the $\mathfrak{F}$-representation is the diagonal matrix $\mathbf{D}$, as we may prove by comparing (16) with (6). It is this fact which sometimes makes it convenient to use as basis vectors the eigenvectors of the optical system or the eigenvectors of some element of the system.

At this point, it is desirable to introduce a few definitions. A vector $\mathcal{E}_{(1)}$ is *normal* if

$$\bar{\mathcal{E}}_{(1)}\mathcal{E}_{(1)}^* \equiv aa^* + bb^* = 1 \qquad (17)$$

and two vectors are *orthogonal* with respect to one another if

$$\bar{\mathcal{E}}_{(1)}\mathcal{E}_{(2)}^* \equiv a_1a_2^* + b_1b_2^* = 0. \qquad (18)$$

Two vectors are *orthonormal* if each of them is normal, and if they are orthogonal. The condition that the eigenvectors of a matrix be normal involves no loss in generality, but the requirement that they be orthogonal places a restriction on the nature of the matrix. A sufficient condition that the eigenvectors of $\mathbf{M}$ be orthogonal is that $\mathbf{M}$ be either Hermitian or unitary, and a necessary condition is that the absolute values of m_3 and m_4 be equal. The necessary *and* sufficient condition that the eigenvectors of $\mathbf{M}$ be orthogonal is that $\mathbf{M}$ commute with its Hermitian adjoint:[3]

$$\mathbf{M}\mathbf{M}^\dagger = \mathbf{M}^\dagger\mathbf{M}.$$

It was pointed out in Part I that the intensity, of the light is proportional to $\bar{\mathcal{E}}\mathcal{E}^*$. If and only if the basis vectors of the $\mathfrak{F}$-representation are orthonormal, however, do we have

$$\bar{\mathcal{E}}\mathcal{E}^* = \bar{\mathfrak{F}}\mathfrak{F}^*. \qquad (19)$$

Equation (19) is a special case of Plancherel's theorem.[2] More generally, if $\mathcal{E}_j$ and $\mathcal{E}_k$ are any two vectors and if the basis vectors are orthonormal, we have

$$\bar{\mathcal{E}}_j\mathcal{E}_k^* = \bar{\mathfrak{F}}_j\mathfrak{F}_k^*. \qquad (20)$$

If the basis vectors are orthonormal, it is no real loss in generality to take them to be

$$\mathcal{E}_{(1)} \equiv \begin{pmatrix} a_1 \\ b_1 \end{pmatrix}; \quad \mathcal{E}_{(2)} \equiv \begin{pmatrix} -b_1^* \\ a_1^* \end{pmatrix} \qquad (21)$$

with

$$a_1a_1^* + b_1b_1^* = 1, \qquad (22)$$

in which case Δ is equal to unity.

Partially Polarized Light

Up to this point, it has always been assumed that the light is in a definite state of polarization. Nearly all of the ordinary sources of light, however, yield light which is almost completely unpolarized. In the following treatment of partially polarized light, it will be assumed that light is passing through the optical system in only one direction, so that it is proper to infer that the intensity, or energy flux vector, is parallel to the axis of the system, and that its length is proportional to the square of the length of the light vector.

Unpolarized light of unit intensity may be described as the combination of two linearly polarized components, one parallel to the x axis, and the other parallel to the y axis, each of intensity one-half, and with the added stipulation that all values of the relative phase of the two components are equally likely, so that, when one computes the total intensity, it is necessary to average over the relative phase. There is, however, nothing sacred about light polarized along the axes of the coordinate system, nor is there anything sacred about linearly polarized light. More generally, unpolarized light of unit intensity may be regarded as the superposition of any two orthogonal components, each of intensity one-half, and with all values of the relative phase equally likely.

In the following discussion we make the assumption, convenient for ease of presentation, that a normal light vector has an intensity of one-half. Such a correspondence may always be obtained by suitable choice of the units in which energy and the electric vector are expressed. Let $\mathcal{E}_{01}$ and $\mathcal{E}_{02}$ be the two orthonormal components in terms of which unpolarized light of unit intensity is described. Then, after passing through an optical system with the matrix $\mathbf{M}^{(n)}$, the corresponding components of the emergent light will be

$$\mathcal{E}_{n1}=\mathbf{M}^{(n)}\mathcal{E}_{01} \quad \text{and} \quad \mathcal{E}_{n2}=\mathbf{M}^{(n)}\mathcal{E}_{02}. \tag{23}$$

In general, $\mathcal{E}_{n1}$ and $\mathcal{E}_{n2}$ will be neither normal nor orthogonal. The emergent light may be described as the combination of the vectors $\mathcal{E}_{n1}$ and $\mathcal{E}_{n2}$, with the understanding that all values of the relative phase are equally likely. If one wishes to know the intensity of the transmitted light, one simply finds the intensity of $\mathcal{E}_{n1}$ and of $\mathcal{E}_{n2}$, and adds these two intensities. Or, more generally, if one wishes to know the intensity of a specified component in the emergent light, one finds the intensity of the given component in $\mathcal{E}_{n1}$ and $\mathcal{E}_{n2}$, and adds.

The result of computing the intensity of the transmitted light, or the intensity of any component of it, is, as it must be, independent of the arbitrary choice of the orthonormal vectors $\mathcal{E}_{01}$ and $\mathcal{E}_{02}$. The intensity transmission factor for initially unpolarized light of an optical system with the matrix $\mathbf{M}^{(n)}$ is easily shown to be

$$(|\mathbf{M}^{(n)}\mathcal{E}_{01}|^2+|\mathbf{M}^{(n)}\mathcal{E}_{02}|^2) =\tfrac{1}{2}(|m_1|^2+|m_2|^2+|m_3|^2+|m_4|^2). \tag{24}$$

If $\mathcal{E}_{n1}$ and $\mathcal{E}_{n2}$ have the same absolute value and if they are orthogonal, the emergent light is also unpolarized. If they differ in intensity, or if they are not orthogonal, the emergent light is said to be *partially polarized*. More specially, the emergent light is *completely polarized* if either $\mathcal{E}_{n1}$ or $\mathcal{E}_{n2}$ is zero, or if they are linearly dependent.

Further Discussion of Γ

In Part II (p. 496), we introduced a quantity $\Gamma(\mathbf{M})$, which was defined as the maximum value of $|\mathbf{M}\mathcal{E}|/|\mathcal{E}|$ with respect to variations of the two components of $\mathcal{E}$. We see that Γ^2 is equal to the maximum value of the intensity transmission factor of the optical system represented by $\mathbf{M}$. If we denote the intensity transmission factor by T, we have

$$T\equiv|\mathbf{M}\mathcal{E}|^2/|\mathcal{E}|^2. \tag{25}$$

In the following paragraphs, we shall find explicit expressions for the maximum value of T, and also for the minimum value, for any matrix $\mathbf{M}$.

For the present purpose, it is no loss of generality to assume that $\mathcal{E}$ has the following normal form:

$$\mathcal{E}=\begin{pmatrix} e^{i\gamma}\sin\theta \\ e^{-i\gamma}\cos\theta \end{pmatrix}. \tag{26}$$

We now find that T is given explicitly by the following expression:

$$\begin{aligned} T\equiv|\mathbf{M}\mathcal{E}|^2=&(|m_1|^2+|m_3|^2)\sin^2\theta \\ &+(|m_2|^2+|m_4|^2)\cos^2\theta \\ &+[(m_1m_4^*+m_2^*m_3)e^{2i\gamma} \\ &+(m_1^*m_4+m_2m_3^*)e^{-2i\gamma}]\sin\theta\cos\theta. \end{aligned} \tag{27}$$

In order to find the maximum value of T, and thus the value of Γ, we must assign to γ and θ the values which make T stationary with respect to variations in γ and θ. These values are found to be

$$\gamma=\frac{1}{2}\arg(m_1^*m_4+m_2m_3^*), \tag{28}$$

$$\tan 2\theta=\frac{2|m_1^*m_4+m_2m_3^*|}{|m_2|^2+|m_4|^2-|m_1|^2-|m_3|^2}, \tag{29}$$

whence

$$\begin{aligned} T_{\text{stat}}=&\tfrac{1}{2}(|m_1|^2+|m_2|^2+|m_3|^2+|m_4|^2 \\ &+\tfrac{1}{2}(|m_2|^2+|m_4|^2-|m_1|^2-|m_2|^2)\sec 2\theta, \end{aligned} \tag{30}$$

where θ is understood to have the value (29), or

$$\begin{aligned} T_{\text{stat}}=&\tfrac{1}{2}(|m_1|^2+|m_2|^2+|m_3|^2+|m_4|^2) \\ &\pm\tfrac{1}{2}[(|m_2|^2+|m_4|^2-|m_1|^2-|m_3|^2)^2 \\ &+4|m_1m_4^*+m_2^*m_3|^2]^{\frac{1}{2}} \end{aligned} \tag{31}$$

if θ is eliminated.

It is clear the $T_{\max}$ is to be associated with the plus sign in (31). We then have $\Gamma=(T_{\max})^{\frac{1}{2}}$.

By (29), we see that the two values of θ which yield the maximum and minimum values of T differ by 90°, and from this fact it follows at once that the vector $\mathcal{E}_{max}$ which maximizes T is orthogonal to the vector $\mathcal{E}_{min}$ which minimizes T. It should be noted, however, that in general $\mathcal{E}_{max}$ and $\mathcal{E}_{min}$ are *not* eigenvectors of the matrix **M**; if and only if the matrix possesses orthogonal eigenvectors will $\mathcal{E}_{max}$ and $\mathcal{E}_{min}$ be the eigenvectors of the matrix.

It is interesting to note, by comparison of (31) with (24), that the arithmetic mean of T_{max} and T_{min} is equal to the transmission factor for unpolarized light.

It has been pointed out earlier in this part (cf. paragraph following Eq. (16)) that if one uses as basis vectors the eigenvectors of **M**, then in the $\mathfrak{F}$-representation, the matrix **M** becomes diagonal, with the diagonal elements d_1 and d_2. The reader should be warned against the error of concluding from this fact, that $\Gamma(\mathbf{M})$ is equal to the larger of $|d_1|$ and $|d_2|$. It follows from the last two paragraphs that $\Gamma(\mathbf{M})$ is equal to the larger of $|d_1|$ and $|d_2|$ if and only if the eigenvectors of **M** are orthogonal.

APPLICATIONS

Circular Birefringence and Circular Dichroism

Fresnel's phenomenological explanation of the optical activity of quartz was that the indices of refraction of quartz were different for right and left circularly polarized light. It is evident physically that such a difference will produce a rotation of the plane of linearly polarized light.

In Part I we found that the matrix of a rotator was $\mathbf{S}(\tilde{\omega})$ by making use of the fact that a rotator produced a rotation of the plane of linearly polarized light which was independent of the angle of the plane of polarization. It is interesting to derive this matrix also by the use of Fresnel's assumption. By assumption, the normal eigenvectors of a rotator are

$$\mathcal{E}_{(1)} = 2^{-\frac{1}{2}}\begin{pmatrix} 1 \\ i \end{pmatrix}, \quad \mathcal{E}_{(2)} = 2^{-\frac{1}{2}}\begin{pmatrix} -1 \\ i \end{pmatrix} \tag{32}$$

and we shall indicate the relative phase retardation of $2\tilde{\omega}$ by writing

$$d_1 = e^{i\tilde{\omega}}, \quad d_2 = e^{-i\tilde{\omega}}. \tag{33}$$

By substituting these expressions in (10), we find at once that the matrix of a rotator is

$$\begin{pmatrix} \cos\tilde{\omega} & -\sin\tilde{\omega} \\ \sin\tilde{\omega} & \cos\tilde{\omega} \end{pmatrix} \equiv \mathbf{S}(\tilde{\omega}). \tag{34}$$

The fact that we obtain the same matrix by considering a rotator *qua* a rotator, and as a crystal possessing circular birefringence, provides a mathematical proof of the physical equivalence of the two points of view.

Circular dichroism differs from circular birefringence only in that the two components are differentially absorbed, rather than differentially retarded. In the case of pure circular dichroism, therefore, we take the eigenvalues to be real. From (10) and (32), we then find for the matrix

$$\begin{pmatrix} \frac{1}{2}(d_1+d_2) & \frac{1}{2}i(d_1-d_2) \\ -\frac{1}{2}i(d_1-d_2) & \frac{1}{2}(d_1+d_2) \end{pmatrix}. \tag{35}$$

Circular dichroism is of interest primarily in connection with liquids, although a measurement of its magnitude in crystalline nickel sulphate has been reported recently.[4]

Elliptical Birefringence and Elliptical Dichroism

The treatment of elliptical birefringence and dichroism is exactly analogous to the foregoing treatment. If the eigenvectors are orthogonal, the matrix for the case of elliptical birefringence turns out to be the general unitary matrix, and we have seen that this matrix is the most general matrix which one encounters with optical elements which have no absorption. A matrix with the eigenvalues (33) and with non-orthogonal eigenvectors has a value of Γ which is greater than unity; such a matrix is therefore not physically realizable.

Crystals of Low Symmetry: Monoclinic and Triclinic

Monoclinic and triclinic crystals are crystals of such low symmetry that the principal axes of absorption are not parallel to the principal axes of refraction. In this section we shall derive by a new approach the matrix which corresponds to a crystalline plate cut from such crystals. Our treatment will be restricted to those crystals which do not possess optical activity—that is to say, to those of the crystal classes C_s and C_{2h} (monoclinic), and C_i (triclinic).

Our explicit treatment will be restricted to the case in which the crystalline plate is cut from a monoclinic crystal with the symmetry axis normal to the surface of the plate. At the conclusion of the derivation, however, we shall show that the results are also valid for triclinic crystals.

The general equations for the propagation of plane waves of light through triclinic crystals have been developed by the use of standard electromagnetic theory.[5] The derivation by this means is entirely rigorous, but it is tedious, and the resulting equations, in the general case, are complicated in form and yield little or no physical insight.

The present derivation of the matrix representing a monoclinic crystal without optical activity and with its axis normal to the surface of the plate is based on the observation that such a crystal is physically equivalent to an optical system whose odd numbered elements are identical partial polarizers each with the same orientation ω_P and whose even numbered elements are identical retardation plates with the orientation ω_G, in the limit in which the number of plates becomes infinite and their thickness zero with the total thickness of the optical system remaining constant. Indeed, the writer finds that this method of representing a monoclinic crystal is the only one which provides him any physical insight into its optical behavior.

In the following derivation, we shall first determine the matrix corresponding to a finite number of pairs of partial polarizers and retardation plates, and we shall then pass to the indicated limit.

The matrix of a retardation plate with the orientation ω_G is

$$\mathbf{S}(\omega_G)\mathbf{G}\mathbf{S}(-\omega_G), \tag{36}$$

where

$$\mathbf{G} \equiv \begin{pmatrix} e^{i\gamma} & 0 \\ 0 & e^{-i\gamma} \end{pmatrix}. \tag{37}$$

Similarly, the matrix of a partial polarizer with the orientation ω_P is

$$\mathbf{S}(\omega_P)\mathbf{P}\mathbf{S}(-\omega_P), \tag{38}$$

where

$$\mathbf{P} \equiv \begin{pmatrix} p_1 & 0 \\ 0 & p_2 \end{pmatrix} = \begin{pmatrix} \kappa e^{\xi} & 0 \\ 0 & \kappa e^{-\xi} \end{pmatrix} \tag{39}$$

$$\left.\begin{aligned} &0 \leq p_1 \leq 1, \\ &0 \leq p_2 \leq 1, \\ &\kappa = (p_1 p_2)^{\frac{1}{2}}, \\ &\xi = \tfrac{1}{2}\log(p_1/p_2). \end{aligned}\right\} \tag{40}$$

If we suppose that the light passes first through the partial polarizer, the matrix of the combination of the partial polarizer and retardation plate is then

$$\mathbf{B} \equiv \mathbf{S}(\omega_G)\mathbf{G}\mathbf{S}(-\omega_G)\mathbf{S}(\omega_P)\mathbf{P}\mathbf{S}(-\omega_P) \tag{41}$$

and the matrix representing the optical system composed of n such pairs, one after the other, is $\mathbf{B}^n$, according to Eq. (15, I).

It will materially reduce the algebraic complexity of the resulting equations, however, if instead of determining $\mathbf{B}^n$ directly, we first determine the nth power of

$$\mathbf{A} \equiv \mathbf{S}(-\omega_G)\mathbf{B}\mathbf{S}(\omega_G) = \mathbf{G}\mathbf{S}(\omega)\mathbf{P}\mathbf{S}(-\omega), \tag{42}$$

where

$$\omega \equiv \omega_P - \omega_G \tag{43}$$

and then determine $\mathbf{B}^n$ from $\mathbf{A}^n$ by the simple relation

$$\mathbf{B}^n = \mathbf{S}(\omega_G)\mathbf{A}^n\mathbf{S}(-\omega_G). \tag{44}$$

The physical significance of this mathematical simplification is that we are first rotating our coordinate axes so that they are parallel to the principal axes of the retardation plate; the transformation (44) then corresponds to rotating the coordinate axes back to their original position.

From Eq. (42) we find that the matrix elements of $\mathbf{A}$ are

$$\left.\begin{aligned} a_1 &= \kappa e^{i\gamma}(\cosh\xi + \cos 2\omega \sinh\xi), \\ a_2 &= \kappa e^{-i\gamma}(\cosh\xi - \cos 2\omega \sinh\xi), \\ a_3 &= \kappa e^{-i\gamma}\sin 2\omega \sinh\xi, \\ a_4 &= \kappa e^{i\gamma}\sin 2\omega \sinh\xi. \end{aligned}\right\} \tag{45}$$

In order to obtain the nth power of the matrix $\mathbf{A}$ by the use of Eqs. (8, III), it is first necessary to determine the characteristic values of $\mathbf{A}$. By (3), we find

$$\left.\begin{aligned} d_1, d_2 &= \alpha \pm (\alpha^2-1)^{\frac{1}{2}} \\ &= (1+\beta^2)^{\frac{1}{2}} \pm \beta \\ &= \alpha \pm \beta, \end{aligned}\right\} \quad (46)$$

where

$$\left.\begin{aligned} \alpha &\equiv \cosh \xi \cos \gamma + i \cos 2\omega \sinh \xi \sin \gamma, \\ \beta &\equiv (\alpha^2-1)^{\frac{1}{2}}, \end{aligned}\right\} \quad (47)$$

and where we shall in the following correlate d_1 with the plus sign in (46).

As a convenient abbreviation, we take $\mathbf{A}' \equiv \mathbf{A}^n$. We then have

$$d_1' = (\alpha+\beta)^n, \quad d_2' = (\alpha-\beta)^n \quad (48)$$

and substitution of (45) and (48) in (8, III) yields the explicit expressions for the matrix elements of $\mathbf{A}' \equiv \mathbf{A}^n$.

As the second step in our determination of the matrix representing a monoclinic crystal, we must take the limit of $\mathbf{A}'$ as n becomes infinite under the condition that γ_0, ξ_0, and κ_0 are held constant:

$$\gamma_0 = n\gamma, \quad \xi_0 = n\xi, \quad \kappa_0 = \kappa^n. \quad (49)$$

We find

$$\left.\begin{aligned} \lim a_1' &= \lim (\kappa^n/2\beta)[(a_1-1+\beta)(1+\beta)^n \\ &\qquad -(a_1-1-\beta)(1-\beta)^n] \\ &= \kappa_0(\cosh z + \Psi_1 \sinh z), \\ \lim a_2' &= \kappa_0(\cosh z - \Psi_1 \sinh z), \\ \lim a_3' &= \lim a_4' = \kappa_0 \Psi_2 \sinh z, \end{aligned}\right\} \quad (50)$$

where

$$\left.\begin{aligned} z^2 &\equiv \xi_0^2 - \gamma_0^2 + 2ix_0\gamma_0 \cos 2\omega = \lim n^2\beta, \\ \Psi_1 &\equiv (\xi_0 \cos 2\omega + i\gamma_0)/z, \\ \Psi_2 &\equiv (\xi_0 \sin 2\omega)/z. \end{aligned}\right\} \quad (51)$$

It is easily confirmed that the matrix elements of $\lim \mathbf{A}'$ are independent of which of the two possible values of z is used, as long as the same root is used throughout.

Finally, we obtain the explicit expression for $\mathbf{\Theta}$, the matrix representing a monoclinic crystal whose two pairs of principal axes have the orientations ω_G and ω_P:

$$\begin{aligned} \lim \mathbf{B}^n &= \mathbf{S}(\omega_G)(\lim \mathbf{A}')\mathbf{S}(-\omega_G) \\ &= \mathbf{\Theta} \equiv \begin{pmatrix} \kappa_0(\cosh z + \Gamma_1 \sinh z) & \kappa_0\Gamma_2 \sinh z \\ \kappa_0\Gamma_2 \sinh z & \kappa_0(\cosh z - \Gamma_1 \sinh z) \end{pmatrix} \end{aligned} \quad (52)$$

in which we have used the abbreviations

$$\left.\begin{aligned} z^2 &\equiv \xi_0^2 - \gamma_0^2 + 2i\xi_0\gamma_0 \cos 2(\omega_P - \omega_G), \\ \Gamma_1 &\equiv (\xi_0 \cos 2\omega_P + i\gamma_0 \cos 2\omega_G)/z, \\ \Gamma_2 &\equiv (\xi_0 \sin 2\omega_P + i\gamma_0 \sin 2\omega_G)/z. \end{aligned}\right\} \quad (53)$$

In this case also, it matters not which of the two possible values of z is used.

If the optical system represented by $\mathbf{\Theta}$ has a thickness t, it is clear by comparison with Eq. (2, I) that the difference of the two principal indices of refraction is given by

$$n_{y'} - n_{x'} = \lambda\gamma_0/\pi t \quad (54)$$

and that the two principal extinction coefficients are

$$\left.\begin{aligned} k_{x''} &= \lambda(\log \kappa_0 + \xi_0)/2\pi t, \\ k_{y''} &= \lambda(\log \kappa_0 - \xi_0)/2\pi t. \end{aligned}\right\} \quad (55)$$

These relations serve to evaluate ξ_0, γ_0, and κ_0 in terms of the principal optical coefficients of the crystal.

The two eigenvectors of $\mathbf{\Theta}$ are

$$\mathcal{E}_{(1)} \doteq \begin{pmatrix} 1+\Gamma_1 \\ \Gamma_2 \end{pmatrix} \doteq \begin{pmatrix} \Gamma_2 \\ 1-\Gamma_1 \end{pmatrix} \quad (56)$$

and

$$\mathcal{E}_{(2)} \doteq \begin{pmatrix} \Gamma_1-1 \\ \Gamma_2 \end{pmatrix} \doteq \begin{pmatrix} -\Gamma_2 \\ 1+\Gamma_1 \end{pmatrix}. \quad (57)$$

Because of the identity,

$$\Gamma_1^2 + \Gamma_2^2 \equiv 1 \quad (58)$$

the eigenvectors satisfy the relation $\bar{\mathcal{E}}_{(1)}\mathcal{E}_{(2)} = 0$, but do not satisfy the orthogonality condition $\bar{\mathcal{E}}_{(1)}\mathcal{E}_{(2)}^* = 0$. The eigenvectors of $\mathbf{\Theta}$ are therefore orthogonal if and only if $\mathcal{E}_{(2)}$ and $\mathcal{E}_{(2)}^*$ are linearly dependent, and this is not true in general.

Triclinic Crystals

Up to this point, our detailed considerations have related exclusively to monoclinic crystals.

There is a sense, however, in which the results are also valid for triclinic crystals, as we shall now show.

It is characteristic of all crystals without optical activity that their effect on light passing through them is the same whether the light is being propagated in one direction or in the reverse direction. From a comparison of Eqs. (14) and (23) of Part I, it follows at once that the matrix representing a plate cut from a crystal without optical activity must be symmetrical.

Secondly, it is not difficult to show that *any* symmetrical matrix with a determinant which is real and non-negative may be written in the form $\mathbf{\Theta}$ with the quantities κ_0, γ_0, ξ_0, ω_G, and ω_P all real. The proof is similar to those used in Part II. It therefore follows from this paragraph and the last that the matrix representing any triclinic crystal without optical activity may also be written in the form $\mathbf{\Theta}$.

In the triclinic case, however, the quantities κ_0, γ_0, ξ_0, ω_G, and ω_P are not so directly related to the principal coefficients as in the monoclinic case (cf. Eqs. (54) and (55)). It is convenient to consider these quantities as being those of the monoclinic crystal which is equivalent to the triclinic crystal under discussion. The equivalence is valid only for a plate cut from the triclinic crystal at a given angle.

ACKNOWLEDGMENT

The writer would like to acknowledge the helpful discussion of various members of the Polaroid Corporation. He is particularly indebted to Dr. Cutler D. West and to Dr. Martin Grabau for their helpful suggestions.

REFERENCES

(1) R. Clark Jones, J. Opt. Soc. Am. **31**, 488 (1941); Henry Hurwitz, Jr., and R. Clark Jones, J. Opt. Soc. Am. **31**, 493 (1941); R. Clark Jones, J. Opt. Soc. Am. **31**, 500 (1941).
(2) M. Plancherel, Rend. di Palermo **30**, 289 (1910).
(3) For the definition of $\mathbf{M}^\dagger$, see the footnote on p. 492 in Part I.
(4) L. R. Ingersoll, P. Rudnick, F. G. Slack, and M. Underwood, Phys. Rev. **57**, 1145 (1940).
(5) M. Berek, Fortschr., Mineral. Krist. Petrog. **22**, 1–104 (1937).

20

Reprinted from *J. Opt. Soc. Am.*, **37**, 107–110 (Feb. 1947)

A New Calculus for the Treatment of Optical Systems

V. A More General Formulation, and Description of Another Calculus

R. CLARK JONES
Research Laboratory, Polaroid Corporation, Cambridge, Massachusetts

(Received October 18, 1946)

After a brief review of the nature of the calculus, a formulation is provided for the application of the calculus to a wider class of optical systems than those postulated in the earlier papers of this series. The application is found, however, to be subject to the drastic limitation that the system must not depolarize the light passing through it. Since homogeneous crystalline plates are almost completely free of depolarization, this limitation does not exist in the previously contemplated application to piles of crystalline plates. A more powerful calculus, whose formulation is due to Soleillet and whose development is due to Perrin and Mueller, is described, and is found to be free of this limitation. Its general properties are compared with those of the calculus discussed in this series of papers.

INTRODUCTION

IN 1941 and 1942 there appeared in this Journal a series of four papers[1–4] with the same main title as this one. These four papers, rather heavily burdened with mathematical equations, represented the elaboration of a very simple idea, that of using a two-by-two matrix to represent the effect of a crystalline plate on the state of polarization of the light passing through it.

If one establishes two mutually perpendicular directions, x and y, both of them perpendicular to the direction of propagation of the light, then the polarized light which is incident on the crystal may be described by giving the phases and amplitudes of the x and y components of its electric vector. Both the phases and the amplitudes of these components, denoted by E_x and E_y, may be indicated by permitting E_x and E_y to assume complex values.

It is a physical fact that, for all crystals, the two components E_x' and E_y' of the electric vector of the light emerging from the far side of the crystal are linear functions of the two components of the incident light. Thus one may write:

$$E_x'=m_1E_x+m_4E_y,\quad E_y'=m_3E_x+m_2E_y, \tag{1}$$

where the m's are constants, in general complex, which are characteristic of the crystal. If one introduces the definitions:

$$\mathcal{E}\equiv\begin{pmatrix}E_x\\E_y\end{pmatrix};\quad \mathbf{M}\equiv\begin{pmatrix}m_1 & m_4\\ m_3 & m_2\end{pmatrix}, \tag{2}$$

then Eqs. (1) may be written in the simple form:

$$\mathcal{E}'=\mathbf{M}\mathcal{E}. \tag{3}$$

This purely formal step of introducing a matrix notation turned out to be fruitful of new and interesting results, because of the large bag of tricks which may be used in the manipulation of matrices. Nearly all of the operations and concepts which are familiar to those working in

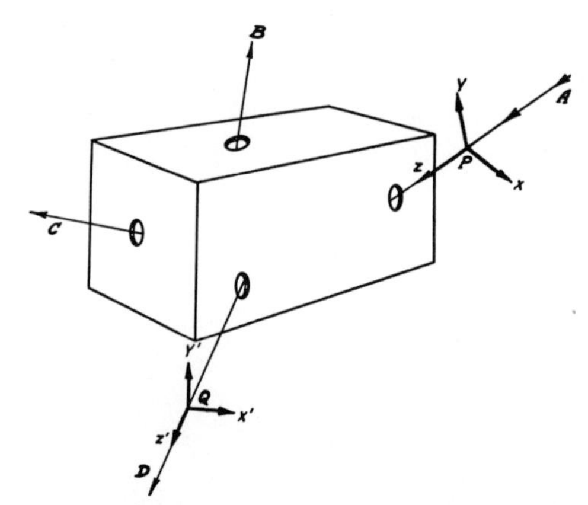

FIG. 1. Showing the box which represents the more general formulation of the calculus contained in this paper. The beam of light A is entering the box, and the beams B, C, and D are emerging from the box. This figure also shows the coordinate systems for the entering and emerging beams described in the text.

[1] I. R. Clark Jones, J. Opt. Soc. Am. **31**, 488–493 (1941).
[2] II. Henry Hurwitz, Jr., and R. Clark Jones, J. Opt. Soc. Am. **31**, 493–499 (1941).
[3] III. R. Clark Jones, J. Opt. Soc. Am. **31**, 500–503 (1941).
[4] IV. R. Clark Jones, J. Opt. Soc. Am. **32**, 486–493 (1942).

quantum mechanics are useful in relation to optical systems: characteristic values and characteristic vectors, coordinate transformation, change of basis vectors, and diagonalization, for example. Indeed the only concept prominent in quantum mechanics that is not useful is that of the Hermitian matrix. Unitary matrices, on the other hand, turn out to be co-extensive with the class of non-absorbing crystals.

A MORE GENERAL FORMULATION

Now it is possible to generalize the physical formulation of the calculus, although in so doing the possibility of a limitation arises which did not exist as long as one was concerned only with homogeneous crystals.

Suppose that the box shown in Fig. 1 represents an optical system—any optical system whatever. Furthermore, suppose that when a beam of light A enters the box, it gives rise to a number of emergent beams, indicated as B, C, D. In general, no two of the beams are supposed to lie in the same plane.

At some point P along the beam A, erect a right-handed set of mutually perpendicular axes x, y, and z, with the z axis parallel to the forward wave normal, and at some point Q along any one of the emergent beams, say D, erect an x', y', and z' coordinate system with z' parallel to the forward wave normal. Place a stop around the beam at P and decrease the maximum diameter of its aperture until the wave front is so small in area that it is essentially flat, and of uniform amplitude and state of polarization.

It is assumed that the light which passes through the aperture in the stop at P is completely polarized and monochromatic.

Then the x' and y' components of the electric vector at Q will be linear functions of the x and y components of the electric vector at P, as indicated by Eqs. (1) and (3), provided that both of the following conditions are satisfied.

Condition **I**. The light at Q is completely polarized—that is to say, the optical system within the box does not depolarize the light passing through it.

Condition **II**. The light at Q is monochromatic and of the same frequency as the light at P. This restriction rules out Raman effect and fluorescence within the optical system.

Thus, subject to these two important conditions, it has been possible to generalize the calculus so that it holds not merely for plane-parallel crystalline plates, but for a much wider class of optical systems. The generalization has been accomplished, however, only at the cost of the simplicity of description which characterizes the application to plane-parallel plates of crystalline material.

The Two Conditions

The second of the two conditions is not important in practice, but the first condition is very important, because nearly all actual optical systems depolarize the light passing through them to some extent.

The first condition exists because a simple statement of the electric vector of the light is unable to express a state of polarization which is not complete. In order to specify a state of partial polarization, it is necessary to specify two electric vectors, with the proviso that the light consists of the addition of the two vectors without any correlation between their phases—that is to say, that they be added incoherently.

The calculus is able to treat the passage of partially polarized light through a non-depolarizing optical system in the manner discussed in IV, but it is completely unable to treat systems which serve to depolarize the light which passes through them.

The restriction of the first condition may be avoided altogether by the use of a more powerful calculus which Professor Hans Mueller has developed in considerable detail in a course of lectures during 1945–1946 at the Massachusetts Institute of Technology.

THE MORE POWERFUL CALCULUS

In order to develop a calculus for the treatment of depolarizing optical systems, it is first of all desirable to have a concise means of representing partially polarized light. Such a representation is provided by the Stokes parameters[5] of a beam of light.

Any beam of partially polarized light may be considered as the incoherent superposition of two beams, the first of which is natural (unpolar-

[5] G. G. Stokes, Trans. Camb. Phil. Soc. 9, 399 (1852).

ized) light, and the second of which is completely polarized light. The resolution into the two beams is unique.

The beam of natural light is characterized completely by the r.m.s. amplitude A of its electric vector. The completely polarized beam may be defined by the r.m.s. amplitudes A_x and A_y of the x and y components of the electric vector, and by their relative phase ϕ.

The Stokes parameters of the beam are defined by

$$\begin{aligned} I &\equiv A_x^2+A_y^2+A^2, \\ M &\equiv A_x^2-A_y^2, \\ C &\equiv 2A_xA_y\cos\phi, \\ S &\equiv 2A_xA_y\sin\phi, \end{aligned} \tag{4}$$

and may be arranged to form a vector, the Stokes vector $\mathfrak{S}$, in the manner

$$\mathfrak{S} \equiv \begin{pmatrix} I \\ M \\ C \\ S \end{pmatrix}. \tag{5}$$

The Stokes vector has the important property that, if two separate beams of light are superposed incoherently, then the Stokes vector of the combination is the sum of the Stokes vectors of the two original beams. On the other hand, if the two beams are added coherently, the Stokes vector of the combination is still calculable from the Stokes vectors of the individual beams, but is not obtained by simple addition of the vectors.

On the other hand, the $\mathcal{E}$ vectors are added if the two beams are combined coherently, whereas if the two beams are superposed incoherently, the result is not expressible as an $\mathcal{E}$ vector, since the incoherent combination of two beams generally results in partially polarized light.

The Stokes vector and the $\mathcal{E}$ vector both require the specification of four numbers to determine them. In the case of the Stokes vector, the four numbers are the four Stokes parameters, whereas in the case of the $\mathcal{E}$ vector, the four numbers are the real and imaginary parts of the two components. In either of the vectors, three of the four numbers may be considered as specifying the state of polarization of the light. With the Stokes vector, the remaining number may be considered as the specification of the degree of polarization of the light, whereas, with the $\mathcal{E}$ vector, the degree of polarization is always 100 percent, and the fourth number determines the absolute phase of the light.

In the case of completely polarized light, the Stokes vector can be calculated from the $\mathcal{E}$ vector, but, if the Stokes vector is given, the absolute phase of the $\mathcal{E}$ vector is left undetermined.

The Mueller Matrices

It has been shown by Soleillet[6] that the four components of the vector $\mathfrak{S}'$ which represents the light emerging from any optical system which satisfies condition **II** are linear functions of the four components of the vector $\mathfrak{S}$ which represents the light entering the optical system. Thus one may write

$$\mathfrak{S}' = \boldsymbol{M}\mathfrak{S}, \tag{6}$$

where $\boldsymbol{M}$, a matrix introduced by Mueller, is a four-by-four matrix with 16 real matrix elements.

Thus the Mueller matrices permit one to treat optical systems which depolarize light in a way which is very similar to that in which the two-by-two matrices permit one to treat non-depolarizing systems.

As already mentioned, Professor Mueller has elaborated the use of these four-by-four matrices in a recent course of lectures.[7] He expects to include a comprehensive treatment of their properties and applications in a forthcoming book on polarized light.

Comparison

The Mueller matrices contain 16 coefficients, whereas the two-by-two matrices contain 8 parameters (real and imaginary part of each of the four matrix elements). The two-by-two matrices, however, specify the absolute phase of the optical system, whereas the Mueller matrices do not. Thus the significant comparison is 8 compared with 17 coefficients if the absolute phase is included, or 7 compared with 16 if it is not.

If the optical system does not depolarize, then

[6] Paul Soleillet, Ann. de physique **12**, 23–97 (1929). See also Francis Perrin, J. Chem. Phys. **10**, 415–427 (1942).

[7] The Mueller matrices are used in a declassified report by Hans Mueller, "Memorandum on the polarization optics of the photo-elastic shutter," Report No. 2 of OSRD, project OEMsr-576, November 15, 1943.

9 identities exist among the 16 Mueller coefficients, so that only 7 of them are independent. Under this condition, the Mueller matrix may be computed from the two-by-two matrix, and the two-by-two matrix, except for a phase factor, may be computed from the Mueller matrix. If the optical system does depolarize, however, none of the 9 identities holds in the most general case.

The primary usefulness of both of the calculi lies in treating light which passes through the successive elements of the optical system in a prescribed way. The existence of multiple paths, particularly those provided by interreflections, increases the complexity of the calculation. When multiple paths do exist, the choice of the calculus may depend on whether the multiple beams are combined coherently, or incoherently.

21

Reprinted from *J. Opt. Soc. Am.*, **37**, 110–112 (Feb. 1947)

A New Calculus for the Treatment of Optical Systems

VI. Experimental Determination of the Matrix*

R. Clark Jones
Research Laboratory, Polaroid Corporation, Cambridge, Massachusetts

(Received October 18, 1946)

A simple, straightforward method is described for determining experimentally the matrix of any crystalline plate. The method involves three measurements of the state of polarization of light transmitted by the plate, and also a measurement of its transmission factor for natural light. These measurements determine the matrix uniquely, except for a phase factor whose practical significance is small. The conditions are stated for the application of the method to the more general type of optical system described in V. A statement of the content of future papers in this series is included.

THE PROBLEM

LET us examine the following problem: Suppose that someone has presented us with a plane-parallel slab of crystalline material, and that he wishes to know the two-by-two matrix which represents the crystal for normal incidence at a given wave-length. He has obligingly scratched on one of the two plane surfaces the x and y coordinate axes with respect to which he wishes the matrix stated, and he has specified on which side the incident light is to fall.

A SIMPLE SOLUTION

There are many ways by which the measurement may be done, but the following is probably one of the most simple. The measurement is made in five steps.

1. Let the incident light be linearly polarized with its electric vector parallel to the x axis of the slab, so that E_y is zero. Then the fundamental pair of equations (Eqs. (1) of V)

$$E_x' = m_1 E_x + m_4 E_y, \quad E_y' = m_3 E_x + m_2 E_y, \tag{1}$$

become for this special case:

$$E_x' = m_1 E_x, \quad E_y' = m_3 E_x, \tag{2}$$

or

$$m_1/m_3 = E_x'/E_y', \tag{3}$$

where, in accordance with conditions **I** and **II** of the preceding paper, it is assumed that the emergent light is monochromatic and completely polarized. Now the measurement of the complex ratio E_x'/E_y' is equivalent to the measurement of the state of the polarization of the emergent light, which may be accomplished in a routine manner with a quartz wedge. Thus one may determine the complex ratio $k_1 \equiv m_1/m_3$.

The determination of the state of polarization of the emergent light involves the determination of the axial ratio and sense of rotation of the vibration ellipse, and the determination of the

* Presented at the Thirty-First Annual Meeting of the Optical Society of America, October 3–5, 1946, New York, New York.

azimuth of the major axis of this ellipse. By Eqs. (27) and (28) of I, these determinations may be used to calculate the ratio r of the amplitude of the x component to the amplitude of the y component of the electric vector of the emergent light, and to calculate the phase ϕ of the x component relative to that of the y component. In terms of r and ϕ, the ratio k_1 defined in the last paragraph is equal to $re^{i\phi}$.

2. The measurement is repeated with the electric vector of the incident light parallel to the y axis of the slab. This determines $k_2 \equiv m_4/m_2$.

The slab is now turned over, so that the light passes through it in the opposite direction. If one continues to use the x and y axes which are scratched on the slab, the matrix to be used for the reverse direction of transmission is the transpose of the matrix for the forward direction, so that the equations which replace (1) are:

$$E_x' = m_1E_x + m_3E_y, \quad E_y' = m_4E_x + m_2E_y. \quad (4)$$

3. With the electric vector of the incident light parallel to the x axis of the inverted slab, one now determines $k_3 \equiv m_1/m_4$.

4. With the electric vector parallel to the y axis, one finally determines $k_4 \equiv m_3/m_2$.

Actually, only three of the four measurements indicated above are independent, since one has

$$k_1k_4 \equiv k_2k_3. \quad (5)$$

The existence of this relation does serve, however, as a check on the internal consistency of the measurements. In fact, this relation really involves two independent checks, since both the real and the imaginary parts of the two products must be equal.

By solving for the m's in terms of the k's, one finds that the matrix **M** for the forward direction of transmission is

$$\mathbf{M} = A\begin{pmatrix} k_1k_4 & k_2 \\ k_4 & 1 \end{pmatrix}, \quad (6)$$

where A is a complex constant which is so far undetermined.

The absolute value of A determines the transmission factor of the slab, and is accordingly determined by any single measurement of the transmission factor. From the point of view of simplicity and symmetry, perhaps the most attractive measurement to make is that of the transmission factor of the plate for unpolarized (natural) light. It has been shown in IV that this transmission factor is given by

$$T_{\text{nat}} = \tfrac{1}{2}[|m_1|^2 + |m_2|^2 + |m_3|^2 + |m_4|^2], \quad (7)$$

so that the absolute value of A in terms of T_{nat} and the k's is

$$|A|^2 = \frac{2T_{\text{nat}}}{1 + |k_2|^2 + |k_4|^2 + |k_1|^2|k_4|^2}. \quad (8)$$

5. Thus the absolute value of A may be determined by a measurement of the transmission factor of the slab for natural light.

Actually, however, because of the practical difficulty of obtaining completely unpolarized monochromatic light, it will be more convenient in practice to measure the transmission factors for the incident light which is used in either step **1** or **2**. One easily finds from (6) that the transmission factor for the incident light used in **1** is given by

$$T_1 = |A|^2|k_4|^2(1 + |k_1|^2), \quad (9)$$

and the corresponding transmission factor for step **2**

$$T_2 = |A|^2(1 + |k_2|^2), \quad (10)$$

whence $|A|$ is given by either of the following two expressions

$$|A|^2 = \frac{T_1}{|k_4|^2(1 + |k_1|^2)} = \frac{T_2}{1 + |k_2|^2}, \quad (11)$$

where the quantities on the right are all measured quantities.

This completes the determination of the matrix of the slab for most practical purposes, since the absolute phase of the matrix is only rarely of interest. Throughout the developments in the previous papers of this series, the phase was chosen so that the determinant of the matrix was real and positive. If this is done, the value of A is completely determined.

If, however, one wishes to determine the value of the phase of A, it is necessary to use some method which is in principle capable of measuring the absolute phase retardation of an optical system. One method is to interpose the slab in one of the two paths of an interferometer, and to determine the shift in the fringe pattern. This measurement must be made with light whose state of polarization is that of one of the two eigenvectors of the slab. The eigenvectors of $\mathbf{M}$ may be determined without knowing the value of A because they are independent of the value of A and depend only on the k's.

THE MORE GENERAL PROBLEM

In the case of the more general type of optical system described in V, the same method of solution may be used, provided first that the conditions **I** and **II** are satisfied, so that the calculus is applicable, and provided second that the system satisfies the principle of reciprocity. The principle of reciprocity is assumed in the use of the transpose of **M** as the matrix for the reverse transmission of the light. In practice, any optical system which satisfies conditions **I** and **II** will satisfy the reciprocity principle, provided no magnetic fields are present. Thus the absence of magnetic fields is the primary condition for the applicability of the solution described above to the more general type of optical system.

SOLUTION WHEN RECIPROCITY FAILS

In the event that the optical system does not satisfy the reciprocity principle, it is still possible to obtain a simple solution of the problem. Instead of the measurements numbered **3** and **4**, the following two measurements should be made.

6. With the electric vector of the incident linearly polarized light parallel to the bisector of the angle between the positive x and y axes, so that $E_x=E_y$, one has from Eq. (1)

$$E_x'=(m_1+m_4)E_x, \quad E_y'=(m_3+m_2)E_x. \tag{12}$$

This permits the determination of the complex ratio

$$k_6\equiv\frac{m_1+m_4}{m_3+m_2}.$$

7. Similarly, with the electric vector of the incident light perpendicular to the bisector of the positive x and y axes, so that $E_x=-E_y$, one determines:

$$k_7\equiv\frac{m_1-m_4}{m_3-m_2}.$$

With the aid of a few lines of algebra, one finds that the constant k_4 is related to k_1, k_2, and k_6 or to k_1, k_2, and k_7 by the relations:

$$k_4=\frac{k_6-k_2}{k_1-k_6}=\frac{k_7-k_2}{k_7-k_1}. \tag{13}$$

Just as with the third and fourth measurements, the sixth and seventh measurements are not independent, as indicated by the second equality in Eq. (13).

With the ratio k_4 obtained by Eq. (13), the determination of the matrix is just as described in the previous section. In the case in which reciprocity fails, it is, of course, necessary to repeat measurements **1**, **2**, **5**, and **6** or **7** with the opposite direction of transmission of the light if one wishes to know the matrix for the opposite direction of transmission.

This completes the description of the method used to obtain the matrix when the system does not satisfy the reciprocity principle.

STATEMENT

Future papers in this series will include a discussion of the matrix as a solution of a differential equation, and the relation between the matrix of a crystalline plate and the complex dielectric tensor of the plate.

22

Reprinted from *J. Opt. Soc. Am.*, **38,** 671–685 (Aug. 1948)

A New Calculus for the Treatment of Optical Systems.
VII. Properties of the N-Matrices

R. Clark Jones
Research Laboratory, Polaroid Corporation, Cambridge, Massachusetts
(Received January 2, 1948)

The preceding papers of this series have examined the properties of an optical calculus which represented each of the separate elements of an optical system by means of a single matrix **M**. This paper is concerned with the properties of matrices, denoted by **N**, which refer not to the complete element, but only to a given infinitesimal path length within the element.

If **M** is the matrix of the optical element up to the point z, where z is measured along the light path, then the **N**-matrix at the point z is defined by

$$\mathbf{N} \equiv (d\mathbf{M}/dz)\mathbf{M}^{-1}. \tag{A}$$

Thus one may write symbolically,

$$\mathbf{N} = d\log\mathbf{M}/dz, \tag{B}$$

and

$$\mathbf{M} = \mathbf{M}_0 \exp(\textstyle\int \mathbf{N} dz). \tag{C}$$

A general introduction is contained in Part I. The definition and general properties of the **N**-matrices are treated in Part II. Part III contains a detailed discussion of the important special case in which the optical medium is homogeneous, so that **N** is independent of z; Part III contains in Eq. (3.26) the explicit relation which corresponds to the symbolic relation (C). Part IV describes a systematic method, based on the **N**-matrices, by which the optical properties of the system at each point may be described uniquely and quantitatively as a combination of a certain amount of linear birefringence, a certain amount of circular dichroism, etc.; the method of resolution is indicated in Table I. Part V treats the properties of the inhomogeneous crystal which is obtained by twisting a homogeneous crystal about an axis parallel to the light path.

1.0 INTRODUCTION

THE preceding papers of this series[1] have been concerned with the properties of a calculus for the treatment of optical systems which modify the state of polarization of the light passing through them.

The calculus is based on the fundamental physical fact that the two principal components of the electric vector $\mathcal{E}'$ of the light emerging from an optical element are linear functions of the components of the electric vector $\mathcal{E}$ of the light entering the optical element, subject to certain conditions discussed in Paper V. Thus one may write

$$\mathcal{E}' = \mathbf{M}\mathcal{E}, \tag{1.1}$$

where the electric vector $\mathcal{E}$ is defined by

$$\mathcal{E} \equiv \begin{pmatrix} X \\ Y \end{pmatrix}, \tag{1.2}$$

and where the matrix **M** is defined by

$$\mathbf{M} \equiv \begin{pmatrix} m_1 & m_4 \\ m_3 & m_2 \end{pmatrix}. \tag{1.3}$$

X and Y are the two principal transverse (com-

[1] I.—J. Opt. Soc. Am. **31**, 488–493 (1941); II.—*ibid.*, **31**, 493–499 (1941); III.—*ibid.*, **31**, 500–503 (1941); IV.—*ibid.*, **32**, 486–493 (1942); V.—*ibid.*, **37**, 107–110 (1947); VI.—*ibid.*, **37**, 110–112 (1947).

plex) components of the light, and m_1, m_2, m_3, and m_4 are complex numbers which are functions only of the optical element in question, and do not depend on X or Y.

In order to contrast the procedure used previously with that to be presented in this paper, let it be emphasized that the calculus made no effort to describe the state of polarization at every point in the optical system, but described the electric vector only at a number of places in the system—namely, in the spaces which separated the units designated as the elements of the optical system. This limitation occurred naturally because the fundamental mode of description by use of the matrix operators assigned a single matrix to describe the over-all behavior of the optical element, and provided no description of the optical behavior within subdivisions of the optical element.

This paper is concerned with a more detailed approach which permits the determination of the electric vector at each point along the path of the light. The matrix operator $\mathbf{M}$ of the complete element is represented as a line integral of a matrix $\mathbf{N}$ which is determinate at each point along the path of the light.

For the sake of simplicity of description, the presentation in this paper will be restricted to the case in which the optical element has plane and parallel surfaces, and in which the element is homogeneous in every sufficiently thin, plane and parallel section which is parallel with the surfaces. The formulation will thus include the case of light which passes normally through a crystal which has been twisted (not necessarily uniformly) about the normal to its surfaces and which has been stretched (not necessarily uniformly) along the normal. It will be evident, however, that the procedure may easily be generalized to hold for arbitrarily deformed crystals, and for the more general types of optical elements discussed in Paper V, by the introduction of suitable curvilinear coordinates.

2.0 DEFINITION AND PROPERTIES OF THE N-MATRICES

2.1 Definition of the N-Matrix

The crystal is assumed to have plane and parallel surfaces, and to be two-dimensionally homogeneous within every plane which is parallel with the surfaces. The origin of coordinates is at the surface at which the light enters, and the positive z axis extends into the crystal.

Consider first the matrix $\mathbf{M}_{z,z'}$ which represents the optical properties of a thin layer of the crystal whose surfaces have the coordinates z and z'. One then has by definition

$$\mathcal{E}_{z'} = \mathbf{M}_{z,z'}\mathcal{E}_z. \tag{2.1}$$

The $\mathbf{N}$-matrix at the coordinate z is then defined by

$$\mathbf{N}_z \equiv \lim_{z'=z} \frac{\mathbf{M}_{z,z'} - 1}{z' - z}. \tag{2.2}$$

Now let $\mathbf{M}_z$ represent the matrix of that part of the crystal which lies between the surface at which the light enters and the parallel surface at the depth z:

$$\mathcal{E}_z = \mathbf{M}_z\mathcal{E}_0, \tag{2.3}$$

where $\mathcal{E}_0$ is the electric vector of the light which enters the crystal. The matrix $\mathbf{M}_{z,z'}$ may then be written

$$\mathbf{M}_{z,z'} = \mathbf{M}_{z'}\mathbf{M}_z^{-1}. \tag{2.4}$$

Substitution of Eq. (2.4) in Eq. (2.2) then yields

$$\mathbf{N}_z \equiv \lim_{z'=z} \frac{\mathbf{M}_{z'} - \mathbf{M}_z}{z' - z}\mathbf{M}_z^{-1} \tag{2.5}$$

$$= (d\mathbf{M}_z/dz)\mathbf{M}_z^{-1},$$

or

$$\mathbf{N} \equiv (d\mathbf{M}/dz)\mathbf{M}^{-1}, \tag{2.6}$$

where the subscript z has been omitted because it is no longer needed. It is to be understood that both $\mathbf{M}$ and $\mathbf{N}$ are functions of z.

Equation (2.6) should be considered as the general definition of the matrix $\mathbf{N}$.

By writing Eq. (2.6) in the form

$$d\mathbf{M}/dz = \mathbf{N}\mathbf{M}, \tag{2.7}$$

it is evident that $\mathbf{N}$ is the operator which determines $d\mathbf{M}/dz$ from $\mathbf{M}$. It will be shown in the next section that $\mathbf{N}$ is also the operator which determines $d\mathcal{E}/dz$ from $\mathcal{E}$.

In Paper I of this series, it was shown that crystals representing various distinctive types of optical behavior corresponded to specific forms of the $\mathbf{M}$-matrices, and the various forms were exhibited. In the same way, there is a correspondence between the type of optical behavior and the form of the **N**-matrices. The form of the **N**-matrices for eight different (exhaustive and mutually exclusive) types of optical behavior is shown in Table I.

As a specific example, consider the matrix

$$\mathbf{M}=\mathbf{S}(\tfrac{1}{2}\alpha z^2);\quad \mathbf{M}^{-1}=\mathbf{S}(-\tfrac{1}{2}\alpha z^2), \tag{2.A}$$

where **S** is the rotation matrix defined by Eq. (2.15). Then by differentiation of Eq. (2.A) one finds

$$d\mathbf{M}/dz=\alpha z\mathbf{S}(\tfrac{1}{2}\alpha z^2)\mathbf{S}(\tfrac{1}{2}\pi) \tag{2.B}$$

and, finally, by use of Eqs. (2.6), (2.A), and (2.B) one finds:

$$\mathbf{N}=\begin{pmatrix}0 & -\alpha z\\ \alpha z & 0\end{pmatrix} \tag{2.C}$$

2.2 Differential Equation for $\mathcal{E}$

Another important property of the matrix **N** is that it is the operator which determines $d\mathcal{E}/dz$ from $\mathcal{E}$. The proof follows.

By differentiation, one finds from Eq. (2.3)

$$d\mathcal{E}/dz=(d\mathbf{M}/dz)\mathcal{E}_0. \tag{2.8}$$

By the use of Eq. (2.6), the last relation becomes

$$d\mathcal{E}/dz=\mathbf{N}\mathbf{M}\mathcal{E}_0, \tag{2.9}$$

and by the use of Eq. (2.3) again, one obtains finally

$$d\mathcal{E}/dz=\mathbf{N}\mathcal{E}. \tag{2.10}$$

Accordingly, the **N**-matrix may be considered as a generalized wave number, since the wave number, defined as k in the expression

$$A=A_0\exp(i(\omega t-kz)), \tag{2.11}$$

is the number by which the amplitude is multiplied if the amplitude is differentiated with respect to z:

$$dA/dz=-ikA. \tag{2.12}$$

As a specific example of the use of Eq. (2.10) consider the matrix:

$$\mathbf{M}=\mathbf{S}(\omega z). \tag{2.D}$$

Substitution of Eq. (2.D) in Eq. (2.6) then yields:

$$\mathbf{N}=\omega\mathbf{S}(\tfrac{1}{2}\pi). \tag{2.E}$$

if the expression (2.E) for **N** is then substituted in the vector equation (2.10), this equation becomes equivalent to the following two scalar equations:

$$\begin{aligned} dX/dz&=-\omega Y,\\ dY/dz&=\omega X.\end{aligned} \tag{2.F}$$

Integration of the Eqs. (2.F) subject to the boundary condition that **M** must be equal to the unit matrix when z is zero, then leads back to the expression (2.D) for **M**.

2.3 Rotation of the Optical Element

It will now be shown that the **N**-matrices are transformed upon rotation of the element, in just the same way as the **M**-matrices.

Let a rotation from the positive x axis toward the positive y axis be termed positive. Then upon rotation of the optical element through the positive angle ω, it was shown in Paper I (p. 190) that the matrix **M** is transformed by the relations

$$\mathbf{M}'=\mathbf{S}(\omega)\mathbf{M}\mathbf{S}(-\omega), \tag{2.13}$$

$$\mathbf{M}'^{-1}=\mathbf{S}(\omega)\mathbf{M}^{-1}\mathbf{S}(-\omega), \tag{2.14}$$

where $\mathbf{S}(\omega)$ is the rotation matrix

$$\mathbf{S}(\omega)\equiv\begin{pmatrix}\cos\omega & -\sin\omega\\ \sin\omega & \cos\omega\end{pmatrix}. \tag{2.15}$$

Upon differentiating Eq. (2.13) with respect to z:

$$d\mathbf{M}'/dz=\mathbf{S}(\omega)(d\mathbf{M}/dz)\mathbf{S}(-\omega), \tag{2.16}$$

and then substituting Eqs. (2.14) and (2.16) in Eq. (2.6), one finds directly

$$\mathbf{N}'=(d\mathbf{M}'/dz)\mathbf{M}'^{-1}=\mathbf{S}(\omega)\mathbf{N}\mathbf{S}(-\omega), \tag{2.17}$$

which relation was to be proved.

2.4 Relations between N and M

It is evident from Eq. (2.6) that if the value of **M** and also its derivative with respect to z is known for a single value of z, then the matrix **N** is also known for that value of z. The converse statement, however, is not at all true. In general, it is necessary to know the value of the **N**-matrix for all values of z lying between z_1 and z_2 if one wishes to determine the matrix **M** which corresponds to that part of the crystal which lies between z_1 and z_2. This is a long-winded way of saying that **M** is an integral function of the **N**-matrix.

Thus, whenever **N** has a prescribed dependence on the coordinate z, it is always possible in principle to express **M** as a function of z. Section 3.0 is devoted to the relations between **N** and **M** when **N** is independent of z, as is the case for a homogeneous crystal.

There is a number of further relations between **N** and **M** which hold quite generally. These relations may be derived from Eq. (2.6) and from the definition of the reciprocal matrix.

Because of the relation

$$\mathbf{M}\mathbf{M}^{-1}=1, \tag{2.18}$$

Eq. (2.6) may also be written

$$\mathbf{N}=-\mathbf{M}(d\mathbf{M}^{-1}/dz). \tag{2.19}$$

Then by use of Eqs. (2.6) through (2.19), one may also derive

$$\begin{aligned}\mathbf{N}^2&=-(d\mathbf{M}/dz)(d\mathbf{M}^{-1}/dz)\\&=-\mathbf{M}(d\mathbf{M}^{-1}/dz)(d\mathbf{M}/dz)\mathbf{M}^{-1}\\&=\tfrac{1}{2}[\mathbf{M}(d^2\mathbf{M}^{-1}/dz^2)+(d^2\mathbf{M}/dz^2)\mathbf{M}^{-1}].\end{aligned} \tag{2.20}$$

The following relations will also be useful:

$$\begin{aligned}\mathbf{N}^2+(d\mathbf{N}/dz)&=(d^2\mathbf{M}/dz^2)\mathbf{M}^{-1},\\\mathbf{N}^2-(d\mathbf{N}/dz)&=\mathbf{M}(d^2\mathbf{M}^{-1}/dz^2).\end{aligned} \tag{2.21}$$

It is significant to note that all of the relations between $\mathbf{N}$ and $\mathbf{M}$ which have been developed so far, as well as others to be developed below, are consistent with the following formal relations

$$\mathbf{M}=\exp(\mathbf{N}z)\quad \mathbf{M}^{-1}=\exp(-\mathbf{N}z), \tag{2.23}$$

provided that the derivatives are defined by

$$d\exp(\mathbf{N}z)/dz\equiv\mathbf{N}\exp(\mathbf{N}z), \tag{2.24}$$

$$d\exp(-\mathbf{N}z)/dz\equiv-\exp(-\mathbf{N}z)\mathbf{N}. \tag{2.25}$$

See also Section 3.2 below.

As a specific example of the meaning of the last three equations, consider the matrix $\mathbf{M}$ defined by Eq. (2.A). Substitution of Eqs. (2.A) and (2.C) in Eqs. (2.19)–(2.21) then indicates that the equations hold for the matrix defined by Eq. (2.A); in particular one finds:

$$\mathbf{N}^2=-\alpha^2z^2\mathbf{1}, \tag{2.G}$$

$$\mathbf{N}^2+\frac{d\mathbf{N}}{dz}=-\alpha^2z^2\mathbf{1}+\alpha\mathbf{S}(\tfrac{1}{2}\pi)=\begin{pmatrix}-\alpha^2z^2 & -\alpha\\ +\alpha & -\alpha^2z^2\end{pmatrix}. \tag{2.H}$$

3.0 THE SPECIAL CASE: N INDEPENDENT OF z

This part is devoted to a study of the special case in which $\mathbf{N}$ is independent of z, as is the case for a homogeneous crystal. The statement that $\mathbf{N}$ is independent of z means simply that each of the four matrix elements is independent of z.

3.1 Derivation of M in Terms of N by Means of Integration

Perhaps the most straightforward way to determine $\mathbf{M}$ as a function of z when $\mathbf{N}$ is independent of z is to integrate Eq. (2.7):

$$d\mathbf{M}/dz=\mathbf{N}\mathbf{M} \tag{3.1}$$

subject to the boundary condition that $\mathbf{M}$ must be equal to the unit matrix when z is zero. The solution by this means involves the integration of four simultaneous first-order differential equations.

It is possible to obtain the solution more simply, however, by integrating Eq. (2.10), which involves only two simultaneous first-order differential equations. The vector equation (2.10) is equivalent to the following two scalar equations:

$$\begin{aligned}dX/dz&=n_1X+n_4Y,\\dY/dz&=n_3X+n_2Y.\end{aligned} \tag{3.2}$$

These two equations may be combined to obtain the following second-order differential equation which is satisfied by both X and Y:

$$d^2X/dz^2+2T(dX/dz)+DX=0, \tag{3.3}$$

where T is an abbreviation for the half-trace of $\mathbf{N}$

$$T\equiv\tfrac{1}{2}(n_1+n_2), \tag{3.4}$$

and where D is the determinant of $\mathbf{N}$:

$$D\equiv n_1n_2-n_3n_4. \tag{3.5}$$

It is convenient to introduce at this point also the definition of the discriminant Q:

$$Q\equiv(T^2-D)^{\frac{1}{2}}=(\tfrac{1}{4}(n_1-n_2)^2+n_3n_4)^{\frac{1}{2}}. \tag{3.6}$$

The Eqs. (3.2) and (3.3) must now be integrated subject to the condition that X and Y are equal, respectively, to X_0 and Y_0 when z is equal to zero. The integration may be accomplished by straightforward elementary methods, and the result is the same as that obtained below by a different method and given in Eq. (3.26).

3.2 Series Relation for M in Terms of N

Starting with the relation (3.1), it follows by successive differentiation, in view of the fact that $\mathbf{N}$ is independent of z, that:

$$d^k\mathbf{M}/dz^k=\mathbf{N}^k\mathbf{M}. \tag{3.7}$$

Substitution of Eq. (3.7) in the Taylor series expansion for $\mathbf{M}$ in powers of $z-z_0$:

$$\mathbf{M}=\mathbf{M}_0+\sum_{k=1}^{\infty}\frac{d^k\mathbf{M}}{dz^k}\frac{(z-z_0)^k}{k!} \tag{3.8}$$

then leads to:

$$\mathbf{M}=\left\{1+\sum_{k=1}^{\infty}\mathbf{N}^k\frac{(z-z_0)^k}{k!}\right\}\mathbf{M}_0, \qquad (3.9)$$

where $\mathbf{M}_0$ is the value of $\mathbf{M}$ at $z=z_0$. This relation may be written symbolically as

$$\mathbf{M}=e^{\mathbf{N}(z-z_0)}\mathbf{M}_0. \qquad (3.10)$$

When z_0 is zero, Eq. (3.9) may be written

$$\mathbf{M}=1+\mathbf{N}z+\tfrac{1}{2}\mathbf{N}^2z^2+\tfrac{1}{6}\mathbf{N}^3z^3+\cdots \qquad (3.11)$$

which relation may be written symbolically

$$\mathbf{M}=\exp(\mathbf{N}z). \qquad (3.12)$$

The last relation should be compared with Eq. (2.23) above.

The conditions for the convergence of these series expansions will not be examined here.

3.3 Identity of the Eigenvectors of N and M

It will now be shown that the eigenvectors of **M** are equal to the eigenvectors of **N** when **N** is independent of z.

The two eigenvectors of **M** are defined as the two linearly independent vectors which satisfy the following vector equation:

$$\mathbf{M}\mathcal{E}_M=\lambda_M\mathcal{E}_M, \qquad (3.13)$$

where λ_M is a constant for each of the two eigenvectors and is called the eigenvalue corresponding to that eigenvector.

It is physically evident that if $\mathcal{E}_M$ is an eigenvector of a given thickness of a homogeneous crystal, then it is also an eigenvector of any other thickness of that crystal. Accordingly, it may be concluded that $\mathcal{E}_M$ is independent of z when **N** is independent of z. By differentiation of Eq. (3.13), it then follows that

$$(d\mathbf{M}/dz)\mathcal{E}_M=(d\lambda_M/dz)\mathcal{E}_M. \qquad (3.14)$$

Upon rewriting Eq. (3.13) in the form

$$\mathcal{E}_M=\lambda_M\mathbf{M}^{-1}\mathcal{E}_M, \qquad (3.15)$$

and substituting Eq. (3.15) in the left-hand member of Eq. (3.14), one finds:

$$(d\mathbf{M}/dz)\mathbf{M}^{-1}\lambda_M\mathcal{E}_M=(d\lambda_M/dz)\mathcal{E}_M, \qquad (3.16)$$

or

$$\mathbf{N}\mathcal{E}_M=(1/\lambda_M)(d\lambda_M/dz)\mathcal{E}_M. \qquad (3.17)$$

Since $\mathcal{E}_M$ and **N** are both independent of z, it now follows from Eq. (3.17) that $\mathcal{E}_M$ is also an eigenvector of **N**:

$$\mathcal{E}_M\doteq\mathcal{E}_N, \qquad (3.18)$$

where the symbol $\doteq$ means that the quantities on either side are equal except for a constant numerical factor. It then follows also that the coefficient of $\mathcal{E}_M$ on the right-hand side of Eq. (3.17) is the eigenvalue of **N** which corresponds to the eigenvector $\mathcal{E}_M$:

$$d\log\lambda_M/dz=\lambda_N. \qquad (3.19)$$

Thus, the identity of the eigenvectors has been established.

The relation between the eigenvalues of **M** and **N** is now easily obtained from Eq. (3.19). By integration one finds

$$\lambda_M=\exp(\lambda_N z), \qquad (3.20)$$

where the constant of integration has been evaluated by the condition that λ_M must be unity when $z=0$.

3.4 Explicit Expression for M in Terms of N

An explicit expression for **M** in terms of z and the matrix elements of **N** will now be obtained on the basis of results derived in the preceding section, when **N** is independent of z.

It was shown in Paper IV that if a matrix has the eigenvectors:

$$\mathcal{E}_1=\begin{pmatrix}X_1\\Y_1\end{pmatrix}\quad \mathcal{E}_2=\begin{pmatrix}X_2\\Y_2\end{pmatrix}, \qquad (3.21)$$

with the corresponding eigenvalues λ_1 and λ_2, then the matrix can be expressed uniquely in the form:

$$\mathbf{M}=\frac{1}{X_1Y_1-X_2Y_2}\times\begin{pmatrix}\lambda_1X_1Y_2-\lambda_2X_2Y_1 & (\lambda_2-\lambda_1)X_1X_2\\(\lambda_1-\lambda_2)Y_1Y_2 & \lambda_2X_1Y_2-\lambda_1X_2Y_1\end{pmatrix}. \qquad (3.22)$$

Also from the material given in Paper IV the eigenvectors of **N** and, therefore, also of **M**, may be expressed in the form

$$\mathcal{E}_N=\mathcal{E}_M=\begin{pmatrix}\tfrac{1}{2}(n_1-n_2)\pm{}_1Q_N\\n_3\end{pmatrix}, \qquad (3.23)$$

and the eigenvalues of **N** may be expressed in the form

$$\lambda_N = T_N \pm_1 Q_N. \qquad (3.24)$$

By substitution of Eq. (3.24) in Eq. (3.20) one finds that the eigenvalues of **N** may be written

$$\lambda_M = \exp((T_N \pm_1 Q_N)z). \qquad (3.25)$$

By substitution of Eqs. (3.23) and (3.25) in Eq. (3.22) one then finds the following expression for the matrix **M**:

$$\mathbf{M} = \exp(T_N z) \begin{bmatrix} \cosh Q_N z + \frac{1}{2}(n_1 - n_2)\dfrac{\sinh Q_N z}{Q_N} & n_4 \dfrac{\sinh Q_N z}{Q_N} \\ n_3 \dfrac{\sinh Q_N z}{Q_N} & \cosh Q_N z - \frac{1}{2}(n_1 - n_2)\dfrac{\sinh Q_N z}{Q_N} \end{bmatrix}. \qquad (3.26)$$

This equation is the explicit form of the relation indicated symbolically in Eq. (3.12). It should be noted that the matrix defined by Eq. (3.26) is symmetrical in the parameter Q_N, so that the ambiguity in the sign of Q_N involved in its definition does not introduce any ambiguity in the expression (3.26).

It is of interest to note that this expression has been obtained without any process of integration except that involved in obtaining Eq. (3.20) from Eq. (3.19).

In order to exemplify the results contained in this and the preceding section by a specific case, consider the corresponding **M**- and **N**-matrices defined by Eqs. (2.D) and (2.E):

$$\mathbf{M} = \mathbf{S}(\omega z), \quad \mathbf{N} = \omega \mathbf{S}(\tfrac{1}{2}\pi). \qquad (3.A)$$

The eigenvectors of any scalar multiple of a rotation matrix are the vectors which correspond to right and left circularly polarized light. Accordingly, since both **M** and **N** are scalar multiples of rotation matrices, it follows at once that they have the same eigenvectors.

One has from Eqs. (3.4)–(3.6):

$$\begin{aligned} T_M &= \cos\omega, & T_N &= 0, \\ D_M &= 1, & D_N &= \omega, \\ Q_M &= \pm i \sin\omega, & Q_N &= \pm i\omega, \end{aligned} \qquad (3.B)$$

from which one finds:

$$\lambda_M = T_M \pm Q_M = \exp(\pm i\omega), \qquad (3.C)$$

$$\lambda_N = T_N \pm Q_N = \pm i\omega. \qquad (3.D)$$

By comparison of Eq. (3.C) with Eq. (3.D) it is evident that the eigenvectors satisfy the relation (3.20). Finally, by substitution of

$$\begin{aligned} n_1 &= n_2 = 0, \\ n_3 &= -n_4 = \omega \end{aligned} \qquad (3.E)$$

in Eq. (3.26), one confirms that the matrix **M** is equal to $\mathbf{S}(\omega z)$.

3.5 Explicit Expression for N in Terms of M

If one knows **M** as a function of z, then the corresponding **N**-matrix can be determined directly from the definition (2.6). If, however, one knows the matrix **M** for a given thickness z but does not know the value of **M** for any other thickness, then it is not possible to determine the matrix **N** uniquely. In the latter case, however, it is possible to provide a multiple-valued solution for **N** in terms of **M**, when **N** is independent of z. The significance of the multiple values will be made clear in a few illustrative examples.

The solution for **N** in terms of **M** will be obtained by solving Eq. (3.26) for the n's in terms of the m's.

First of all, one finds immediately from Eq. (3.26) the following simple expression for the determinant of **M**:

$$D_M = \exp(2T_N z). \qquad (3.27)$$

This relation may be inverted to yield T_N as a function of D_M:

$$T_N \equiv \tfrac{1}{2}(n_1 + n_2) = \tfrac{1}{2}\log(\pm_2 D_M^{\frac{1}{2}}), \qquad (3.28)$$

where the $\pm_2$ sign has been introduced to give explicit recognition of the fact that the square root of D_M is double valued.

The four scalar equations which are equivalent to the matrix equation (3.26) may be written:

$$T_M \equiv \tfrac{1}{2}(m_1 + m_2) = \exp(T_N z)\cosh Q_N z, \qquad (3.29)$$

$$m_1 - m_2 = (n_1 - n_2)\exp(T_N z)\frac{\sinh Q_N z}{Q_N}, \qquad (3.30)$$

$$m_3 = n_3 \exp(T_N z)\frac{\sinh Q_N z}{Q_N}, \qquad (3.31)$$

$$m_4 = n_4 \exp(T_N z)\frac{\sinh Q_N z}{Q_N}. \qquad (3.32)$$

Then from Eqs. (3.29) and (3.27) one finds directly with the help of the definition (3.6):

$$Q_M = \pm_1 \exp(T_N z)\sinh Q_N z, \qquad (3.33)$$

where the $\pm_1$ sign has been introduced because both Q_N and Q_M are ambiguous with regard to sign; once this sign has been specified, however, the same value should be used throughout the calculation. This relation may be rewritten in the form:

$$Q_M = \pm_1 Q_N \exp(T_N z)\frac{\sinh Q_N z}{Q_N}. \qquad (3.34)$$

By comparing Eqs. (3.30), (3.31), (3.32), and (3.34) the following relations may be obtained by inspection:

$$\frac{n_1 - n_2}{m_1 - m_2} = \frac{n_3}{m_3} = \frac{n_4}{m_4} = \pm_1 \frac{Q_N}{Q_M}. \qquad (3.35)$$

Equations (3.28) and (3.33) are easily combined to yield

$$Q_M = \pm_1 D_M^{\frac{1}{2}} \sinh Q_N z, \qquad (3.36)$$

and this relation becomes upon inversion:

$$\frac{Q_N}{Q_M} = \pm_1 \frac{\sinh^{-1}[Q_M / D_M^{\frac{1}{2}}]}{z Q_M}. \qquad (3.37)$$

The Eqs. (3.35), (3.28), and (3.37) now permit an explicit solution for the n's in terms of the m's. The result may be written in the form:

$$\mathbf{N} = \begin{pmatrix} F_M + \frac{1}{2}(m_1 - m_2)G_M & m_4 G_M \\ m_3 G_M & F_M - \frac{1}{2}(m_1 - m_2)G_M \end{pmatrix}, \qquad (3.38)$$

where the functions F_M and G_M are defined by:

$$F_M \equiv \frac{1}{z}\log(\pm_2 D_M^{\frac{1}{2}}),$$

$$G_M \equiv \frac{1}{zQ_M}\sinh^{-1}\frac{Q_M}{(\pm_2 D_M^{\frac{1}{2}})}$$

$$= \frac{1}{zQ_M}\log\frac{Q_M \pm_3 T_M}{(\pm_2 D_M^{\frac{1}{2}})}, \qquad (3.39)$$

where both values of $\pm_2$ and of $\pm_3$ must be used if all of the possible solutions for $\mathbf{M}$ are to be obtained.

It should be noted that both F_M and G_M are multiple-valued functions.

3.6 Illustrative Examples

In using Eq. (3.38) to determine the matrix $\mathbf{N}$ when the matrix $\mathbf{M}$ is known only for a given thickness, the chief difficulty lies in making sure that one has obtained all of the formal solutions given by Eq. (3.38). The best way to make this situation clear is by means of a few examples.

Consider first the matrix

$$\mathbf{M} = \begin{pmatrix} i & 0 \\ 0 & -i \end{pmatrix}. \qquad (3.40)$$

It is evident upon inspection that this is the matrix of a half-wave plate with its axes parallel with the coordinate axes. The solution of Section 3.5 will now be employed to determine all of the $\mathbf{N}$-matrices which are consistent with Eq. (3.40).

One has from Eq. (3.40):

$$T_M = 0,$$
$$D_M^{\frac{1}{2}} = 1, \qquad (3.41)$$
$$Q_M = i,$$

where only one of the two possible values of Q_M and of $D_M^{\frac{1}{2}}$ has been specified. Either value of Q_M, if used consistently, leads to the same final result.

In computing the general solution of Eq. (3.38), the only safe procedure is to compute separately the solutions for each of the four different combinations of signs in Eq. (3.39), and then to combine the separate solutions.

In the special case of Eq. (3.40), however, there are only two different choices of sign because T_M is zero. With the first choice, $\pm_2 = +$, one finds

$$zF_M = \log 1 = 2\pi i k,$$
$$zG_M = -i\log i = 2\pi l + \tfrac{1}{2}\pi, \qquad (3.42)$$

whence

$$\mathbf{N}z = \tfrac{1}{2}\pi i \begin{pmatrix} 4k + 4l + 1 & 0 \\ 0 & 4k - 4l - 1 \end{pmatrix}, \qquad (3.43)$$

where k and l are any two integers.

With the second choice, $\pm_2 = -$, one finds

$$zF_M = \log(-1) = 2\pi i k + \pi i,$$
$$zG_M = -i\log(-i) = 2\pi l - \tfrac{1}{2}\pi, \qquad (3.44)$$

whence

$$\mathbf{N}z=\tfrac{1}{2}\pi i\begin{pmatrix}4k+4l+1 & 0\\ 0 & 4(k+1)-4l-1\end{pmatrix}. \quad (3.45)$$

Both Eqs. (3.43) and (3.45) are consistent with the following general solution

$$\mathbf{N}z=\begin{pmatrix}\pi i(2p+\tfrac{1}{2}) & 0\\ 0 & \pi i(2q-\tfrac{1}{2})\end{pmatrix}, \quad (3.46)$$

where p and q are any two integers.

The case just discussed is one in which the two separate calculations led to the same general solution. Slightly different is the second case:

$$\mathbf{M}=\begin{pmatrix}0 & -1\\ 1 & 0\end{pmatrix}. \quad (3.47)$$

The values of T_M, D_M, and Q_M for this matrix are also given by Eq. (3.41); accordingly, the functions F_M and G_M as given by Eq. (3.42) or Eq. (3.44) hold also for this case. One then finds from Eq. (3.38) for the two separate choices of $\pm_2$:

$$\mathbf{N}z=\tfrac{1}{2}\pi\begin{pmatrix}4ki & -4l-1\\ 4l+1 & 4ki\end{pmatrix}, \quad (3.48)$$

$$\mathbf{N}z=\tfrac{1}{2}\pi\begin{pmatrix}(4k+2)i & -4l+1\\ 4l-1 & (4k+2)i\end{pmatrix}, \quad (3.49)$$

where k and l are any two integers.

These two solutions may be combined to obtain the following general solution

$$\mathbf{N}z=\begin{pmatrix}\pi ip & -(2q+p+\tfrac{1}{2})\pi\\ (2q+p+\tfrac{1}{2})\pi & \pi ip\end{pmatrix}, \quad (3.50)$$

where p and q are any two integers, and where even values of p correspond to the first choice, and odd values to the second.

The third and last example is one in which there are four different choices of the signs involved in Eq. (3.39). Consider the matrix

$$\mathbf{M}=\tfrac{1}{2}\begin{pmatrix}11 & 5\\ 7 & 9\end{pmatrix}. \quad (3.51)$$

One finds

$$Q_M=3,\quad D_M^{\frac{1}{2}}=4,\quad T_M=5. \quad (3.52)$$

Four different solutions will now be found which correspond, respectively, to the following four choices

	$\pm_2$	$\pm_3$
I	+	+
II	+	−
III	−	+
IV	−	−

The functions F_M for choices I and II, and choices III and IV, respectively, are

$$F_M=\log 4=2\epsilon+2\pi ik, \quad (3.53)$$

$$F_M=\log(-4)=2\epsilon+\pi i(2k+1), \quad (3.54)$$

where k is any integer, and ϵ is the principal value of log 2:

$$\epsilon=0.69315\cdots. \quad (3.55)$$

Correspondingly, the functions G_M are given by

$$3zG_M=\log 2=\epsilon+2\pi il, \quad (3.56)$$

$$3zG_M=\log(-\tfrac{1}{2})=-\epsilon+\pi i(2l+1), \quad (3.57)$$

$$3zG_M=\log(-2)=\epsilon+\pi i(2l+1), \quad (3.58)$$

$$3zG_M=\log\tfrac{1}{2}=-\epsilon+2\pi il, \quad (3.59)$$

respectively, where l is any integer.

The four separate solutions for the matrix $\mathbf{N}$ may now be written

$$6\mathbf{N}z=\begin{pmatrix}13\epsilon+2\pi i(6k+l) & 5\epsilon+10\pi il\\ 7\epsilon+14\pi il & 11\epsilon+2\pi i(6k-l)\end{pmatrix}, \quad (3.60)$$

$$6\mathbf{N}z=\begin{pmatrix}11\epsilon+\pi i(12k+2l+1) & -5\epsilon+5\pi i(2l+1)\\ -7\epsilon+7\pi i(2l+1) & 13\epsilon+\pi i(12k-2l-1)\end{pmatrix}, \quad (3.61)$$

$$6\mathbf{N}z=\begin{pmatrix}13\epsilon+\pi i(12k+2l+7) & 5\epsilon+5\pi i(2l+1)\\ 7\epsilon+7\pi i(2l+1) & 11\epsilon+\pi i(12k-2l+5)\end{pmatrix}, \quad (3.62)$$

$$6\mathbf{N}z=\begin{pmatrix}11\epsilon+2\pi i(6k+l+3) & -5\epsilon+10\pi il\\ -7\epsilon+14\pi il & 13\epsilon+2\pi i(6k-l+3)\end{pmatrix}. \quad (3.63)$$

The general solution for **N** is now to be obtained by combining the solutions exhibited in the last four equations.

The desired combination may be expressed as follows:

$$6\mathbf{N}z=\begin{pmatrix}\epsilon(12+(-1)^p)+\pi i(12r+7q+p) & 5(-1)^p\epsilon+5\pi i(p+q)\\ 7(-1)^p\epsilon+7\pi i(p+q) & \epsilon(12-(-1)^p)+\pi i(12r+5q-p)\end{pmatrix}, \tag{3.64}$$

where p, q, and r are any three integers. The relation between Eq. (3.64) and the partial solutions (3.60)–(3.63) is as follows:

	p	q
I	even	even
II	odd	even
III	even	odd
IV	odd	odd

The physical significance of the solutions obtained in these three examples will be discussed in Section 4.4.

4.0 LAMELLAR REPRESENTATION OF A CRYSTAL

In the latter part (pp. 490–493) of Paper IV a treatment was presented whose object was to show that any crystal without optical activity was optically equivalent to a laminated crystal, whose alternate laminae were identical and identically oriented. One of the alternate laminae was a retardation plate of prescribed retardation and orientation, and the other lamina was a partial polarizer of prescribed orientation and principal absorption coefficients.

One of the purposes of the treatment was to provide a means for the mind to grasp as a whole the optical behavior of such a crystal. To the writer, at least, it was helpful to realize that any section of any homogeneous crystal (with zero optical activity) could be considered, as far as its optical properties for normal incidence were concerned, as an interleaved pile of retardation plates and partial polarizers, in the limit in which the thickness of each lamina approaches zero.

A second purpose of the treatment was to provide a straightforward method of breaking down the complete optical description of the crystal section in terms of the matrix **M**, into a number of relatively simple component properties.

In this section, the treatment will be similar in spirit to that just mentioned, but will be more general in that it will hold for any homogeneous crystal whatever, and different in that the treatment will be based on the **N**-matrices.

Although the treatment given in this section relates explicitly to homogeneous crystals, the method of resolution may also be used for inhomogeneous crystals with the understanding that the resolution holds only at the given point for a given direction of propagation of the light, whereas for homogeneous crystals the resolution holds at every point in the crystal for a given direction of propagation of the light.

4.1 Method of Attack

It is desired to find some method of representing the properties of an arbitrary homogeneous crystal as a combination of a number of simple properties. As a first attempt, one might try to find a simple way of factoring the matrix **M** of the crystal into the product of a finite number of simple **M**-matrices, each of which would represent a simple crystal property, such as circular dichroism, linear birefringence, or isotropic absorption. This effort fails, because the constants which specify the component matrices depend on the order in which the matrices are multiplied.

The resolution of this difficulty may be realized by noting that the **M**-matrix of a very thin section of the crystal differs only slightly from the unit matrix, and that such matrices have a product which is independent of the order in which the matrices are multiplied. Accordingly, it is possible to represent the **M**-matrix of a thin section of the crystal as a product of factors which meaningfully represent the crystal as a combination of simple properties. It turns out that the representation is most conveniently presented in terms of the **N**-matrices.

Consider a sandwich of eight thin laminae. Let the **N**-matrix of each of the laminae be denoted by $\mathbf{N}_k$, where k assumes integral values from one to eight, and let the corresponding thicknesses of the laminae be τ_k. Each lamina is to be called "thin" if and only if its thickness satisfies the conditions

$$|n_{ik}|\tau_k \ll \tfrac{1}{8}, \tag{4.1}$$

where n_{ik} is the ith matrix element of the kth lamina.

Because of the conditions just stated, Eq. (3.11) may be written

$$\mathbf{M}_k = 1 + \mathbf{N}_k\tau_k + O(\tau_k^2), \tag{4.2}$$

where the term $O(\tau_k^2)$ should be read "a term of the order τ_k^2." Then if the light passes through the laminae in the order in which they are numbered, the **M**-matrix of the sandwich, $\mathbf{M}_s$, is

$$\mathbf{M}_s = \mathbf{M}_8\mathbf{M}_7 \cdots \mathbf{M}_2\mathbf{M}_1. \tag{4.3}$$

By use of Eq. (4.2) this may be written:

$$\mathbf{M}_s = 1 + \textstyle\sum_k \mathbf{N}_k\tau_k + O((\sum_k \tau_k)^2). \tag{4.4}$$

It is now convenient to introduce a matrix $\overline{\mathbf{N}}$ which is a weighted average of the $\mathbf{N}_k$'s:

$$\overline{\mathbf{N}} \equiv \frac{\sum_k \mathbf{N}_k\tau_k}{\sum_k \tau_k}. \tag{4.5}$$

Each of the $\mathbf{N}_k$'s is weighted in proportion to the thickness of the lamina. Furthermore, let τ be the total thickness of the sandwich

$$\tau \equiv \textstyle\sum_k \tau_k. \tag{4.6}$$

By the use of the last two abbreviations Eq. (4.4) may be written in the following convenient form:

$$\mathbf{M}_s = 1 + \overline{\mathbf{N}}\tau + O(\tau^2). \tag{4.7}$$

It will now be shown that a pile containing a large number of the octuple sandwiches behaves as though it were a homogeneous crystal with an **N**-matrix equal to $\overline{\mathbf{N}}$, in the limit in which the thickness of the sandwiches approaches zero while the number becomes infinite.

Consider such a pile of the octuple sandwiches. If the number of the sandwiches is q, the total thickness z of the pile is $q\tau$, and the **M**-matrix of the pile is

$$\mathbf{M} = \mathbf{M}_s^{\,q} = (1 + \overline{\mathbf{N}}\tau + O(\tau^2))^{z/\tau}. \tag{4.8}$$

Now, while the total thickness z of the pile remains constant, let the number q of the octuple sandwiches increase without limit, so that τ approaches zero. Accordingly, one is faced with the problem of calculating the limit of (4.8) as τ approaches zero. This calculation involves finding the qth power of $\mathbf{M}_s$ and may be performed in the manner exemplified at length in Paper III (p. 501); the result of the calculation is that **M** is related to $\overline{\mathbf{N}}$ by Eq. (3.26) provided that $\overline{\mathbf{N}}$ is substituted for **N** in (3.26). Thus in the given limit ($\tau \rightarrow 0$) the pile behaves just as though it were a homogeneous crystal with an **N**-matrix equal to $\overline{\mathbf{N}}$.

This result is independent of the ordering of the eight laminae in each sandwich, since no matter what order is used the **M**-matrix of the sandwich is still given by:

$$\mathbf{M}_s = 1 + \overline{\mathbf{N}}\tau + O(\tau^2),$$

as is evident by inspection of the derivation of (4.4) from (4.3).

Thus one has arrived at the concept that any given homogeneous crystal is optically equivalent to a laminated crystal as described above, provided only that the constants of the laminae be so chosen that $\overline{\mathbf{N}}$ is equal to the **N**-matrix of the given crystal. Furthermore, the equivalence is an invariant one in the sense that the equivalence is independent of the ordering of the laminae within each sandwich.

Since eight real constants are required to specify an **N**-matrix (the real and imaginary parts of each of the four matrix elements), it is evident that by suitably choosing the form of each of the eight $\mathbf{N}_k$ matrices so that each contains only one adjustable constant, a unique solution for the constants may always be obtained.

In the next section the form of the eight $\mathbf{N}_k$ matrices will be chosen as just described, and furthermore in such a way that each corresponds to a particularly simple type of optical behavior. For example, the matrix $\mathbf{N}_3$ will represent circular birefringence; $\mathbf{N}_5$ will represent linear birefringence with axes parallel to the coordinate axes, etc.

Before specifying the $\mathbf{N}_k$'s, however, it is convenient to introduce a simplification and a change in notation. Henceforth, it will be assumed that all of the laminae have the same thickness:

$$\tau_k = \tfrac{1}{8}\tau.$$

Furthermore, let $\mathbf{\Theta}_k$ be defined as one-eighth of $\mathbf{N}_k$:

$$\mathbf{\Theta}_k \equiv \tfrac{1}{8}\mathbf{N}_k.$$

One then has

$$\bar{\mathbf{N}} = \sum_k \mathbf{\Theta}_k.$$

The $\mathbf{\Theta}_k$'s are such that if every $\mathbf{\Theta}_k$ is zero except one, denoted by $\mathbf{\Theta}_l$, then in the limit $(\tau \to 0)$ the pile of octuple sandwiches will behave as though it were a homogeneous crystal with a **N**-matrix equal to $\mathbf{\Theta}_l$. Thus it seems reasonable to postulate that the magnitude of each $\mathbf{\Theta}_k$ represents the amount of the corresponding optical property in a homogeneous crystal whose **N**-matrix is $\sum_k \mathbf{\Theta}_k$. The examination of several **M**-matrices from this point of view is carried out in Section 4.4.

4.2 Specification of the Eight Θ Matrices

The eight matrices denoted by $\mathbf{\Theta}_k$ in the preceding section will now be determined so that each of them represents a particularly simple type of crystal. The eight matrices are listed in Table I along with a description of the corresponding optical behavior and the quantitative significance of the associated constants. For example, $\mathbf{\Theta}_1$ introduces a progressive change in phase corresponding to the mean index of refraction of the material, $\mathbf{\Theta}_2$ introduces an absorption corresponding to the mean absorption coefficient of the material, $\mathbf{\Theta}_3$ introduces circular birefringence (optical activity), and so on. It will be noted that each of the matrices involves a single real number so that, as must be the case, the matrix $\bar{\mathbf{N}}$ involves eight independent real numbers.

A number of conventions are involved in Table I. The time factor, the index of refraction n, and the extinction coefficient k are as they appear in the following relation for an isotropic medium:

$$\mathcal{E} = \mathcal{E}_0 \exp(i\omega t - 2\pi(k+in)z/\lambda).$$

The x, y, and z coordinate system is assumed to be right handed, and the light to travel in the direction of the positive z axis. Positive rotations about the z axis are defined as those which rotate from the positive x axis toward the positive y axis. Then the vectors representing right and left circularly polarized lights are*

$$\mathcal{E}_r = \begin{pmatrix} -i \\ 1 \end{pmatrix},$$

$$\mathcal{E}_l = \begin{pmatrix} i \\ 1 \end{pmatrix}.$$

In a stationary plane normal to the z axis, circularly polarized light whose vector rotates in the positive direction is left circularly polarized. Furthermore, with these conventions a rotation of the plane of polarization in the positive direction is a levorotation.

A rather unusual type of resolution is involved in the last four of the matrices listed in Table I: The linear birefringence is split into two components, one that part which is parallel with the coordinate axes, and the other parallel with the bisectors of the coordinate axes. The sum of $\mathbf{\Theta}_5$ and $\mathbf{\Theta}_7$ is

$$\mathbf{\Theta}_5 + \mathbf{\Theta}_7 = \begin{pmatrix} ig_0 & ig_{45} \\ ig_{45} & -ig_0 \end{pmatrix}. \tag{4.9}$$

If one compares the last expression with the matrix of a linearly birefringent plate with its fast axis rotated through the angle α from the x axis:

$$\begin{pmatrix} ig\cos 2\alpha & ig\sin 2\alpha \\ ig\sin 2\alpha & -ig\cos 2\alpha \end{pmatrix}, \tag{4.10}$$

where g is positive, it follows that (4.9) is equivalent to a wave plate of retardation specified by

$$g^2 = g_0^2 + g_{45}^2, \tag{4.11}$$

and with a fast axis at the angle

$$\tan 2\alpha = g_{45}/g_0. \tag{4.12}$$

---

* In the writer's opinion *the* definition of right circularly polarized light is that an instantaneous picture of the space distribution of its electric vector represents a right spiral. This definition is adhered to by Pockels (*Lehrbuch der Kristalloptik* (B. G. Teubner, Berlin, 1906), pp. 7–8) and by Born (*Optik* (Verlag Julius Springer, Berlin, 1933), p. 24), and is the convention in terms of which all of the experimental data on optical activity is presented in the literature. Born, however, seems dissatisfied with the convention, as indicated by the following quotation: "Gemäss der Tradition, an die man sich halten muss, bezieht man den Drehsinn 'rechts' und 'links' nicht auf die Fortpflanzungsrichtung, sondern auf die entgegengesetzte, die 'Blickrichtung'." In the opinion of the writer, however, the definition should not be founded upon the time variation of the electric vector in a given plane as seen from a given direction, but should rather be founded on the spatial configuration of the electric vector at a given moment.

TABLE I.

This table shows the form of the **N**-matrices for eight different and independent types of crystalline behavior.

$\mathbf{\Theta}_1=-\eta\begin{pmatrix} i & 0 \\ 0 & i \end{pmatrix}$	The parameter η is the propagation constant, or the phase retardation per unit thickness, and is thus related to the index of refraction n by $\eta=2\pi n/\lambda.$
$\mathbf{\Theta}_2=-\kappa\begin{pmatrix} 1 & 0 \\ 0 & 1 \end{pmatrix}$	The parameter κ is the amplitude absorption coefficient to the base e per unit thickness, and is thus related to the extinction coefficient k by $\kappa=2\pi k/\lambda.$
$\mathbf{\Theta}_3=\omega\begin{pmatrix} 0 & -1 \\ 1 & 0 \end{pmatrix}$	The parameter ω is a measure of the circular birefringence, and is equal to the rotation (in the positive direction) of the plane of linearly polarized light, in radians per unit thickness. It is equal to half of the difference of propagation constants for right and left circularly polarized lights, $\omega=\frac{1}{2}(\eta_r-\eta_l),$ and is positive for crystals which are levorotatory.
$\mathbf{\Theta}_4=\delta\begin{pmatrix} 0 & -i \\ i & 0 \end{pmatrix}$	The parameter δ is a measure of the circular dichroism, and is equal to half of the difference of the absorption coefficients for left and right circularly polarized lights: $\delta=\frac{1}{2}(\kappa_l-\kappa_r).$ The parameter is positive for crystals which are more transparent for right polarized light.
$\mathbf{\Theta}_5=g_0\begin{pmatrix} i & 0 \\ 0 & -i \end{pmatrix}$	The parameter g_0 is a measure of that part of the linear birefringence which is parallel with the coordinate axes. It is equal to one-half of the difference between the two principal propagation constants $g_0=\frac{1}{2}(\eta_y-\eta_x),$ and is thus positive when the fast (smaller index) axis is parallel with the x axis.
$\mathbf{\Theta}_6=p_0\begin{pmatrix} 1 & 0 \\ 0 & -1 \end{pmatrix}$	The parameter p_0 is a measure of that part of the linear dichroism which is parallel with the coordinate axes. It is equal to one-half of the difference of the two principal absorption coefficients $p_0=\frac{1}{2}(\kappa_y-\kappa_x),$ and is thus positive when the more highly transmitting axis is parallel with the x axis.
$\mathbf{\Theta}_7=g_{45}\begin{pmatrix} 0 & i \\ i & 0 \end{pmatrix}$	The parameter g_{45} is a measure of that part of the linear birefringence which is parallel with the bisectors of the coordinate axes. It is equal to one-half of the difference between the two principal propagation constants $g_{45}=\frac{1}{2}(\eta_{-45}-\eta_{45}),$ and is thus positive when the fast axis bisects the positive x and y axes.
$\mathbf{\Theta}_8=p_{45}\begin{pmatrix} 0 & 1 \\ 1 & 0 \end{pmatrix}$	The parameter p_{45} is a measure of that part of the linear dichroism which is parallel with the bisectors of coordinate axes. It is equal to one-half of the difference between the two principal absorption coefficients $p_{45}=\frac{1}{2}(\kappa_{-45}-\kappa_{45}),$ and is thus positive when the more highly transmitting axis bisects the positive x and y axes.

The linear dichroism is correspondingly expressed by $\mathbf{\Theta}_6$ and $\mathbf{\Theta}_8$.

By Table I and Eq. (4.7), the matrix $\overline{\mathbf{N}}$ may be written explicitly as follows:

$$\overline{\mathbf{N}}=\begin{pmatrix} -\kappa+p_0-i\eta+ig_0 & -\omega+p_{45}-i\delta+ig_{45} \\ \omega+p_{45}+i\delta+ig_{45} & -\kappa-p_0-i\eta-ig_0 \end{pmatrix}. \tag{4.13}$$

It is evident from the form of this matrix that $\overline{\mathbf{N}}$ may be made equal to any given **N**-matrix by suitable choice of the eight constants, and that this choice is always unique.

From this expression for $\overline{\mathbf{N}}$ one finds at once:

$$T_{\overline{N}}=-\kappa-i\eta,$$
$$Q_{\overline{N}}^2=(p_0+ig_0)^2+(p_{45}+ig_{45})^2-(\omega+i\delta)^2. \tag{4.14}$$

Thus, $T_{\overline{N}}$ and $Q_{\overline{N}}$ together depend on all of the eight parameters.

The general expression for the matrix $\overline{\mathbf{M}}$ which corresponds to $\overline{\mathbf{N}}$ may be obtained by substituting Eqs. (4.13) and (4.14) in Eq. (3.26).

4.3 Special Case of Non-Absorbing Crystals

In the special case in which the crystal does not absorb, the even numbered $\mathbf{\Theta}_k$'s will all be zero, so that the matrix $\bar{\mathbf{N}}$ may be written in the following relatively simple form:

$$\bar{\mathbf{N}}=\begin{pmatrix} -i\eta+ig_0 & -\omega+ig_{45} \\ \omega+ig_{45} & -i\eta-ig_0 \end{pmatrix}, \qquad (4.15)$$

from which one finds

$$\begin{aligned} T_{\bar{N}} &= -i\eta, \\ Q_{\bar{N}}^2 &= -(g^2+\omega^2) = -\Gamma^2, \end{aligned} \qquad (4.16)$$

where g is defined by Eq. (4.11). Substitution of Eqs. (4.15) and (4.16) in Eq. (3.26) then yields:

$$\bar{\mathbf{M}}=\exp(-i\eta z)\begin{bmatrix} \cos\Gamma z+ig_0\dfrac{\sin\Gamma z}{\Gamma} & (-\omega+ig_{45})\dfrac{\sin\Gamma z}{\Gamma} \\ (\omega+ig_{45})\dfrac{\sin\Gamma z}{\Gamma} & \cos\Gamma z-ig_0\dfrac{\sin\Gamma z}{\Gamma} \end{bmatrix}. \qquad (4.17)$$

Perhaps the most important simple observation that can be made about this unitary matrix is that, except for the factor written before the matrix which represents the absolute phase, the matrix $\mathbf{M}$ is periodic in z with the period:

$$z_0=2\pi/(g^2+\omega^2)^{\frac{1}{2}}=2\pi/\Gamma. \qquad (4.18)$$

This fact can also be represented by stating that the wave number (the number of periods in unit length) is given by:

$$k=1/z_0=(g^2+\omega^2)^{\frac{1}{2}}/2\pi. \qquad (4.19)$$

If one introduces the wave numbers corresponding to g and ω when either of these types of retardation exists alone,

$$k_g=g/2\pi;\quad k\omega=\omega/2\pi, \qquad (4.20)$$

one has

$$k=(k_g{}^2+k_\omega{}^2)^{\frac{1}{2}}. \qquad (4.21)$$

This relation indicates that a small amount of circular birefringence superposed on a large amount of linear birefringence will affect the value of the wave number k only by a quantity of the second order in their ratio.

The existence of the periodic behavior exhibited by Eq. (4.17) can also be predicted by the representation involving Poincaré's sphere.[2] In this representation the distance z_0 is the distance in which the sphere experiences a rotation through 360°.

[2] H. Poincaré, *Théorie Mathématique de la Lumière* (Gauthier-Villars et Fils, Paris, 1892).

4.4 Illustrative Examples

The formulae developed in this section will now be used to discuss the physical significance of the mathematical solutions obtained in Section 3.6.

Consider the first example, in which it is found that Eq. (3.46) represented the most general value of $\mathbf{N}$ which is consistent with the $\mathbf{M}$-matrix defined by Eq. (3.40). By comparing Eq. (3.46) with the general expression (4.13), one finds

$$\begin{aligned} \eta &= -\pi(p+q)/z, \\ g_0 &= \pi(p-q+\tfrac{1}{2})/z, \end{aligned} \qquad (4.22)$$

with the other six constants all zero. Thus the crystal is linearly birefringent with its axes parallel to the coordinate axes. It is an $(n+\frac{1}{2})$-wave plate, where n is even if $p+q$ is even, and odd otherwise. The quantity η is the propagation constant the crystal would have if it were replaced with an isotropic material with an index of refraction equal to the mean of the two principal indices of the crystal. The number of wave-lengths within the thickness of the isotropic material would be equal to $-\frac{1}{2}(p+q)$.

The second example is represented by Eqs. (3.47) and (3.50). One finds

$$\begin{aligned} \eta &= -\pi p/z, \\ \omega &= (2q+p+\tfrac{1}{2})\pi/z, \end{aligned} \qquad (4.23)$$

with the other six constants all zero. Thus the crystal is optically active, and rotates the plane 90°, plus or minus a number of complete revolutions if p is even, and plus or minus an odd number of 180° rotations if p is odd.

The last example is much more complex. By comparing the general solution (3.64) with Eq.

(4.13), one finds

$$\begin{aligned} \eta &= -\pi(q+2r)/z, \\ \kappa &= -2\epsilon/z, \\ \omega &= \tfrac{1}{6}(-1)^p\epsilon/z = p_0, \\ \delta &= \tfrac{1}{6}\pi(p+q)/z = g_0, \\ g_{45} &= \pi(p+q)/z, \\ p_{45} &= (-1)^p\epsilon. \end{aligned} \quad (4.24)$$

Thus, in general, none of the constants is zero. In the simplest possible case, in which p, q, and r are all zero, the constants become

$$\begin{aligned} -\tfrac{1}{2}\kappa = 6\omega = 6p_0 = p_{45} = \epsilon, \\ \eta = \delta = g_0 = g_{45} = 0. \end{aligned} \quad (4.25)$$

In this simple case, the crystal is described as a combination of a negative isotropic absorption, a positive optical activity, and a linear dichroism with its fast axis at the angle defined by

$$\tan 2\alpha = p_{45}/p_0 = 6, \quad (4.26)$$

or

$$\alpha = 40.8°. \quad (4.27)$$

The circular dichroism, and the linear birefringence are both zero.

The only one of the eight constants which is independent of p, q, and r is κ, and its value is $-1.385/z$. Since a physically possible crystal cannot have a negative absorption coefficient, the matrix defined by Eq. (3.51) does not represent a physically realizable crystal. The last statement is easily confirmed by substitution of Eq. (3.51) in Eq. (31) of Paper IV, which substitution leads to a value of Γ (as defined in Paper IV) of 8.4. It follows that the matrix (3.51) must be divided by a number whose absolute value is not less than 8.4 in order to convert it to a matrix which represents a physically realizable crystal.

5.0 THE TWISTED CRYSTAL

5.1 The Uniformly Twisted Crystal

Sections 3.0 and 4.0 have treated the important special case in which $\mathbf{N}$ is independent of z. There is another special dependence of $\mathbf{N}$ upon z which permits a simple solution. This second special case is the case in which an originally homogeneous crystal is twisted uniformly about an axis parallel to the direction of transmission of the light.

Let $\mathbf{N}_0$ be the matrix corresponding to the untwisted crystal. Then, according to Eq. (2.17), the matrix $\mathbf{N}$ which corresponds to the twisted crystal is given by

$$\mathbf{N} = \mathbf{S}(kz)\mathbf{N}_0\mathbf{S}(-kz), \quad (5.1)$$

where $\mathbf{S}$ is the rotation matrix defined by Eq. (2.15) and where k is the angular twist per unit thickness. The differential equation (2.10) may now be written

$$\frac{d\mathcal{E}}{dz} = \mathbf{S}(kz)\mathbf{N}_0\mathbf{S}(-kz)\mathcal{E}. \quad (5.2)$$

The general solution of this differential equation will now be found.

Let $\mathcal{E}'$ be defined by

$$\mathcal{E}' \equiv \mathbf{S}(-kz)\mathcal{E}. \quad (5.3)$$

Substitution of this relation in Eq. (5.2) then yields

$$\mathbf{S}(kz)\frac{d\mathcal{E}'}{dz} + \frac{d\mathbf{S}(kz)}{dz}\mathcal{E}' = \mathbf{S}(kz)\mathbf{N}_0\mathcal{E}', \quad (5.4)$$

or

$$d\mathcal{E}'/dz = \{\mathbf{N}_0 - k\mathbf{S}(\tfrac{1}{2}\pi)\}\mathcal{E}', \quad (5.5)$$

where the relation

$$(d/dz)\mathbf{S}(kz) \equiv \omega\mathbf{S}(kz)\mathbf{S}(\tfrac{1}{2}\pi) \quad (5.6)$$

has been used. The substitution

$$\mathbf{N}' \equiv \mathbf{N}_0 - k\mathbf{S}(\tfrac{1}{2}\pi) \quad (5.7)$$

then permits Eq. (5.5) to be written in the folfowing form:

$$d\mathcal{E}'/dz = \mathbf{N}'\mathcal{E}', \quad (5.8)$$

where $\mathbf{N}'$ is a matrix which is independent of z. Accordingly, the solution of Eq. (5.8) may be written down directly from the solution obtained in Section 3.4:

Let the relation (3.26) by which $\mathbf{M}$ is expressed in terms of $\mathbf{N}$ by written symbolically as

$$\mathbf{M} = \exp(\mathbf{N}z) \quad (5.9)$$

in accordance with Eq. (3.12). Then the solution of (5.8) may be written

$$\mathcal{E}' = \exp(\mathbf{N}'z)\mathcal{E}_0', \quad (5.10)$$

where $\mathcal{E}_0'$ is the value of the vector $\mathcal{E}_0'$ at $z=0$, as in Section 2.1. By the use of (5.3), the last equation becomes

$$\mathbf{S}(-kz)\mathcal{E} = \exp(\mathbf{N}'z)\mathcal{E}_0, \quad (5.11)$$

or

$$\mathcal{E} = \mathbf{S}(kz)\exp(\mathbf{N}'z)\mathcal{E}_0, \quad (5.12)$$

from which one infers that the matrix $\mathbf{M}$ which

corresponds to the $\mathbf{N}$-matrix (5.1) is

$$\mathbf{M} = \mathbf{S}(kz)\exp(\{\mathbf{N}_0 - k\mathbf{S}(\tfrac{1}{2}\pi)\}z). \quad (5.13)$$

Thus the matrix $\mathbf{M}$ of the twisted crystal is equal to $\mathbf{S}(kz)$ multiplied by the $\mathbf{M}$-matrix corresponding to a homogeneous crystal with an $\mathbf{N}$-matrix equal to $\mathbf{N}_0 - k\mathbf{S}(\frac{1}{2}\pi)$.

Equation (5.13) is the general integral of Eq. (5.2) which satisfies the boundary condition that $\mathbf{M}$ is equal to the unit matrix when z is equal to zero.

5.2 The Arbitrarily Twisted Crystal

The transformation effected in the preceding section may also be employed with an arbitrarily twisted crystal. If $\mathbf{N}_0$ is the matrix of the untwisted crystal, the $\mathbf{N}$ matrix of the twisted crystal is

$$\mathbf{N} = \mathbf{S}(\omega(z))\mathbf{N}_0\mathbf{S}(-\omega(z)), \quad (5.14)$$

where $\omega(z)$ is the arbitrary function of z which specifies the angle of twist at the coordinate z.

Then by the same transformation used in Section 5.1, one finds that the $\mathbf{M}$ matrix which corresponds to Eq. (5.14) is given by

$$\mathbf{M} = \mathbf{S}(\omega(z))\mathbf{M}', \quad (5.15)$$

where $\mathbf{M}'$ is the $\mathbf{M}$ matrix which corresponds to

$$\mathbf{N}' \equiv \mathbf{N}_0 - (d\omega(z)/dz)\mathbf{S}(\tfrac{1}{2}\pi). \quad (5.16)$$

Only in the case of the uniformly twisted crystal is $d\omega(z)/dz$ a constant, and thus only in this case is the matrix $\mathbf{N}'$ independent of z.

In general, however, it is easier to determine the matrix $\mathbf{M}'$ than it is to determine the matrix $\mathbf{M}$ directly from Eq. (5.14), since the trigonometrical factors which occur in $\mathbf{N}$ do not occur in $\mathbf{N}'$.

23

Reprinted from *J. Opt. Soc. Am.*, **46**, 126–131 (Feb. 1956)

New Calculus for the Treatment of Optical Systems. VIII. Electromagnetic Theory

R. Clark Jones
Research Laboratory, Polaroid Corporation, Cambridge, Massachusetts

(Received June 23, 1955)

The preceding seven papers of this series present a systematic procedure for computing the effect of an optical system on the state of polarization of the light that passes through it. The **M**-matrices discussed in the first six papers represent the over-all effect of an optical system; the **N**-matrices described in Paper VII are essentially path derivatives of the **M**-matrices and represent the local optical properties at a given point along the light-path.

In this paper we suppose that the medium is an anisotropic crystal and note that the description of the local optical properties by the **N**-matrices must be closely related to the description of the local properties by the dielectric and gyration tensors that are employed in standard crystal optics. We find the exact relation between the **N**-matrix and the above-mentioned tensors. It is shown how one can compute the dielectric and gyration tensors from a knowledge of the **N**-matrices for several different directions of the light-path in the crystal. It is also shown how one can compute the **N**-matrix for any given light direction from a knowledge of the dielectric and gyration tensors; the computation entails finding the square root of a two-by-two complex matrix.

Taken together, the eight papers of this series present a compact and systematic procedure for the solution of problems in crystal optics. The **N**-matrices have the advantage that circular birefringence and circular dichroism are treated in the same framework used for linear birefringence and linear dichroism.

INTRODUCTION

THE first six papers of this series[1] were concerned with developing the properties of certain matrices (**M**-matrices) which are a compact statement of the over-all effect of a given optical system on the polarization of the light that passes through the system. The **M**-matrix of the optical system contains the coefficients of the linear equations by which one computes from the electric vector of the entering light the electric vector of the emergent light.

When the optical system is a homogeneous material, such as a crystal, Paper VII showed how one can introduce an **N**-matrix that describes the properties of the homogeneous material. The relations between the **M**-matrices and the **N**-matrices were the subject of that paper. Since the **N**-matrices describe the optical properties of the material, they necessarily have a close relationship to the dielectric and gyration tensors that are conventionally used to describe the properties of optical media. The present paper derives explicit formulas by which given the dielectric and gyration tensors of the medium one can calculate the **M**-matrix for any given direction of propagation.

The eight papers of this series form a unified mathematical equivalent for the geometrical procedures commonly employed in crystal optics. Given the **M**-matrix for several directions of propagation through a crystal, as measured by the simple procedure described in Paper VI, one can compute the corresponding **N**-matrices by the methods developed in Paper VII, and then compute the dielectric and gyration tensors by the relations developed in this paper. Conversely, if one is given the dielectric and gyration tensors of a crystal, one can compute in the opposite sequence to find the **M**-matrix for light passing in any given direction through a given section of the crystal.

[1] I, J. Opt. Soc. Am. **31**, 488–493 (1941); II, **31**, 493–499 (1941); III, **31**, 500–503 (1941); IV, **32**, 486–493 (1942); V, **37**, 107–110 (1947); VI, **37**, 110–112 (1947); VII, **38**, 671–684 (1948).

After a brief section that introduces the terminology and a few definitions, Sec. 2.0 introduces the dielectric and gyration tensors, combines them, and then reduces them to a two-dimensional form suitable for direct comparison with the corresponding **N**-matrix.

The general relation between the dielectric and gyration tensors and the **N**-matrix is shown in different forms in Sec. 3.0 in Eqs. (3.3), (3.5), (3.9), and (3.13). Section 3.0 is thus the heart of the paper.

If one wishes to determine the **N**-matrix for a given direction of propagation from the dielectric and gyration tensors, it is necessary to find the square root of a two-by-two matrix. Systematic methods for this operation are described in Sec. 4.0.

Finally, Sec. 5.0 reviews briefly the forms that the gyration tensor can assume, as limited by the symmetry elements of the various crystal groups.

1.0 DEFINITIONS AND TERMINOLOGY

In a right-handed Cartesian coordinate system x, y, z, let the wave front of a monochromatic light wave lie in the x-, y-plane, and let the wave be traveling in the positive z-direction. Let the vector $\mathcal{E}$ be defined as the column vector.

$$\mathcal{E}\equiv\begin{pmatrix}E_x\\E_y\end{pmatrix}=\mathbf{i}E_x+\mathbf{j}E_y, \tag{1.1}$$

where E_x and E_y are the x- and y-components of the electric vector $\mathcal{E}$ associated with the light wave.

The **N**-matrix at a given point of the medium, for a given direction of propagation, and for a given wavelength, may be defined as the operator that determines from $\mathcal{E}$ the derivative $d\mathcal{E}/dz$:

$$d\mathcal{E}/dz=\mathbf{N}\mathcal{E}. \tag{1.2}$$

The matrix elements of **N** are denoted as follows:

$$\mathbf{N}\equiv\begin{pmatrix}n_1 & n_4\\ n_3 & n_2\end{pmatrix}. \tag{1.3}$$

The matrix elements n are in general complex, as are also the vector components E_x and E_y. The time factor is $e^{i\omega t}$.

The matrix $\mathbf{N}$ was defined in Paper VII in terms of the $\mathbf{M}$-matrices, whose properties were the subject of the first six papers of this series. If $\mathbf{M}_z$ is the matrix of the optical element up to the position z, then the $\mathbf{N}$-matrix at the position z may be defined by:

$$\mathbf{N}_z \equiv (d\mathbf{M}_z/dz)\mathbf{M}_z^{-1}. \tag{1.4}$$

As implied by the above notation, two-component vectors will be indicated by script capitals. Three-component vectors will be indicated by German letters; the only exception is that the three-dimensional gradient operator is indicated by ∇. Two-by-two matrices are indicated by boldface letters, and three-by-three matrices by boldface italic letters.

2.0 ELECTROMAGNETIC THEORY

The electromagnetic theory of light propagation in crystals is beset by a fundamental difficulty when one considers the theory to that degree of completeness requisite for the inclusion of optical activity: It is easy to show[2] that the magnetic effects are of the same order of importance as electrical effects in contributing to circular dichroism and circular birefringence. But the simultaneous treatment of both the electrical and magnetic effects in crystals leads to mathematical relations of unmanageable complexity.[3] Accordingly, it is customary in treating crystals to ignore completely the magnetic effects, and to place all of the burden upon the electrical effects for the explanation of experimental results.

This neglect is not quite so bad as it looks upon first glance, because the chief experimental interest lies in the relation between the directional dependence of the optical rotation and the symmetry class of the crystal, and this relation is not changed by the inclusion of magnetic effects. For this reason, the present treatment will follow Born[2] in ignoring the magnetic effects. The neglect does have, however, the result that the constitutive relations to be used do not satisfy rigorously the the principle of conservation of energy.

The treatment will be based on the two Maxwell field equations

$$c\,\mathrm{curl}\,\mathfrak{E} = -\partial\mathfrak{B}/\partial t \tag{2.1}$$

$$c\,\mathrm{curl}\,\mathfrak{H} = \partial\mathfrak{D}/\partial t + 4\pi\mathfrak{J} \tag{2.2}$$

and the two constitutive relations

$$\partial\mathfrak{D}/\partial t + 4\pi\mathfrak{J} = \boldsymbol{K}\partial\mathfrak{E}/\partial t \tag{2.3}$$

$$\mathfrak{B} = \mathfrak{H}, \tag{2.4}$$

[2] Max Born, *Optik* (Verlag Julius Springer, Berlin, Germany, 1933), pp. 403–413.

[3] Reference 2, p. 414.

where $\mathfrak{E}$ is the three-dimensional column vector

$$\mathfrak{E} \equiv \begin{bmatrix} E_x \\ E_y \\ E_z \end{bmatrix} = \mathbf{i}E_x + \mathbf{j}E_y + \mathbf{k}E_z, \tag{2.5}$$

with corresponding definitions for $\mathfrak{D}$, $\mathfrak{H}$, and $\mathfrak{B}$. The equating of $\mathfrak{B}$ and $\mathfrak{H}$ represents the ignoring of magnetic effects. The operator $\boldsymbol{K}$ is a linear multiplicative operator in the absence of optical activity, but in the presence of optical activity it is necessary to assume that $\boldsymbol{K}$ contains also differential operators.

By the usual transformations, one finds

$$\frac{\partial^2\mathfrak{D}}{\partial t^2} = \boldsymbol{K}\frac{\partial^2\mathfrak{E}}{\partial t^2} = c^2(\nabla^2 - \mathrm{grad\ div})\mathfrak{E}. \tag{2.6}$$

By use of the postulate that all of the components vary with time according to the factor $e^{i\omega t}$, one obtains finally

$$\boldsymbol{K}\mathfrak{E} = \lambda\!\!\!^{-2}\,(\mathrm{grad\ div} - \nabla^2)\mathfrak{E}, \tag{2.7}$$

where

$$\lambda\!\!\!^{-} \equiv c/\omega. \tag{2.8}$$

THE OPERATOR $\boldsymbol{K}$

In the absence of optical activity, the operator $\boldsymbol{K}$ is a linear multiplicative operator, the complex dielectric tensor of the medium

$$\boldsymbol{K} = \boldsymbol{Q} = \begin{bmatrix} q_{11} & q_{12} & q_{13} \\ q_{21} & q_{22} & q_{23} \\ q_{31} & q_{32} & q_{33} \end{bmatrix}, \tag{2.9}$$

where in order to satisfy the principle of conservation of energy, it is necessary to assume that $\boldsymbol{Q}$ is symmetrical:

$$q_{ij} = q_{ji}. \tag{2.10}$$

The matrix elements q_{ij} are real only when the medium is nonconducting at the frequency in question.

In order to introduce the effects which constitute optical activity, it is necessary to assume that $\mathfrak{D}$ depends not only upon the components of $\mathfrak{E}$, but also upon the space derivatives of the electric vector. If this generalization of Eq. (2.9) is carried through completely, each of the quantities q_{ij} in Eq. (2.9) must be replaced by

$$q_{ij} \rightarrow q_{ij} + \phi_{ij1}\frac{\partial}{\partial x} + \phi_{ij2}\frac{\partial}{\partial y} + \phi_{ij3}\frac{\partial}{\partial z}, \tag{2.11}$$

with the result that 27 new quantities ϕ_{ijk} are introduced into the operator $\boldsymbol{K}$.

The phenomena constituting optical activity depend only on the antisymmetrical part of the operator $\boldsymbol{K}$. Since the operator $\mathfrak{U}\times$, where $\times$ indicates the vector cross product, may be written as an antisymmetric matrix:

$$\mathfrak{U}\times\mathfrak{V}=\begin{bmatrix}0 & -U_z & U_y\\ U_z & 0 & -U_x\\ -U_y & U_x & 0\end{bmatrix}\begin{bmatrix}V_x\\ V_y\\ V_z\end{bmatrix}, \qquad (2.12)$$

it is plausible to suppose that for the purpose of treating optical activity in crystals, it is sufficient to assume that $\mathfrak{D}$ depends upon $\mathfrak{E}$ in the following way:

$$\mathfrak{D}=\boldsymbol{K}\mathfrak{E}=Q\mathfrak{E}+\bar{\lambda}(G\nabla)\times\mathfrak{E}, \qquad (2.13)$$

where G is a three-by-three matrix, and where ∇ is the gradient operator:

$$\nabla\equiv\mathbf{i}\frac{\partial}{\partial x}+\mathbf{j}\frac{\partial}{\partial y}+\mathbf{k}\frac{\partial}{\partial z}. \qquad (2.14)$$

Born has derived Eq. (2.13) rigorously from molecular theory, and has given the name "gyration tensor" to the matrix G. Accordingly, the relation (2.13) will henceforth be considered as firmly established.

It follows that the operator $\boldsymbol{K}$ may be written

$$\boldsymbol{K}=\begin{bmatrix}q_{11} & q_{12}-\bar{\lambda}g_{31}\nabla_x-\bar{\lambda}g_{32}\nabla_y-\bar{\lambda}g_{33}\nabla_z & q_{13}+\bar{\lambda}g_{21}\nabla_x+\bar{\lambda}g_{22}\nabla_y+\bar{\lambda}g_{23}\nabla_z\\ q_{21}+\bar{\lambda}g_{31}\nabla_x+\bar{\lambda}g_{32}\nabla_y+\bar{\lambda}g_{33}\nabla_z & q_{22} & q_{23}-\bar{\lambda}g_{11}\nabla_x-\bar{\lambda}g_{12}\nabla_y-\bar{\lambda}g_{13}\nabla_z\\ q_{31}-\bar{\lambda}g_{21}\nabla_x-\bar{\lambda}g_{22}\nabla_y-\bar{\lambda}g_{23}\nabla_z & q_{32}+\bar{\lambda}g_{11}\nabla_x+\bar{\lambda}g_{12}\nabla_y+\bar{\lambda}g_{13}\nabla_z & q_{33}\end{bmatrix}, \qquad (2.15)$$

where ∇_x is an obvious abbreviation for $\partial/\partial x$, etc.

The matrix G is neither symmetrical nor antisymmetrical, in general. The matrix elements g_{ij} are real only when the crystal is nonabsorbing.

TWO-DIMENSIONAL FORM

It is convenient to suppose that the electric vector represents a plane wave with the wave normal parallel with the z-axis. It is then possible to write the operator $\boldsymbol{K}$ in a two-dimensional form, and the components of $\mathfrak{E}$ depend only on z. One has from Eq. (2.7),

$$\boldsymbol{K}\mathfrak{E}=-\bar{\lambda}^2\frac{\partial^2}{\partial z^2}\begin{pmatrix}E_x\\ E_y\\ 0\end{pmatrix} \qquad (2.16)$$

$$=-\bar{\lambda}^2\partial^2\mathcal{E}/\partial z^2.$$

Thus the z-component of $\mathfrak{D}$ is zero:

$$D_z=(\boldsymbol{K}\mathfrak{E})_z=0. \qquad (2.17)$$

For this special case, furthermore, the operator $\boldsymbol{K}$ may be written

$$\boldsymbol{K}=\begin{bmatrix}q_{11} & q_{12}-\bar{\lambda}g_{33}\nabla_z & q_{13}+\bar{\lambda}g_{23}\nabla_z\\ q_{21}+\bar{\lambda}g_{33}\nabla_z & q_{22} & q_{23}-\bar{\lambda}g_{13}\nabla_z\\ q_{31}-\bar{\lambda}g_{23}\nabla_z & q_{32}+\bar{\lambda}g_{13}\nabla_z & q_{33}\end{bmatrix}. \qquad (2.18)$$

The condition (2.17) that D_z be zero may now be written explicitly:

$$(\boldsymbol{K}\mathfrak{E})_z=(q_{31}-\bar{\lambda}g_{23}\nabla_z)E_x+(q_{32}+\bar{\lambda}g_{13}\nabla_z)E_y+q_{33}E_z=0. \qquad (2.19)$$

By solving this expression for E_z, and substituting the result in the corresponding expressions for $(\boldsymbol{K}\mathfrak{E})_x$ and $(\boldsymbol{K}\mathfrak{E})_y$, one obtains

$$(\boldsymbol{K}\mathfrak{E})_x=[q_{11}-(q_{13}+\bar{\lambda}g_{23}\nabla_z)(q_{31}-\bar{\lambda}g_{23}\nabla_z)/q_{33}]E_x+[q_{12}-\bar{\lambda}g_{33}\nabla_z-(q_{13}+\bar{\lambda}g_{23}\nabla_z)\times(q_{32}+\bar{\lambda}g_{13}\nabla_z)/q_{33}]E_y, \qquad (2.20)$$

$$(\boldsymbol{K}\mathfrak{E})_y=[q_{21}+\bar{\lambda}g_{33}\nabla_z-(q_{31}-\bar{\lambda}g_{23}\nabla_z)\times(q_{23}-\bar{\lambda}g_{13}\nabla_z)/q_{33}]E_x+[q_{22}-(q_{23}-\bar{\lambda}g_{13}\nabla_z)(q_{32}+\bar{\lambda}g_{13}\nabla_z)/q_{33}]E_y, \qquad (2.21)$$

$$(\boldsymbol{K}\mathfrak{E})_z=0. \qquad (2.22)$$

The first two of the last three relations may be written

$$\mathfrak{D}=\mathbf{K}\mathcal{E}, \qquad (2.23)$$

where $\mathbf{K}$ is the two-by-two matrix operator:

$$\mathbf{K}\equiv\begin{pmatrix}q_{11}-(q_{13}+\bar{\lambda}g_{23}\nabla_z)(q_{31}-\bar{\lambda}g_{23}\nabla_z)/q_{33} & q_{12}-\bar{\lambda}g_{33}\nabla_z-(q_{13}+\bar{\lambda}g_{23}\nabla_z)(g_{32}+\bar{\lambda}g_{13}\nabla_z)/q_{33}\\ q_{21}+\bar{\lambda}g_{33}\nabla_z-(q_{31}-\bar{\lambda}g_{23}\nabla_z)(g_{23}-\bar{\lambda}g_{13}\nabla_z)/q_{33} & q_{22}-(q_{23}-\bar{\lambda}g_{13}\nabla_z)(g_{32}+\bar{\lambda}g_{13}\nabla_z)/q_{33}\end{pmatrix}. \qquad (2.24)$$

Explicit cognizance will now be given to the symmetry of the q_{ij}'s, and the important assumption will be made that the g_{ij}'s are small compared with unity. The squares and cross products of the g_{ij}'s may then be ignored, with the result that one finds:

$$\mathbf{K}=\begin{bmatrix}q_{11}-q_{13}^2/q_{33} & q_{12}-q_{13}q_{23}/q_{33}-\bar{\lambda}\Omega_{12}\dfrac{\partial}{\partial z}\\ q_{12}-q_{13}q_{23}/q_{33}+\bar{\lambda}\Omega_{12}\dfrac{\partial}{\partial z} & q_{22}-q_{23}^2/q_{33}\end{bmatrix}, \qquad (2.25)$$

where Ω_{12} is an abbreviation for

$$\Omega_{12} \equiv (q_{13}g_{13}+q_{23}g_{23}+q_{33}g_{33})/q_{33}. \qquad (2.26)$$

The parameter Ω_{12} that appears in (2.25) is the special form taken by the invariant

$$\Omega = (\boldsymbol{Gs}\cdot\boldsymbol{Qs})/(\boldsymbol{s}\cdot\boldsymbol{Qs}), \qquad (2.27)$$

when the wave normal $\boldsymbol{s}$ is parallel with the z-axis.

If one makes the approximation that the dielectric tensor is isotropic—that is to say, is a scalar multiple of the unit tensor, then Ω becomes

$$\Omega = \boldsymbol{s}\cdot\boldsymbol{Gs}. \qquad (2.28)$$

For this special case, Ω clearly depends only on the symmetrical part of the tensor $\boldsymbol{G}$. Since most crystals are nearly isotropic, it follows that the optical activity depends primarily on the symmetrical part of $\boldsymbol{G}$. In the last equation, Ω is sometimes called the *scalar parameter of gyration.*[4]

3.0 GENERAL RELATIONS BETWEEN N AND K

In the preceding section a two-dimensional operator $\mathbf{K}$ was derived from electromagnetic theory such that

$$\mathbf{K}\mathcal{E} = -(\lambda^2 d^2/dz^2)\mathcal{E}, \qquad (3.1)$$

where $\partial/\partial x$ has been replaced by d/dx, because henceforth z is the only independent variable. But from Eq. (1.2),

$$d\mathcal{E}/dz = \mathbf{N}\mathcal{E}, \qquad (3.2)$$

whence

$$\mathbf{K} = -\lambda^2(\mathbf{N}^2 + d\mathbf{N}/dz). \qquad (3.3)$$

This is the desired relation between $\mathbf{N}$ and the matrix $\mathbf{K}$ which is related to the dielectric tensor. For the important special case in which $\mathbf{N}$ is independent of z, this relation becomes:

$$\mathbf{K} = -(\lambda\mathbf{N})^2. \qquad (3.4)$$

$\mathbf{N}$ is independent of z for every homogeneous medium, including the case of single crystals.

By the use of Eq. (2.21) of Paper VII, Eq. (3.3) may be written in terms of $\mathbf{M}$ instead of $\mathbf{N}$:

$$\mathbf{K} = -\lambda^2(d^2\mathbf{M}/dz^2)\mathbf{M}^{-1}. \qquad (3.5)$$

Let $\mathbf{K}$ be written as the sum of a symmetrical part $\mathbf{S}$ and an antisymmetrical part $\mathbf{A}$:

$$\mathbf{K} = \mathbf{S} + \mathbf{A}. \qquad (3.6)$$

By use of Eq. (3.1) and Eq. (2.25) one obtains

$$\mathbf{K} = \mathbf{S} + \lambda\Omega_{12}\mathbf{R}(\tfrac{1}{2}\pi)d/dz = -\lambda^2 d^2/dz^2, \qquad (3.7)$$

where $\mathbf{R}(\frac{1}{2}\pi)$ is defined by

$$\mathbf{R}(\tfrac{1}{2}\pi) \equiv \begin{pmatrix} 0 & -1 \\ 1 & 0 \end{pmatrix}. \qquad (3.8)$$

[4] Reference 2, p. 415.

Use of Eq. (3.2) then yields

$$\mathbf{S} + \lambda\Omega_{12}\mathbf{R}(\tfrac{1}{2}\pi)\mathbf{N} + \lambda^2\mathbf{N}^2 + \lambda^2 d\mathbf{N}/dz = 0. \qquad (3.9)$$

Equation (3.9) is the general relation between $\mathbf{N}$ on one hand, and $\boldsymbol{Q}$ and $\boldsymbol{G}$ on the other hand.

For the important special case in which $\mathbf{N}$ is independent of z, the last equation reduces to

$$\mathbf{S} + \lambda\Omega_{12}\mathbf{R}(\tfrac{1}{2}\pi)\mathbf{N} + \lambda^2\mathbf{N}^2 = 0. \qquad (3.10)$$

DETERMINATION OF N FROM Q AND G

Let us direct our attention first to the problem of determining $\mathbf{N}$ if $\boldsymbol{Q}$ and $\boldsymbol{G}$ are given. In the general situation where Ω_{12} differs from zero, Eq. (3.10) is a quadratic equation for $\mathbf{N}$.

I have not been able to find the general solution of Eq. (3.10) in closed form. In practice, the coefficient Ω_{12} will always be very small compared with unity; this fact permits one to obtain a very rapidly converging solution by successive approximation.

Let $\mathbf{N}^{(0)}$ be the solution of Eq. (3.9) when Ω_{12} is zero:

$$\mathbf{N}^{(0)} = -ik\mathbf{S}^{\frac{1}{2}}, \qquad (3.11)$$

where $k \equiv \omega/c$ is the reciprocal of λ. The solution for $\mathbf{N}^{(0)}$ given by (3.11) is of course multiple valued, since in general every two-by-two matrix has four distinct square roots. Only one of these four square roots is physically meaningful, however, as will be discussed in detail in Sec. 4.0. Henceforth in this section it will be supposed that the physically acceptable root is always employed.

If the value of $\mathbf{N}$ given by (3.11) is substituted in the linear term of (3.10), one obtains

$$\mathbf{S} - i\Omega_{12}\mathbf{R}(\tfrac{1}{2}\pi)\mathbf{S}^{\frac{1}{2}} + \lambda^2\mathbf{N}^2 = 0. \qquad (3.12)$$

Let $\mathbf{N}^{(1)}$ be the solution of this equation:

$$\mathbf{N}^{(1)} = -ik(\mathbf{S} - i\Omega_{12}\mathbf{R}(\tfrac{1}{2}\pi)\mathbf{S}^{\frac{1}{2}})^{\frac{1}{2}}. \qquad (3.13)$$

Similarly, the $(m+1)$th approximation is given by

$$\mathbf{N}^{(m+1)} = -ik(\mathbf{S} + \lambda\Omega_{12}\mathbf{R}(\tfrac{1}{2}\pi)\mathbf{N}^{(m)})^{\frac{1}{2}}. \qquad (3.14)$$

The solution given by (3.13) is probably the most generally useful solution. This solution may be simplified by expanding the square root in powers of Ω_{12}. The first two terms of this expansion are

$$\mathbf{N}^{(1)} = -ik\mathbf{S}^{\frac{1}{2}} - \tfrac{1}{2}k\Omega_{12}\mathbf{R}', \qquad (3.15)$$

where $\mathbf{R}'$ is the solution of

$$\mathbf{R}' + \mathbf{S}^{\frac{1}{2}}\mathbf{R}'\mathbf{S}^{-\frac{1}{2}} = 2\mathbf{R}(\tfrac{1}{2}\pi). \qquad (3.16)$$

In this approximation, it is clear that the elements of the gyration tensor $\boldsymbol{G}$ contribute *linearly* to the $\mathbf{N}$-matrix of the medium.

4.0 THE SQUARE ROOT OF A MATRIX

Introduction

In order to evaluate the matrix **N** to any given order of approximation it is necessary to be able to find the square root of a matrix.

Let **F** be a two-by-two complex matrix, and let **Y** be its square root

$$\mathbf{Y}^2=\mathbf{F}. \tag{4.1}$$

Our problem is, given **F**, to find **Y**.

In the discussion of physical acceptability, it is specifically assumed that the matrix **F** is equal either to **S** or to one of the $\mathbf{S}+\lambda\Omega_{12}\mathbf{R}(\frac{1}{2}\pi)\mathbf{N}^{(m)}$.

There are usually four distinct matrices **Y** whose square is **F**, but only one of the four mathematically possible **Y**'s is physically acceptable. Consider, for example, a transparent, linearly birefringent crystalline plate without optical activity; the principal crystal axes are oriented so that two of them lie in the plane of the surface, and one is normal to the surface; the x, y, z coordinate axes are parallel to the crystal axes, with the z-axis normal to the surface of the plate. Suppose that the principal dielectric constants are

$$q_{11}=9,\quad q_{22}=4. \tag{4.2}$$

Then the matrix **S** is

$$\mathbf{S}=\begin{pmatrix}9 & 0\\ 0 & 4\end{pmatrix} \tag{4.3}$$

and the matrix **N** to all orders of approximation is

$$\mathbf{N}=-ik\mathbf{S}^{\frac{1}{2}}. \tag{4.4}$$

The four square roots of Eq. (4.3) are

$$\mathbf{S}^{\frac{1}{2}}=\begin{pmatrix}\pm 3 & 0\\ 0 & \pm 2\end{pmatrix}. \tag{4.5}$$

Inspection of Table I of Paper VII indicates that the diagonal elements of $\mathbf{S}^{\frac{1}{2}}$ are the principal indices of refraction of the crystal. Since these are intrinsically positive, the only physically acceptable value of $\mathbf{S}^{\frac{1}{2}}$ is

$$\mathbf{S}^{\frac{1}{2}}=\begin{pmatrix}3 & 0\\ 0 & 2\end{pmatrix}. \tag{4.6}$$

To the general rule that a matrix has four distinct roots, an important exception is the case in which the matrix **F** is a unit matrix, or a scalar multiple of a unit matrix. The unit matrix has the four square roots

$$\mathbf{Y}=\begin{pmatrix}\pm 1 & 0\\ 0 & \pm 1\end{pmatrix}, \tag{4.7}$$

but has also an infinite number of roots

$$\mathbf{Y}=\begin{pmatrix}a & b\\ b & -a\end{pmatrix}, \tag{4.8}$$

where

$$a^2+b^2=1. \tag{4.9}$$

For example,

$$\mathbf{Y}=\frac{1}{2^{\frac{1}{2}}}\begin{pmatrix}1 & 1\\ 1 & -1\end{pmatrix} \tag{4.10}$$

is a square root of **1**. None of the roots (4.8), however, meets the criterion for physical acceptability.

General Solution

We now have the problem of choosing that one of the four mathematical values of **Y** which is physically acceptable. The most fruitful approach involves the use of the eigenvalues and eigenvectors of the matrix **F**.

The eigenvalues of **F** are the two distinct values of ξ which satisfy

$$(f_1-\xi)(f_2-\xi)=f_3f_4. \tag{4.11}$$

Let the two distinct values of ξ be denoted by ξ_1 and ξ_2. The eigenvectors $\mathcal{E}_1$ and $\mathcal{E}_2$ which correspond, respectively, to ξ_1 and ξ_2, are the vectors which satisfy

$$(\mathbf{F}-\xi_k\mathbf{1})\mathcal{E}_k=0. \quad k=1 \text{ or } 2. \tag{4.12}$$

It is easy to show that **Y** has the same eigenvectors as **F**, and that the eigenvalues of **Y**, denoted by ϕ_1 and ϕ_2, are related to the eigenvalues of **F** by

$$\phi_1^2=\xi_1,\quad \phi_2^2=\xi_2. \tag{4.13}$$

Then by use of the general expression for a matrix with given eigenvalues and given eigenvectors [Eq. (10) of Paper IV] one finds the following general expression for the matrix **Y**:

$$\mathbf{Y}=\frac{1}{\phi_1+\phi_2}\begin{pmatrix}f_1+\phi_1\phi_2 & f_4\\ f_3 & f_2+\phi_1\phi_2\end{pmatrix}, \tag{4.14}$$

where the ϕ's are defined by Eq. (4.14). There are four different pairs ϕ_1 and ϕ_2 which satisfy Eq. (4.14), and these correspond to the four mathematical square roots of **F**.

If, now, the antisymmetrical term in Eq. (3.12) be ignored, the eigenvalues ϕ_1 and ϕ_2 are the indices of refraction of the electric vectors of those electromagnetic waves which are transmitted through the crystal without change of state of polarization. There are four such waves, two $\mathcal{E}_1^{(+)}$ and $\mathcal{E}_2^{(+)}$ which move in the positive z-direction, and two $\mathcal{E}_1^{(-)}$ and $\mathcal{E}_2^{(-)}$ which move in the opposite direction. From these four waves, four different pairs of waves may be chosen, as follows:

$$\begin{aligned}&\mathcal{E}_1^{(+)},\ \mathcal{E}_2^{(+)};\\ &\mathcal{E}_1^{(+)},\ \mathcal{E}_2^{(-)};\\ &\mathcal{E}_1^{(-)},\ \mathcal{E}_2^{(+)};\\ &\mathcal{E}_1^{(-)},\ \mathcal{E}_2^{(-)}.\end{aligned} \tag{4.15}$$

These four pairs are the eigenvectors of the four

different matrices **Y**, and correspond, respectively, to the four different choices of ϕ_1 and ϕ_2:

$$\begin{aligned} &\bar{\phi}_1, \quad \bar{\phi}_2; \\ &\bar{\phi}_1, \quad -\bar{\phi}_2; \\ -&\bar{\phi}_1, \quad \bar{\phi}_2; \\ -&\bar{\phi}_1, \quad -\bar{\phi}_2; \end{aligned} \tag{4.16}$$

where $\bar{\phi}_1$ and $\bar{\phi}_2$ are the values of ϕ_1 and ϕ_2 which satisfy Eq. (4.13) *and which have positive real parts.*

The calculus involving the **N**- and **M**-matrices is specifically designed for handling light which is traveling in one direction only; accordingly, only the first and last of the pairs in Eq. (4.15) are acceptable on this ground. Furthermore, it has been assumed throughout this series of papers that the beam of light is traveling in the positive z-direction; accordingly, only the first of the pairs in Eq. (4.15) is acceptable.

Thus, in conclusion, the single value of **Y** which is acceptable physically is that one obtained from Eq. (4.13) and Eq. (4.14) by the added condition that ϕ_1 and ϕ_2 must have positive real parts.

A second way of writing the matrix **Y** in terms of **F** is

$$\mathbf{Y}=\begin{pmatrix} (f_1-\chi)^{\frac{1}{2}} & \chi^{\frac{1}{2}} \\ \chi^{\frac{1}{2}} & (f_2-\chi)^{\frac{1}{2}} \end{pmatrix}, \tag{4.17}$$

where χ is a root of

$$4Q^2\chi^2+4T\chi+P^2=0, \tag{4.18}$$

and where

$$\begin{aligned} T&=\tfrac{1}{2}(f_1+f_2), \\ P&=f_3f_4, \\ Q^2&=f_3f_4+\tfrac{1}{4}(f_1-f_2)^2. \end{aligned} \tag{4.19}$$

This solution, however, yields 32 different values of **Y**, only 4 of which are actual square roots of **F**, and only one of which is the physically acceptable root.

Numerical Example

The results of this section will be exemplified by a consideration of the specific matrix

$$\mathbf{F}=\begin{pmatrix} 9 & 1 \\ 1 & 4 \end{pmatrix}. \tag{4.20}$$

In order to find the square root of this matrix, we shall first employ Eq. (4.14). Equation (4.11) becomes

$$\xi^2-13\xi+35=0, \tag{4.21}$$

whose solution is

$$\xi_1=3.8074, \; \xi_2=9.1926. \tag{4.22}$$

One then has

$$\bar{\phi}_1=1.9513, \; \bar{\phi}_2=3.0319, \tag{4.23}$$

whose substitution in (4.14) yields

$$\mathbf{Y}=\begin{pmatrix} 2.9933 & 0.20067 \\ 0.20067 & 1.9899 \end{pmatrix}. \tag{4.24}$$

TABLE I.

C_1	g_{11}	g_{22}	g_{33}	g_{12}	g_{21}	g_{13}	g_{31}	g_{23}	g_{32}
C_2	g_{11}	g_{22}	g_{33}	g_{12}	g_{21}	0	0	0	0
C_3, C_4, C_6	g_{11}	g_{11}	g_{33}	g_{12}	$-g_{12}$	0	0	0	0
D_2	g_{11}	g_{22}	g_{33}	0	0	0	0	0	0
D_3, D_4, D_6	g_{11}	g_{11}	g_{33}	0	0	0	0	0	0
T, O	g_{11}	g_{11}	g_{11}	0	0	0	0	0	0
C_s	0	0	0	0	0	g_{13}	g_{31}	g_{23}	g_{32}
S_4	g_{11}	$-g_{11}$	0	g_{12}	g_{12}	0	0	0	0
D_{2d}	g_{11}	$-g_{11}$	0	0	0	0	0	0	0
C_{2v}	0	0	0	g_{12}	g_{21}	0	0	0	0
C_{3v}, C_{4v}, C_{6v}	0	0	0	g_{12}	$-g_{12}$	0	0	0	0
C_{3h}, D_{3h}, T_d	0	0	0	0	0	0	0	0	0

We shall now find the same result by means of the alternative solution (4.17). Equation (4.18) becomes

$$29\chi^2-26\chi+1=0, \tag{4.25}$$

whose solution is

$$\chi=0.040268, \; 0.85628. \tag{4.26}$$

Only the first of these two values of χ leads to a physically acceptable solution. If one substitutes the first of these two values in Eq. (4.17), one finds (4.24), except that the sign of each of the four matrix elements is undetermined. The signs must be determined by the condition that the eigenvalues of **Y** must have positive real parts, and by the condition that the matrix **Y** must be a square root of **F**. These two conditions yield (4.24) uniquely.

5.0 CRYSTALLOGRAPHY

We conclude this paper with a brief discussion of the limitations that crystal symmetry imposes on the form of the matrix $\boldsymbol{G}$.

It is well known that a crystal with a center of symmetry cannot possess optical activity. All crystals with centers of symmetry therefore have $\boldsymbol{G}$-tensors that are identically zero. Voigt[5] has shown that the tensor $\boldsymbol{G}$ is also zero for the crystal classes C_{3h}, D_{3h}, and T_d, even though these classes do not have centers of symmetry.

For the classes C_{3v}, C_{4v}, C_{6v}, Voigt shows that the only nonzero elements are $g_{12}=-g_{21}$. These classes are all uniaxial, however, and this has the result that the parameter Ω is identically zero for all directions of propagation through the crystal. Thus crystals of these classes also show no optical activity.

Table I shows the general form of the gyration tensor for all of the 21 crystal classes that do not have a center of symmetry. In this table, the distinguished axis is always the z-axis, in accordance with the usage of Voigt,[5] of Wooster,[6] and of the IRE Standards.[7]

[5] W. Voigt, *Lehrbuch der Kristallphysik* (B. G. Teubner, Leipzig, 1928), pp 313–315.

[6] W. A. Wooster, *Textbook on Crystal Physics* (Cambridge University Press, New York, 1938).

[7] Standards on Piezoelectric Crystals, Proc. Inst. Radio Engrs. 37, 1378–1395 (1949).

24

Written for this Benchmark volume by
Dr. Robert Clark Jones, Polaroid Corporation,
Cambridge, Mass.

A Comedy of Errors

R. CLARK JONES

In the theory of the optical properties of crystals, the anisotropy of the refraction is introduced by the dielectric tensor, and the anisotropy of the optical activity is introduced by the gyration tensor. When elements of these tensors are constants of the crystal, and in particular do not depend on the direction of the wave normal in the crystal, the tensors are called "material." Otherwise they are "nonmaterial."

The errors revolve about the assertions that the gyration tensors of Born,[1] Fedorov,[2] and of Jones[3] (the writer) are material. I believe there is now agreement that the gyration tensors of Fedorov and of Jones are material, that of Born, not.

My own formulation of the gyration tensor in Reference 3 in 1956 was based on Eq. (9) in an article by Condon and Seitz,[4] who in turn refer to earlier work by Voigt in 1905 and Gibbs in 1882. At the top of the second column on page 128, I clearly implied that my gyration tensor was identical with Born's; this is an error. In 1958 Pancharatnam[5] showed that the relation between my gyration tensor and Born's involved the index of refraction of the wave normal, and thus the two tensors could not both be material. Pancharatnam[5] accepted Born's erroneous statement that Born's tensor was material, and in a section entitled "An error in Jones' paper" concluded (erroneously) that my gyration tensor was nonmaterial.

In 1959 Fedorov[2] was able to show that the assumption that an equation of energy conservation exists is sufficient to reduce the number of constants from 27 to 9, and to obtain the equation of energy. His formulation follows that of Gibbs, Voigt, and of Condon and Seitz.[4] Fedorov compares his results with those of several others, and concludes that the Born gyration tensor of Reference 1 is nonmaterial. Fedorov indicates some puzzlement about this; Fedorov finds, from statements in a 1923 book by Born, that it is easy to show that Born's gyration tensor in Reference 1 is nonmaterial.

In summary, two actual errors were made. (I erroneously implied that Born's tensor and mine were identical; Born erroneously stated that his tensor was material.)

These errors led Pancharatnam to make erroneous statements. I conclude that the gyration tensors of Fedorov and of Jones are material; the gyration tensor of Born is nonmaterial.

The material in this letter has been known to me since 1959. The republication of my series of eight papers[3] by William Swindell of the University of Arizona seems an appropriate occasion to set the record straight.

References

1. Max Born, *Optik, Ein Lehrbuch der electromagnetischen Lichttheorie,* Springer, Berlin, 1933. The gyration tensor is introduced by Eqs. (1) and (2) on page 414.
2. F. I. Fedorov, *Optika i Spektroskopiia,* **6,** 85–93 (1959); English translation in *Optics and Spectroscopy,* **6,** 49–53 (1959). This is a monumental article, in which the author derives for the first time the correct equation of energy in an anisotropic, optically active, absorbing crystal. In the original Russian article time derivatives are indicated by dots that are of variable height above the letters. When the equations are reduced in the English translation, the reader should know that the dots are sometimes a bit obscure. (I am indebted to Fedorov for a reprint of the Russian article.) The gyration tensor is not so much introduced as it is derived. See Eqs. (1) to (13). There is a typographical error in the important Eq. (13): The boldface H should be a boldface E; this error affects both the Russian and the English versions.
3. R. Clark Jones, *J. Opt. Soc. Am.,* **46,** 126–131 (1956). This is the last of a series of eight papers. The gyration tensor is introduced in Eq. (2.13).
4. E. U. Condon and F. Seitz, *J. Opt. Soc. Am.,* **22,** 393–401 (1932).
5. S. Pancharatnam, *Proc. Indian Acad. Sci.,* **48,** 227–244 (1958). The indicated section is on page 244.

V
Other Descriptions of Polarization

Editor's Comments on Papers 25 Through 30

25 Perrin: *Polarization of Light Scattered by Isotropic Opalescent Media*

26 Parke: *Optical Algebra*

27 Wolf: *Coherence Properties of Partially Polarized Electromagnetic Radiation*

28 Parrent and Roman: *On the Matrix Formulation of the Theory of Partial Polarization in Terms of Observables*

29 Barakat: *Theory of the Coherency Matrix for Light of Arbitrary Spectral Bandwidth*

30 Takenaka: *A Unified Formalism for Polarization Optics by Using Group Theory*

Hans Mueller formulated a matrix method of describing optical instruments. Using a 4 × 4 matrix, the system is capable of handling scattering and depolarizing systems as well as the simpler types of polarization elements. Mueller did not publish his work on this subject in the open literature.

Mueller matrix methods and applications are well described by Edward Collett ["The Description of Polarization in Classical Physics," *Amer. J. Phys.*, **36,** 713–725 (1968); **39,** 517–528 (1971)] and William H. McMaster ["Matrix Representation of Polarization," *Rev. Modern Phys.*, **33,** 8–28 (1961)].

Prior to Mueller, Paul Soleillet ["Sur les paramètres caractérisant la polarisation partielle de la lumière dans les phénomènes de fluorescence," *Ann. Phys.*, **12,** 23 (1929)] showed that the Stokes parameters of a beam of fluorescent radiation were linearly related to the Stokes parameters of the exciting beam.

Francis Perrin developed a general theory of scattering using a 4 × 4 matrix now commonly known as the Mueller matrix. This paper is reprinted here and may be considered to be the foundation paper in the area of 4 × 4 matrix descriptions of polarized light and polarizing devices.

Francis Perrin, son of the Nobel Laureate Jean Perrin, was born in Paris in 1901. He held the position of Professor at the Collège de France from 1946 to 1972. He was also High Commissioner of the French Atomic Energy Commission from 1951 to 1970 and has been honored with Membership of the French Academy of Sciences (1951) and several other national academies.

Following the pioneering work of Stokes and the later studies by Wiener, Jones, and Mueller, there have been many articles in the literature concerning advances in the theoretical descriptions of polarized light and polarization-related devices. Space forbids the inclusion of more than a representative sample of the more recent contributions to the field.

Papers 26 through 30 are in the areas of matrix algebras (Parke, Takenaka), coherence and partial polarization (Wolf, Barakat), and partial polarization and instrumental matrices (Parrent and Roman).

Nathan Grier Parke, III (1927–) studied under Hans Mueller at MIT (1945–1948). The paper included here is condensed from part of his Ph.D. dissertation. Since 1951, he has been president of Parke Mathematical Laboratories, Inc.

Emil Wolf (1922–) is professor of physics at the University of Rochester. He is well known for introducing to the English language and updating Max Born's book *Optik.* Wolf has published extensively in the areas of physical optics and electromagnetic theory.

George B. Parrent, Jr. (1931–) has been with Technical Operations Research, Inc., since 1965 and is known for his work in coherence theory and information theory as applied to optics.

Paul Roman (1925–) is professor of physics at Boston University. His major research interests have been theory of elementary particles and quantum field theory.

Richard Barakat has been senior scientist at Bolt, Beranek and Newman, Inc., since 1970, and lecturer in applied mathematics at Harvard University since 1969. He has published considerably in the area of mathematical optics.

Hiroshi Takenaka (1941–) is with the Nippon-Kogaku Co., Ltd., in Tokyo. His principal research areas include polarization microscopy and electro-optical pattern recognition.

25

Reprinted from *J. Chem. Phys.*, **10**, 415–427 (July 1942)

Polarization of Light Scattered by Isotropic Opalescent Media*

FRANCIS PERRIN

University of Paris, Paris, France, and Department of Chemistry, Columbia University, New York, New York

(Received March 23, 1942)

A general study is given of the polarization of light scattered by isotropic media whose elements of heterogeneity are not very small in comparison with the wave-length, (suspensions, colloidal solutions, solutions of large molecules, . . .). This includes an extension of a theory by R. S. Krishnan, who, considering certain particular states of polarization of the incident light and applying the law of reciprocity, had proved the equality of two of the four coefficients which are to be considered in these cases. Using Stokes' linear representation of the polarization of light beams, it is shown that the scattering through a given angle and for a given wave-length is characterized by the 16 coefficients of the linear forms which express the four polarization parameters of the scattered beam in terms of the four corresponding parameters of the incident beam and that the law of reciprocity leads to six relations between these sixteen coefficients. For an isotropic asymmetrical medium (having rotatory power), the scattering is thus characterized by *ten* independent coefficients. In the case of a symmetrical medium, four of these coefficients must be zero, leaving only *six* scattering coefficients, and if the scattering particles are spherical, there are two additional relations between these coefficients. The comparison with dipolar scattering by very small elements shows that the best test to prove multipolar scattering is the existence of some ellipticity in the scattered light when the incident beam is linearly polarized in a direction oblique to the scattering plane.

I

THE scattering of light by a macroscopically homogeneous medium is caused by some microscopical structure. If the dimensions of the elements of this structure are very small in comparison with the wave-length of the light, the scattering has the well-known simple characteristics of secondary dipolar emission.[17–20] But if

* Publication assisted by the Ernest Kempton Adams Fund for Physical Research of Columbia University.

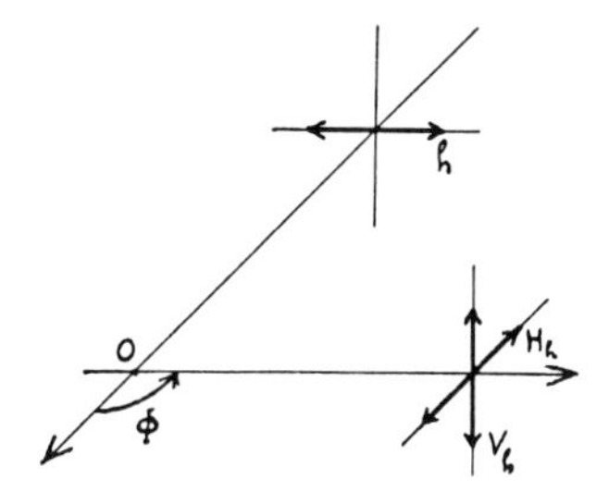

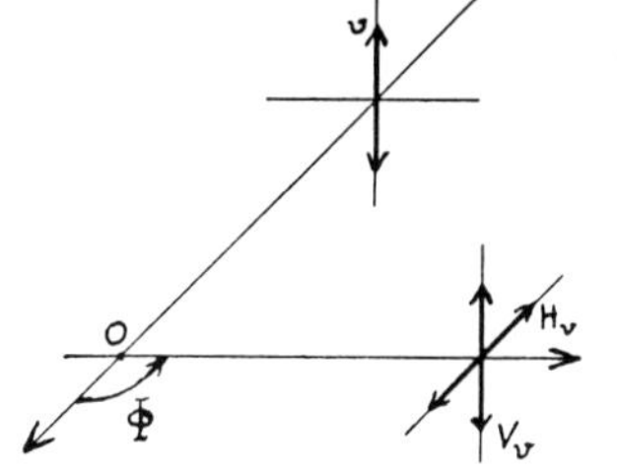

FIG. 1.

these elements have dimensions comparable to the wave-length, the phenomenon is more complicated, and has been experimentally studied only in particular conditions of excitation or observation, and theoretically only for spherical scattering particles.[7, 21]

Our purpose is to extend a method used by R. S. Krishnan, and to point out the independent parameters which are necessary for specifying, in general, the intensity and polarization of the light scattered by any isotropic medium for given scattering angle and wave-length. A summary of this research has been read before the French Society of Physical Chemistry in May, 1939.[22]

We shall have to distinguish between *symmetrical* media, for which the center of any large spherical volume is a center of symmetry and any plane through this center a plane of symmetry, and the asymmetrical media, which usually have some optical rotatory power. The isotropy may be only statistical, as the result of an isotropic distribution of small anisotropic elements.

We shall see that for isotropic media, whose scattering elements are not very small compared with the wave-length, media which are more or or less turbid or opalescent, it is necessary to introduce several new parameters, whose values will contribute to the determination of the magnitude, shape, and optical properties of these elements. This method will be applicable in the study of smokes, fogs, suspensions, emulsions, colloidal solutions, solutions of large molecules, and also of media with widely extended fluctuations such as pure fluids, or liquid mixtures, near their critical state, glasses, etc.

My thanks are due to Dr. R. Wurmser with whom I had several discussions on these questions of scattering, which were the origin of this research.

II. THE RELATION OF R. S. KRISHNAN

In several papers published in 1938, R. S. Krishnan[1–5] has established theoretically, and has given the experimental verification of, a relation between the intensities of certain components of the light scattered by an isotropic symmetrical medium for some particular conditions of polarization of the incident light.

At a point of a horizontal light beam linearly polarized and of given intensity, he considered the scattering in a horizontal direction making an angle ϕ with the incident beam. He denoted by H_h and V_h the intensities of the horizontal and vertical vibrations of the scattered beam when the direction of vibration of the incident beam is horizontal, and by H_v and V_v these intensities when this direction is vertical (Fig. 1). He defined the corresponding depolarization factors of the scattered light by the ratios (generally smaller than 1)

$$\rho_h = V_h/H_h, \quad \rho_v = H_v/V_v. \tag{1}$$

The superposition without any phase relation, of the two considered incident beams, polarized at right angles and of the same intensity, gives an unpolarized incident beam to which corresponds a scattered beam whose horizontal and vertical intensities of vibration are $H_u = H_h + H_v$ and $V_u = V_h + V_v$. The depolarization factor of this scattered beam is defined by the ratio

$$\rho_u = H_u/V_u$$

and has thence the value

$$\rho_u = (H_h + H_v)/(V_h + V_v). \tag{2}$$

The measurements of the three depolarization

factors ρ_v, ρ_h, and ρ_u thus give the ratios of the four quantities H_h, V_h, H_v, V_v.

Using a general law of reciprocity due to Lord Rayleigh, R. S. Krishnan obtained the relation

$$H_v = V_h \quad (3)$$

for any symmetrical isotropic medium, and proved that this relation must be true also for a medium with only axial symmetry around the vertical direction perpendicular to the plane of scattering.[2]

This relation is verified in two particular cases already known: (1) For very small scattering particles (dipolar scattering) V_v, H_v, V_h are independent of ϕ, and (Lord Rayleigh)[6]

$$H_v = V_h,$$
$$H_h = V_v \cos^2 \phi + H_v \sin^2 \phi,$$

which gives for transverse scattering

$$H_v = V_h = H_h \quad (\text{dipoles}, \phi = \pi/2). \quad (4)$$

(2) For spherical scattering particles of any dimension, and for all values of ϕ (G. Mie)[7]

$$H_v = V_h = 0 \quad (\text{spheres}). \quad (5)$$

R. S. Krishnan gave, moreover, the experimental proof of Eq. (3) for various non-spherical large particles, for which the observed intensities H_v and V_h are always equal, but generally different from 0 and from H_h.

From the Eqs. (1), (2), and (3), the relation (6) results:

$$\rho_u = (1 + 1/\rho_h)/(1 + 1/\rho_v). \quad (6)$$

As a consequence of reciprocity it is thus unnecessary to measure the depolarization factor ρ_u for unpolarized excitation, if the depolarization factors ρ_h and ρ_v for horizontally and vertically polarized excitation have been measured.

Finally, in a paper published in November 1939, R. S. Krishnan[8] considered the case in which the direction of vibration of the linearly polarized incident beam makes an angle θ with the normal to the plane of scattering. Neglecting the correlation of phase which then exists between the horizontal and vertical components of vibration of the incident beam, he obtained for the ratio of the intensities of the horizontal and vertical components of vibration of the scattered light the equation,

$$\rho_\theta = H_\theta / V_\theta = (1 + \text{tg}^2 \theta / \rho_h)/(\text{tg}^2 \theta + 1/\rho_v). \quad (7)$$

This equation, though in agreement with the particular theoretical results of Lord Rayleigh (any small particles) and of G. Mie (large spheres) did not seem to him to be generally true, because of the arbitrariness of the hypothesis he had made to obtain it. He reported even a few measurements he had made of the scattering by large non-spherical particles showing some disagreement with it. However, we shall show that this equation must be valid for any symmetrical medium.

III. THE LAW OF RECIPROCITY IN OPTICS

In his book on the *Theory of Sound*, Lord Rayleigh established a theorem of reciprocity for the forces and displacements in the neighborhood of an equilibrium state of a mechanical system governed by linear equations.[9] Later on he extended to optics, without a new demonstration, the law of reciprocity; he merely indicated in a footnote the necessity of specifying the states of polarization.[10] R. S. Krishnan referred to this statement by Lord Rayleigh of the theorem of reciprocity to establish, in the case of light scattering, the relation we have given above. But the conditions he considered are particular, because he did not take into account the possible correlation of phases between the two components of vibration of each beam of light.

To apply the law of reciprocity to the most general phenomenon of scattering by an isotropic medium, it is necessary to start from a precise statement of this law.

Any monochromatic beam of light may be considered, in an infinite number of ways, as the superposition, with more or less phase correlation, of two completely polarized beams of complementary characters, for instance rectangular linear polarizations, or inverse circular polarizations. We shall choose as reference polarization states, for each direction of propagation, the states of linear polarization along two fixed rectangular axes.

Let us consider, given any system in which light can be scattered and absorbed, for an incident linearly polarized beam F_1 having an intensity I_1, coming from a linear polarizer N, a particular emerging beam from which we may separate a linearly polarized component F'_1, having an intensity I'_1, by means of a linear polarizer N'. Let us associate with these beams the inverse beams, that is to say, an incident polarized beam F_2 coming from the polarizer N', with an intensity equal to I_1, in the direction opposite to that of the emerging beam F'_1, and the corresponding emerging beam F'_2 coming out of the polarizer N in the direction opposite to that of the incident beam F_1. The law of reciprocity states that the intensity I'_2 of this last beam F'_2 is equal to the intensity I'_1 of the beam F'_1: *If two incident polarized beams have equal intensities, the inverse emerging beams of the same polarization, which are associated with them, also have equal intensities.*

This law is true only if the considered optical system is not affected by a reversal of time, so that the sense of propagation of light be immaterial. There must be no movements, no electrical currents, no magnetic fields. To extend it to more general cases it is necessary to reverse, together with the direction of light propagation, all movements, electrical currents, and magnetic fields. For instance, it is well known that the law of reciprocity, as stated above, is not true for a system in which magnetic rotatory power comes into account, if the magnetic fields are not reversed with the sense of propagation of light.

Moreover, only monochromatic beams of the same frequency must be considered. At least it is necessary that the mechanisms which modify the frequency can be reversed with the propagation of light, as for instance a change of frequency caused by scattering by a moving body. The law of reciprocity is not valid for fluorescence or for Raman effect, in which the change of frequency is irreversible. In scattering phenomenon it is only relevant for Rayleigh scattering, with no or small symmetrical frequency changes.

It is also interesting to give the corpuscular statement of the law of reciprocity: If a photon associated with the incident polarized beam F_1 has a probability p to come out of the optical system associated with the ,polarized beam F'_1, then inversely, a photon associated with the beam F_2, reverse, to F'_1, has the same probability p to come out associated with the beam F'_2 reverse to F_1.

The law of reciprocity is thus seen to be connected with the general principle of quantum mechanics asserting the equal probability of inverse transitions between two states of the same energy.

IV. STOKES' LINEAR REPRESENTATION OF STATES OF POLARIZATION

Let us first consider a completely polarized monochromatic beam of light, whose electrical vibration may be represented by its components along two rectangular axes

$$E_x = p_1 \cos(\omega t + \varphi_1),\quad E_y = p_2 \cos(\omega t + \varphi_2), \tag{8}$$

the amplitudes p_1, p_2, and the frequency $\omega/2\pi$ being positive. Let δ be the phase difference of these components

$$\delta = \varphi_1 - \varphi_2 \tag{9}$$

and I_e the total intensity of vibration

$$I_e = p_1^2 + p_2^2. \tag{10}$$

The terminal point of the oscillating vector $\mathbf{E}$ thus specified, describes, in the direct or reverse sense according to the positive or negative sign of $\sin\delta$, an ellipse with semi-axes a and b ($b \leqslant a$), whose major axis makes an angle α with the x axis. Let us set

$$\operatorname{tg}\beta = \pm b/a, \qquad -\pi/4 \leqslant \beta \leqslant \pi/4, \tag{11}$$

taking the sign + or − according to the sense of rotation, i.e., so that always

$$\operatorname{tg}\beta \sin\delta > 0.$$

With such definitions, it is known that

$$\begin{aligned} p_1^2 - p_2^2 &= I_e \cos 2\beta \cos 2\alpha, \\ 2p_1p_2 \cos\delta &= I_e \cos 2\beta \sin 2\alpha, \\ 2p_1p_2 \sin\delta &= I_e \sin 2\beta. \end{aligned} \tag{12}$$

These three quantities, which we shall name M_e, C_e, S_e, determine the elliptic vibration (except its phase). In Poincaré's representation they are considered as the rectangular components of a vector in space, whose length is I_e, longitude 2α, and latitude 2β.

No actual light is strictly monochromatic. The amplitudes and phases of the components of any light vibration undergo slow variations without strict correlation. The ellipse of vibration, which is still determined at each moment, changes its shape and magnitude, slowly in comparison with the period of vibration but extremely swiftly in comparison with the duration of any measurement. It is thus possible to measure only mean values.

The study of the polarization of a light beam requires the use of analyzers, each giving the mean intensity of a vibration E_a obtained as a linear combination, with given changes in phase, of the two components E_x and E_y of the initial vibration

$$E_a = c_1 p_1 \cos(\omega t + \varphi_1 + \eta_1) + c_2 p_2 \cos(\omega t + \varphi_2 + \eta_2). \quad (13)$$

This mean intensity has the value

$$\begin{aligned} I_a = &\tfrac{1}{2}(c_1^2 + c_2^2)(\langle p_1^2\rangle_{Av} + \langle p_2^2\rangle_{Av}) \\ &+ \tfrac{1}{2}(c_1^2 - c_2^2)(\langle p_1^2\rangle_{Av} - \langle p_2^2\rangle_{Av}) \\ &+ c_1 c_2 \cos(\eta_1 - \eta_2)\langle 2p_1 p_2 \cos\delta\rangle_{Av} \\ &- c_1 c_2 \sin(\eta_1 - \eta_2)\langle 2p_1 p_2 \sin\delta\rangle_{Av}, \end{aligned}$$

the notation $\langle\ \rangle_{Av}$ denoting the mean value with respect to time. By the use of four different (linearly independent) analyzers it is thus possible to calculate the quantities

$$\begin{aligned} I &= \langle p_1^2\rangle_{Av} + \langle p_2^2\rangle_{Av} = \langle I_e\rangle_{Av}, \\ M &= \langle p_1^2\rangle_{Av} - \langle p_2\rangle_{Av} = \langle M_e\rangle_{Av}, \\ C &= 2\langle p_1 p_2 \cos\delta\rangle_{Av} = \langle C_e\rangle_{Av}, \\ S &= 2\langle p_1 p_2 \sin\delta\rangle_{Av} = \langle S_e\rangle_{Av}, \end{aligned} \quad (14)$$

and if these four quantities are known, it is possible to calculate the intensity that will be measured with any analyzer corresponding to certain values of the coefficients c_1 and c_2 and of the phase shift $(\eta_1 - \eta_2)$. That is to say, the four quantities I, M, C, S give a complete description of the polarization properties of the light beam (Stokes).[11]

It is easy to prove that for any beam of light the parameters I, M, C, S, verify the inequality

$$I \geqslant (M^2 + C^2 + S^2)^{\frac{1}{2}}, \quad (15)$$

since the equality is true only for completely polarized light; and if four quantities satisfy this condition they may be considered the polarization parameters of a light beam.

Any light beam, having a partial polarization specified by the values I, M, C, S of the Stokes' parameters, may be considered as the superposition, without any phase correlation, of a beam of natural light having an intensity

$$I_N = I - (M^2 + C^2 + S^2)^{\frac{1}{2}} \quad (16)$$

and of a beam of completely polarized elliptic light having an intensity

$$I_E = (M^2 + C^2 + S^2)^{\frac{1}{2}} \quad (17)$$

and whose ellipse of vibration is defined by the angles α and β given by the relations

$$\begin{aligned} I_E \cos 2\beta \cos 2\alpha &= M, \\ I_E \cos 2\beta \sin 2\alpha &= C, \\ I_E \sin 2\beta &= S. \end{aligned} \quad (18)$$

The ratio

$$p = I_E / I = (M^2 + C^2 + S^2)^{\frac{1}{2}} / I, \quad (0 \leqslant p \leqslant 1) \quad (19)$$

is called the degree of polarization.

The essential property of the Stokes' parameters is their additivity in the superposition of two independent beams of light, i.e., without any correlation between the perturbations of their phases or amplitudes. This additivity corresponds to the absence of any interference.

When a beam of light passes through some optical arrangement, or more generally, produces a secondary beam of light, the intensity and the state of polarization of the emergent beam are functions of those of the incident beam. If two independent incident beams are superposed the new emergent beam will be, if the process is linear, the superposition without interference of the two emergent beams corresponding to the separate incident beams. Consequently, in such a linear process, from the additivity properties of the Stokes' parameters, the parameters I', M', C', S' which define the polarization of the emergent beam must be homogeneous linear functions of the parameters I, M, C, S corresponding to the incident beam; the sixteen coefficients of these linear functions will completely characterize the corresponding optical phenomenon. This fundamental remark is due in its general formulation to P. Soleillet.[12] We shall use it in the next section in the study of scattering. Let us give here only the linear transformation formulas of the Stokes' parameters in two simple cases which we shall have to consider:

When the light beam is rotated through an angle ψ around its direction of propagation, for instance by passing through a crystal plate with simple rotatory power, we have

$$\begin{aligned} I' &= I, \\ M' &= M \cos 2\psi - C \sin 2\psi, \\ C' &= M \sin 2\psi + C \cos 2\psi, \\ S' &= S, \end{aligned} \tag{20}$$

and these equations also give the transformation of the Stokes' parameters when the reference axes are rotated through an angle $-\psi$.

When a difference in phase φ is introduced between the components of the vibration along the axes, for instance by a birefringent crystal plate with axes parallel to the reference axes (axes of maximum speed along Ox for $\varphi > 0$), we have

$$\begin{aligned} I' &= I, \\ M' &= M, \\ C' &= C \cos \varphi - S \sin \varphi, \\ S' &= C \sin \varphi + S \cos \varphi. \end{aligned} \tag{21}$$

It is interesting to note how the method used by Stokes' to characterize a state of polarization may be generalized and connected with the wave statistics of J. von Neumann:[13] Let us consider a system of n harmonic oscillations of the same frequency subjected to small random perturbations; we may represent them by complex expressions

$$E_k = P_k \exp(i\omega t), \quad P_k = p_k \exp(i\varphi_k), \tag{22}$$

the modulus p_k and the arguments φ_k varying in course of time, slowly in comparison with the period of oscillation, but quickly in comparison with the duration of any measurement. Let us suppose that we can measure the mean intensity of an oscillation E linearly dependent on these oscillations

$$E = \sum_k C_k E_k, \quad C_k = c_k \exp(i\eta_k). \tag{23}$$

The value of this mean intensity is (the asterisk indicating the change to the complex conjugate quantity)

$$\langle EE^* \rangle_{Av} = \sum_{kl} C_k C_l^* \langle P_k P_l^* \rangle_{Av}. \tag{24}$$

The mean intensity depends on the particular oscillations involved only through the von Neumann's matrix

$$\Gamma_{kl} = \langle P_k P^*_l \rangle_{Av}, \tag{25}$$

the knowledge of which determines all we can know about these oscillations by such measurements. Since this matrix is hermitic, we can set

$$\Gamma_{kk} = \mu_k, \quad \Gamma_{kl} = \gamma_{kl} + i\sigma_{kl}, \quad (k \neq l) \tag{26}$$

μ_k, $\gamma_{kl} = \gamma_{lk}$, $\sigma_{kl} = -\sigma_{lk}$ being real quantities. The diagonal terms μ_k are the mean intensities of the oscillations:

$$\mu_k = \langle p_k^2 \rangle_{Av} \tag{27}$$

and the other terms give the correlations between the oscillations:

$$\begin{aligned} \gamma_{kl} &= \langle p_k p_l \cos(\varphi_k - \varphi_l) \rangle_{Av}, \\ \sigma_{kl} &= \langle p_k p_l \sin(\varphi_k - \varphi_l) \rangle_{Av}. \end{aligned} \tag{28}$$

The state of excitation of n oscillators having the same frequency is thus defined by n^2 real quantities. For instance, a vectorial vibration in space having three components must be defined by *nine* quantities (P. Soleillet),[12] and the possibilities of interference between two beams of light of the same frequency depend on *sixteen* parameters, since there are four components (two for each beam).

When the different oscillations are the components of an oscillating vector **P** $\exp(i\omega t)$, the matrix Γ_{kl} has, for any change of the reference axes, the variance of a tensor of the second order, since its elements are the mean values of the products of the components of two vectors (**P** and **P***).

It is easy to prove that the determinant of the matrix Γ_{kl} and all its diagonal minors are always positive or zero.

When the ratios of the P_k's are independent of time, the oscillation state of the system is said to be *pure* (complete polarization in the case of a light beam); all the diagonal minors of the matrix Γ_{kl} are then zero, and conversely. When it is not so, the state is said to be *mixed* (partial polarization). In general, any mixed state of oscillation of a system of n oscillators may be considered, in an infinite number of ways, as the superposition of n pure states without correlation (for a state to be equivalent to the superposition of less pure states, it is necessary that the determinant of the Γ_{kl} be zero).

Still more generally, it is possible, in a similar way, to find the quantities which will appear in the linear investigation of a non-harmonic system whose motion is described by n statistical func-

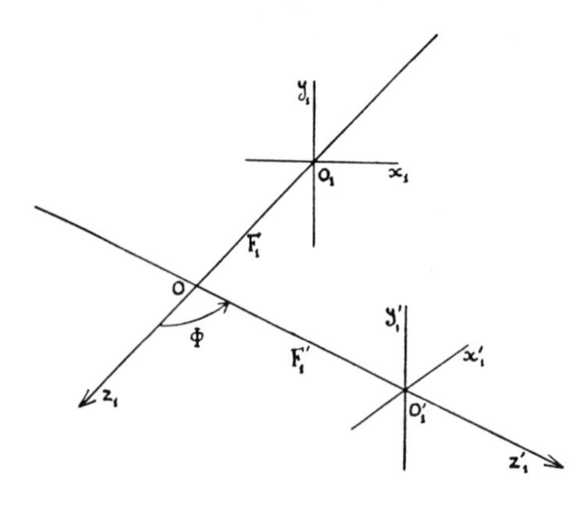

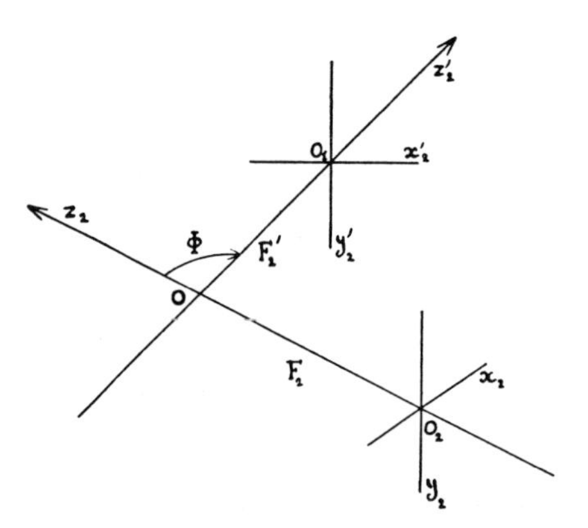

FIG. 2. (Above.) FIG. 3. (Below.)

tions of the time $E_k(t)$: These will be the *correlation functions* of M. Courtines[14] and J. Bernamont[15] defined by the relation

$$f_{kl}(\tau) = \langle E_k(t)E_l(t+\tau)\rangle_{Av} = f_{lk}(-\tau) \qquad (29)$$

or the functions corresponding to them by the Laplacian transformation

$$\Gamma_{kl}(\nu) = \int_{-\infty}^{+\infty} f_{kl}(\tau) \exp(-2\pi i\nu\tau)d\tau, \qquad (30)$$

which are the amplitudes of their representation by Fourier's integrals

$$f_{kl}(\tau) = \int_{-\infty}^{+\infty} \Gamma_{kl}(\nu) \exp(2\pi i\nu\tau)d\nu \qquad (31)$$

and verify the condition of hermiticity

$$\Gamma_{lk}(\nu) = \Gamma_{kl}(\nu)^*. \qquad (32)$$

It is thus seen that the state of motion of a system having n degrees of freedom is characterized, when linear methods of analysis are used, by n^2 continuous spectra giving, for each frequency, n spectral densities of intensity and $n(n-1)$ spectral densities of correlation.

V. THE SIXTEEN SCATTERING COEFFICIENTS OF AN ARBITRARY ISOTROPIC MEDIUM

If a monochromatic parallel beam of light F_1 is propagated along an axis O_1z_1, we consider the light scattered at a point O of this axis in a direction Oz'_1, making an angle ϕ (between 0 and π) with Oz_1. Let us specify the state of polarization of the incident beam F_1 by the values I_1, M_1, C_1, S_1 of its Stokes' parameters for the axes $O_1x_1y_1$, and that of the scattered beam F'_1 by the values I'_1, M'_1, C'_1, S'_1 of its Stokes' parameters for the axes $O'_1x'_1y'_1$. The two sets of rectangular axes $O_1x_1y_1z_1$ and $O'_1x'_1y'_1z'_1$ are right handed; the planes $O_1z_1x_1$ and $O'_1z'_1x'_1$ coincide; and the parallel axes O_1y_1 and $O'_1y'_1$ are orientated so that the rotation ϕ smaller than π which brings Oz_1 on Oz'_1 is positive around their common direction (Fig. 2).

We assume the linear character of light scattering for the superposition of non-coherent beams: If two independent light beams F and G which can be propagated along O_1z_1 give, when separate, the scattered beams F' and G' along Oz'_1, the scattered beam when the incident beams F and G are superposed without any phase relation, may be obtained by the superposition without phase relation of the two beams F' and G'. Since the Stokes' parameters are additive for the superposition of non-coherent beams, this requires that the quantities I'_1, M'_1, C'_1, S'_1 be linear homogeneous functions of the quantities I_1, M_1, C_1, S_1, i.e., that

$$\begin{aligned} I'_1 &= a_{11}I_1 + a_{12}M_1 + a_{13}C_1 + a_{14}S_1, \\ M'_1 &= a_{21}I_1 + a_{22}M_1 + a_{23}C_1 + a_{24}S_1, \\ C'_1 &= a_{31}I_1 + a_{32}M_1 + a_{33}C_1 + a_{34}S_1, \\ S'_1 &= a_{41}I_1 + a_{42}M_1 + a_{43}C_1 + a_{44}S_1. \end{aligned} \qquad (33)$$

If the scattering is produced by an isotropic medium, the sixteen *scattering coefficients* a_{ik} will depend only, for a given medium, on the frequency of the light and on the scattering angle ϕ.

We shall see in the next sections that symmetry and reciprocity arguments prove that the number of independent scattering coefficients is actually less than sixteen.

VI. SYMMETRICAL MEDIUM

If the scattering medium is symmetrical, and consequently without any rotatory power, the plane of scattering $Oz_1z'_1$ determined by the directions of excitation and observation, is a plane of symmetry for the medium. Therefore, if the incident beam (I_1, M_1, C_1, S_1) is replaced by the symmetrical beam with respect to this plane, a beam whose parameters are (I_1, M_1, $-C_1$, $-S_1$), the new scattered beam must be symmetrical to the first scattered beam, and hence its parameters will be (I'_1, M'_1, $-C'_1$, $-S'_1$). In other words the relations (33) must hold if the sign of the parameters C_1, S_1, C'_1, S'_1 is changed whatever the values of I_1, M_1, C_1, S_1. This requires that for an isotropic symmetrical medium

$$a_{13}=a_{14}=a_{23}=a_{24}=a_{31}=a_{32}=a_{41}=a_{42}=0,$$

and consequently that for such a medium the relations (33) reduce to

$$\begin{aligned} I'_1&=a_{11}I_1+a_{12}M_1,\\ M'_1&=a_{21}I_1+a_{22}M_1,\\ C'_1&=a_{33}C_1+a_{34}S_1,\\ S'_1&=a_{43}C_1+a_{44}S_1. \end{aligned} \tag{33A}$$

For a symmetrical medium the number of scattering coefficients is only eight.

VII. RECIPROCITY

We must now apply the law of reciprocity, and for this purpose consider an incident beam being propagated along the axis Oz_2 opposite to the direction Oz'_1 of the first scattered beam, and the corresponding scattered beam F'_2 in the direction Oz'_2 opposite to the direction Oz_1 of the first incident beam (Fig. 3). We shall take as reference axes for the states of polarization of these new beams two sets of axes having, with respect to F_2 and F'_2 the same orientation as the axes previously used with respect to F_1 and F'_1: For beam F_2 the axis O_2x_2 in coincidence with $O'_1x'_1$ and the axis O_2y_2 opposite to $O'_1y'_1$, for beam F'_2 the axis $O'_2x'_2$ in coincidence with O_1x_1 and the axis $O'_2y'_2$ opposite to O_1y_1. With these reference axes the Stokes' parameters I'_2, M'_2, C'_2, S'_2 of the scattered beam F'_2 will be expressed as linear functions of the parameters I_2, M_2, C_2, S_2 of the incident beam F_2 by relations identical to the relations (33):

$$\begin{aligned} I'_2&=a_{11}I_2+a_{12}M_2+a_{13}C_2+a_{14}S_2,\\ M'_2&=a_{21}I_2+a_{22}M_2+a_{23}C_2+a_{24}S_2,\\ C'_2&=a_{31}I_2+a_{32}M_2+a_{33}C_2+a_{34}S_2,\\ S'_2&=a_{41}I_2+a_{42}M_2+a_{43}C_2+a_{44}S_2. \end{aligned} \tag{34}$$

The coefficients a_{ik} have the same values, since the scattering angle has the same value ϕ and since the medium is supposed to be isotropic.*

To be able to express reciprocity, we must introduce on the paths of the incident and scattered beams some polarizers transforming an initial beam having a fixed linear polarization into an incident beam of any elliptical polarization, and any component of the scattered beam into a final beam having, like the initial beam a fixed linear polarization.

For this purpose we may introduce on the path of the incident beam F_1: (1) a linear polarizer N with fixed orientation such that the electrical vibration of the light emerging from it is along O_1x_1, this light having thus the Stokes' parameters (1, 1, 0, 0) if its intensity is unity; (2) a crystal plate R with rotatory power turning the light vibration through an angle ψ; according to formulas (20) the Stokes' parameters of the beam after this plate will be (1, $\cos 2\psi$, $\sin 2\psi$, 0); (3) a birefringent crystal plate B with its axes parallel to the reference axes $O_1x_1y_1$ and producing a change in phase φ between the vibrations along O_1x_1 and O_1y_1. The incident beam thus obtained may have any complete elliptical polarization; its Stokes' parameters will be, according to Eq. (21),

$$\begin{aligned} I_1&=1,\\ M_1&=\cos 2\psi,\\ C_1&=\sin 2\psi \cos \varphi,\\ S_1&=\sin 2\psi \sin \varphi. \end{aligned} \tag{35}$$

* It is even sufficient that the direction Oz_1 and Oz_2 be equivalent in the scattering medium, which might, for instance, have only axial symmetry around the perpendicular Oy to the scattering plane.

Similarly, on the path of the scattered beam F'_1 whose Stokes' parameters I'_1, M'_1, C'_1, S'_1 will be obtained by relations (33) in which I_1, M_1, C_1, S_1 have the values given by (35), we shall introduce: (1) a birefringent crystal plate B' with its axes parallel to the axes $O'_1x'_1y'_1$ and producing a change in phase φ' between the vibrations along $O'_1x'_1$ and $O'_1y'_1$; after this plate the Stokes' parameters of the beam will be (Eq. 21)

$$\begin{aligned}&I'_1,\\&M'_1,\\&C'_1\cos\varphi'-S'_1\sin\varphi',\\&C'_1\sin\varphi'+S'_1\cos\varphi';\end{aligned}$$

(2) a crystal plate R' with rotatory power turning the light vibration through an angle ψ'; after this plate the Stokes' parameters will become (Eq. 20)

$$\begin{aligned}&I'_1,\\&M'_1\cos 2\psi'-(C'_1\cos\varphi'-S'_1\sin\varphi')\sin 2\psi',\\&M'_1\sin 2\psi'+(C'_1\cos\varphi'-S'_1\sin\varphi')\cos 2\psi',\\&C'_1\sin\varphi'+S'_1\cos\varphi';\end{aligned}$$

(3) a linear polarizer N' with fixed orientation selecting the electrical vibration along $O'_1x'_1$. The intensity J_1 of the light coming out of this polarizer will be equal to half the sum of the first two Stokes' parameters of the light entering it; thus

$$J_1=\tfrac{1}{2}[I'_1+M'_1\cos 2\psi'-C'_1\sin 2\psi'\cos\varphi' +S'_1\sin 2\psi'\sin\varphi']. \quad (36)$$

Let us consider now the light being propagated in the reverse direction through this arrangement, when the intensity of the beam F_2 coming out of the polarizer N' toward the scattering medium has an intensity equal to unity. The values of the Stokes' parameters are then for this beam (1, 1, 0, 0) and after the plates R' and B' they will be

$$\begin{aligned}I_2&=1,\\M_2&=\cos 2\psi',\\C_2&=\sin 2\psi'\cos\varphi',\\S_2&=\sin 2\psi'\sin\varphi'.\end{aligned}\quad (37)$$

The parameters I'_2, M'_2, C'_2, S'_2 of the corresponding scattered beam will be given by Eq. (34) I_2, M_2, C_2, S_2 having the values (37), and the intensity of the light coming out of the arrangement through the polarizer N, obtained as for the first scattered beam, will be

$$J_2=\tfrac{1}{2}[I'_2+M'_2\cos 2\psi-C'_2\sin 2\psi\cos\varphi +S'_2\sin 2\psi\sin\varphi]. \quad (38)$$

Since the incident beams F_1 and F_2 have the same intensity, and the initial and final states of polarization are the same complete linear polarization, the intensities J_1 and J_2 of the emergent beams must be equal, according to the law of reciprocity, whatever crystal plates are introduced in the path of the light.

We have, consequently,

$$I'_1+M'_1\cos 2\psi'-C'_1\sin 2\psi'\cos\varphi' +S_1'\sin 2\psi'\sin\varphi'=I'_2+M'_2\cos 2\psi -C'_2\sin 2\psi\cos\varphi+S'_2\sin 2\psi\sin\varphi, \quad (39)$$

whatever the values of the angles ψ, φ, ψ', φ'. Using Eqs. (33), (34), (35), and (37), we obtain

$$\begin{aligned}&(a_{12}-a_{21})(\cos 2\psi-\cos 2\psi')\\&+(a_{13}+a_{31})(\sin 2\psi\cos\varphi-\sin 2\psi'\cos\varphi')\\&+(a_{14}-a_{41})(\sin 2\psi\sin\varphi-\sin 2\psi'\sin\varphi')\\&+(a_{23}+a_{32})(\sin 2\psi\cos\varphi\cos 2\psi'-\sin 2\psi'\cos\varphi'\cos 2\psi)\\&+(a_{24}-a_{42})(\sin 2\psi\sin\varphi\cos 2\psi'-\sin 2\psi'\sin\varphi'\cos 2\psi)\\&-(a_{34}+a_{43})\sin(\varphi-\varphi')\sin 2\psi\sin 2\psi'=0.\end{aligned}\quad (40)$$

This identity can be maintained only if all the coefficients of the trigonometrical expressions in it are zero: For $\psi=0$, $\psi'=\pi/2$ it reduces to

$$a_{12}-a_{21}=0; \quad (41)$$

for $\psi=-\psi'=\pi/4$, $\varphi=\varphi'=0$, to

$$a_{13}+a_{31}=0; \quad (42)$$

for $\psi=-\psi'=\pi/4$, $\varphi=\varphi'=\pi/2$, to

$$a_{14}-a_{41}=0. \quad (43)$$

After the suppression of the three first terms in the identity (40), as a consequence of the zero value of their coefficients, the new identity obtained reduces, for $\psi=\pi/4$, $\psi'=\varphi=\varphi'=0$ to

$$a_{23}+a_{32}=0; \quad (44)$$

for $\psi=\pi/4$, $\psi'=0$, $\varphi=\varphi'=\pi/2$, to

$$a_{24}-a_{42}=0; \quad (45)$$

and for $\psi=\psi'=\pi/4$, $\varphi=\pi/2$, $\varphi'=0$, to

$$a_{34}+a_{43}=0. \quad (46)$$

The sixteen scattering coefficients must therefore obey six conditions of symmetry or antisymmetry, and the linear relations which give the Stokes' parameters of the scattered beam in terms of those of the incident beam, may be written

$$\begin{aligned} I' &= a_1 I + b_1 M - b_3 C + b_5 S, \\ M' &= b_1 I + a_2 M - b_4 C + b_6 S, \\ C' &= b_3 I + b_4 M + a_3 C + b_2 S, \\ S' &= b_5 I + b_6 M - b_2 C + a_4 S. \end{aligned} \tag{47}$$

The scattering of light, through a given angle, by an asymmetrical isotropic medium is characterized by ten independent coefficients.

If the scattering medium is symmetrical, and has therefore no rotatory power, the coefficients b_3, b_4, b_5, b_6 are necessarily zero, since in this case the relations (47) must have the form of the relations (33A), and the linear relations between the Stokes' parameters are reduced to

$$\begin{aligned} I' &= a_1 I + b_1 M, \\ M' &= b_1 I + a_2 M, \\ C' &= a_3 C + b_2 S, \\ S' &= -b_2 C + a_4 S. \end{aligned} \tag{47A}$$

The scattering of light, through a given angle, by a symmetrical isotropic medium is characterized by six independent coefficients.

R. S. Krishnan in his publications dated 1938[1-5] considered only incident beams with linear polarization either horizontal ($I=1$, $M=1$, $C=S=0$) or vertical ($I=1$, $M=-1$, $C=S=0$). These excitation conditions involve only the four coefficients a_{11}, a_{12}, a_{21}, a_{22}, and the relation obtained by him $H_v = V_h$ is equivalent to the first relation proved here $a_{12}=a_{21}$.

In the more general case considered in his paper of 1939, the incident beam has a linear polarization in a direction making an angle θ with the perpendicular to the scattering plane. The Stokes' parameters of the incident beam are then

$$I=1, \quad M=-2\cos 2\theta, \quad C=\sin 2\theta, \quad S=0$$

and, applying Eq. (47) we obtain

$$\rho_\theta = \frac{H_\theta}{V_\theta} = \frac{I'+M'}{I'-M'} = \frac{a_1 + b_1(1-\cos 2\theta) - a_2\cos 2\theta - (b_3+b_4)\sin 2\theta}{a_1 - b_1(1+\cos 2\theta) + a_2 \cos 2\theta - (b_3-b_4)\sin 2\theta}.$$

When θ is 0 or $\pi/2$ this formula gives

$$\rho_v = \frac{H_v}{V_v} = \frac{a_1-a_2}{a_1+2b_1+a_2}, \quad \rho_h = \frac{V_h}{H_h} = \frac{a_1-a_2}{a_1-2b_1+a_2},$$

which shows that the expression of ρ_θ may be written

$$\rho_\theta = \frac{1+\operatorname{tg}^2\theta/\rho_h - 2(b_3+b_4)(a_1-a_2)^{-1}\operatorname{tg}\theta}{\operatorname{tg}^2\theta + 1/\rho_v - 2(b_3-b_4)(a_1-a_2)^{-1}\operatorname{tg}\theta}.$$

This proves that Eq. (7) proposed with some doubt by R. S. Krishnan, must be true for any symmetrical medium, since then $b_3=b_4=0$ but not for an asymmetrical medium.

VIII. FORWARD AXIAL SCATTERING

The light scattered in the same direction as the incident beam cannot be distinguished from the remaining light of this beam. However, if the incident beam has a small aperture ω, it is possible to define, as a limit when $\omega \to 0$, a transmitted beam T, with axis Oz_1 and aperture ω, and a scattered beam including all rays between two cones of apertures η ($\eta > \omega$) and $\eta + d\eta$, having also Oz_1 as axis. The scattered beam F'_0 thus defined may vary with the small angle η, but whatever this angle, it will have axial symmetry around Oz_1.

The law of reciprocity applies to the transmitted beam T as well as to the axially scattered beam F'_0. If reference axes parallel to those chosen for the incident beam are used for the polarization states of these beams, the linear relations between Stokes' parameters will have in both cases, the form obtained above. Writing them for beam F'_0, we will indicate the zero value of the scattering angle ϕ by marking the coefficients with a superscript 0

$$\begin{aligned} I' &= a_1^0 I + b_1^0 M - b_3^0 C + b_5^0 S, \\ M' &= b_1^0 I + a_2^0 M - b_4^0 C + b_6^0 S, \\ C' &= b_3^0 I + b_4^0 M + a_3^0 C + b_2^0 S, \\ S' &= b_5^0 I + b_6^0 M - b_2^0 C + a_4^0 S. \end{aligned} \tag{48}$$

In this case we must also express the axial symmetry around the common direction of the incident and scattered beams: If the incident beam is turned around itself through any angle, the axially scattered beam F'_0 will simply turn around itself through the same angle. It is sufficient to consider an infinitesimal rotation

$\frac{1}{2}d\alpha$ which changes the beam (I, M, C, S) into a beam F_1 with parameters

$$\begin{aligned} I_1 &= I, \\ M_1 &= M - C d\alpha, \\ C_1 &= M d\alpha + C, \\ S_1 &= S, \end{aligned} \tag{49}$$

and the beam $F'_0(I', M', C', S')$ into a beam F'_{01} with the parameters

$$\begin{aligned} I'_1 &= I', \\ M'_1 &= M' - C' d\alpha, \\ C'_1 &= M' d\alpha + C', \\ S'_1 &= S'. \end{aligned} \tag{50}$$

The parameters I'_1, M'_1, C'_1, S'_1 must be expressed in terms of I_1, M_1, C_1, S_1 by the relations identical to relations (48). Eliminating I, M, C, S and I', M', C', S' between the relations (48), (49), and (50), we obtain

$$\begin{aligned} I'_1 &= a_1{}^0 I_1 + (b_1{}^0 + b_3{}^0 d\alpha) M_1 \\ &\qquad - (b_3{}^0 - b_1{}^0 d\alpha) C_1 + b_5{}^0 S_1, \\ M'_1 &= (b_1{}^0 - b_3{}^0 d\alpha) I_1 + a_2{}^0 M_1 \\ &\qquad - [b_4{}^0 + (a_3{}^0 - a_2{}^0) d\alpha] C_1 \\ &\qquad + (b_6{}^0 - b_2{}^0 d\alpha) S_1, \\ C'_1 &= (b_3{}^0 + b_1{}^0 d\alpha) I_1 + (b_4{}^0 - (a_3{}^0 - a_2{}^0) d\alpha] M_1 \\ &\qquad + a_3{}^0 C_1 + (b_2{}^0 + b_6{}^0 d\alpha) S_1, \\ S_1' &= b_5{}^0 I_1 + (b_6{}^0 + b_2{}^0 d\alpha) M_1 \\ &\qquad - (b_2{}^0 - b_6{}^0 d\alpha) C_1 + a_4{}^0 S_1. \end{aligned} \tag{51}$$

So that this linear system be identical with system (48) it is necessary and sufficient that

$$b_1{}^0 = b_2{}^0 = b_3{}^0 = b_6{}^0 = 0, \quad a_2{}^0 = a_3{}^0.$$

The linear transformation for the Stokes' parameter is thus, for axial scattering:

$$\begin{aligned} I' &= a_1{}^0 I + b_5{}^0 S, \\ M' &= a_2{}^0 M - b_4{}^0 C, \\ C' &= b_4{}^0 M + a_2{}^0 C, \\ S' &= b_5{}^0 I + a_4{}^0 S. \end{aligned} \tag{53}$$

These equations apply to any asymmetrical medium. They show that forward axial scattering involves in general *five* independent coefficients.

For a symmetrical medium $b_4{}^0$ and $b_5{}^0$ are necessarily zero; therefore, the linear transformation is reduced to

$$\begin{aligned} I' &= a_1{}^0 I, \\ M' &= a_2{}^0 M, \\ C' &= a_2{}^0 C, \\ S' &= a_4{}^0 S. \end{aligned} \tag{53A}$$

The forward axial scattering involves in this case only *three* independent coefficients.

For the transmitted beam T the linear transformation will have also the form (53) or (53A) but with other values of the coefficients, which will in this case correspond to simple classical properties: The quantity

$$1 - a_1{}^0$$

will measure the absorption for natural light; the quantity

$$b_4{}^0 : a_2{}^0$$

the tangent of the rotation (rotatory power); the quantity

$$1 - (a_2{}^{02} + b_4{}^{02} + b_5{}^{02})^{\frac{1}{2}} : a_1{}^0$$

the depolarization for linear light; the quantity

$$1 - a_4{}^0 : a_1{}^0$$

the depolarization for circular light; and the quantity

$$b_5{}^0 : a_1{}^0$$

the circular dichroism.

IX. BACKWARD AXIAL SCATTERING

A similar argument may be applied to the scattering in the direction opposite to that of the incident beam, since there is also in this case axial symmetry. But then any rotation of the incident beam around its direction must produce an *inverse* rotation of the same magnitude of the scattered beam around its direction.

Marking with a superscript π the coefficients of the linear transformation (47) for the value $\phi = \pi$ which correspond to backward scattering, the invariance for an infinitesimal rotation leads to the conditions

$$b_1{}^\pi = b_2{}^\pi = b_3{}^\pi = b_4{}^\pi = b_6{}^\pi = 0, \quad a_2{}^\pi = -a_3{}^\pi \tag{54}$$

and the linear transformation is

$$\begin{aligned} I' &= a_1{}^\pi I + b_5{}^\pi S, \\ M' &= a_2{}^\pi M, \\ C' &= -a_2{}^\pi C, \\ S' &= b_5{}^\pi I + a_4{}^\pi S. \end{aligned} \tag{55}$$

The backward scattering by an asymmetrical medium involves only *four* independent coefficients.

For a symmetrical medium $b_5{}^\pi$ must be zero, so that the linear transformation is

$$\begin{aligned} I' &= a_1{}^\pi I, \\ M' &= a_2{}^\pi M, \\ C' &= -a_2{}^\pi C, \\ S' &= a_4{}^\pi S. \end{aligned} \tag{55A}$$

The backward scattering involves then *three* independent coefficients, like the forward scattering.

X. SCATTERING BY PARTICLES HAVING SPHERICAL SYMMETRY

When the incident light is completely polarized, the light scattered by one particle of any shape is completely polarized. Any depolarization of the light scattered by an isotropic distribution of similar particles is the result of the difference in polarization of light waves scattered by particles of different orientation; but if the scattering is caused by identical particles having spherical symmetry, the polarization will be the same for all the partial waves, and there will be no depolarization, if we suppose the emulsion sufficiently diluted, so that double scattering be negligible.

Thus, for such a diluted emulsion of identical spherical particles, whenever the incident light is completely polarized, the scattered light must be also completely polarized; consequently, whenever the Stokes' parameters of the incident beam verify the condition

$$I^2 - M^2 - C^2 - S^2 = 0$$

the Stokes' parameters of the scattered beam must verify the similar condition

$$I'^2 - M'^2 - C'^2 - S'^2 = 0.$$

In other words, the linear transformation (47) on the Stokes' parameters must correspond in this case to a rotation and a similarity in a Minkowsky four-dimensional space. Expressing this we obtain the relations

$$\begin{aligned} a_1b_1 - a_2b_1 - b_3b_4 - b_5b_6 &= 0, \\ -a_1b_3 + b_1b_4 - a_3b_3 + b_2b_5 &= 0, \\ a_1b_5 - b_1b_6 - b_2b_3 - a_4b_5 &= 0, \\ -b_1b_3 + a_2b_4 - a_3b_4 + b_2b_6 &= 0, \\ b_1b_5 - a_2b_6 - b_2b_4 - a_4b_6 &= 0, \\ -b_3b_5 + b_4b_6 - a_3b_2 + a_4b_2 &= 0, \end{aligned} \tag{56}$$

and

$$\begin{aligned} a_1^2 &= a_2^2 + b_3^2 + b_4^2 + b_5^2 + b_6^2, \\ a_1^2 &= a_3^2 + b_1^2 + b_2^2 + b_4^2 + b_5^2, \\ a_1^2 &= a_4^2 + b_1^2 + b_2^2 + b_3^2 + b_6^2. \end{aligned} \tag{57}$$

The three relations (57) are a consequence of the six relations (56), which are linear with respect to the four quantities a_1, a_2, a_3, a_4 and are coherent if *

$$b_2b_3b_4 + b_2b_5b_6 + b_1b_4b_6 - b_1b_3b_5 = 0. \tag{58}$$

They give then

$$\begin{aligned} 2a_1 &= \frac{b_3b_4 + b_5b_6}{b_1} + \frac{b_2b_5 + b_1b_4}{b_3} + \frac{b_2b_3 + b_1b_6}{b_5}, \\ 2a_2 &= -\text{———} + \text{———} + \text{———}, \\ 2a_3 &= -\text{———} + \text{———} - \text{———}, \\ 2a_4 &= \text{———} + \text{———} - \text{———}, \end{aligned} \tag{59}$$

from which follows a linear relation between the a's

$$a_1 - a_2 + a_3 - a_4 = 0. \tag{60}$$

There are thus only five independent parameters in the scattering by identical spherical particles without mirror symmetry.

In the case of an emulsion containing spherical particles differing in magnitude or optical properties, each scattering coefficient will be the sum of the corresponding coefficients for the various types of spherical particles in the mixture. The only relation between the scattering coefficients which will hold after this adding process, will be the linear relation (60), which is thus characteristic of scattering by any mixture of spherical particles without mirror symmetry.

For identical spherical particles with mirror symmetry we must have $b_3 = b_4 = b_5 = b_6 = 0$ and the general solution of Eqs. (56), (57) gives

$$a_1 = a_2, \quad a_3 = a_4 \tag{60A}$$

and

$$a_1^2 = a_3^2 + b_1^2 + b_2^2. \tag{61}$$

The two relations (60A) being linear will remain true for a mixture of different spherical particles

* It is also possible to solve the Eqs. (56) when the b's obey the two conditions

$$b_1b_2 = b_3b_6 = b_4b_5,$$

but the singular solutions then obtained do not correspond to scattering by spherical particles.

with mirror symmetry: There are then four independent scattering coefficients.

The first of the relations (60A) is equivalent to the relation $H_v = V_h = 0$ resulting, in the case of homogeneous symmetrical spheres, from the theory of G. Mie.[7]

XI. COMPARISON WITH DIPOLAR SCATTERING

The general polarization properties of dipolar secondary light emission have been determined by P. Soleillet.[12] His theory results in the fact that the linear relations between the Stokes' parameters of the incident and scattered beams, must be in the case of dipolar scattering

$$\begin{aligned} I' &= (a - b\sin^2\phi)I - b\sin^2\phi M, \\ M' &= -b\sin^2\phi I + b(1+\cos^2\phi)M, \\ C' &= 2b\cos\phi C, \\ S' &= 2c\cos\phi S, \end{aligned} \tag{62}$$

in which a, b, c are independent of the angle of scattering ϕ. There is then no distinction between symmetrical and asymmetrical media.

For any angle of scattering, dipolar scattering is qualitatively characterized by the condition

$$b_2 = 0, \tag{63}$$

which expresses the non-existence of any ellipticity in the scattered light when there is no ellipticity in the polarization of the incident light ($S' = 0$ when $S = 0$).

For transverse dipolar scattering ($\phi = \pi/2$) we have, moreover,

$$a_3(\pi/2) = a_4(\pi/2) = 0 \tag{64}$$

and

$$b_1(\pi/2) + a_2(\pi/2) = 0, \tag{65}$$

which proves that whatever the polarization of the incident light, there is then neither obliquity of polarization ($C' = 0$) nor ellipticity ($S' = 0$), and that for an incident beam polarized in the plane of scattering ($I = M$, $C = S = 0$) the scattered light is not at all polarized ($M' = C' = S' = 0$).

This last test was used by R. S. Krishnan to prove the multipolar character of light scattering by some media, particularly by liquid mixtures in the neighborhood of the critical state. But this proof has been criticized,[16, 23] because convergence in the incident beam or secondary scattering is, at least qualitatively, a possible cause of the observed polarizations.

The existence of some ellipticity in the scattered light for an incident beam linearly obliquely polarized ($C \neq 0$, $S = 0$) which would prove that the coefficient b_2 is not zero, would be a much more sure test of the multipolar character of the scattering, and consequently, of the non-negligible magnitude, in comparison with the wave-length, of the elements of heterogeneity of the scattering medium.

LITERATURE

(1) R. S. Krishnan, Proc. Ind. Acad. Sci. **fA**, 21 (1938).
(2) R. S. Krishnan, Proc. Ind. Acad. Sci. **fA**, 91 (1938).
(3) R. S. Krishnan, Proc. Ind. Acad. Sci. **fA**, 98 (1938).
(4) R. S. Krishnan, Kolloid. Zeits. **84**, 2 (1938).
(5) R. S. Krishnan, Kolloid. Zeits. **84**, 18 (1938).
(6) Lord Rayleigh, Phil. Mag. **35**, 373 (1918).
(7) G. Mie, Ann. d. Physik **25**, 377 (1908).
(8) R. S. Krishnan, Proc. Ind. Acad. Sci. **A10**, 395 (1939).
(9) Lord Rayleigh, *Theory of Sound* (edition 1877). volume 1, §109, p. 93.
(10) Lord Rayleigh, Phil. Mag. **49**, 324 (1900).
(11) G. G. Stokes, Trans. Camb. Phil. Soc. **9**, 399 (1852).
(12) P. Soleillet, Ann. de physique **12**, 23 (1929).
(13) J. von Neumann, *Mathematische Grundlagen der Quantenmechanik* (Berlin, 1932).
(14) M. Courtines, Congrès international d'Électricité **2**, 545 (Paris, 1932).
(15) J. Bernamont, Ann. de physique **7**, 71 (1937).
(16) B. K. Mookerjee, Ind. J. Phys. **12**, 15 (1938).
(17) Lord Rayleigh, Phil. Mag. **47**, 375 (1899).
(18) R. Gans, Ann. d. Physik **37**, 881 (1912).
(19) R. Gans, Ann. d. Physik **62**, 331 (1920).
(20) Y. Rocard, Ann. de physique **10**, 116 (1928).
(21) Lord Rayleigh, Proc. Roy. Soc. **A84**, 25 (1910).
(22) Francis Perrin, J. de Chimie physique **36**, 234 (1939).
(23) S. Parthsarathy, Phil. Mag. **29**, 148 (1940).
(24) R. S. Krishnan, Phil. Mag. **29**, 515 (1940).

26

Reprinted from *J. Math. Physics*, **28**(2), 131–139 (1949)

OPTICAL ALGEBRA*

BY NATHAN GRIER PARKE III

There have been three major lines of attack on the problem of obtaining a useful and satisfying mathematical representation of light and the instruments through which it passses. Wiener[1] developed a generalized harmonic analysis and a statistical description of light. He introduced the coherency matrix. His work follows the lead of Rayleigh, Schuster and others. Jones[2] developed a matrix-vector calculus of monochromatic plane waves based on the electric vector. Mueller[3] developed a phenomenological matrix-vector calculus based on the Stokes[4] vector which is defined in terms of observable light intensities. The work of Jones and Mueller is characterized by the explicit recognition of the role of the instrument and its representation by a matrix. All these researches are primarily concerned with polarization or coherence, not with images.

The central purpose of the present research was to determine the mathematical relation between the work of Wiener, Jones and Mueller. The interrelation was discovered. The result is an algebraic formulation of optical statistics in which the results of Wiener, Jones and Mueller play natural and essential roles. As a result of the connection it is now possible to compute theoretically the expected values of the observables defined phenomenologically by Mueller. Indeed, one has what appears to be the beginning of a statistical optics which bears the same relation to phenomenological optics that statistical mechanics bears to thermodynamics.[4]

The overall structure of optical algebra is most readily apparent from Fig. 1. The algebras have been associated with the names of Jones and Mueller, the original investigators. However, at the beginning of the present research, only the Jones frequency algebra and the elementary Mueller frequency algebra existed. The remainder of the structure is new.

Jones Frequency Algebra. Jones frequency algebra is a mathematical system in which the complex vector function

(1) $$E(\omega) = (F_1(\omega), F_2(\omega)) \;.=.\; \text{Maxwell vector}^5$$

* This work has been supported in part by the Signal Corps, the Air Materiel Command and O.N.R. It is a condensation of the mathematical part of a thesis, "Matrix Optics," submitted in partial fulfillment of the requirements for the degree of Doctor of Philosophy at the Massachusetts Institute of Technology (1948).

[1] Wiener, N.: Acta Math. **55:** 117 (1930); Extensive bibliography.

[2] Jones, R. C.: Jour. Opt. Soc. Am., **31:** 488 (1941); **37:** 707 (1947); **38:** 671 (1948).

[3] Mueller, H.: M.I.T., course 8.262; Fall 1946, and an unpublished manuscript.

[4] The subject is treated in full in: Parke, N. G.: "Matrix Optics," Ph.D. Thesis, Course VIII, M.I.T., 1948. 200 pp. Ch. I, Generalized Optical Algebra; Ch. II, Illustrations and Examples; Ch. III, Statistical Harmonic Analysis; Ch. IV, Quasi-Stationary Scattering. The present paper summarizes the mathematical aspects of the algebraic structure. Space limitations preclude any motivation or interpretation. A slightly more extensive summary will be found in Report No. 70, Research Laboratory of Electronics, M.I.T., 1948.

[5] The symbol ".=." is read "called" or "represents."

characterizes the radiation. The complex matrix function

(2) $$J(\omega) = [J_{ij}(\omega)] \mathrel{.=.} \text{Jones matrix}$$

characterizes the instrument. The definition of J is the basic transfer relation,

(3) $$E' = 2\pi JE, \text{ for all } E;\ E' \mathrel{.=.} \text{outgoing radiation.}$$

The following operations occur,

(4) $$E = E_1 + E_2 \mathrel{.=.} \text{vector addition} \mathrel{.=.} \text{coherent superposition}$$

			ELEMENTARY ALGEBRAS	GENERALIZED ALGEBRAS
JONES ALGEBRAS	Theoretical Quantities	ω-Algebra	Jones Frequency Algebra Eq. (3) $E' = 2\pi JE$	Symbols Maxwell Quantities .=. E, e, f .=. Radiation Stokes Quantities .=. L, ϕ, S .=. Radiation Jones Quantities .=. J, j .=. Instrument Mueller Quantities .=. M, m .=. Instrument
		t-Algebra	Jones Time Algebra Eq. (25) $e'(t) = \int_{-\infty}^{\infty} j(t_0)e(t - t_0)\, dt_0$ Eq. (34) $f'_\beta(t) = j_\beta^\alpha(\underline{t_0}) f_\alpha(t - \underline{t_0})$	Generalized Jones Time Algebra Eq. (30) $f'^{ij}_\alpha(t) = j^{i\beta}_\alpha(t_0) f^j_\beta(t - \underline{t_0})$
			Wiener's Generalized Harmonic Analysis	
MUELLER ALGEBRAS	Observables	t-Algebra	Mueller Time Algebra Eq. (38) $\varphi'_{\alpha\beta}(t) = m^{\gamma\delta}_{\alpha\beta}(\underline{t_1}, \underline{t_2}) \varphi_{\gamma\delta}(t - \underline{t_1} + \underline{t_2})$	Generalized Mueller Time Algebra Eq. (50) $\varphi'_{\alpha\beta}(t) = m^{jk\gamma\delta}_{\alpha\beta}(\underline{t_1}, \underline{t_2}) \varphi_{jk\gamma\delta}(t - \underline{t_1} + \underline{t_2})$
		ω-Algebra	Mueller Frequency Algebra Eq. (40) $S'_{\alpha\beta}(\omega) = M^{\gamma\delta}_{\alpha\beta}(\omega) S_{\gamma\delta}(\omega)$	Generalized Mueller Frequency Algebra Eq. (52) $S'_{\alpha\beta}(\omega) = M^{jk\gamma\delta}_{\alpha\beta}(\omega) S_{jk\gamma\delta}(\omega)$

FIG. 1. THE STRUCTURE OF OPTICAL ALGEBRA

(5) $$J = J_2 J_1 \mathrel{.=.} \text{matrix multiplication} \mathrel{.=.} \text{instruments in series}$$

(6) $$J = J_1 + J_2 \mathrel{.=.} \text{instruments in parallel.}$$

Mueller Frequency Algebra. Mueller frequency algebra is a mathematical system in which the complex vector function

(7) $$L(\omega) = (I(\omega), M(\omega), C(\omega), S(\omega)) \mathrel{.=.} \text{Stokes vector}$$

characterizes the radiation. The complex matrix function

(8) $$M(\omega) = [M_{ij}(\omega)] \mathrel{.=.} \text{Mueller matrix}$$

characterizes the instrument. The definition of M is the basic transfer relation,

(9) $$L' = ML, \text{ for all } L;\ L' \mathrel{.=.} \text{outgoing radiation.}$$

The following operations occur,

(10) $L = L_1 + L_2$.=. vector addition .=. incoherent superposition

(11) $M = M_2 M_1$.=. matrix multiplication .=. instruments in series

(12) $M = M_1 + M_2$.=. matrix addition .=. instruments in parallel.

Wiener's Generalized Harmonic Analysis. Wiener has pointed out that, in the electromagnetic theory of light, the field vectors E and B are not observables as they appear to be at lower frequencies.[6] Optical observations always end in intensity measurements using the eye, a photographic plate, the photoelectric effect, a bolometer, etc. The quantities of the Maxwell theory most nearly corresponding with these observations are,

$$\tfrac{1}{2}(E.D + B.H) \text{ .=. energy density}$$

$$E \times H \text{ .=. Poynting vector .=. energy flow}$$

quantities depending "quadratically" on the field quantities E, B, D, H. This linear-quadratic duality is exactly the relation between Jones and Mueller algebra. The correspondence is supplied by Wiener's generalized harmonic analysis.

As introduced by Wiener, the generalized harmonic analysis of a function $f(t)$, takes place in two steps,

$$\begin{aligned} \varphi(t) &= \lim_{T\to\infty} \frac{1}{2T} \int_{-T}^{T} f(t + t_0) f^*(t_0)\, dt_0 \\ &= \langle f(t + t_0) f^*(t_0) \rangle_{\text{Avg.}} \text{ .=. auto-correlation} \end{aligned} \tag{13}$$

$$\begin{aligned} S(\omega) &= \frac{1}{2\pi} \int_{-\infty}^{\infty} \varphi(t) e^{i\omega t}\, dt \\ &= F\varphi(t) \text{ .=. Fourier transformation.} \end{aligned} \tag{14}$$

If this is applied to

(15) $f(t) = a_n f_n(t)$; a_n complex; (Summation convention),

then

$$\varphi(t) = a_n \varphi_{nm}(t) a_m^* \tag{16}$$

where

$$\varphi_{nm}(t) = \langle f_n(t + t_0) f_m^*(t_0) \rangle \text{ .=. Interference matrix.} \tag{17}$$

The Fourier transformation of eq. (16) yields,

$$S(\omega) = a_n S_{nm}(\omega) a_m^* \tag{18}$$

where

$$S_{nm}(\omega) = F\varphi_{nm}(t) \text{ .=. Spectral matrix.} \tag{19}$$

[6] Wiener, N.: Jour. Franklin Institute **207:** 525 (1929).

Wiener prefers the matrix

$$(20) \qquad C_{nm}(\omega) = \frac{1}{2\pi} \int_{-\infty}^{\infty} \varphi_{nm}(t) \frac{e^{i\omega t} - 1}{it} \, dt \; .=. \text{ coherency matrix.}$$

When differentiation is meaningful,

$$(21) \qquad S_{nm}(\omega) = (d/d\omega) C_{nm}(\omega).$$

For a monochromatic plane wave of frequency ω, it turns out that the spectral matrix is

$$(22) \qquad S_{nm}(\omega) = \frac{1}{2} \begin{bmatrix} I + M & C - iS \\ C + iS & I - M \end{bmatrix}$$

$$\delta \; .=. \text{ unit impulse function}$$

$$(I, M, C, S) \; .=. \text{ monochromatic Stokes vector.}$$

Jones Time Algebra. Jones time algebra is the more fundamental. Its Fourier transform (when defined on a finite time interval) is Jones frequency algebra. Briefly,

$$(23) \qquad e(t) = (f_1(t), f_2(t)) \; .=. \text{ Maxwell vector } .=. \text{ radiation,}$$

$$(24) \qquad E(\omega) = Fe(t),$$

$$(25) \qquad e'(t) = \int_{-\infty}^{\infty} j(t_0) e(t - t_0) \, dt_0 \; .=. \text{ convolution } .=. \text{ transfer relation,}$$

$$(26) \qquad E'(\omega) = Fe'(t)$$

$$(27) \qquad J(\omega) = Fj(t)$$

$$(28) \qquad e(t) = e_1(t) + e_2(t) \; .=. \text{ vector addition } .=. \text{ superposition.}$$

J is the response of an optical instrument to polarized monochromatic waves. j is the response of an optical instrument to polarized impulses. The correlation applies to functions of time; hence the introduction of Jones time algebra. The quantum theory of scattering, or the Mie theory of scattering, refers to monochromatic plane waves and gives the J of Jones frequency algebra. It turns out that it is only necessary to make conceptual use of Jones time algebra. The J's can be used directly to compute the M's of Mueller frequency algebra, e.g., eq. (53).

Generalized Jones Algebra. The problem which motivates the generalization of Jones algebra is scattering from a statistical assemblage of centers, Fig. 2. The following notation is useful,

$$f_\beta^j(t) \; .=. \text{ set of } N \text{ incoming wavelets; } j = 1, 2, \cdots, N$$

$$j_\alpha^{i\beta}(t) \; .=. \text{ set of } M \text{ scattering centers; } i = 1, 2, \cdots, M$$

These multi-index quantities are direct sums of the corresponding elementary quantities. Using the summation convention one has,

$$f'^{ij}_{\alpha}(t) = \int_{-\infty}^{\infty} j^{i\beta}_{\alpha}(t_0) f^{j}_{\beta}(t - t_0)\, dt_0 \;.=.\; MN \text{ outgoing wavelets.} \tag{29}$$

The last operation is a combined Kronecker product, summation and convolution. In order to free the calculation of integral signs one introduces the "convolution convention": *Repetition of a time variable with a single bar under both occurrences implies convolution.* Thus,

$$f'^{ij}_{\alpha}(t) = j^{i\beta}_{\alpha}(\underline{t_0}) f^{j}_{\beta}(t - \underline{t_0}). \tag{30}$$

If only a single wave comes in,

$$f'^{i}_{\alpha}(t) = j^{i\beta}_{\alpha}(t_0) f_{\beta}(t - \underline{t_0}). \tag{31}$$

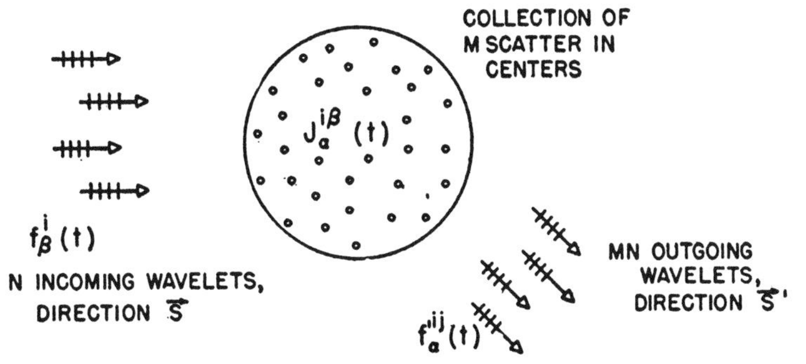

FIG. 2. SCATTERING, A BASIC OPTICAL PATTERN

The geometry of the relative retardations of the various scattered wavelets $f'^{i}_{\alpha}(t)$ must be considered before summation on i is carried out. Let $r_i(t)$ be the position of the M scattering centers, Fig. 3.
At the point of observation P, eq. (31) becomes

$$f'^{i}_{\alpha}\left(t - \frac{s' \cdot r}{v}\right) = j^{i\beta}_{\alpha}(\underline{t_0}) f_{\beta}\left(t - \frac{s' \cdot (r - r_i(t))}{v} - \frac{s \cdot r_i(t)}{v} - \underline{t_0}\right), \tag{32}$$

i not summed, v .=. velocity of radiation. After change of epoch, eq. (32) becomes,

$$f'_{\alpha}(t) = j^{i\beta}_{\alpha}(t_0) f_{\beta}\left(t - \frac{(s - s') \cdot r_i(t - \underline{t_0})}{v}\right); \qquad (i \text{ summed}). \tag{33}$$

This result involves a quasi-stationary assumption on the scattering centers but leads to many interesting results in agreement with a widely scattered literature, including scattering of x-rays by crystals, liquids and gases. Elementary Jones algebra could not describe depolarizing instruments and partially polarized radiation, the generalized algebra can.

Elementary Mueller Algebra. The relation between Jones and Mueller algebra is most easily understood if one restricts attention to the effect of a single elementary instrument,

$$f'_\beta(t) = j_\beta^\alpha(\underline{t_0}) f_\alpha(t - \underline{t_0}). \tag{34}$$

The Wiener correlation of the incoming radiation is

$$\varphi_{\alpha\beta}(t) = \langle f_\alpha(t + t_0) f_\beta^*(t_0) \rangle. \tag{35}$$

One can introduce a "correlation convention": *Repetition of a time variable with a double bar under both occurrences implies correlation,* and rewrite eq. (35),

$$\varphi_{\alpha\beta}(t) = f_\alpha(t + \underline{\underline{t_0}}) f_\beta^*(\underline{\underline{t_0}}). \tag{36}$$

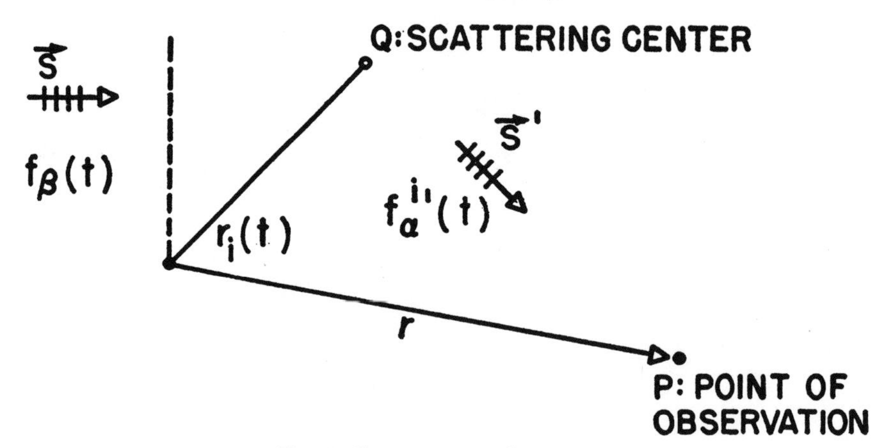

Fig. 3. Retardation Geometry

Also,

$$\varphi'_{\alpha\beta}(t) = f'_\alpha(t + \underline{\underline{t_0}}) f'^*_\beta(\underline{\underline{t_0}}) = j_\alpha^\gamma(\underline{t_1}) j_\beta^{\delta *}(\underline{t_2}) f_\gamma(t - \underline{t_1} + \underline{t_2} + \underline{\underline{t_0}}) f_\delta^*(\underline{\underline{t_0}}). \tag{37}$$

From this one obtains the important formula,

$$\varphi'_{\alpha\beta}(t) = m_{\alpha\beta}^{\gamma\delta}(t_1, \underline{t_2}) \varphi_{\gamma\delta}(t - t_1 + \underline{t_2}) \ .=. \ \text{transfer relation}, \tag{38}$$

in which

$$m_{\alpha\beta}^{\gamma\delta}(t_1, t_2) = j_\alpha^\gamma(t_1) j_\beta^{\delta *}(t_2) \ .=. \ \text{Mueller function}, \tag{39}$$

the Kronecker product of two Jones matrices. The Fourier transform of eq. (39) gives

$$S'_{\alpha\beta}(\omega) = M_{\alpha\beta}^{\gamma\delta}(\omega) S_{\gamma\delta}(\omega) \tag{40}$$

where,

$$M_{\alpha\beta}^{\gamma\delta}(\omega) = (2\pi J_\delta^\gamma(\omega))(2\pi J_\beta^\delta(\omega))^* \ .=. \ \text{Mueller function}. \tag{41}$$

If one introduces the "index reduction transformation,"

$$i = \alpha + 2(\beta - 1); j = \gamma + 2(\delta - 1) \tag{42}$$

one obtains,

$$M_{\alpha\beta}^{\gamma\delta}(\omega) = M_{ij} \;.\!=\!.\; \text{complex Mueller matrix,} \tag{43}$$

$$S_{\gamma\delta}(\omega) = L_j \;.\!=\!.\; \text{complex Stokes vector,} \tag{44}$$

$$S'_{\alpha\beta}(\omega) = L'_i \;.\!=\!.\; \text{complex Stokes vector.} \tag{45}$$

On dropping indices and using matrix notation,

$$L'(\omega) = M(\omega)L(\omega). \tag{46}$$

These matrices and vectors differ from those used by Mueller. In particular, the complex Stokes vector turns out to be,

$$L = (I + M, C + iS, C - iS, I - M). \tag{47}$$

The complex Mueller matrix differs correspondingly from the real Mueller matrix used by Mueller. The advantage of the complex Mueller matrix is its simple relation to the Jones matrix, eq. (41). The real Stokes vector and Mueller matrix have easier phenomenological definitions.

Generalized Mueller Algebra. One begins with the transfer relation for generalized Jones algebra,

$$f'_\alpha(t) = j_\alpha^{i\beta}(\underline{t_0})f_{j\beta}(t - \underline{t_0}) \tag{33}$$

where,

$$f_{j\beta}(t) = f_\beta\left(t - \frac{(s - s')\cdot r_j(t)}{v}\right). \tag{48}$$

Wiener correlation gives,

$$\varphi'_{\alpha\beta}(t) = [j_\alpha^{\cdot j\gamma}(\underline{t_1})j_\beta^{\cdot k\delta *}(\underline{t_2})]f_{j\gamma}(t + \underline{\underline{t_0}} - \underline{t_1})f^*_{k\delta}(\underline{\underline{t_0}} - \underline{t_2}). \tag{49}$$

This reduces to,

$$\varphi'_{\alpha\beta}(t) = m_{\alpha\beta}^{jk\gamma\delta}(\underline{t_1}, \underline{t_2})\varphi_{jk\gamma\delta}(t - \underline{t_1} + \underline{t_2}), \tag{50}$$

where

$$m_{\alpha\beta}^{jk\gamma\delta}(t_1, t_2) = j_\alpha^{\cdot j\gamma}(t_1)j_\beta^{\cdot k\delta *}(t_2) \;.\!=\!.\; \text{Mueller function.} \tag{51}$$

The Fourier transform of eq. (5) yields,

$$S'_{\alpha\beta}(\omega) = M_{\alpha\beta}^{jk\gamma\delta}(\omega)S_{jk\gamma\delta}(\omega), \tag{52}$$

where,

$$M_{\alpha\beta}^{jk\gamma\delta}(\omega) = (2\pi J_\alpha^{j\gamma}(\omega))(2\pi J_\beta^{k\delta}(\omega))^* \;.\!=\!.\; \text{Mueller function.} \tag{53}$$

When $j = k$, the M's are called auto-functions, when $j \neq k$ they are called cross-functions.

The difficulties of the scattering problem are reduced to the evaluation of the interference,

$$\varphi_{jk\gamma\delta}(\omega) = \underline{\underline{f_{j\gamma}(t+t_0)f^*_{k\delta}(t_0)}}. \tag{54}$$

For an important class of cases, the Fourier transform of eq. (54) separates, i.e.,

$$S_{jk\gamma\delta}(\omega) = N_{jk}(\omega)S_{\gamma\delta}(\omega). \tag{55}$$

This leads to a simplification in which $N_{jk}(\omega)$ is characteristic of the statistical distribution of scattering centers. In this case,

$$M^{\gamma\delta}_{\alpha\beta}(\omega) = N_{jk}(\omega)M^{jk\gamma\delta}_{\alpha\beta}(\omega), \tag{56}$$

a general law for the composition of Mueller matrices, for instruments in parallel. Each scattering center is thought of as an elementary instrument.

The Form of $N_{jk}(\omega)$ in Typical Cases. The importance of $N_{jk}(\omega)$ is well illustrated by three examples which embody a unification of the treatment of scattering from liquids, solids and gases, i.e., coherent scattering (solids), partially coherent scatterings (liquids, electrolytes), and incoherent scattering (gases).

1. Coherent Scattering,

$$N_{jk}(\omega) = e^{i(s-s')\cdot(r_j-r_k)\omega/v}. \tag{57}$$

2. Incoherent Scattering,

$$N_{jk} = \delta_{jk}. \tag{58}$$

3. Partially Coherent Scattering (Radial Symmetry),

$$N_{jk}(s) = \begin{cases} 1, & \text{when } j = k \\ \dfrac{i(s)}{nV} & \text{when } j \neq k, \end{cases} \tag{59}$$

$$s = \frac{\omega\,|\,s - s'\,|}{v}; \qquad i(s) = F(\rho(r) - n) \tag{60}$$

n .=. concentration

$\rho(r)$.=. radial concentration density.

4. Special Case of an Electrolyte

$$i(s) = \pm\frac{1}{2(1+(s/\kappa)^2)} \tag{61}$$

$+$.=. unlike-signed ions

$-$.=. like-signed ions

$$\kappa^2 = \frac{2nZ^2\,|\,e\,|}{\epsilon kT} \tag{62}$$

$N_{jk}(\omega)$ is called a Debye statistical matrix because it is closely related to some of Debye's results in electrolyte theory and in x-ray scattering.

Composition of Mueller Matrices. Written in full, eq. (54) becomes,

$$\varphi_{jk\gamma\delta}(t) = f_\gamma\left(t + \underline{\underline{t_0}} - \frac{(s - s'\cdot r_j(t - \underline{\underline{t_0}})}{v}\right) f_\delta^*\left(\underline{\underline{t_0}} - \frac{(s - s')\cdot r_k(\underline{\underline{t_0}})}{v}\right). \tag{63}$$

Making the quasi-stationary approximation,

$$\begin{aligned} \varphi_{jk\gamma\delta}(t) &= f_\gamma\left(t + \underline{\underline{t_1}} - \frac{(s - s')\cdot(r_j(\underline{\underline{t_2}}) - r_k(\underline{\underline{t_2}}))}{v}\right) f_\delta^*(\underline{\underline{t_1}}) \\ &= \varphi_{\gamma\delta}\left(t - \frac{(s - s')\cdot(r_j(\underline{\underline{t_2}}) - r_k(\underline{\underline{t_2}}))}{v}\right). \end{aligned} \tag{64}$$

The Fourier transform of the last expression gives,

$$S_{jk\gamma\delta}(\omega) = S_{\gamma\delta}(\omega)e^{i\omega(s-s')(r_j(\underline{\underline{t_2}})-r_k(\underline{\underline{t_2}}))/v} \tag{65}$$

If one defines,

$$N_{jk}(\omega) = \lim_{T\to\infty} \frac{1}{2T}\int_{-T}^{T} e^{i\omega(s-s')\cdot(r_j(t)-r_k(t))/v}\, dt \tag{66}$$

then

$$S_{ij\gamma\delta}(\omega) = N_{jk}(\omega)S_{\gamma\delta}(\omega) \tag{67}$$

which is eq. (55), as required. Under these circumstances one has a general rule for the composition of Mueller matrices, eq. (56).

Acknowledgement. I am indebted to Hans Mueller for pointing out the problem, guiding the research and encouraging an independent investigation of the mathematical aspects. I am also indebted to Y. W. Lee for helpful discussions of the close relation between my results and his results in network theory. Lee uses a system matrix analogous to the Jones matrix, and a squared system matrix corresponding to the Mueller matrix. This type of analysis appears to be applicable to any linear system with a transfer function and a significant linear-quadratic duality such as voltage-power.

39 Main Street
Concord, Mass.

(Received November 15, 1948)

27

Reprinted from *Il Nuovo Cimento,* Ser. 10, **13,** 1165–1181 (Sept. 1959)

Coherence Properties of Partially Polarized Electromagnetic Radiation (*).

E. WOLF (**)

Department of Theoretical Physics - University of Manchester

(ricevuto il 4 Giugno 1959)

Summary. — This paper is concerned with the analysis of partial polarization from the standpoint of coherence theory. After observing that the usual analytic definition of the Stokes parameters of a quasi-monochromatic wave are not unique, a simple experiment is analysed, which brings out clearly the observable parameters of a quasi-monochromatic light wave. The analysis leads to a unique coherency matrix and to a unique set of Stokes parameters, the latter being associated with the representation of the coherency matrix in terms of Pauli's spin matrices. In this analysis the concept of Gabor's analytic signal proves to be basic. The degree of coherence between the electric vibrations in any two mutually orthogonal directions of propagation of the wave depends in general on the choice of the two orthogonal directions. It is shown that its maximum value is equal to the degree of polarization of the wave. It is also shown that the degree of polarization may be determined in a new way from relatively simple experiments which involve a compensator and a polarizer, and that this determination is analogous to the determination of the degree of coherence from Young's interference experiment.

1. – Introduction.

Although it is generally known that there is an intimate connection between partial polarization and partial coherence (see, for example, WIENER [1], p. 187), a systematic analysis of the properties of partially polarized radiation from

(*) The research described in this paper has been partially sponsored by the Air Force Cambridge Research Center of the Air Research and Development Command, United States Air Force, through its European Office, under Contract No. AF 61(052)-169.

(**) Now at the Institute of Optics, University of Rochester, Rochester, N.Y.

([1]) N. WIENER: *Acta Math.*, **55**, § 9 (1930).

the standpoint of coherence theory does not appear to have been made so far. There are several reasons why it seems desirable to carry out such an analysis. The numerous investigations made in recent years in connection with partially coherent light (ZERNIKE [2], HOPKINS [3], WOLF [4a,b,c], BLANC-LAPIERRE and DUMONTET [5]) have clearly shown that there is still a great deal to be learnt about the statistical properties of high frequency electromagnetic radiation. Moreover, as we shall briefly indicate, the usual treatments of partial polarization are not entirely satisfactory.

The most systematic treatments of partial polarization utilize the concept of Stokes parameters (introduced by G. G. STOKES in 1852 [6]) which are usually defined as follows: Consider a plane, quasi-monochromatic electromagnetic wave and let the components of the electric vector in two mutually perpendicular directions at right angles to the direction of propagation of the wave be represented in the form

$$\left\{\begin{aligned} E_x(t) &= A_1(t)[\cos \Phi_1(t) - \bar{\omega}t]\,, \\ E_y(t) &= A_2(t)[\cos \Phi_2(t) - \bar{\omega}t]\,, \end{aligned}\right. \tag{1.1}$$

where $\bar{\omega}$ denotes the mean frequency and t the time. The Stokes parameters are the four quantities

$$\left\{\begin{aligned} s_0 &= \langle A_1^2\rangle + \langle A_2^2\rangle\,, \\ s_1 &= \langle A_1^2\rangle - \langle A_2^2\rangle\,, \\ s_2 &= 2\langle A_1 A_2 \cos(\Phi_1 - \Phi_2)\rangle\,, \\ s_3 &= 2\langle A_1 A_2 \sin(\Phi_1 - \Phi_2)\rangle\,, \end{aligned}\right. \tag{1.2}$$

where the sharp brackets denote time average. Although to-day there exists an extensive literature in which the Stokes parameters play a central role (see, for example CHANDRASEKHAR [7]), it does not appear to have been noticed that the above relations do not define the parameters uniquely. For in (1.1) only $E_x(t)$ and $E_y(t)$ can be regarded as uniquely associated with the wave, whereas the A's and Φ's may evidently be chosen in many ways, leading to different sets of Stokes parameters. Only a careful analysis of experiment may be expected to lead to a unique set. Such an analysis is carried out in

---

[2] F. ZERNIKE: *Physica*, **5**, 785 (1938).

[3] H. H. HOPKINS: *Proc. Roy. Soc.*, A **208**, 263 (1951).

[4] E. WOLF: (*a*) *Proc. Roy. Soc.*, A **225**, 96 (1954); (*b*) *Nuovo Cimento*, **12**, 884 (1954); (*c*) *Proc. Roy. Soc.*, A **230**, 96 (1955).

[5] A. BLANC-LAPIERRE and P. DUMONTET: *Revue d'Optique*, **34**, 1 (1955).

[6] G. G. STOKES: *Trans. Camb. Phil. Soc.*, **9**, 399 (1852); also his *Mathematical and Physical Papers*, vol. **3** (Cambridge, 1901), p. 233.

[7] S. CHANDRASEKHAR: *Radiative Transfer* (Oxford, 1950), § 15.

Section **2** of this paper and shows that the unique set obtained is intimately related to the appropriate « degree of coherence » of the electric vibrations in the two orthogonal directions.

Another unsatisfactory feature of the usual treatments of partial polarization has been clearly pointed out in an interesting recent paper by PANCHARANTHAM ([8], expecially p. 399-340): A partially polarized beam is usually described in terms of incoherent superposition of polarized and unpolarized beams and the interference phenomena arising from the superposition of these beams are analysed by using the concept of coherence and incoherence alone. However, the decomposition may be carried out in many different ways and it is by no means evident that the different decompositions will always lead to identical results. In any case this approach masks completely the *invariant characteristics* of the different representations. These unsatisfactory features can only be expected to be removed by the introduction of intermediate states (partial coherence).

In the present paper the basic properties of a quasi-monochromatic partially polarized electromagnetic wave are discussed from the standpoint of coherency theory and the invariant characteristics of such a wave are clearly brought out. The basic tool used for this purpose is the coherency matrix introduced in a previous paper (WOLF ([4b])), specialized to the problem in question; however, unlike in the previous paper, the coherency matrix is introduced here from the analysis of a simple experiment.

2. – The coherency matrix of a plane, quasi-monochromatic electromagnetic wave.

Consider a plane, quasi-monochromatic wave and let $E_x^{(r)}(t)$ and $E_y^{(r)}(t)$ represent (*) the components of the electric vector $\boldsymbol{E}^{(r)}$ at a typical point in the wave field in two mutually orthogonal directions at right angles to the directions of propagation of the wave. We assume that $\boldsymbol{E}^{(r)}$ may be represented as a Fourier integral and write

$$(2.1)\qquad \left\{\begin{aligned} E_x^{(r)}(t) &= \int_0^\infty a_1(\omega)\cos[\varphi_1(\omega)-\omega t]\,\mathrm{d}\omega\,, \\ E_y^{(r)}(t) &= \int_0^\infty a_2(\omega)\cos[\varphi_2(\omega)-\omega t]\,\mathrm{d}\omega\,. \end{aligned}\right.$$

([8]) S. PANCHARATNAM: *Proc. Ind. Acad. Sci.*, A **44**, 398 (1956),

(*) The superscript « r » is introduced because the (real) electric wave function will shortly be regarded as the real part of a suitably chosen complex wave function.

Since the wave is assumed to be quasi-monochromatic, the spectral amplitudes $a_1(\omega)$ and $a_2(\omega)$ will be appreciable only in a narrow range

$$\bar{\omega} - \tfrac{1}{2}\Delta\omega \leqslant \omega \leqslant \bar{\omega} + \tfrac{1}{2}\Delta\omega \,, \tag{2.2}$$

where $\Delta\omega$ is small compared with the mean frequency $\bar{\omega}$.

Suppose that the wave is passed through a device (compensator) which introduces retardations in $E_x^{(r)}$ and $E_y^{(r)}$. Let $\varepsilon_1(\omega)$ and $\varepsilon_2(\omega)$ be the phase delays in the Fourier components of frequency ω of $E_x^{(r)}$ and $E_y^{(r)}$ respectively. The electric wave emerging from the compensator has the components

$$\left\{\begin{aligned} \mathscr{E}_x^{(r)}(t) &= \int_0^\infty a_1(\omega)\cos\left[\varphi_1(\omega) - \varepsilon_1(\omega) - \omega t\right] \mathrm{d}\omega \,, \\ \mathscr{E}_y^{(r)}(t) &= \int_0^\infty a_2(\omega)\cos\left[\varphi_2(\omega) - \varepsilon_2(\omega) - \omega t\right] \mathrm{d}\omega \,, \end{aligned}\right. \tag{2.3}$$

it being assumed that reflection and absorption losses are negligible. If we use the identity $\cos[A - \varepsilon] = \cos A \cos \varepsilon + \sin A \sin \varepsilon$ in (2.3) and assume that for any two frequencies ω' and ω'' in the range (2.2)

$$|\varepsilon_1(\omega') - \varepsilon_1(\omega'')| \ll 2\pi \,, \qquad |\varepsilon_2(\omega') - \varepsilon_2(\omega'')| \ll 2\pi \,, \tag{2.4}$$

(2.3) may be re-written as

$$\left\{\begin{aligned} \mathscr{E}_x^{(r)} &= E_x^{(r)}(t)\cos\bar{\varepsilon}_1 + E_x^{(i)}(t)\sin\bar{\varepsilon}_1 \,, \\ \mathscr{E}_y^{(r)} &= E_y^{(r)}(t)\cos\bar{\varepsilon}_2 + E_y^{(i)}(t)\sin\bar{\varepsilon}_2 \,, \end{aligned}\right. \tag{2.5}$$

where $\bar{\varepsilon}_1 = \varepsilon_1(\bar{\omega})$, $\bar{\varepsilon}_2 = \varepsilon_2(\bar{\omega})$ and $E_x^{(i)}$ and $E_y^{(i)}$ are the Fourier integrals *conjugate* to $E_x^{(r)}$ and $E_y^{(r)}$ respectively, *i.e.*

$$\left\{\begin{aligned} E_x^{(i)} &= \int_0^\infty a_1(\omega)\sin\left[\varphi_1(\omega) - \omega t\right] \mathrm{d}\omega \,, \\ E_y^{(i)} &= \int_0^\infty a_2(\omega)\sin\left[\varphi_2(\omega) - \omega t\right] \mathrm{d}\omega \,. \end{aligned}\right. \tag{2.6}$$

As is well known (cf. TITCHMARSH [9]), any two conjugate functions are related

[9] E. C. TITCHMARSH: *Introduction to the Theory of Fourier Integrals* (Oxford, 1948), 2nd ed., chap. v.

by Hilbert's reciprocity relations, *i.e.* by relations of the form

$$E_x^{(i)}(t) = \frac{P}{\pi}\int_{-\infty}^{\infty} \frac{E_x^{(r)}(t')}{t'-t}\,\mathrm{d}t'\,, \qquad E_x^{(r)}(t) = -\frac{P}{\pi}\int_{-\infty}^{\infty} \frac{E_x^{(i)}(t')}{t'-t}\,\mathrm{d}t'\,, \tag{2.7}$$

P denoting the Cauchy principal value at $t' = t$.

Suppose now that the wave emerging from the compensator is sent through a polarizer, which only transmits the component which makes an angle θ with the x-direction. This component is given by

$$\mathscr{E}^{(r)}(t;\, \theta, \varepsilon_1, \varepsilon_2) = \mathscr{E}_x^{(r)}(t) \cos\theta + \mathscr{E}_y^{(r)}(t) \sin\theta\,, \tag{2.8}$$

so that the intensity of the light emerging from the polarizer is

$$\begin{aligned} I(\theta_1, \varepsilon_1, \varepsilon_2) &= 2\langle \mathscr{E}^{(r)^2}(t, \theta, \varepsilon_1, \varepsilon_2)\rangle \\ &= 2\langle \mathscr{E}_x^{(r)^2}\rangle \cos^2\theta + 2\langle \mathscr{E}_y^{(r)^2}\rangle \sin^2\theta + 4\langle \mathscr{E}_x^{(r)} \mathscr{E}_y^{(r)}\rangle \cos\theta \sin\theta\,, \end{aligned} \tag{2.9}$$

where sharp brackets denote time average. Here the wave field was assumed to be stationary (*), so that the intensity I is independent of the time instant at which the average is taken and the factor 2 on the right of the first equation of (2.9) was introduced to simplify later calculations. Next we substitute from (2.5) into (2.9) and use the following relations which may be readily proved from the properties of Hilbert transforms (**):

$$\left\{\begin{aligned} &\langle E_x^{(r)^2}\rangle = \langle E_x^{(i)^2}\rangle\,, \qquad \langle E_y^{(r)^2}\rangle = \langle E_y^{(i)^2}\rangle\,, \\ &\langle E_x^{(r)} E_y^{(r)}\rangle = \langle E_x^{(i)} E_y^{(i)}\rangle\,, \\ &\langle E_x^{(r)} E_y^{(i)}\rangle = -\langle E_x^{(i)} E_y^{(r)}\rangle\,, \\ &\langle E_x^{(r)} E_x^{(i)}\rangle = \langle E_y^{(i)} E_y^{(r)}\rangle = 0\,. \end{aligned}\right. \tag{2.10}$$

(*) Stationarity in the strict sense of the theory of random functions would imply that the field vectors are not square integrable and hence we would not be justified in using Fourier integral analysis. This difficulty may be avoided in the usual way by assuming that the field exists only for a finite time interval $-T \leqslant t \leqslant T$ and proceeding to the limit $T \to \infty$ at the end of the calculations. The final results are the same whether or not this refinement is made.

(**) The formulae (2.10) are valid quite generally. When the field is quasi-monochromatic as here assumed, they may be proved in a very simple way by the following

Further, if we set

$$\delta = \bar{\varepsilon}_1 - \bar{\varepsilon}_2 \,, \tag{2.11}$$

and write $I(\theta, \delta)$ in place of $I(\theta, \bar{\varepsilon}_1, \bar{\varepsilon}_2)$ [since $\bar{\varepsilon}_1$ and $\bar{\varepsilon}_2$ enter the expression for the intensity only through their difference] we obtain the following expression for the time averaged intensity:

$$I(\theta, \delta) = 2\langle E_x^{(r)2}\rangle \cos^2\theta + 2\langle E_y^{(r)2}\rangle \sin^2\theta + \\ + 4\cos\theta \sin\theta\{\cos\delta\langle E_x^{(r)} E_y^{(r)}\rangle - \sin\delta\langle E_x^{(r)} E_y^{(i)}\rangle\}\,. \tag{2.12}$$

The formulae (2.12) may be expressed in a more convenient form, by using in place of the real wave functions the associated analytic signals of GABOR [11],

---

argument due to BRACEWELL [[10], p. 102]. We set

$$E_x(t) = E_x^{(r)}(t) + iE_x^{(i)}(t) = A_1(t)\exp[-i\bar{\omega}t]\,,$$

$$E_y(t) = E_y^{(r)}(t) + iE_y^{(i)}(t) = A_2(t)\exp[-i\bar{\omega}t]\,.$$

Then, if the field is quasi-monochromatic, the (generally complex) quantities A_1 and A_2 will vary slowly with t in comparison with the periodic term, and we have, for example,

$$\langle E_x^{(r)2}\rangle = \left\langle\left(\frac{A_1\exp[-i\omega t] + A_1^*\exp[i\omega t]}{2}\right)^2\right\rangle = \\ = \frac{1}{4}\langle A_1^2\exp[-2i\bar{\omega}t]\rangle + \frac{1}{2}\langle A_1 A_1^*\rangle + \frac{1}{4}\langle A_1^{*2}\exp[2i\bar{\omega}t]\rangle\,.$$

The first and the last term on the right vanish because of the rapidly varying terms $\exp[-2i\bar{\omega}t]$ and $\exp[2i\bar{\omega}t]$, so that

$$\langle E_x^{(r)2}\rangle = \tfrac{1}{2}\langle A_1 A_1^*\rangle\,.$$

Similarly

$$\langle E_x^{(i)2}\rangle = \left\langle\left(\frac{A_1\exp[-i\bar{\omega}t] - A_1^*\exp[i\bar{\omega}t]}{2i}\right)^2\right\rangle = \frac{1}{2}\langle A_1 A_1^*\rangle\,.$$

Comparison of the last two formulae gives the first relation in (2.10); the other relations may be proved in a similar way.

[10] R. N. BRACEWELL: *Proc. I.R.E.*, **46**, 97 (1958).

[11] D. GABOR: *Journ. Inst. Elect. Engrs.*, **93**, part III, 429 (1946).

i.e. by using in place of $E_x^{(r)}$ and $E_y^{(r)}$, the functions (*)

$$
(2.13)\quad \begin{cases} E_x(t) = E_x^{(r)}(t) + i\,E_x^{(i)}(t) = \int_0^\infty a_1(\omega) \exp\left[i[\varphi_1(\omega) - \overline{\omega}t]\right] \mathrm{d}\omega\,, \\ E_y(t) = E_y^{(r)}(t) + i\,E_y^{(i)}(t) = \int_0^\infty a_2(\omega) \exp\left[i[\varphi_2(\omega) - \omega t]\right] \mathrm{d}\omega\,. \end{cases}
$$

Using (2.10) we have the relations

$$(2.14a)\qquad \langle E_x E_x^* \rangle = 2\langle E_x^{(r)2} \rangle = 2\langle E_x^{(i)2} \rangle\,,$$

$$(2.14b)\qquad \langle E_x E_y^* \rangle = 2\langle E_x^{(r)} E_y^{(r)} \rangle - 2i\langle E_x^{(i)} E_y^{(i)} \rangle\,,$$

etc. With the help of (2.14) the formula (2.12) becomes

$$
(2.15)\quad I(\theta, \delta) = J_{xx} \cos^2\theta + J_{yy} \sin^2\theta + J_{xy} \cos\theta \sin\theta \exp[-i\delta] + J_{yx} \sin\theta \cos\theta \exp[i\delta]\,,
$$

where the J's are the elements of the *coherency matrix*

$$
(2.16)\qquad \boldsymbol{J} = \begin{bmatrix} J_{xx} & J_{xy} \\ J_{yx} & J_{yy} \end{bmatrix} \doteq \begin{bmatrix} \langle E_x E_x^* \rangle & \langle E_x E_y^* \rangle \\ \langle E_y E_x^* \rangle & \langle E_y E_y^* \rangle \end{bmatrix}.
$$

The formula (2.15) expresses in a compact form the intensity of the wave after transmission through the compensator (which introduces a phase delay δ) and the polarizer (oriented so as to transmit the component which makes an angle θ with the x-axis) in terms of the coherency matrix $\boldsymbol{J}$ which characterizes the incident wave.

Since $J_{yx} = J_{xy}^*$ the coherency matrix is *Hermitian*. Its trace represents the intensity of the incident wave,

$$
(2.17)\qquad \mathrm{Tr}\,\boldsymbol{J} = J_{xx} + J_{yy} = \langle E_x E_x^* \rangle + \langle E_y E_y^* \rangle = 2\langle E_x^{(r)2} \rangle + 2\langle E_y^{(r)2} \rangle\,,
$$

and its non-diagonal elements express the correlation between the x and y-components of the complex vector $\boldsymbol{E}$. Further it follows from Schwarz' inequality

(*) An analytic signal is a complex function characterized by the property that its Fourier integral contains no spectral components of positive (or negative) frequencies. This fact alone implies that the real and imaginary parts of the signal are conjugate functions and hence Hilbert transforms of each other.

for integrals that $|J_{xy}| \leqslant \sqrt{J_{xx}}\sqrt{J_{yy}}$, $|J_{yx}| \leqslant \sqrt{J_{yy}}\sqrt{J_{xx}}$; hence, since $J_{yx} = J_{xy}$,

$$|\boldsymbol{J}| = J_{xx}J_{yy} - J_{xy}J_{yx} \geqslant 0 \,, \tag{2.18}$$

i.e. the discriminant of the coherency matrix is non-negative.

Let $A_1(t)$, $A_2(t)$ be the amplitudes and $\Psi_1(t)$ and $\Psi_2(t)$ the phases of $E_x(t)$ and $E_y(t)$ respectively, *i.e.*

$$E_x(t) = A_1(t)\exp[i\Psi_1(t)]\,, \qquad E_y(t) = A_2(t)\exp[i\Psi_2(t)]\,. \tag{2.19}$$

Then, from (2.13),

$$\begin{cases} E_x^{(r)}(t) = A_1(t)\cos[\Psi_1(t)]\,, & E_x^{(i)}(t) = A_1(t)\sin[\Psi_1(t)]\,, \\ E_y^{(r)}(t) = A_2(t)\cos[\Psi_2(t)]\,, & E_y^{(i)}(t) = A_2(t)\sin[\Psi_2(t)]\,, \end{cases} \tag{2.20}$$

and, if we introduce quantities $\Phi_1(t)$ and $\Phi_2(t)$ by the relations

$$\Psi_1(t) = \Phi_1(t) - \bar{\omega}t\,, \qquad \Psi_2(t) = \Phi_2(t) - \bar{\omega}t\,, \tag{2.21}$$

where $\bar{\omega}$ is the mean frequency, the components $E_x^{(r)}$, $E_y^{(r)}$ of the (real) electric vector are represented by expressions of the form (1.1), but the representation is now *unique*. In terms of the A's and Ψ's the elements of the coherency matrix are

$$\begin{cases} J_{xx} = \langle A_1^2\rangle\,, \\ J_{yy} = \langle A_2^2\rangle\,, \\ J_{xy} = \langle A_1A_2\exp[i(\Psi_1-\Psi_2)]\rangle\,, \\ J_{yx} = \langle A_1A_2\exp[-i(\Psi_1-\Psi_2)\rangle\,. \end{cases} \tag{2.22}$$

We may now introduce a set of Stokes parameters by the relations

$$\begin{cases} s_0 = \langle A_1^2\rangle + \langle A_2^2\rangle & = J_{xx} + J_{yy}\,, \\ s_1 = \langle A_1^2\rangle - \langle A_2^2\rangle & = J_{xx} - J_{yy}\,, \\ s_2 = 2\langle A_1A_2\cos(\Psi_1-\Psi_2)\rangle & = J_{xy} + J_{yx}\,, \\ s_3 = 2\langle A_1A_2\sin(\Psi_1-\Psi_2)\rangle & = i(J_{yx} - J_{xy})\,. \end{cases} \tag{2.23}$$

We see that this set of Stokes parameters is unique and that uniqueness has been achieved with the help of analytic signals, the introduction of which was suggested by the appearance of conjugate functions in the analysis of our experiment.

The relation between the Stokes parameters and the coherency matrix may also be expressed in the form

$$\mathbf{J} = \tfrac{1}{2}\sum_{i=0}^{3} s_i \boldsymbol{\sigma}_i \,, \tag{2.24}$$

where $\boldsymbol{\sigma}_0$ is the unit matrix

$$\boldsymbol{\sigma}_0 = \begin{bmatrix} 1 & 0 \\ 0 & 1 \end{bmatrix},$$

and $\boldsymbol{\sigma}_1$, $\boldsymbol{\sigma}_2$, $\boldsymbol{\sigma}_3$, are the Pauli spin matrices

$$\boldsymbol{\sigma}_1 = \begin{bmatrix} 1 & 0 \\ 0 & -1 \end{bmatrix}, \quad \boldsymbol{\sigma}_2 = \begin{bmatrix} 0 & 1 \\ 1 & 0 \end{bmatrix}, \quad \boldsymbol{\sigma}_3 = \begin{bmatrix} 0 & i \\ -i & 0 \end{bmatrix}. \tag{2.26}$$

The connection between a coherency matrix, Stokes parameters and Pauli's spin matrices has been noted previously (FANO ([12])); however, as already mentioned, the non-uniqueness of the usual analytic definition of the Stokes parameters appears to have escaped attention.

Finally we mention that the coherency matrix (2.16) was introduced in an earlier paper (WOLF ([4b])) from more formal considerations. Our present analysis shows that this matrix appears in a natural way from the analysis of a simple experiment.

3. – Some consequences of the basic intensity formula.

To see the physical significance of the intensity formula (2.15) we re-write it in a somewhat different form. We set

$$\frac{J_{xy}}{\sqrt{J_{xx}}\sqrt{J_{yy}}} = \mu_{xy} = |\mu_{xy}| \exp[i\beta_{xy}] \,. \tag{3.1}$$

([12]) U. FANO: *Phys. Rev.*, **93**, 121 (1954).

It follows from (2.18) that

$$|\mu_{xy}| \leqslant 1 .$$

By analogy with the theory of partially coherent scalar fields we may call μ_{xy} *the complex degree of coherence* of the electric vibrations in the x and y directions. It absolute value $|\mu_{xy}|$ is a measure of the degree of correlation of the vibrations and its phase represents their « effective phase difference ».

If we substitute from (3.1) into the intensity formula (2.15) and use the relation $J_{yx} = J^*_{xy}$ we obtain the following expression for the intensity:

$$(3.3) \qquad I(\theta, \delta) = J_{xx} \cos^2\theta + J_{yy} \sin^2\theta + 2\sqrt{J_{xx}}\sqrt{J_{yy}} \cos\theta \sin\theta \, |\mu_{xy}| \cos(\beta_{xy} - \delta) .$$

This expression is formally identical with the basic interference law of partially coherent fields [WOLF ([4a]), p. 102, THOMPSON and WOLF ([13]), p. 896]. It shows that the intensity $I(\theta, \delta)$ may be regarded as arising from the interference of two beams of intensities:

$$(3.4) \qquad I^{(1)} = J_{xx} \cos^2\theta , \qquad I^{(2)} = J_{yy} \sin^2\theta ,$$

and with complex degree of coherence μ_{xy}, after a phase difference δ has been introduced between them.

Returning to (2.15) we see that the elements of the coherency matrix of a quasi-monochromatic plane wave may be determined from very simple experiments. It is only necessary to measure the intensity for several different values of θ (orientation of polarizer) and δ (delay introduced by a compensator), and solve the corresponding relations obtained from (2.15). Let $\{\theta, \delta\}$, denote the measurement corresponding to a particular pair θ, δ. A convenient set of measurements is the following:

$$(3.5) \qquad \{0°, 0\}, \quad \{45°, 0\}, \quad \{90°, 0\}, \quad \{135°, 0\}, \quad \left\{45°, \frac{\pi}{2}\right\}, \quad \left\{135°, \frac{\pi}{2}\right\} .$$

It follows from (2.15) that, in terms of the intensities determined from these six measurements, the elements of the coherency matrix are given by

$$(3.6) \qquad \left\{ \begin{aligned} J_{xx} &= I(0°, 0) , \\ J_{yy} &= I(90°, 0) , \\ J_{xy} &= \frac{1}{2}\{I(45°, 0) - I(135°, 0)\} + \frac{1}{2} i \left\{ I\left(45°, \frac{\pi}{2}\right) - I\left(135°, \frac{\pi}{2}\right) \right\} , \\ J_{yx} &= \frac{1}{2}\{I(45°, 0) - I(135°, 0)\} - \frac{1}{2} i \left\{ I\left(45°, \frac{\pi}{2}\right) - I\left(135°, \frac{\pi}{2}\right) \right\} . \end{aligned} \right.$$

([13]) B. J. THOMPSON and E. WOLF: *Journ. Opt. Soc. Amer.*, **47**, 895 (1957).

Thus we see that the elements of the coherency matrix represent ***measurable*** physical quantities.

In the theory of partially coherent scalar fields, the concept of Michelson's visibility (*) of fringes plays a central role (cf. ZERNIKE (2)). We will now derive an expression for a quantity defined in a similar way and we shall see later that this quantity has a simple physical meaning.

It follows from (2.15) by a straightforward calculation, that the maxima and minima of the intensity (with respect to both θ and δ) are

$$
\left\{
\begin{aligned}
I_{\max} &= \frac{1}{2}(J_{xx}+J_{yy})\left[1+\sqrt{1-\frac{4|\boldsymbol{J}|}{(J_{xx}+J_{yy})^2}}\right], \\
I_{\min} &= \frac{1}{2}(J_{xx}+J_{yy})\left[1-\sqrt{1-\frac{4|\boldsymbol{J}|}{(J_{xx}+J_{yy})^2}}\right].
\end{aligned}
\right. \tag{3.7}
$$

Hence

$$
\frac{I_{\max}-I_{\min}}{I_{\max}+I_{\min}} = \sqrt{1-\frac{4|\boldsymbol{J}|}{(J_{xx}+J_{yy})^2}}\,. \tag{3.8}
$$

Now if the x and y axes are rotated about the direction of propagation of the wave, the coherency matrix will change. There are, however, two invariants for such rotations, namely the discriminant $|\boldsymbol{J}|$ and the trace $\mathrm{Tr}\boldsymbol{J} = J_{xx}+J_{yy}$ of the matrix. Since on the right hand side of (3.8) the elements of $\boldsymbol{J}$ enter only in these combinations, it follows that the expression is invariant with respect to rotations of the axes and hence may be expected to have a physical significance. We shall see shortly (eq. (5.14) below) that it represents the degree of polarization of the wave.

4. – Coherency matrices of natural and of monochromatic radiation.

Light which is most frequently encountered in nature has the property that the intensity of its components in any direction perpendicular to the direction of propagation is the same; and, moreover, the intensity is not af-

(*) The visibility $\mathscr{V}$ of fringes at a point P in the fringe pattern is defined by the formula

$$
\mathscr{V} = \frac{I_{\max}-I_{\min}}{I_{\max}+I_{\min}},
$$

where $I_{\max}$ and $I_{\min}$ are the maximum and minimum intensities in the immediate neighbourhood of P.

fected by any previous retardation of one of the rectangular components relative to the other, into which the light may have been resolved. In other words

$$I(\theta, \delta) = \text{constant} \tag{4.1}$$

for all values of θ and δ. Such light is called *natural light*; and we may define « natural » electromagnetic radiation of any other spectral range in a strictly similar way.

It is evident from (3.3) that $I(\theta, \delta)$ is independent of δ and θ, if and only if

$$|\mu_{xy}| = 0\,, \qquad \text{and} \qquad J_{xx} = J_{yy}\,. \tag{4.2}$$

The first condition implies that the electric vibrations in the x and y directions are mutually incoherent. According to (3.1) and the relation $J_{yx} = J^*_{yx}$, (4.2) may also be written as

$$J_{xy} = J_{yx} = 0\,, \qquad J_{xx} = J_{yy}\,, \tag{4.3}$$

and it follows that *the coherency matrix of natural radiation* of intensity $J_{xx} + J_{yy} = I$ is

$$\frac{1}{2} I \begin{bmatrix} 1 & 0 \\ 0 & 1 \end{bmatrix}. \tag{4.4}$$

Next let us consider the coherency matrix of monochromatic radiation. In this case the amplitudes A_1 and A_2 and the phases Ψ_1 and Ψ_2 in (2.22) are independent of time and the coherency matrix has the form

$$\begin{bmatrix} A_1^2 & A_1 A_2 \exp[i(\Psi_1 - \Psi_2)] \\ A_1 A_2 \exp[-i(\Psi_1 - \Psi_2)] & A_2^2 \end{bmatrix}. \tag{4.5}$$

We see that in this case

$$|\mathbf{J}| = J_{xx} J_{yy} - J_{xy} J_{yx} = 0\,, \tag{4.6}$$

i.e. the discriminant of the coherency matrix is zero. The complex degree of coherence now is

$$\mu_{xy} = \frac{J_{xy}}{\sqrt{J_{xx}}\sqrt{J_{yy}}} = \exp[(\Psi_1 - \Psi_2)\,, \tag{4.7}$$

i.e. its absolute value is unity (complete coherence) and its phase is equal to the difference between the phases of the two components.

5. – The degree of polarization.

Before deriving an expression for the degree of polarization in terms of the coherency matrix we shall establish a simple theorem relating to the coherency matrix of a wave resulting from the superposition of a number of mutually independent waves.

Consider N mutually independent quasi-monochromatic waves propagated in the same direction (z-say) and let $E_x^{(n)}$, $E_y^{(n)}$ ($n = 1, 2, ..., N$) be the analytic signals associated with the components of the electric vibrations of the n-th wave in the directions of the x and y-axes. The components of the resulting wave then are

$$E_x = \sum_{n=1}^{N} E_x^{(n)}, \qquad E_y = \sum_{n=1}^{N} E_y^{(n)}, \tag{5.1}$$

so that the elements of the coherency matrix are

$$\begin{cases} J_{kl} = \langle E_k E_l^* \rangle = \sum_{n=1}^{N} \sum_{m=1}^{N} \langle E_k^{(n)} E_l^{(m)*} \rangle \,, \\ \quad = \sum_{n=1}^{N} \langle E_k^{(n)} E_l^{(n)*} \rangle + \sum_{n \neq m} \langle E_k^{(n)} E_l^{(m)*} \rangle \,. \end{cases} \tag{5.2}$$

Since the waves are assumed to be independent, each term under the last summation sign is zero, and it follows that

$$J_{kl} = \sum_{n=1}^{N} J_{kl}^{(n)}, \tag{5.3}$$

where $J_{kl}^{(n)} = \langle E_k^{(n)} E^{(n)*} \rangle$ are the elements of the coherency matrix of the n-th wave. The formula (5.3) shows that the coherency matrix of a wave resulting from the superposition of a number of independent waves is the sum of the coherency matrices of the individual waves.

To find an expression for the degree of polarization of a wave, we first represent the wave as a superposition of a wave of natural radiation and a wave of monochromatic radiation, independent of the former. Let $\boldsymbol{J}$ be the coherency matrix of the given wave and let $\boldsymbol{J}^{(1)}$ and $\boldsymbol{J}^{(2)}$ be the coherency matrices of the two independent waves into which we decompose it. Then according to (4.4) and (4.6) $\boldsymbol{J}^{(1)}$ and $\boldsymbol{J}^{(2)}$ must be of the form

$$\boldsymbol{J}^{(1)} = \begin{bmatrix} A & 0 \\ 0 & A \end{bmatrix}, \qquad \boldsymbol{J}^{(2)} = \begin{bmatrix} B & D \\ D^* & C \end{bmatrix}, \tag{5.4}$$

where $A \geqslant 0$, $B \geqslant 0$, $C \geqslant 0$ and

$$BC - DD^* = 0 . \tag{5.5}$$

In order to show that such a decomposition is possible we must determine quantities A, B, C, D, subject to the above conditions, such that the given coherency matrix $\mathbf{J} = [J_{ik}]$ is equal to the sum of two matrices of the form (5.4),

$$\mathbf{J} = \mathbf{J}^{(1)} + \mathbf{J}^{(2)} . \tag{5.6}$$

The relation (5.6) implies that

$$\begin{cases} J_{xx} = A + B , & J_{xy} = D , \\ J_{yx} = D^* , & J_{yy} = A + C . \end{cases} \tag{5.7}$$

On substituting for B, C, D and D^* from (5.7) into (5.5) we find that (*)

$$A = \tfrac{1}{2}(J_{xx} + J_{yy}) \pm \tfrac{1}{2}\sqrt{(J_{xx} + J_{yy})^2 - 4|\mathbf{J}|} . \tag{5.8}$$

Since $J_{yx} = J^*_{yx}$ the product $J_{xy}J^*_{yx}$ is non-negative, and it follows from (2.18) that

$$|\mathbf{J}| \leqslant J_{xx}J_{yy} \leqslant \tfrac{1}{4}(J_{xx} + J_{yy})^2 , \tag{5.9}$$

so that both the roots (5.8) are non-negative. Consider first the solution with the negative sign in front of the square root. We then have from (5.7),

$$\begin{cases} A = \tfrac{1}{2}(J_{xx} + J_{yy}) - \tfrac{1}{2}\sqrt{(J_{xx} + J_{yy})^2 - 4|\mathbf{J}|} , \\ B = \tfrac{1}{2}(J_{xx} - J_{yy}) + \tfrac{1}{2}\sqrt{(J_{xx} + J_{yy})^2 - 4|\mathbf{J}|} , \\ C = \tfrac{1}{2}(J_{yy} - J_{xx}) + \tfrac{1}{2}\sqrt{(J_{xx} + J_{yy})^2 - 4|\mathbf{J}|} , \\ D = J_{xy} , \\ D^* = J_{yx} . \end{cases} \tag{5.10}$$

Now

$$\sqrt{(J_{xx} + J_{yy})^2 - 4|\mathbf{J}|} = \sqrt{(J_{xx} - J_{yy})^2 + 4J_{xy}J_{yx}} \geqslant |J_{xx} - J_{yy}| . \tag{5.11}$$

Hence B and C are also non-negative as required. The other root given by (5.8) (with the positive sign in front of the square root) leads to negative values

(*) A is seen to be a characteristic root (eigenvalue) of the coherency matrix $\mathbf{J}$.

of B and C and must therefore be rejected. We have thus obtained a unique decomposition of the required kind.

The total intensity of the wave is

$$I_{\text{tot}} = \operatorname{Tr} \boldsymbol{J} = J_{xx} + J_{yy} \; ; \tag{5.12}$$

and the intensity of the monochromatic (and hence *polarized*) part is

$$I_{\text{pol}} = \operatorname{Tr} \boldsymbol{J}^{(2)} = B + C = \sqrt{(J_{xx} + J_{yy})^2 - 4|\boldsymbol{J}|} \, . \tag{5.13}$$

Hence the *degree of polarization* P of the original wave is

$$P = \frac{I_{\text{pol}}}{I_{\text{tot}}} = \sqrt{1 - \frac{4|\boldsymbol{J}|}{(J_{xx} + J_{yy})^2}} \, . \tag{5.14}$$

Since this expression involves only the two rotational invariants of the coherency matrix $\boldsymbol{J}$, the degree of polarization is independent of the particular choice of the x and y axes, as might have been expected.

Comparison of (5.14) with (3.8) shows that the quantity $(I_{\text{max}} - I_{\text{min}})/(I_{\text{max}} + I_{\text{min}})$ is precisely the degree of polarization P of the wave.

Unlike the degree of polarization, the degree of coherence depends on the choice of the x and y directions. We shall not investigate in detail the changes in the degree of coherence as the x, y axes are rotated; we shall only consider an extreme case, which is of special physical interest.

If in the expression (5.14) for the degree of polarization P we write out in full the discriminant $\boldsymbol{J}$, and use the expression (3.1) for the degree of coherence μ_{xy} we find that the following relation holds between P and $|\mu_{xy}|$:

$$1 - P^2 = \frac{J_{xx} J_{yy}}{[\frac{1}{2}(J_{xx} + J_{yy})]^2} [1 - |\mu_{xy}|^2] \, . \tag{5.15}$$

Since the geometric mean of any two positive numbers cannot exceed their arithmetic mean it follows that $1 - P^2 \leqslant 1 - |\mu_{xy}|^2$, *i.e.*

$$P \geqslant |\mu_{xy}| \, . \tag{5.16}$$

The equality sign in (5.16) will hold if an only if $J_{xx} = J_{yy}$, *i.e.* if the (time averaged) intensities in the two orthogonal directions are equal. We shall now show that a pair of directions always exists for which this is the case.

Suppose that we take a new pair of orthogonal directions x', y' perpendicular to the direction of propagation of the wave and let φ be the angle between x and x'. The components E_x, E_y, of the electric vector (in the complex representations (2.13)) in the new directions are

$$
\begin{cases} E_{x'} = \quad E_x \cos\varphi + E_y \sin\varphi\,, \\ E_{y'} = -E_x \sin\varphi + E_y \cos\varphi\,. \end{cases} \tag{5.17}
$$

Hence the elements of the transformed coherency matrix $\boldsymbol{J}' = [J_{k'l'}] = [\langle E_{k'} E^*_{l'}\rangle]$ are

$$
\begin{cases} J_{x'x'} = J_{xx}c^2 + J_{yy}s^2 + (J_{xy} + J_{yx})cs\,, \\ J_{y'y'} = J_{xx}s^2 + J_{yy}c^2 - (J_{xy} + J_{yx})\,cs\,, \\ J_{x'y'} = (J_{yy} - J_{xx})cs + J_{xy}c^2 - J_{yx}s^2\,, \\ J_{y'x'} = (J_{yy} - J_{xx})cs + J_{yx}c^2 - J_{xy}s^2\,, \end{cases} \tag{5.18}
$$

where

$$
c = \cos\varphi\,, \qquad s = \sin\varphi\,. \tag{5.19}
$$

The intensities in the x' and y' directions will be equal (*i.e.* $J_{x'x'} = J_{y'y'}$) if

$$
J_{xx}c^2 + J_{yy}s^2 + (J_{xy} + J_{yx})\,cs = J_{xx}s^2 + J_{yy}c^2 - (J_{xy} + J_{yx})\,cs\,.
$$

Solving this equation for φ we obtain

$$
\operatorname{tg} 2\varphi = \frac{J_{yy} - J_{xx}}{J_{xy} + J_{yx}}\,. \tag{5.20}
$$

Since $J_{yx} = J^*_{xy}$ and J_{xx} and J_{yy} are real, this equation always has a real root. Thus *there always exists a pair of directions for which the two intensities are equal. For this pair of directions the degree of coherence* $|\mu_{xy}|$ *of the electric vibrations has its maximum value and this value is equal to the degree of polarization of the wave.*

This special pair of directions has a simple geometrical significance. If, as in (5.6) we represent the wave as incoherent mixture of a wave of natural radiation and a wave of monochromatic (and therefore completely polarized) radiation, the angle χ which the major axes of the vibrational ellipse of the polarized portion makes with the x-direction is given by (see CHANDRASEKHAR [7], p. 33, eq. (180))

$$
\operatorname{tg} 2\chi = \frac{s_2}{s_1} = \frac{J_{xy} + J_{yx}}{J_{xx} - J_{yy}}\,. \tag{5.21}
$$

It follows from (5.20) and (5.21) that $(\operatorname{tg} 2\chi)\cdot(\operatorname{tg} 2\varphi) = -1$ so that $\chi - \varphi = 45°$ or $135°$. This implies that *the directions for which* $P = |\mu_{xy}|$ *are the bisectors of the principal directions (directions of the major and minor axes) of the vibrational ellipse of the polarized portion of the wave.*

It is evident from the foregoing discussion that the introduction of coherence concepts into the theory of partial polarization leads to a clearer understanding of the behaviour of partially polarized radiation and suggests new ways for the measurement of its degree of polarization.

RIASSUNTO (*)

In questo articolo si fa l'analisi della polarizzazione parziale dal punto di vista della teoria della coerenza. Dopo aver rilevato che la usuale definizione analitica dei parametri di Stokes per un'onda quasi monocromatica non è univoca, si prende in esame un semplice esperimento il quale introduce, in maniera chiara, i parametri osservabili per un'onda luminosa quasi monocromatica. L'analisi conduce ad un'unica matrice di coerenza e ad un unico gruppo di parametri di Stokes; quest'ultimo è associato alla rappresentazione della matrice di coerenza in funzione delle matrici di spin di Pauli. Tale analisi mostra la fondamentale importanza del concetto di segnale analitico di Gabor. Il grado di coerenza fra le vibrazioni elettriche in due direzioni qualunque, reciprocamente ortogonali, di propagazione dell'onda dipende, in generale, dalla scelta delle due direzioni ortogonali. Si dimostra che il valore massimo di esso uguaglia il grado di polarizzazione dell'onda. Si dimostra altresì che il grado di polarizzazione può essere dedotto, seguendo una via nuova, da esperimenti relativamente semplici che richiedono l'uso di un compensatore e di un polarizzatore, e che tale determinazione è analoga a quella del grado di coerenza ricavata dall'esperimento di interferenza di Young.

(*) *Traduzione a cura della Redazione.*

28

Reprinted from *Il Nuovo Cimento,* Ser. 10, **15,** 370–388 (Feb. 1960)

On the Matrix Formulation of the Theory of Partial Polarization in Terms of Observables.

G. B. PARRENT, Jr. (*) and P. ROMAN

Department of Theoretical Physics, The University - Manchester

(ricevuto il 19 Ottobre 1959)

Summary. — The coherency matrix of a quasi-monochromatic plane wave is deduced from a matrix representation of the analytic signal associated with the electric field. It is shown that if the radiation passes through a physical device, such as a compensator, absorber, rotator, or polarizer, the effect of this interaction can be fully described in terms of appropriately choosen operators which transform directly the coherency matrix. The complex degree of coherence is defined in terms of the operators mentioned above, and from this is deduced an expression characterizing the degree of polarization. It is shown that the quantity deduced in this manner is identical with that obtained from the more conventional definition. An experiment is described which can serve to measure the components of the correlation matrix.

1. – Introduction.

The study of partial polarization has a great deal in common with the study of partial coherence, since both are necessitated by the statistical nature of « natural » radiation. Thus for example a rigorously monochromatic beam of electromagnetic radiation would be completely coherent and completely polarized; and the behaviour of such a beam is simply determined. However, if one considers a more realistic model of a beam of light, *e.g.* quasi-monochro-

(*) Permanent address: Electromagnetic Radiation Lab., Electronics Research Directorate, Air Force Cambridge Research Center, L. G. Hanscom, A. F. B. Bedford, Mass., U.S.A.

matic (which results from the superposition of a large number of randomly timed statistically independent pulses with the same central frequency), the problem of describing its behaviour is essentially different. What was a completely determined problem becomes a statistical one in which even the concepts of amplitude and phase become ambiguous, requiring redefinition.

It is only when dealing with such statistical radiation that the terms partial coherence and partial polarization have any meaning.

In view of this genetic relationship between these two aspects of the study of natural light one would expect the techniques and concepts which prove important for one to be helpful in understanding the other. Consequently, the recent emphasis on the formulation of coherence theory in terms of measurable quantities (correlation functions), and the success which this approach has enjoyed, suggest that the more general problems of partial polarization might also be better understood if they were formulated in a similar way.

In fact, considerable progress with this approach has already been made through the introduction by WOLF ([1]) of the correlation tensor. The matrix of this tensor, the correlation matrix, is the appropriate entity for the description of partially polarized beams. Its physical significance was, however, left somewhat obscure. In a later paper WOLF ([2]) identified the elements of the coherency matrix $\mathscr{J}$ with the factors occurring in the expression for the intensity in a beam after transmission through a compensator and a polarizer. In his paper ref. ([1]) he pointed out that $\mathscr{J}$ is formally equivalent to the density matrix in the study of scattering phenomena.

In spite of the attention focussed upon the coherency matrix, the uniformity which this entity introduces has gone largely unnoticed. The formulation of the subject as introduced here stresses the analogy between this field of research and modern scattering theory.

We shall use here a matrix representation of the electric field and derive formulae for the solution of the problem arising in the description of partially polarized fields and their interaction with physical devices. We mention that in the last decade several attempts have been made to describe in terms of matrices the action of various physical devices on radiation passed through them (see, for example, CLARK JONES ([3]) or WESTFOLD ([4]), where also further literature is quoted). Our approach will be, however, very different from these and also more general. In particular, we shall avoid the use of the « Stokes vector »

$$\boldsymbol{s} = (s_0, s_1, s_2, s_3)\,,$$

([1]) E. WOLF: *Nuovo Cimento*, **12**, 884 (1954).
([2]) E. WOLF: *Nuovo Cimento*, **13**, 1165 (1959).
([3]) R. CLARK JONES: *Journ. Opt. Soc. Am.*, **46**, 126 (1956).
([4]) K. C. WESTFOLD: *Journ. Opt. Soc. Am.*, **49**, 717 (1959).

which, in fact, has not the transformation property of a four-vector. Furthermore, we shall not need 4×4 matrices to characterize the interactions, but only 2×2 matrices. An attractive feature of the present approach is that *one obtains matrix operators characteristic of the physical devices (interactions) which operate directly on the coherency matrix.* Further, the physical characteristic to be measured (*e.g.* the intensity I) is always given by an expression of the form

$$F = \mathrm{Sp}\,[\mathscr{F}\mathscr{I}]\,, \tag{1}$$

where F is the characteristic to be measured, $\mathscr{I}$ is the coherency matrix, $\mathscr{F}$ is an operator describing the experiment (the device), and Sp indicates the trace. In (1) *the operator* $\mathscr{F}$ *depends only on the physical devices and* $\mathscr{I}$ *depends only on the measurable properties of the beam.* That is, $\mathscr{I}$ describes the state of the field.

2. – Scope, definitions and notation.

Throughout this discussion we shall limit our attention to quasi-monochromatic fields. In essence this approximation requires that the spectral width of the radiation is negligible compared to the mean frequency, *i.e.*

$$\Delta\nu \ll \bar{\nu}\,. \tag{2a}$$

The introduction of this approximation enables us to evaluate frequency-dependent quantities at the mean frequency, but also limits the validity of the development to phenomena involving relatively small path differences Δl, *i.e.* the theory is valid when

$$\Delta l \ll c/\Delta\nu\,, \tag{2b}$$

where c is the velocity of light. For a full discussion of the consequences of this approximation the reader is referred to BORN and WOLF [5]. Further, we limit our attention to plane waves in which case the electric vector has two components, perpendicular to the direction of propagation. While these approximations are appropriate for many problems of interest, the analysis given here may be extended to polychromatic non-planar waves. This generalization will be treated in a later communication.

---

[5] M. BORN and E. WOLF: *Principles of Optics* (London, 1959), p. 502.

The real electric field will be denoted by the column matrix

$$\mathscr{E}^r(\boldsymbol{x}, t) = \begin{pmatrix} E_x^r(\boldsymbol{x}, t) \\ E_y^r(\boldsymbol{x}, t) \end{pmatrix}. \tag{3}$$

Here, as throughout this paper, script letters denote matrices; bold face type denotes vectors; and a superscript r denotes a real function of a real variable.

Our analysis will not deal directly with $\mathscr{E}^r$ but rather with the associated analytic signal representation of the field. The advantage of this representation for problems involving correlation functions become apparent from the investigations of various authors ([6-8]); therefore we shall simply review briefly the method of obtaining the analytic signal from the real field.

It is assumed (see p. 1169 of ref. ([2])) that the field possesses a Fourier transform; thus

$$E_x^r(\boldsymbol{x}, t) = \int_0^\infty a(\boldsymbol{x}, \nu) \sin\,[\varphi(\boldsymbol{x}, \nu) - 2\pi\nu t]\, \mathrm{d}\nu\,.$$

We associate with E_x^r its « conjugate function » $E_x^i(\boldsymbol{x}, t)$ defined by

$$E_x^i(\boldsymbol{x}, t) = \int_0^\infty a(\boldsymbol{x}, \nu) \cos\,[\varphi(\boldsymbol{x}, \nu) - 2\pi\nu t]\, \mathrm{d}\nu\,.$$

It is easily shown that E_x^i is the Hilbert transform of E_x^r, *i.e.*

$$E_x^i(\boldsymbol{x}, t) = \frac{1}{\pi} P \int_{-\infty}^{+\infty} \frac{E_x^r(\boldsymbol{x}, t')}{t' - t}\, \mathrm{d}t'\,,$$

where P denotes Cauchy's principal value. The analytic signal E_x is then defined as

$$E_x(\boldsymbol{x}, t) = E_x^r(\boldsymbol{x}, t) + iE_x^i(\boldsymbol{x}, t)\,.$$

Similar considerations hold for E_y.

([6]) Cf. ref. ([5]), p. 492.
([7]) G. B. Parrent jr: *Journ. Opt. Soc. Am.*, **49**, 787 (1959).
([8]) P. Roman and E. Wolf: *Ann. Phys.*, in press.

In terms of these functions our representation becomes

$$\mathscr{E}(\boldsymbol{x}, t) = \begin{pmatrix} E_x(\boldsymbol{x}, t) \\ E_y(\boldsymbol{x}, t) \end{pmatrix} = \mathscr{E}^r(\boldsymbol{x}, t) + i\mathscr{E}^i(\boldsymbol{x}, t) \,. \tag{4}$$

In this representation the intensity I at the point $\boldsymbol{x}$ may be expressed as

$$I = \mathrm{Sp}\, \langle \mathscr{E} \times \mathscr{E}^\dagger \rangle \,. \tag{5}$$

Here $\mathscr{E}^\dagger$ is the Hermitian conjugate of $\mathscr{E}$, *i.e.* the row matrix

$$\mathscr{E}^\dagger = \tilde{\mathscr{E}}^* = [E_x^* \; E_y^*];$$

further, $\langle ... \rangle$ indicates the time average, and $\times$ denotes the Kronecker product of matrices. That this definition is equivalent to that normally given, may be seen by writing (5) in full, when we obtain

$$I = \langle \boldsymbol{E} \cdot \boldsymbol{E}^* \rangle = 2 \langle (\boldsymbol{E}^r)^2 \rangle \,.$$

(In the last step some simple properties of Hilbert transforms have been utilized.)

We introduce now the coherency matriw $\mathscr{I}$ by the definition

$$\mathscr{I} = \langle \mathscr{E} \times \mathscr{E}^\dagger \rangle \,, \tag{6}$$

which is clearly Hermitian:

$$\mathscr{I}^\dagger = \mathscr{I} \,.$$

$\mathscr{I}$ is in fact the matrix of the correlation tensor introduced by WOLF (1) who proved that the elements of this matrix are observables of the radiation field. In terms of $\mathscr{I}$ the intensity (5) becomes simply

$$I = \mathrm{Sp}\, \mathscr{I} \,. \tag{7}$$

One further remark about notation. At one stage in the analysis the Pauli matrices will be used. Since our representation differs slightly from that normally used, we shall summarize it at this point. These matrices obey the algebra

$$\begin{cases} \sigma_\alpha \sigma_\beta = -i\sigma_\gamma \,, & (\alpha, \beta, \gamma) = (1, 2, 3) \text{ and cycl.;} \\ (\sigma_i)^2 = \sigma_0 \,, & (i = 0, 1, 2, 3) \,, \\ \sigma_i \sigma_0 = \sigma_0 \sigma_i = \sigma_i \,. & \end{cases} \tag{8}$$

It follows from (8) that

$$\mathrm{Sp}\,(\sigma_i\sigma_j) = 2\delta_{ij}\,, \qquad (i,\,j = 0,\,1,\,2,\,3)\,. \tag{9}$$

The matrices may be represented by putting

$$\sigma_0 = \begin{pmatrix} 1 & 0 \\ 0 & 1 \end{pmatrix}, \quad \sigma_1 = \begin{pmatrix} 1 & 0 \\ 0 & -1 \end{pmatrix}, \quad \sigma_2 = \begin{pmatrix} 0 & 1 \\ 1 & 0 \end{pmatrix}, \quad \sigma_3 = \begin{pmatrix} 0 & i \\ -i & 0 \end{pmatrix}. \tag{10}$$

In the following sections we shall obtain transformation equations for $\mathscr{I}$ corresponding to the passage of partially polarized radiation through the various physical devices which are of interest in the study of such fields. In Sections **8** and **9** we discuss the various parameters used to specify the state of such statistical radiation and their relation to the coherency matrix. The relation between $\mathscr{I}$ and the density matrix will be briefly discussed in the concluding remarks.

3. – Compensator.

The compensator is a device which introduces a phase change ε_x in the x-component and ε_y in the y-component in each spectral component of the field vector. This results in a relative phase difference $\delta = \varepsilon_x - \varepsilon_y$ between the x- and y-components. Thus the compensator produces a selective rotation of the electric field in Fourier space. In the general case δ is of course a function of frequency; however, consistent with the quasi-monochromatic approximation we take the phase shift for each spectral component to be equal to that for the mean frequency. The compensator may, therefore, be represented or characterized by a unitary rotation matrix

$$\mathscr{C} = \begin{pmatrix} \exp\,[\frac{i}{2}\,\delta] & 0 \\ 0 & \exp\,[-\frac{i}{2}\,\delta] \end{pmatrix}. \tag{11}$$

If $\mathscr{E}$ is the field incident on the compensator, the emergent field $\mathscr{E}_C$ may be written as

$$\mathscr{E}_C = \mathscr{C}\mathscr{E}\,; \tag{12}$$

and the coherency matrix for the emergent field is given, according to the definition (6), by

$$\mathscr{I}_C = \langle \mathscr{E}_C \times \mathscr{E}_C^\dagger \rangle = \langle \mathscr{C}\mathscr{E} \times \mathscr{E}^\dagger \mathscr{C}^\dagger \rangle\,. \tag{13}$$

Hence we may write (*)

$$\mathscr{I}_C = \mathscr{C}\langle \mathscr{E} \times \mathscr{E}^\dagger \rangle \mathscr{C}^\dagger ; \tag{14}$$

and substituting from (6) we obtain, using also the unitary nature of $\mathscr{C}$,

$$\mathscr{I}_C = \mathscr{C}\mathscr{I}\mathscr{C}^\dagger = \mathscr{C}\mathscr{I}\mathscr{C}^{-1} . \tag{15}$$

Thus the effect of the compensator may be characterized by a rotation operator in Fourier space acting directly on the coherency matrix $\mathscr{I}$, without recourse to the unmeasurable quantities of the $\mathscr{E}$ field itself.

To compute the intensity I_C at a point in the emergent beam we take, according to (7), the trace on both sides of (15); and since the argument of the trace may be cyclically permuted, we obtain

$$I_C = I .$$

That the intensity is unchanged under the passage of the beam through a compensator, could have been anticipated from the fact that a rotation leaves the trace invariant and absorption and reflection were ignored. However, absorption is easily incorporated into the scheme as will be shown in the next section.

4. – Absorption.

Absorption is characterized by a decrease in field strength and may therefore be represented by a matrix of the form

$$\mathscr{A} = \begin{pmatrix} \exp[-\frac{1}{2}\eta_x] & 0 \\ 0 & \exp[-\frac{1}{2}\eta_y] \end{pmatrix} . \tag{16}$$

Here η_x and η_y are the absorption coefficients for the x- and y-components respectively, evaluated at the mean frequency $\bar{\nu}$. The effect of absorption is to produce a field $\mathscr{E}_A$ given by

$$\mathscr{E}_A = \mathscr{A}\mathscr{E} , \tag{17}$$

(*) The associative property used in obtaining (14) is of course not permissible in general; indeed in the general case this operation is meaningless. However, when it is possible to perform such an operation, as in the present and in the following cases, the associative property is valid as may be easily verified.

and the coherency matrix becomes

$$\mathcal{I}_A = \mathcal{A}\mathcal{I}\mathcal{A}^{\dagger}\,. \tag{18}$$

The reduced intensity is clearly

$$I_A = \mathrm{Sp}\,[\mathcal{A}\mathcal{I}\mathcal{A}^{\dagger}] = \mathrm{Sp}\,[\mathcal{A}^2\mathcal{I}]\,. \tag{19}$$

Note that the measurable intensity is given by an expression of the form (1).

5. - Rotator.

Various materials and physical devices produce a rotation of the electric vector, *i.e.* they rotate the plane of polarization. Such a device is termed a rotator and may clearly be characterized by a rotation operator

$$\mathcal{R}(\alpha) = \begin{pmatrix} \cos\alpha & \sin\alpha \\ -\sin\alpha & \cos\alpha \end{pmatrix}, \tag{20}$$

where α is the angle through which the field is rotated. Note that $\mathcal{R}$ is a real antisymmetric unimodular unitary matrix.

The emergent field $\mathcal{E}_R$ following a rotator is given by

$$\mathcal{E}_R = \mathcal{R}(\alpha)\mathcal{E}\,, \tag{21}$$

and the coherency matrix becomes

$$\mathcal{I}_R = \mathcal{R}(\alpha)\mathcal{I}\mathcal{R}^{\dagger}(\alpha) = \mathcal{R}(\alpha)\mathcal{I}\mathcal{R}^{-1}(\alpha) = \mathcal{R}(\alpha)\mathcal{I}\mathcal{R}(-\alpha)\,. \tag{22}$$

Thus the action of the rotator may also be represented as an operation directly on the coherency matrix which contains only observable quantities. The intensity of course remains unchanged.

6. - Polarizer.

The last device to be considered in this discussion before turning to general considerations is a polarizer, such as a Nichol prism, which passes only a particular component of the field, say the component making an angle θ with the x-direction. That is, the polarizer takes the projection of the $\mathcal{E}$ field on the direction θ. The polarizer may thus be characterized by a projection operator $\mathscr{P}_+(\theta)$. Projection operators are singular and satisfy the idempotency condition

$$\mathscr{P}_+(\theta)\mathscr{P}_+(\theta) = \mathscr{P}_+(\theta)\,. \tag{23}$$

Associated with every projection operator $\mathscr{P}_+(\theta)$ is an orthogonal proejction operator $\mathscr{P}_-(\theta)$ representing a projection on the direction orthogonal to θ and satisfying the conditions

$$\begin{cases} \mathscr{P}_-(\theta)\mathscr{P}_-(\theta) = \mathscr{P}_-(\theta)\,, \\ \mathscr{P}_+(\theta)\mathscr{P}_-(\theta) = \mathscr{P}_-(\theta)\mathscr{P}_+(\theta) = 0\,, \\ \mathscr{P}_+(\theta) + \mathscr{P}_-(\theta) = 1\,. \end{cases} \tag{24}$$

Since the polarizer takes a projection of the $\mathscr{E}$ field, it may be represented by the operator

$$\mathscr{P}_+(\theta) = \begin{pmatrix} \cos^2\theta & \sin\theta\cos\theta \\ \sin\theta\cos\theta & \sin^2\theta \end{pmatrix}. \tag{25}$$

The associated operator $\mathscr{P}_-(\theta)$ may be written as

$$\mathscr{P}_-(\theta) = \mathscr{P}_+\left(\theta + \frac{\pi}{2}\right) = \begin{pmatrix} \sin^2\theta & -\sin\theta\cos\theta \\ -\sin\theta\cos\theta & \cos^2\theta \end{pmatrix}. \tag{26}$$

That these Hermitian operators satisfy the above conditions (23) and (24) is readily verified.

The field $\mathscr{E}_P$ emerging from the polarizer is given as

$$\mathscr{E}_P = \mathscr{P}_+(\theta)\mathscr{E}\,; \tag{27}$$

and the coherency matrix becomes

$$\mathscr{J}_P = \mathscr{P}_+\mathscr{J}\mathscr{P}_+^\dagger = \mathscr{P}_+\mathscr{J}\mathscr{P}_+\,, \tag{28}$$

since $\mathscr{P}_+$ is Hermitian. Thus the effect of the polarizer is also represented by an operation directly on the observable entity $\mathscr{J}$. Unlike the devices previously considered, however, the polarizer does not leave the intensity unaltered. The intensity is given by

$$I_P = \mathrm{Sp}\,\mathscr{J}_P = \mathrm{Sp}[\mathscr{P}_+\mathscr{J}\mathscr{P}_+] = \mathrm{Sp}[\mathscr{P}_+\mathscr{J}]\,, \tag{29}$$

where the property (23) of $\mathscr{P}_+$ was used. The reduced intensity is again an expression of the form (1).

For further application we note here that, as follows directly from (25) and (20),

$$\mathscr{R}^\dagger(\alpha)\,\mathscr{P}_+(\theta)\mathscr{R}(\alpha) = \mathscr{P}_+(\theta+\alpha)\,, \tag{30}$$

and, in particular, using also (26),

$$\mathscr{R}^\dagger\left(-\frac{\pi}{2}\right)\mathscr{P}_+(0)\mathscr{R}\left(-\frac{\pi}{2}\right) = \mathscr{P}_+\left(-\frac{\pi}{2}\right) = \mathscr{P}_-(0)\,. \tag{30a}$$

7. – Cascaded systems.

The result of cascading the various devices discussed thus far is of course simply obtained by the consecutive application of the respective operators. One special case of such a cascaded system of particular interest is the compensator followed by a polarizer. Computing the intensity at a point in the beam emerging from such an arrangement we obtain

$$\begin{aligned} I_K(\theta, \delta) &= \mathrm{Sp}[\mathscr{P}_+(\theta)\mathscr{C}(\delta)\mathscr{I}\mathscr{C}^\dagger(\delta)\,\mathscr{P}_+(\theta)] = \mathrm{Sp}[\mathscr{P}_+^2(\theta)\mathscr{C}(\delta)\mathscr{I}\mathscr{C}^\dagger(\delta)] = \\ &= \mathrm{Sp}[\mathscr{C}^\dagger(\delta)\,\mathscr{P}_+(\theta)\mathscr{C}(\delta)\mathscr{I}] = \mathrm{Sp}[\mathscr{K}(\theta, \delta)\mathscr{I}]\,, \end{aligned} \tag{31}$$

where we have set

$$\mathscr{K} = \mathscr{C}^\dagger\mathscr{P}_+\mathscr{C} = \mathscr{C}^{-1}\mathscr{P}_+\mathscr{C}\,. \tag{32}$$

The matrix $\mathscr{K}$ which, incidentally, arises from $\mathscr{P}_+$ through a rotation in Fourier space, is useful enough to be written out in full here for later reference. We find

$$\mathscr{K}(\theta, \delta) = \begin{pmatrix} \cos^2\theta & \exp[\frac{i}{2}\delta]\sin\theta\cos\theta \\ \exp[-\frac{i}{2}\delta]\sin\theta\cos\theta & \sin^2\theta \end{pmatrix}. \tag{33}$$

Note that (31) is also of the form (1), and that the effect of the combination is represented by an operation directly on $\mathscr{I}$. If (31) is written in detail, we obtain the familiar result for such an arrangement, as discussed by WOLF [2]:

$$I_K(\theta, \delta) = J_{xx}\cos^2\theta + J_{yy}\sin^2\theta + (J_{xy}\exp[-i\delta] + J_{yx}\exp[i\delta])\sin\theta\cos\theta\,. \tag{34}$$

Here J_{xx} etc., are the components of the coherency matrix.

8. – The state of the field.

The state of polarization of an electromagnetic field has been described is several ways: in terms of Stokes parameters, the degree of coherence between the x- and y-components of the field, and the degree of polarization. We shall discuss each of these modes of description in terms of the coherency matrix. Most of the relationships discussed in this section were pointed out by WOLF (2). They are, however, obtained here from a quite different point of view which is being stressed as physically more meaningful in this discussion.

The customary set of Stokes parameters has been recently the subject of considerable discussion and it was pointed out by WOLF (2) that as usually defined they are not unique. This ambiguity was, however, removed by the introduction of the analytic signal representation. It was also shown recently by ROMAN (9) that the concept of Stokes parameters can be extended to the case of non-planar waves.

While it is the opinion of the authors that the elements of $\mathscr{I}$ are physically a more meaningful set of parameters, we will digress briefly and discuss the relation between $\mathscr{I}$ and the Stokes parameters.

Anticipating the discussion in Section **10**, we note that the coherency matrix is formally identical to the density matrix, and therefore the expansion of the density matrix given by FANO (10) may be used, *i.e.* we may set

$$\mathscr{I} = \tfrac{1}{2}\sum_{k=0}^{3} s_k \sigma_k , \tag{35}$$

where the σ_k are the Pauli matrices (10) and the s_k are the Stokes parameters. In keeping with the theme of this paper, we wish to express the Stokes parameters as derivable from $\mathscr{I}$, since they are observables of the field. Multiplying (35) by σ_i and taking the trace we obtain, using (9), the solution

$$s_i = \mathrm{Sp}[\sigma_i \mathscr{I}] , \qquad (i = 0, 1, 2, 3) . \tag{36}$$

Note that the Stokes parameters, as all measurables of the field, are obtained from an expression of the form (1).

Another parameter often used in the description of partially polarized fields is the degree of polarization, which may be defined as the ratio of the intensity of the completely polarized part of the radiation to the total intensity. It was shown by WOLF (2) that an unambiguous meaning can be given to the

---

(9) P. ROMAN: *Nuovo Cimento*, **13**, 974 (1959).
(10) U. FANO: *Phys. Rev.*, **93**, 121 (1954).

terms « completely polarized part » and « unpolarized part » and a unique expression for the degree of polarization can be obtained in terms of the invariants of $\mathscr{I}$.

The situation is analogous to that which existed in the closely related theory of partial coherence. Here the ratio of the intensity of the « coherent part » of the radiation to the total intensity was proposed as a definition of the degree of coherence. However, a much clearer understanding of the physical situation was obtained by defining the degree of coherence in terms of a cross-correlation function (cfr. WOLF [11]). Following this lead from scalar coherence theory we shall propose here a new definition of the degree of polarization which will be seen to have the advantage that it involves only *directly* observable quantities.

Our definition is suggested by the following considerations. A strictly monochromatic field is completely polarized and completely coherent. Further, as pointed out by WOLF [1,2], the intensity formula (34) is formally identical with the expression for the intensity resulting from the superposition of two partially coherent beams. Now, the significant entity for describing interference phenomena involving partially coherent radiation is the normalised cross-correlation of the interfering disturbances, *i.e.* the degree of coherence. Thus our considerations suggest that *a significant quantity for the specification of the state of a partially polarized field is the degree of coherence between the x- and y- components of the field.* This quantity μ_{xy} may then be defined as the normalized cross-correlation of the x- and y-components of the field.

The x-component of the field is of course given by $\mathscr{P}_+(0)\mathscr{E}$, and the y-component *which is rotated as to interfere with the x-component* can be expressed as $\mathscr{P}_+(0)\mathscr{R}(-\pi/2)\mathscr{E}$. Since, as a natural extension of the intensity formula (5), we can express the cross-correlation of two fields as

$$K = \mathrm{Sp}\, \langle \mathscr{E}_1 \times \mathscr{E}_2^\dagger \rangle\,,$$

the normalized cross-correlation between the aforementioned two field components may be written in the form

$$(37)\qquad \mu_{xy} = \frac{\mathrm{Sp}\, \langle [\mathscr{P}_+(0)\mathscr{E}] \times [\mathscr{P}_+(0)\mathscr{R}(-\pi/2)\mathscr{E}]^\dagger \rangle}{\{\mathrm{Sp}\, \langle [\mathscr{P}_+(0)\mathscr{E}] \times [\mathscr{P}_+(0)\mathscr{E}]^\dagger \rangle \cdot \mathrm{Sp}\, \langle [\mathscr{P}_+(0)\mathscr{R}(-\pi/2)\mathscr{E}] \times [\mathscr{P}_+(0)\mathscr{R}(-\pi/2)\mathscr{E}]^\dagger \rangle\}^{\frac{1}{2}}}\,.$$

Here the normalizing denominator represents the square root of the product of the intensities associated with the two interfering fields.

---

(11) E. WOLF: *Proc. Roy. Soc.*, A **230**, 96 (1955).

Permuting cyclically the arguments of the traces, utilizing the properties (23), (24) and (30a) of the projection operators, the unitarity of $\mathscr{R}$ and the Hermiticity of $\mathscr{P}$, and applying the « associative law » that was used first in the derivation of (14), this expression can be considerably simplified.

We obtain, in view of (6)

$$\mu_{xy} = \frac{\text{Sp}\,[\mathscr{R}(\pi/2)\,\mathscr{P}_+(0)\mathscr{J}]}{\{\text{Sp}\,[\mathscr{P}_+(0)\mathscr{J}]\cdot\text{Sp}\,[\mathscr{P}_-(0)\mathscr{J}]\}^{\frac{1}{2}}}\,. \tag{38}$$

By the application of the Schwarz inequality it may be shown that the nodulus of μ_{xy} is bounded by zero and one and these extreme values are characteristic of incoherence and coherence respectively.

Since, apart from $\mathscr{J}$, the matrices in (38) are real, we may write

$$|\mu_{xy}|^2 = \frac{\text{Sp}\,[\mathscr{R}(\pi/2)\,\mathscr{P}_+(0)\mathscr{J}]\cdot\text{Sp}\,[\mathscr{R}(\pi/2)\,\mathscr{P}_+(0)\,\mathscr{J}^*]}{\text{Sp}\,[\mathscr{P}_+(0)\mathscr{J}]\cdot\text{Sp}\,[\mathscr{P}_-(0)\mathscr{J}]}\,. \tag{39}$$

Our expression for μ_{xy} is somewhat arbitrary, because for a different selection of the x- and y-directions we would of course obtain a different value. Clearly μ_{xy} is not yet a satisfactory measure for the state of polarization. However, we shall now show that there always exists a coordinate frame for which a maximum value of $|\mu_{xy}|^2$ is obtained.

The value of $|\mu_{xy}|^2$ referred to a coordinate system (XY) making an angle θ with the original one may be obtained from (39) by simply noting that in the new system the projection operator $\mathscr{P}(0)$ must be replaced by $\mathscr{P}(\theta)$. Thus

$$|\mu_{xy}(\theta)|^2 = \frac{\text{Sp}\,[\mathscr{R}(\pi/2)\,\mathscr{P}_+(\theta)\mathscr{J}]\cdot\text{Sp}\,[\mathscr{R}(\pi/2)\,\mathscr{P}_+(\theta)\mathscr{J}^*]}{\text{Sp}\,[\mathscr{P}_+(\theta)\mathscr{J}]\cdot\text{Sp}\,[\mathscr{P}_-(\theta)\mathscr{J}]}\,. \tag{40}$$

The straightforward but rather lengthy maximization of (40) with respect to θ yields the following condition on θ:

$$\text{tg}\,2\theta_m = \frac{J_{yy} - J_{xx}}{J_{xy} + J_{yx}}\,, \tag{41}$$

where J_{xy} etc. are the elements of the coherency matrix with respect to the original choice of axes. Since $\mathscr{J}$ is Hermitian, it follows from (41) that there always exists a *real* θ such $|\mu_{xy}|$ is maximal.

Before proceeding we point out that the angle θ_m which maximises $|\mu_{xy}|$ is also the angle through which the coordinates must be rotated in order that the intensities associated with the x- and y-components are equal. This is easily shown by simply demanding that

$$\mathrm{Sp}\,[\mathscr{P}_+(\theta)\mathscr{I}] = \mathrm{Sp}\,[\mathscr{P}_-(\theta)\mathscr{I}]\,,$$

and solving for θ.

We are now in a position to define the degree of polarization of the beam. *The degree of polarization is defined as the maximum value of the modulus of the degree of coherence,* $|\mu_{xy}|$, maximized with respect to θ. According to this definition we obtain the degree of polarization, P, by substituting the value of θ_m given by (41) into (40). We find after some calculation that

$$P = \sqrt{1 - \frac{4\det\mathscr{I}}{(\mathrm{Sp}\,\mathscr{I})^2}}\,. \tag{42}$$

Since $\det\mathscr{I}$ and $\mathrm{Sp}\,\mathscr{I}$ are invariants, it is clear that P as given by (42) is independent of the choice of axes.

We note that equ. (42) is the result derived by WOLF [2] starting from the definition

$$P = \frac{I_{\mathrm{pol}}}{I_{\mathrm{tot}}}\,. \tag{43}$$

Since it will be of help in deriving an experimental scheme for measuring $\mathscr{I}$ (see Section **9**), we will digress now briefly to consider the determination of P from the definition (43)

Since $\mathscr{I}$ is Hermitian, it is possible to diagonalize it with a unitary matrix $\mathscr{D}$, *i.e.* there exists a $\mathscr{D}$ such that $\mathscr{D}^\dagger = \mathscr{D}^{-1}$ and

$$\mathscr{D}\mathscr{I}\mathscr{D}^{-1} = \begin{pmatrix} \lambda_+ & 0 \\ 0 & \lambda_- \end{pmatrix}, \tag{44}$$

where the eiginvalues are given by

$$\lambda_\pm = \frac{1}{2}\,\mathrm{Sp}\,\mathscr{I}\left\{1 \pm \sqrt{1 - \frac{4\det\mathscr{I}}{(\mathrm{Sp}\,\mathscr{I})^2}}\right\}. \tag{45}$$

Physically (44) implies that there exists a frame of reference (coordinate system and relative phase difference) such that the cross-correlation terms vanish. Thus the operator $\mathscr{D}$ must be of the form

$$\mathscr{D}(\alpha, \delta) = \mathscr{R}(\alpha)\mathscr{C}(\delta)\,. \tag{46}$$

The appropriate diagonalizing angles α and δ are readily found to be given by

$$\exp[2i\delta] = J_{yx}/J_{xy}\,, \tag{47a}$$

and

$$\operatorname{tg} 2\alpha = \frac{1}{\cos\delta}\frac{J_{xy}+J_{yx}}{J_{xx}-J_{yy}} = \frac{J_{xy}+J_{yx}}{J_{xx}-J_{yy}}\sqrt{\frac{2}{1+\operatorname{Re}(J_{yx}/J_{xy})}}\,. \tag{47b}$$

Proceeding, we may now write

$$\begin{pmatrix}\lambda_+ & 0\\ 0 & \lambda_-\end{pmatrix} = \lambda_-\begin{pmatrix}1 & 0\\ 0 & 1\end{pmatrix} + (\lambda_+ - \lambda_-)\begin{pmatrix}1 & 0\\ 0 & 0\end{pmatrix}. \tag{48}$$

The first term on the right of (48) represents evidently a completely unpolarized beam and the second a completely polarized beam. Furthermore, this decomposition is unique. Substituting from (48) into the definition (43) we obtain

$$P = \frac{\lambda_+ - \lambda_-}{\lambda_+ + \lambda_-}\,; \tag{49}$$

and substituting from (45) into (49) we obtain the same expression (42) for P as derived above from our definition. Hence the equivalence of the two definiions is established.

9. – Measurement of $\mathscr{I}$.

In order to give a simple and direct physical interpretation to each of the four measurements involved in the determination of $\mathscr{I}$, we consider first a decomposition of $\mathscr{I}$.

Setting

$$J_{xy} = \beta + i\gamma\,, \qquad \text{hence} \qquad J_{yx} = \beta - i\gamma\,,$$

we may separate $\mathscr{I}$ into a real and imaginary part as

$$\mathscr{I} = \begin{pmatrix}J_{xx} & \beta\\ \beta & J_{yy}\end{pmatrix} + i\gamma\begin{pmatrix}0 & 1\\ -1 & 0\end{pmatrix} \equiv \mathscr{I}^r + \gamma\sigma_3\,.$$

We note that

$$\operatorname{Sp}[\mathscr{P}_+(\theta)\mathscr{R}(\alpha)\sigma_3\mathscr{R}^\dagger(\alpha)] = 0$$

for *all* θ and α. In view of the discussion in the previous sections this implies that *the imaginary part of* $\mathscr{I}$ *does not contribute to the intensity unless a compensator is involved in the device through which the radiation has passed.* We may therefore describe the interaction of a partially polarized beam with a polarizer and a rotator solely in terms of the real (and symmetric) matrix $\mathscr{I}^r$.

Thus the intensity in a beam passed by a polarizer is given by

$$I_P(\theta) = \mathrm{Sp}\,[\mathscr{P}_+(\theta)\mathscr{I}^r]\,. \tag{51}$$

Since $\mathscr{I}^r$ is symmetric, it follows from (47*a*) and (46) that the matrix $\mathscr{D}$ which diagonalizes $\mathscr{I}^r$ is a pure rotation operator $\mathscr{R}$. Hence, denoting the eigenvalues of $\mathscr{I}^r$ by $\lambda^r_\pm$, we may write

$$\mathscr{R}(\alpha_d)\mathscr{I}^r\mathscr{R}^{-1}(\alpha_d) = \begin{pmatrix} \lambda^r_+ & 0 \\ 0 & \lambda^r_- \end{pmatrix}, \tag{52}$$

where the diagonalizing angle α_d is, according to (47*b*), determined by the condition

$$\mathrm{tg}\; 2\alpha_d = \frac{2\beta}{J_{xx} - J_{yy}}\,. \tag{53}$$

Using (52) and cyclically permuting the argument of the trace, (51) may be written as

$$I_P(\theta) = \mathrm{Sp}\left\{\mathscr{R}(\alpha_d)\mathscr{P}_+(\theta)\mathscr{R}^+(\alpha_d)\begin{pmatrix} \lambda^r_+ & 0 \\ 0 & \lambda^r_- \end{pmatrix}\right\} = \mathrm{Sp}\left\{\mathscr{P}_+(\theta-\alpha_d)\begin{pmatrix} \lambda^r_+ & 0 \\ 0 & \lambda^r_- \end{pmatrix}\right\}, \tag{54}$$

where use was made of (30). Taking now in particular $\theta = \alpha_d$, we obtain

$$I_P(\alpha_d) = \lambda^r_+\,, \tag{55a}$$

and similarly, for $\theta = \alpha_d + \pi/2$ we get

$$I_P\left(\alpha_d + \frac{\pi}{2}\right) = \lambda^r_-\,. \tag{55b}$$

We now notice that *the same angle* α_d *which diagonalizes* $\mathscr{I}^r$ *makes also the intensity* $I_P(\theta)$ *an extremum.* This can be shown simply by differentiating (51) with respect to θ. We obtain then the condition

$$\mathrm{tg}\; 2\theta_e = \frac{2\beta}{J_{xx} - J_{yy}}\,, \tag{56}$$

which is indeed identical with (53). Whether this extremum is a maximum or a minimum, depends on the sign of β. Combining this result with (55*a*, *b*) we can set

$$\begin{cases} \lambda_+^r = I_{e_1}, \\ \lambda_-^r = I_{e_2}; \end{cases} \tag{57}$$

where I_{e_1} and I_{e_2} denote the first and second extremal intensities (a maximum and a minimum if $\beta > 0$, and the opposite way round if $\beta < 0$). These occur at an angle $\theta_{e_1} = \alpha_d$ and $\theta_{e_2} = \alpha_d + \pi/2$ respectively, where α_d is given by (53).

The physical significance of the above mathematical considerations is obvious. To measure the elements J_{xx}, J_{yy}, and $\beta = \operatorname{Re} J_{xy}$ of $\mathscr{J}$, we use a polarizer and find first the smallest angle α_d for which we obtain an extremal intensity I_{e_1}. We measure this intensity and then turn the polarizer to the setting $\alpha_d + \pi/2$ and measure the new extremal intensity I_{e_2}. By (57) the two intensities give us directly the two eigen-values $\lambda_\pm^r$; then, knowing also the angle α_d we compute

$$\mathscr{J}^r = \begin{pmatrix} J_{xx} & \beta \\ \beta & J_{yy} \end{pmatrix} = \mathscr{R}^{-1}(\alpha_d) \begin{pmatrix} I_{e_1} & 0 \\ 0 & I_{e_2} \end{pmatrix} \mathscr{R}(\alpha_d),$$

where $\mathscr{R}$ is given by (20).

If we like, we can now compute the first three Stokes parameters. According to (36) and (10)

$$\begin{cases} s_0 = J_{xx} + J_{yy}, \\ s_1 = J_{xx} - J_{yy}, \\ s_2 = 2\beta. \end{cases} \tag{58}$$

It follows that if one is interested only in experiments which do not involve compensators, then these three parameters specify the beam completely.

For a complete description we also require, however, the fourth parameter $\gamma = \operatorname{Im} J_{xy}$. This can be obtained by an additional measurement involving a compensator and a polarizer. The intensity is then given by (31). Using a half-wave plate for which $\delta = \pi$ and setting the polarizer at an angle $\theta = \pi/4$ with respect to the x-axis, we obtain, using (33) and the notation of equ. (50)

$$I_K = \tfrac{1}{2}(J_{xx} + J_{yy}) + \gamma;$$

hence

$$\gamma = I_K - \tfrac{1}{2} I_{\text{inc}}, \tag{59}$$

where $I_{\text{inc}} \equiv J_{xx} + J_{yy} = \text{Sp}\,\mathscr{I}$ is the incident intensity. Since J_{xx} and J_{yy} have been measured already in the previous experiment, the sole determination of the intensity I_K suffices to determine γ. As a matter of fact, settings other than $\theta = \pi/4$, $\delta = \pi$ could serve our purpose just as well. Knowing γ, the fourth Stokes parameter can be expressed as

$$s_3 = 2\gamma \,. \tag{60}$$

10. – Concluding remarks.

In the above discussion we have seen that a suitable and comprehensive description of the properties and the interactions of statistical radiation can be obtained in terms of the coherency matrix $\mathscr{I}$. It is interesting to note that both the definition and the properties of $\mathscr{I}$ resemble very much those of the statistical density matrix ϱ introduced for the characterization of statistical mixtures of quantum mechanical systems by VON NEUMANN (*). The application of the density matrix in the description of phenomena connected with electron- and photon-polarization has been discussed in great detail by TOLHOEK (12).

The main resemblance between the use of ϱ and of $\mathscr{I}$ is reflected in the fact that the expectation value of a physical observable is given in terms of the density matrix by

$$F = \langle \mathscr{F} \rangle = \text{Sp}\,[\mathscr{F}\varrho] \,,$$

while the observable value of a physical characteristic connected with a measurement is given in the present formalism by the analagous expression (1). In particular, as one finds easily by combining (15), (18), (21), (27), the « operator of the intensity » is given, in view of (5) and (1), by the operator

$$\mathscr{I} = \mathscr{A}^2(\eta)\, \mathscr{K}(\theta + \alpha, \delta) \,, \tag{61a}$$

where use was made of (30) and the notation (32). Similarly, the « operator of the Stokes parameter s_i » is, according to (36),

$$\mathscr{S}_i = \sigma_i \,. \tag{61b}$$

(*) For a detailed account see, for example, R. C. TOLMAN: *The principles of statistical mechanics* (Oxford, 1938), p. 327.

(12) H. A. TOLHOEK: *Rev. Mod. Phys.*, **28**, 277 (1956).

Furthermore, specifying the statistical ensemble in a new representation of variables, *i.e.*, physically, performing a certain measurement on the system by letting it interact with some device, amounts to a similarity transformation on ϱ:

$$\varrho \to \varrho' = \mathcal{O}\varrho\mathcal{O}^{-1} = \mathcal{O}\varrho\mathcal{O}^{\dagger}\,. \tag{62a}$$

On the other hand, we have seen in Sect. **3** and **6** that the interaction of the statistical beam amounts to a transformation

$$\mathcal{J} \to \mathcal{J}' = \mathcal{O}\mathcal{J}\mathcal{O}^{\dagger}\,. \tag{62b}$$

There are, however, some important differences between the two formalisms. For one thing, ϱ is defined in terms of ensemble-averages, while the definition of $\mathcal{J}$ involves time-averages. Even though it is true that — by virtue of the ergodic hypothesis — the two averaging procedures are normally equivalent, this equivalence will break down in the limiting case of a strictly monochromatic beam.

Another difference is that the transforming operator in (62*a*) is always unitary, but in (62*b*) not necessarily so. For example, the operator $\mathcal{P}_{\pm}$ characterizing a polarizer is not. This circumstance arises from the fact that in our discussions we have confined our attention to describing the emergent beam only, while the complete specification of the interaction should involve also the non-transmitted (reflected or absorbed) component of the radiation as well.

RIASSUNTO (*)

La matrice di coerenza di un'onda piana quasi-monocromatica viene dedotta da una rappresentazione matriciale del segnale analitico associato al campo elettrico. Si mostra che se la radiazione passa attraverso un apparecchio fisico, quale un compensatore, un assorbitore, un rotatore o un polarizzatore, l'effetto di questa interazione può essere completamente descritto in termini di operatori opportunamente scelti che trasformano direttamente la matrice di coerenza. Il grado complesso di coerenza viene definito in termini degli operatori suddetti, e da questo viene dedotta un'espressione che caratterizza il grado di polarizzazione. Si mostra che la quantità così dedotta è uguale a quella ottenuta con la definizione più usuale. Si descrive un esperimento che può servire a misurare le componenti della matrice di correlazione.

(*) *Traduzione a cura della Redazione.*

Reprinted from *J. Opt. Soc. Am.*, **53**(3), 317–323 (1963)

Theory of the Coherency Matrix for Light of Arbitrary Spectral Bandwidth*

RICHARD BARAKAT
Optics Department, Itek Corporation Lexington, Massachusetts
(Received 24 January 1962)

The coherency matrix is introduced in the context of vector-valued stationary stochastic processes with orthogonal increments. The ensemble-averaging approach is used instead of the customary time-averaging approach; this allows us to use the spectral-representation theorem. The Stokes' parameters are interpreted as power spectral densities and are shown to be related to the cospectra and quadspectra of the processes. Generalizations of the Stokes concept for $N\times N$ systems are also studied. Non image-forming optical instruments are treated as four-pole networks and the formalism of generalized transfer-function matrices is applied.

1. INTRODUCTION

IN an important series of papers, Wiener[1–3] was the first to introduce the coherency matrix of a light beam and show its relation to the density matrix of quantum mechanics. Indeed, one of the main reasons why Wiener invented his generalized harmonic analysis was to give a rigorous interpretation of stochastic phenomena and, in particular, white light. Unfortunately his fundamental papers are not easy to read and consequently have not received the recognition they deserve. Unknown to Wiener, Stokes[4] had introduced four parameters (Stokes parameters) for the description of a light beam. The Stokes parameters are obtainable directly from the coherency matrix and are the expansion coefficients of the Pauli spin matrices as Wiener showed (although he did not, of course, term his parameters "Stokes parameters"). In doing so Wiener predated Fano[5] whose paper is commonly believed to be the first to show the connection between the Stokes parameters, coherency matrix, and the Pauli spin matrices. Furthermore Wiener gave an explicit operational procedure for the experimental determination of the Stokes parameters. The fundamental importance of Wiener's work can hardly be overestimated.

Independently of Wiener, Wolf[6,7] has reintroduced the concept of the coherency matrix. Wolf has chosen to take as his fundamental quantities the covariance functions, unlike Wiener, who used the integrated power spectra. These functions are reciprocally related via a Fourier–Stieltjes transform, thus the Wiener and Wolf coherency matrices are functionally related. When the power spectral bandwidth is small (and only then) is it advantageous to use the Wolf coherency matrix; for general theoretical studies it is best to use the Wiener coherency matrix (see Sec. 2 for a complete discussion of the relative merits of each approach). We will use an ensemble-average modification of the Wiener coherency matrix in the present paper.

In 1941, Jones[8–10] studied the effect of various non image-forming optical instruments on an incident-plane electromagnetic wave using the formalism of matrix calculus. What he essentially does is to treat the optical instrument as a four-pole network, the instrument being represented by a 2×2 complex matrix. His analysis is restricted to incident waves that are sinu-

* Presented at the Los Angeles meeting of the Optical Society of America, 1961.
[1] N. Wiener, J. Math. and Phys. **7**, 109 (1928).
[2] N. Wiener, J. Franklin Inst. **207**, 525 (1929).
[3] N. Wiener, Acta Math. **55**, 117 (1930).
[4] G. Stokes, Trans. Cambridge Phil. **9**, 399 (1952). See also: E. Verdet, *Vorlesungen uber die Wellentheorie des Lichtes, Band 2*, German translation by K. Exner (Friedrich Vieweg und Sohn, Braunschweig, 1887) Chap. 24, and Lord Rayleigh, *Scientific Papers* (Cambridge University Press, Cambridge, 1902) Vol. 3, p. 140.
[5] U. Fano, Phys. Rev. **93**, 121 (1954).
[6] E. Wolf, Nuovo cimento **12**, 884 (1954).
[7] E. Wolf, Nuovo cimento **13**, 1165 (1959).
[8] R. C. Jones, J. Opt. Soc. Am. **31**, 488 (1941).
[9] H. Hurwitz and R. C. Jones, J. Opt. Soc. Am. **31**, 493 (1941).
[10] R. C. Jones, J. Opt. Soc. Am. **31**, 493 (1941).

soidal in time so that the work is not directly applicable to the physical problem which is statistical in nature. Nevertheless the work of Jones is the foundation upon which any statistical theory developed must ultimately rest. For this reason it must be considered, along with the studies of Stokes and Wiener, as the classic papers in the field.

A previous attempt at an abstract-operator approach is due to Parrent and Roman,[11] but their elegant work cannot be considered entirely satisfactory as it is restricted to situations where the power spectral bandwidth is narrow.

So much for the very brief historical resume. It has been included to put the necessary future steps into proper perspective.

Both the Wiener and Wolf approaches to the coherency matrix involve time-averaging, we adopt ensemble-averaging because we wish to utilize the spectral representation theorem and the theory of stochastic processes with orthogonal increments. An outline of the mathematics and the reasons for using it are given in Sec. 2. The coherency matrix is introduced and discussed in Sec. 3. The inclusion of the optical instrument into the analysis is carried out in Sec. 4.

2. STOCHASTIC PROCESSES WITH ORTHOGONAL INCREMENTS

We briefly review some aspects of the theory of stationary stochastic processes in this section. There are two approaches to the study of stochastic processes. The usual approach initiated by Wiener[3,12] is termed the harmonic analysis of individual functions and is a time-averaging theory. We do not follow this method but instead utilize the ensemble approach made popular by Khintchine[13] and Doob[14,15] among others. The choice of the latter is not entirely arbitrary but rests upon the fact that the spectral representation theorem of Cramér-Kolomogorov (stated below) which is to play a key role in subsequent sections is always true for the ensemble approach, in contrast to the time-average approach where it is not true in general.

We consider two real stationary stochastic processes $e_j(t)\,(j=1,2)$ which are postulated to be weakly stationary; that is in addition to having zero mean

$$E\{e_j(t)\}=0; \tag{2.1}$$

the second moments

$$E\{e_j(t)e_j(t+s)\}=R_{jj}(s), \qquad (j,k=1,2) \tag{2.2}$$
$$E\{e_j(t)e_k(t+s)\}=R_{jk}(s),$$

depend only on the interval s. Here $R_{jj}(s)$ is termed the autocovariance and $R_{jk}(s)$ the cross covariance. The curly brackets denote *ensemble* properties, and E is the expectation operator. We assume that $R_{jj}(s)$ and $R_{jk}(s)$ are continuous at $s=0$.

On the basis of the above hypotheses (and some additional mild conditions which will always be satisfied in optics) it can be proved that the second moments have the integral representations

$$R_{jj}(s)=\int_{-\infty}^{\infty} e^{i2\pi fs}\varphi_{jj}(f)df, \tag{2.3}$$
$$(j,k=1,2)$$
$$R_{jk}(s)=\int_{-\infty}^{\infty} e^{i2\pi fs}\varphi_{jk}(f)df, \tag{2.4}$$

where $\varphi_{jj}(f)$ is the power spectral density, and $\varphi_{jk}(f)$ the cross-power spectral density. We tacitly assume the processes $e_j(t)$ to possess only absolutely continuous spectra[15]; this is generally the case in optics.

Now (2.3) and (2.4) can be inverted by the Fourier inversion theorem

$$\varphi_{jj}(f)=\int_{-\infty}^{\infty} e^{-i2\pi fs}R_{jj}(s)ds, \tag{2.5}$$
$$(j,k=1,2)$$
$$\varphi_{jk}(f)=\int_{-\infty}^{\infty} e^{-i2\pi fs}R_{jk}(s)ds. \tag{2.6}$$

Equations (2.3) and (2.5) constitute the Wiener–Khintchine theorem in its original interpretation, while (2.4) and (2.6) will be called the generalized Wiener–Khintchine theorem. It can be shown that $R_{jj}(s)$ and $\varphi_{jj}(f)$ are real even functions.

In general the cross power spectral density $\varphi_{jk}(f)$ is complex unlike the power spectral density $\varphi_{jk}(f)$ which is always real. Now $R_{jk}(s)$ being the ensemble average of two real functions must itself be real and by (2.3)

$$R_{jk}(s)=R_{kj}(-s). \tag{2.7}$$

Thus the reality of $R_{jk}(s)$ implies $\varphi_{jk}(f)=\varphi_{kj}^*(f)$.

We effect the decomposition

$$\varphi_{jk}(f)=c_{jk}(f)+iq_{jk}(f), \tag{2.8}$$

where we define (for reasons to be made clear shortly) $c_{jk}(f)$ the cospectral density function, and $q_{jk}(f)$ the quadrature spectral density function.

A direct application of Schwartz's inequality to (2.8) yields the very important relation $0\leq P(f)\leq 1$, where

$$P(f)=[c_{jk}(f)]^2+[q_{jk}(f)]^2/\varphi_{jj}(f)\varphi_{kk}(f), \tag{2.9}$$

---

[11] G. Parrent and P. Roman, Nuovo cimento **15**, 370 (1960).

[12] N. Wiener, *The Fourier Integral and Certain of Its Applications* (Cambridge University Press, Cambridge, 1932), Chap. IV. See also: N. Wiener, *Extrapolation, Interpolation, and Smoothing of Stationary Time Series* (Technology Press, Cambridge, Massachusetts, 1949).

[13] A. Khintchine, Math. Ann. **109**, 604 (1934).

[14] J. L. Doob, "Time Series and Harmonic Analysis," in *Proceedings of the First Berkeley Symposium on Mathematical Statistics and Probability* (University of California Press, Berkeley, 1949).

[15] J. L. Doob, *Stochastic Processes* (John Wiley & Sons, Inc., New York, 1953).

When dealing with stationary stochastic processes it does not matter in principle whether the covariance functions or power spectral densities are used as they are reciprocally related via the Wiener–Khintchine theorems. Although Wolf *et al.*, have treated the covariance functions as the primary observables it is the author's belief that the spectral density functions are physically more meaningful as the primary observables. Now $\varphi_{jj}(f)$ is the power associated with the interval $2df$; the total power of the process $E_j(t)$ is

$$E\{e_j(t)\}^2=\int_{-\infty}^{\infty}\varphi_{jj}(f)df=\sigma^2, \tag{2.10}$$

where σ^2 is the variance of $e_j(t)$. The important point which is to be stressed is that the spectral density function can be regarded as being analogous to an analysis-of-variance table used in statistics in the sense that variability components are studied.[16] It would seem that variability components (spectral density function) in terms of frequency are more meaningful than the covariance expressed in terms of time intervals. This is especially true in optics where theory and experiment are usually stated in terms of frequency. *In dealing with light, whose power spectrum is not narrow, the frequency dependence of the intensity is of primary importance; for this reason the theory is developed in terms of the spectral functions.*

Corresponding to (2.3) and (2.4), there is an equivalent relation for $e_j(t)$ called the spectral representation theorem.[15]

Every stationary stochastic process $e_j(t)$ which is mean-square continuous has the spectral representation

$$e_j(t)=\int_{-\infty}^{\infty}e^{i2\pi ft}dX_j(f), \tag{2.11}$$

where $X_j(f)$ is a complex-valued stochastic process of orthogonal increments:

$$E\{X_j(f)\}=0 \tag{2.12}$$

$$E\{\Delta X_j(f)\Delta X_j^*(f')\}=\delta_{ff'}\varphi_{jj}(f)df(f,f'\geq 0), \tag{2.13}$$

where

$$\Delta X_j(f)=X_j(f+\Delta f)-X_j(f), \tag{2.14}$$

$$\Delta X_j^*(f')=X_j^*(f'-\Delta f')-X^*_j(f'),$$

and

$$\delta_{ff'}=\begin{cases}0, & f\neq f',\\ 1, & f=f'.\end{cases} \tag{2.15}$$

Since $e_j(t)$ is real, then $X(f)=X^*(-f)$ in order that the right-hand side of (2.11) be real.

The spectral integral (2.11) expresses the process $e_j(t)$ as a superposition of exponential functions $\exp(i2\pi ft)$ with stochastic weight factor $X_j(f)$. Note that the ensemble average of $X_j(f)$, (2.13) is related to the power spectral density. The importance of the spectral representation theorem will be made more evident in the next section where we deal with vector-stochastic processes and the coherency matrix. Incidentally, the integral (2.11) is properly a stochastic integral, but can be thought of as a Riemann–Stieltjes integral for purposes of formal manipulations.

3. COHERENCY MATRIX AND ITS GENERALIZATION

Consider a plane electromagnetic wave and let $e_1(t)$ and $e_2(t)$ be the two mutually perpendicular components of the electric vector $e(t)$ which are orthogonal to the direction of propagation.

Now take $\mathbf{e}(t)$ to be a spinor with real components $e_1(t)$ and $e_2(t)$, thus

$$\mathbf{e}(t)=\begin{bmatrix}e_1(t)\\ e_2(t)\end{bmatrix}. \tag{3.1}$$

Next take the direct product of $\mathbf{e}(t)$ and $\mathbf{e}^+(t+s)$

$$\mathbf{e}(t)\otimes\mathbf{e}^+(t+s)=\begin{bmatrix}e_1(t)e_1(t+s) & e_1(t)e_2(t+s)\\ e_2(t)e_1(t+s) & e_2(t)e_2(t+s)\end{bmatrix}, \tag{3.2}$$

where $\otimes$ indicates the direct product (or Kronecker product), and the superscript $+$ indicates that the transpose of the spinor is to be taken. Taking the ensemble average of (3.2) yields (remembering that $e_1(t)$ and $e_2(t)$ are real)

$$E\{\mathbf{e}(t)\otimes\mathbf{e}^+(t+s)\}\equiv\mathbf{R}(s)=\begin{bmatrix}R_{11}(s) & R_{12}(s)\\ R_{21}(s) & R_{22}(s)\end{bmatrix} \tag{3.3}$$

where

$$R_{jk}(s)=E\{e_j(t)e_k(t+s)\}\,(j,k=1,2). \tag{3.4}$$

The matrix $R(s)$ is called the covariance matrix of the random vector processes $e(t)$, and is essentially the coherency matrix that Wolf, Parrent, and Roman have used in their papers. There are however significant differences between (3.3) and their coherency matrix in spite of the fact that both are covariance matrices. The first is that (3.3) is a real matrix because the spinors in (3.2) are real, whereas the Wolf coherency matrix is complex. The second has to do with the fact that their analysis is only operationally meaningful for a narrow spectral bandwidth by virtue of imposing the Hilbert transform relations on the covariance functions rather than on the random process itself.[17] Third, our covariance matrix is obtained by ensemble averaging rather than by time averaging.

Since the frequency domain is to be taken as the fundamental domain, it is only necessary to take the

[16] R. B. Blackman and J. W. Tukey, *The Measurement of Power Spectra from the Point of View of Communications Engineering* (Dover Publications, Inc., New York, 1960).

[17] This problem has been examined by the author in a paper entitled "On the Theory of Quasi-Monochromatic Light," to be submitted for publication shortly.

Fourier transform of (3.3) by using (2.5) and (2.6) to obtain the spectral density matrix

$$\mathbf{\Phi}(f)=\begin{bmatrix}\varphi_{11}(f) & \varphi_{12}(f)\\ \varphi_{21}(f) & \varphi_{22}(f)\end{bmatrix}. \tag{3.5}$$

But since $\varphi_{12}(f)=\varphi_{21}{}^*(f)$, we have that (3.5) is Hermitian. We rewrite (3.5) in terms of the cospectra and quadspectra

$$\mathbf{\Phi}(f)=\begin{bmatrix}\varphi_{11}(f) & c_{21}(f)-iq_{21}(f)\\ c_{21}(f)+iq_{21}(f) & \varphi_{22}(f)\end{bmatrix} \tag{3.6}$$

and term (3.6) the coherency matrix.

Before proceeding further we now utilize the spectral representation theorem (2.11) for $e_1(t)$ and $e_2(t)$ written in real form

$$e_j(t)=2\int_0^\infty [\cos 2\pi ft dU_j(f)-\sin 2\pi ft dV_j(f)], \tag{3.7}$$

where

$$X_j(f)=U_j(f)+iV_j(f). \tag{3.8}$$

The parity relations

$$U_j(f)=U_j(-f), \tag{3.8}$$

$$V_j(f)=-V_j(-f), \tag{3.9}$$

follow from the reality condition on $e_j(t)$.

The real orthogonal processes $U_j(f)$ and $V_j(f)$ satisfy the relations

$$E\{\Delta U_j(f)\Delta V_j(f')\}=0, \tag{3.10a}$$

$$E\{\Delta U_j(f)\Delta U_j(f')\}=E\{\Delta V_j(f)\Delta V_j(f')\}=\delta_{ff'}\varphi_{jj}(f)df, \tag{3.10b}$$

$$E\{\Delta U_j(f)\Delta U_k(f')\}=E\{\Delta V_j(f)\Delta V_k(f')\}=\delta_{ff'}c_{jk}(f)df, \tag{3.10c}$$

$$E\{\Delta U_j(f)\Delta V_k(f')\}=E\{\Delta V_j(f)\Delta U_k(f')\}=\delta_{ff'}q_{jk}(f)df, \tag{3.10d}$$

for f, $f'\geq 0$. The first equation is especially important since it states that the two components of the random wave are incoherent.

The meaning of the cospectral and quadrature spectral densities is now clear. The pair of random amplitudes $U_j(f)(j=1,2)$ is associated with the $\cos(2\pi fs)$ terms while the pair $V_j(f)$ are associated with the sine terms. Noting that $\sin(2\pi fs)$ is $\frac{1}{2}\pi$ out of phase with $\cos 2\pi fs$, we say that the sine terms are "in quadrature" with the cosine terms. The cospectral density measures the power between the cosine ("in phase") random amplitudes, while the quadrature spectral density measures the power between the sine ("in quadrature") random amplitudes.

Any 2×2 matrix, such as (3.6), can always be written in terms of four parameters[18] x_1, x_2, x_3, x_4 (say)

$$\mathbf{\Phi}(f)=\begin{bmatrix}x_4+x_3 & x_1-ix_2\\ x_1+ix_2 & x_4-x_3\end{bmatrix}, \tag{3.11}$$

where $x_i=x_i(f)$. In the special case where the matrix is Hermitian, the x coefficients are real and are given in terms of the original matrix by

$$\begin{aligned}x_1&=\tfrac{1}{2}(\varphi_{21}+\varphi_{21}{}^*)=c_{21},\\ x_2&=\tfrac{1}{2}(\varphi_{21}-\varphi_{21}{}^*)=q_{21},\\ x_3&=\tfrac{1}{2}(\varphi_{11}-\varphi_{22}),\\ x_4&=\tfrac{1}{4}(\varphi_{11}+\varphi_{22}).\end{aligned} \tag{3.12}$$

A knowledge of these parameters (which are the Stokes' parameters) characterizes the vector stochastic process. However these four parameters do not completely determine the process because they involve only second-order statistics and are not sufficient to describe the hierarchy of the probabilistic structure of the stochastic process. Of course if the process were Gaussian then the description would be complete. We wish to stress the point that x_1, x_2, x_3, x_4 represent observables and hence the coherency matrix is a density matrix, a result pointed out by Wiener.

Two invariants of the coherency matrix are its determinant and trace (or spur)

$$\begin{aligned}\det\mathbf{\Phi}(f)&=\varphi_{11}(f)\varphi_{22}(f)-\varphi_{12}(f)\varphi_{21}(f)\\ &=x_4{}^2-x_3{}^2-x_2{}^2-x_1{}^2\geq 0,\end{aligned} \tag{3.13}$$

$$\mathrm{Sp}\mathbf{\Phi}(f)=\varphi_{11}(f)+\varphi_{22}(f)=2x_4\geq 0. \tag{3.14}$$

Particular importance is attached to the second form of (3.13), which has the form of a Lorentz line element. This fact allows us to apply group-theoretic methods employing the Lorentz group to discuss the coherency matrix.[19] It seems surprising that no one has called attention to this point.

At this point we introduce a generalized coherency matrix for N beams instead of 2. It is obtained in exactly the same manner as the 2×2 coherency matrix. We sample a wavefront (not necessarily plane) at N points and call the values of the wave $e_1(t)$, $e_2(t)$, . . ., $e_N(t)$. Take $\mathbf{e}(t)$ to be an Nth order spinor

$$\mathbf{e}(t)=\begin{bmatrix}e_1(t)\\ e_2(t)\\ \cdots\\ \cdots\\ e_N(t)\end{bmatrix}. \tag{3.15}$$

---

[18] G. Temple, *Cartesian Tensors* (Methuen and Company, Ltd., London, England, 1960), p. 67.

[19] This topic has been analyzed by the author and will be published in the near future. A preliminary account of the work was presented at the Los Angeles meeting of the Optical Society of America, October 1961.

Taking the direct product of $\mathbf{e}(t)$ and $\mathbf{e}^+(t+s)$ and ensemble averaging yield the covariance matrix

$$\mathbf{R}(s)=\begin{bmatrix} R_{11}(s)\cdots R_{1N}(s) \\ \cdots \\ R_{N1}(s)\cdots R_{NN}(s) \end{bmatrix}, \quad (3.16)$$

where

$$R_{jk}(s)=E\{e_j(t)e_k(t+s)\}\,(j,\,k=1,\,2,\,\cdots,\,N). \quad (3.17)$$

The Fourier transform of (3.16) is the spectral density matrix or coherency matrix of the vector process $\mathbf{e}(t)$

$$\mathbf{\Phi}(f)=\begin{bmatrix} \varphi_{11}(f)\cdots\varphi_{1N}(f) \\ \cdots \\ \varphi_{N1}(f)\cdots\varphi_{NN}(f) \end{bmatrix}. \quad (3.18)$$

The matrix is Hermitian. In terms of the cospectra and quadspectra, $\mathbf{\Phi}(f)$ reads

$$\mathbf{\Phi}(f)=\begin{bmatrix} \varphi_{11}\cdots C_{21}-iq_{21}\cdots C_{N1}-iq_{N1} \\ C_{21}+iq_{21}\cdots\varphi_{22}\cdots \\ \cdots \\ C_{N1}+iq_{N1}\cdots\varphi_{NN}\cdot \end{bmatrix}. \quad (3.19)$$

The two obvious invariants of (3.19) are its determinant which can be shown to be nonnegative and its trace (spur)

$$\mathrm{Sp}[\mathbf{\Phi}(f)]=\sum_{j=1}^{N}\varphi_{jj}(f)\geq 0. \quad (3.20)$$

The generalization of the Stokes parameters of the 2×2 coherency to the $N\times N$ coherency matrix can be carried out within the framework of expansions in terms of orthogonal operators.[20] Consider a set of matrix operators $\{\mathbf{O}_j\}$ which are Hermitian and obey the orthonormality condition

$$\mathrm{Sp}[\mathbf{O}_j\mathbf{O}_k]=\delta_{jk}. \quad (3.21)$$

Expand the coherency matrix (3.18) into a sum of $\mathbf{O}_j$,

$$\mathbf{\Phi}(f)=\sum_{j=1}^{N^2}\mathrm{Sp}(\mathbf{\Phi}\mathbf{O}_j)\cdot\mathbf{O}_j. \quad (3.22)$$

The $\{\mathbf{O}_j\}$ must form a *complete* orthonormal set of basic operators if the expansion (3.22) is to be unique; hence there must be N^2 of them since $\mathbf{\Phi}(f)$ is $N\times N$ and thus contains N^2 elements. The expansion coefficients of (3.22) [i.e., $\mathrm{Sp}(\mathbf{\Phi}\mathbf{O}_j)$] being linear combinations of the power spectra are themselves observables and are the generalized Stokes parameters.

[20] U. Fano, Revs. Modern Phys. **29**, 74 (1957).

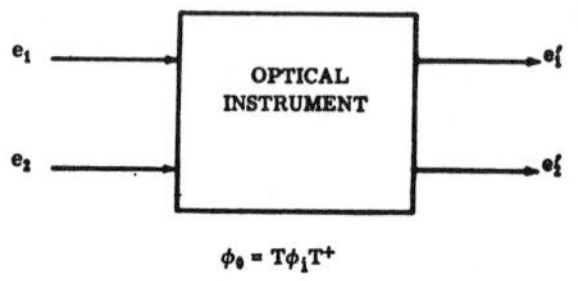

FIG. 1. Schematic diagram of the four-pole network description of non-image-forming optical system.

In the special case of the 2×2 coherency matrix, the Stokes parameters are the expansion coefficients of (3.22) and the Pauli spin matrices σ_j are the requisite matrix operators. Thus $\mathbf{\Phi}(f)$ can be expressed uniquely in the form

$$\mathbf{\Phi}(f)=x_1\sigma_1+x_2\sigma_2+x_3\sigma_3+x_4\sigma_4, \quad (3.23)$$

where

$$\sigma_1=\begin{bmatrix} 0 & 1 \\ 1 & 0 \end{bmatrix},\quad \sigma_2=\begin{bmatrix} 0 & -i \\ i & 0 \end{bmatrix},\quad \sigma_3=\begin{bmatrix} 1 & 0 \\ 0 & -1 \end{bmatrix} \quad (3.24)$$

are the Pauli matrices of quantum mechanics and

$$\sigma_4=\begin{bmatrix} 1 & 0 \\ 0 & 1 \end{bmatrix} \quad (3.25)$$

is the unit matrix. Here the x_j are given by (3.12). This expansion of the coherency matrix is due to Wiener,[2,3] although Fano[5] seems to be the first to have popularized its use.

There is not just one set of generalized Stokes parameters as Roman[21] seems to believe (namely, the 3×3 coherency matrix expansion coefficients), but a set of generalized Stokes parameters corresponding to each value of N taken. However this is of more importance from the *formal* point of view than from the *operational* point of view. With the possible exception of the 2×2 coherency matrix; it is easier to measure the cospectra, quadspectra, and power spectra directly than to form the various generalized Stokes parameters.

It is not possible in this paper to delve into the general problem of measurement by using the coherency matrix, as the subject is somewhat ancillary to our main topic. A lucid discussion is contained in Fano[20] and certain of the general results have been particularized to the optical case by Barakat.[17] There is however one important point which must be stressed: *Namely, that the coherency matrix represents the minimum set of observable inputs necessary to characterize the optical beam.*

4. TRANSFER FUNCTION MATRIX OF NON-IMAGE-FORMING OPTICAL INSTRUMENT

The results of the previous section are a description of the beam only and one must still introduce the optical instrument into the analysis. The fundamental assumption is made that the non image-forming optical instrument can be represented as a four-pole network (see Fig. 1).

[21] P. Roman, Nuovo cimento **13**, 974 (1959).

In the fundamental papers of Jones[8–10] it is postulated that the relation between the input amplitudes (of the electromagnetic vector components) and the output amplitudes is linear. Jones has exploited this linearity to develop his matrix theory. Since only intensities (proportional to the square of the amplitudes) are observables at optical frequencies, we must generalize the work of Jones to the stochastic domain and deal with power spectra.

The basic theorem which will be utilized in our description of the optical instrument is to be found in Grenander and Rosenblatt[22] and in our notation reads[23]

$$\mathbf{\Phi}_0(f)=\mathbf{T}(f)\mathbf{\Phi}_i(f)\mathbf{T}^+(f), \tag{4.1}$$

where $\mathbf{\Phi}_0(f)$ is the output coherency matrix, $\mathbf{\Phi}_i(f)$ the input coherency matrix, and $\mathbf{T}(f)$ the transfer function matrix. In order that this formula be valid, it can be shown that the average output power must be finite. For any conceivable optical situation, this will be true. Equation (4.1) is a generalization of the well-known one-dimensional result

$$\varphi_0(f)=|T(f)|^2\varphi_i(f). \tag{4.2}$$

A proof of (4.1) is outlined below, an alternative proof can also be constructed via use of the spectral representation theorem.[22] The Fourier transform of (3.1), the input-amplitude spectrum, will be termed

$$\boldsymbol{\varepsilon}(f)=\begin{bmatrix}\varepsilon_1(f)\\ \varepsilon_2(f)\end{bmatrix} \tag{4.3}$$

with a dash to represent the output-amplitude spectrum. By virtue of the linearity and stationarity of the instrument, we have the matrix equation

$$\boldsymbol{\varepsilon}'(f)=\mathbf{T}(f)\boldsymbol{\varepsilon}(f), \tag{4.4}$$

where $\mathbf{T}(f)$ is the transfer-function matrix

$$\mathbf{T}(f)=\begin{bmatrix}t_{11}(f) & t_{12}(f)\\ t_{21}(f) & t_{22}(f)\end{bmatrix}. \tag{4.5}$$

In general, $\mathbf{T}(f)$ is a complex matrix. Next take the direct product of (4.4) with its Hermitian conjugate

$$\boldsymbol{\varepsilon}\otimes\boldsymbol{\varepsilon}'^+=\mathbf{T}\boldsymbol{\varepsilon}\otimes(\mathbf{T}\boldsymbol{\varepsilon})^+=\mathbf{T}\boldsymbol{\varepsilon}\otimes\boldsymbol{\varepsilon}^+\mathbf{T}^+. \tag{4.6}$$

Upon forming the ensemble average of (4.6) and defining (see 3.5)

$$E\{\boldsymbol{\varepsilon}'\otimes\boldsymbol{\varepsilon}'^+\}=\mathbf{\Phi}_0(f) \tag{4.7}$$

$$E\{\boldsymbol{\varepsilon}\otimes\boldsymbol{\varepsilon}^+\}=\mathbf{\Phi}_i(f) \tag{4.8}$$

we immediately obtain (4.1).

[22] U. Grenander and M. Rosenblatt, *Statistical Analysis of Stationary Time Series* (John Wiley & Sons, Inc., New York, 1957).

[23] N. G. Parke, J. Math. Phys. **28**, 188, (1949). After this manuscript had been completed, the author found that Parke had derived essentially (4.1) although by entirely different methods. Parke's notation is very complicated and it is difficult to see the relation between his work and this section of the present paper.

If $\mathbf{T}$ is an arbitrary 2×2 complex matrix with $\det\mathbf{T}=\pm 1$, we can show that $\mathbf{\Phi}_0$ is Hermitian on the assumption that $\mathbf{\Phi}_i$ is also Hermitian; furthermore,

$$\det\mathbf{\Phi}_0=\det\mathbf{\Phi}_i \tag{4.9}$$

or by (3.13),

$$x_4^2-x_3^2-x_2^2-x_1^2=x_4'^2-x_3'^2-x_2'^2-x_1'^2. \tag{4.10}$$

Consequently the induced transformation on $\mathbf{\Phi}_0$ is a Lorentz transformation.

A general question arises as to the criteria to be placed upon the elements of $\mathbf{T}$ so that it can represent a physically realizable system. In order that $\mathbf{T}(f)$ represent a physically realizable optical system it must have a gain $g(f)\leq 1$, if we assume that the systems are passive. This is in addition to the finite power restriction. The gain g is defined by the relation

$$g=\frac{\text{output intensity}}{\text{input intensity}}. \tag{4.11}$$

Now

$$g=\mathrm{Sp}\mathbf{\Phi}_0/\mathrm{Sp}\mathbf{\Phi}_i=\mathrm{Sp}[\mathbf{T}^+\mathbf{T}\mathbf{\Phi}_i]/\mathrm{Sp}\mathbf{\Phi}_i. \tag{4.12}$$

In (3.3) we have written $\mathbf{\Phi}$ as a decomposition of Pauli spin matrices, we now write $\mathbf{T}^+\mathbf{T}$ in the same manner. Simple manipulations yield

$$\mathbf{T}^+\mathbf{T}=\begin{bmatrix}t_{11}^2+|t_{12}|^2 & t_{11}t_{21}^*+t_{22}t_{12}\\ t_{11}t_{21}+t_{22}t_{12}^* & t_{22}^2+|t_{21}|^2\end{bmatrix} \tag{4.13}$$

$$=\tau_1\sigma_1+\tau_2\sigma_2+\tau_3\sigma_3+\tau_4\sigma_4,$$

where

$$\begin{aligned}\tau_1&=\tfrac{1}{2}[t_{11}(t_{21}+t_{21}^*)+t_{22}(t_{12}+t_{12}^*)],\\ i\tau_2&=\tfrac{1}{2}[t_{11}(t_{21}-t_{21}^*)-t_{22}(t_{12}-t_{12}^*)],\\ \tau_3&=\tfrac{1}{2}[t_{11}^2-t_{22}^2+|t_{12}|^2-|t_{21}|^2],\\ \tau_4&=\tfrac{1}{2}[t_{11}^2+t_{22}^2+|t_{12}|^2+|t_{21}|^2].\end{aligned} \tag{4.14}$$

The τ coefficients are real, because $\mathbf{T}^+\mathbf{T}$ is Hermitian. The gain can be written

$$g=\frac{1}{x_4}\,\mathrm{Sp}[(\sum_{j=1}^{4}\tau_j\sigma_j)(\sum_{l=1}^{4}x_j\sigma_j)]. \tag{4.15}$$

It is straightforward to show that

$$g=(1/x_4)[\tau_1x_1+\tau_2x_2+\tau_3x_3+\tau_4x_4] \tag{4.16}$$

by using the following properties of the Pauli spin matrices

$$\begin{aligned}\sigma_1^2=\sigma_2^2=\sigma_3^2=\sigma_4,\quad \sigma_2\sigma_3=i\sigma_1=-\sigma_3\sigma_2,\\ \sigma_3\sigma_1=i\sigma_2=-\sigma_1\sigma_3,\quad \sigma_1\sigma_2=i\sigma_3=-\sigma_2\sigma_1.\end{aligned} \tag{4.17}$$

It is interesting to note that g is a symmetric function in τ and x.

The maximum value of the gain will occur when (in vector language) $\tau=(\tau_1,\tau_2,\tau_3)$ is parallel to $x=(x_1,x_2,x_3)$; that is, when

$$\tau_1x_1+\tau_2x_2+\tau_3x_3=x_4[|\tau_1|^2+|\tau_2|^2+|\tau_3|^2]^{\frac{1}{2}}. \tag{4.18}$$

Consequently,

$$g_{max}=\tau_4+[|\tau_1|^2+|\tau_2|^2+|\tau_3|^2]^{\frac{1}{2}}. \tag{4.19}$$

It is now a simple matter to take a hypothetical frequency response matrix and substitute into (4.19) to see whether $g_{max}\leq 1$. *The fact that an instrument has a transfer-function matrix which is unity does not necessarily imply that it is physically realizable.* The Stokes' vector is invariant under such a circumstance but the intensity can, of course, vary.

It has been tacitly assumed that the determinant of **T** does not vanish so that the transfer function matrix is nonsingular. The special but important singular case of $\det\mathbf{T}=0$ occurs in many problems. According to the definition adopted by Jones,[8] a non image-forming optical instrument whose determinant vanishes is termed a (perfect) polarizer. A discussion of the singular case requires a different analytical apparatus and will not be discussed here.

ACKNOWLEDGMENT

I am indebted to Dr. R. Clark Jones for a series of interesting discussions on these topics as well as for his encouragement in the publication of this material.

30

Reprinted from *Nouvelle Revue d'Optique,* **4**(1), 37–41 (1973)

A UNIFIED FORMALISM FOR POLARIZATION OPTICS BY USING GROUP THEORY

by

Hiroshi TAKENAKA

MOTS CLÉS :
Lumière
Polarisée

KEY WORDS :
Polarized
light

SUMMARY

A unified treatment of three polarization calculus methods — Jones matrix, Mueller matrix and Poincaré sphere — is accomplished by using the concept of the Lie group. In the case of the totally transparent system, it is related to the compact Lie group, and in the case of the partially transparent system, it is related to the homogeneous Lorentz group. The physical meaning of the Pauli's spin matrices is given and the generator is introduced and used extensively.

RÉSUMÉ

Formalisme unifié pour la polarisation, à partir de la théorie des groupes.

Le traitement unifié de trois méthodes de calcul pour la polarisation — matrices de Jones, Mueller, et la sphère de Poincaré — est fait en utilisant la notion du groupe Lie. Dans le cas du système complètement transparent, le traitement est relié au groupe compact Lie, tandis qu'il est relié au groupe homogène Lorentz en cas du système partiellement transparent. La signification optique des matrices de spin Pauli est donnée et le générateur est introduit pour l'analyse.

INTRODUCTION

Various methods are used when we analyze or synthesize complicated optical systems related to polarized light. The most well-known methods are the Jones 2×2 matrix, the Mueller 4×4 matrix, and the Poincaré sphere [1]. The relation between these is treated in many papers [2], [3]. But most papers use only algebraic calculation and do not refer to group theory and group theoretical aspects of polarization calculus. The aim of this paper is to show that the relation between these methods can be understood clearly by using group theory in polarization calculus [4] and to introduce the generator representation.

TOTALLY TRANSPARENT SYSTEM

1. — Lie group

The distinguishing characteristic of a Lie group is that the parameters of a product element are analytic functions of the parameters of the factors. This is true of O_3^+, SU(2) and the Lorentz group which will be considered later.

The concept of the generator is developed by this differentiability and it allows us to reduce the study of the whole group to a study of the group elements in the neighbourhood of the identity element. If the parameter varies in a closed interval, it is said to be compact and its important nature is that all representations of the compact groups are equivalent to the unitary repre sentation. O_3^+ and SU(2) are compact. On the other hand, the Lorentz group is not compact so that its representation is not unitary [5].

2. — Orthogonal group O_3^+

A 3×3 real orthogonal matrix forms a group and its independent parameters are 3. The matrix whose determinant is $+1$ is called O_3^+. The rotations around the orthogonal axes are

$$\mathbf{R}_X(\psi) = \begin{pmatrix} 1 & 0 & 0 \\ 0 & \cos\psi & \sin\psi \\ 0 & -\sin\psi & \cos\psi \end{pmatrix}$$

$$(1) \qquad \mathbf{R}_Y(\psi) = \begin{pmatrix} \cos\psi & 0 & -\sin\psi \\ 0 & 1 & 0 \\ \sin\psi & 0 & \cos\psi \end{pmatrix}$$

$$\mathbf{R}_Z(\psi) = \begin{pmatrix} \cos\psi & \sin\psi & 0 \\ -\sin\psi & \cos\psi & 0 \\ 0 & 0 & 1 \end{pmatrix}.$$

The general member of O_3^+ is represented by

(2) $$\mathbf{R}(\alpha, \beta, \gamma) = \mathbf{R}_Z(\gamma)\,\mathbf{R}_X(\beta)\,\mathbf{R}_Z(\alpha)\,.$$

This is the Euler angles in classical mechanics [6].

3. — Special unitary group SU(2)

The Unitary matrix with determinant 1 also forms a group. This is called the special unitary group SU(2). The 2 × 2 unitary, unit determinant matrix has 3 parameters just as O_3^+ does. The general member of SU(2) is written by

(3) $$\mathbf{U} = \begin{pmatrix} a & b \\ -\bar{b} & \bar{a} \end{pmatrix}, \quad a\bar{a} + b\bar{b} = 1$$

where a and b are complex numbers and – means complex conjugate.

4. — SU(2) — O_3^+ homomorphism

O_3^+ represents the rotation in the ordinary three dimensional space, and it has the characteristic of leaving $X^2 + Y^2 + Z^2$ invariant.

The operation of SU(2) on a matrix is given by a unitary transformation

(4) $$\mathbf{M}' = \mathbf{U}\mathbf{M}\mathbf{U}^+$$

where + means ajoint.

Taking **M** to be a 2 × 2 matrix, it can be represented by the linear combination of the unit matrix **1** and the Pauli's spin matrices $\boldsymbol{\sigma}_k$ (k = 1, 2, 3)

(5) $$\mathbf{1} = \begin{pmatrix} 1 & 0 \\ 0 & 1 \end{pmatrix}, \quad \boldsymbol{\sigma}_1 = \begin{pmatrix} 0 & 1 \\ 1 & 0 \end{pmatrix},$$

$$\boldsymbol{\sigma}_2 = \begin{pmatrix} 0 & -\mathrm{i} \\ \mathrm{i} & 0 \end{pmatrix}, \quad \boldsymbol{\sigma}_3 = \begin{pmatrix} 1 & 0 \\ 0 & -1 \end{pmatrix}.$$

Let **M** be a zero trace matrix, then

(6) $$\mathbf{M} = X\boldsymbol{\sigma}_3 + Y\boldsymbol{\sigma}_1 + Z\boldsymbol{\sigma}_2 = \begin{pmatrix} X & Y - \mathrm{i}Z \\ Y + \mathrm{i}Z & -X \end{pmatrix}.$$

Since the trace is invariant under the unitary transformation, the form of **M′** must be

(7) $$\mathbf{M}' = X'\boldsymbol{\sigma}_3 + Y'\boldsymbol{\sigma}_1 + Z'\boldsymbol{\sigma}_2 = \begin{pmatrix} X' & Y' - \mathrm{i}Z' \\ Y' + \mathrm{i}Z' & -X' \end{pmatrix}.$$

As the determinant also remains invariant under the unitary transformation, therefore

$$-(X^2 + Y^2 + Z^2) = -(X'^2 + Y'^2 + Z'^2)\,.$$

In other words $X^2 + Y^2 + Z^2$ is invariant under the operation of SU(2). This is the same as the case of O_3^+ and so SU(2) represents the rotation.

Here let us consider a special case and see how SU(2) describes the rotation. If we put $a = \mathrm{e}^{\mathrm{i}\psi/2}$ $b = 0$ in equation (3), then

$$\mathbf{U}_X\left(\frac{\psi}{2}\right) = \begin{pmatrix} \mathrm{e}^{\mathrm{i}\psi/2} & 0 \\ 0 & \mathrm{e}^{-\mathrm{i}\psi/2} \end{pmatrix}.$$

Calculating $\mathbf{U}_X \mathbf{M} \mathbf{U}_X^+$ and using equations (4), (6), and (7) we get

$$\begin{aligned} X' &= X \\ Y' &= Y\cos\psi + Z\sin\psi \\ Z' &= -Y\sin\psi + Z\cos\psi\,. \end{aligned}$$

We see that the 2 × 2 unitary transformation of the form $\mathbf{U}_X(\psi/2)$ is equivalent to $\mathbf{R}_X(\psi)$ of equation (1)

$$\mathbf{U}_X\left(\frac{\psi}{2}\right) \leftrightarrow \mathbf{R}_X(\psi)\,. \quad (8)$$

Similarly

(8) $$\mathbf{U}_Y\left(\frac{\psi}{2}\right) = \begin{pmatrix} \cos\frac{\psi}{2} & \mathrm{i}\sin\frac{\psi}{2} \\ \mathrm{i}\sin\frac{\psi}{2} & \cos\frac{\psi}{2} \end{pmatrix} \leftrightarrow \mathbf{R}_Y(\psi)$$

$$\mathbf{U}_Z\left(\frac{\psi}{2}\right) = \begin{pmatrix} \cos\frac{\psi}{2} & \sin\frac{\psi}{2} \\ -\sin\frac{\psi}{2} & \cos\frac{\psi}{2} \end{pmatrix} \leftrightarrow \mathbf{R}_Z(\psi)\,.$$

The correspondence is two to one and SU(2) and O_3^+ are homomorphic. From this the representation of SU(2) give O_3^+ automatically.

Corresponding to equation (2) the general member of SU(2) is

(9) $$\mathbf{U}(\alpha, \beta, \gamma) = \mathbf{U}_Z\left(\frac{\gamma}{2}\right)\mathbf{U}_X\left(\frac{\beta}{2}\right)\mathbf{U}_Z\left(\frac{\alpha}{2}\right).$$

5. — The combination of polarization matrix and group theory

Considering the previous content, the optical meaning is clear. The Jones matrix of birefringence or optical activity represents the operation in SU(2) and the Mueller matrix of birefringence or optical activity represents the operation in O_3^+. The Poincaré sphere is the spatial representation of the Mueller matrix. The correspondence between the Jones matrix and the Mueller matrix is two to one.

6. — Pauli's spin matrices and generators for SU(2)

We consider the meaning of Pauli's spin matrix (5). Let $\delta\psi$ be a small angle, then $\mathbf{U}_X(\delta\psi/2)$ is developed as

$$\mathbf{U}_X\left(\frac{\delta\psi}{2}\right) = \begin{pmatrix} 1 + \mathrm{i}\frac{\delta\psi}{2} & 0 \\ 0 & 1 - \mathrm{i}\frac{\delta\psi}{2} \end{pmatrix} = \mathbf{1} + \mathrm{i}\frac{\delta\psi}{2}\boldsymbol{\sigma}_3\,.$$

Similarly

$$\mathbf{U}_Y\left(\frac{\delta\psi}{2}\right) = \mathbf{1} + \mathrm{i}\frac{\delta\psi}{2}\boldsymbol{\sigma}_1 ,$$

$$\mathbf{U}_Z\left(\frac{\delta\psi}{2}\right) = \mathbf{1} + \mathrm{i}\frac{\delta\psi}{2}\boldsymbol{\sigma}_2 .$$

From the above result it is seen that $\boldsymbol{\sigma}_3$, $\boldsymbol{\sigma}_1$ and $\boldsymbol{\sigma}_2$ correspond to the infinitesimal rotations around the X, Y, and Z axes, respectively. Next we show that a finite rotation can be obtained by successive infinitesimal rotation by picking up $\mathbf{U}_X(\psi/2)$ as an example. Let $\delta\psi_1$, $\delta\psi_2$ be small angles, then

$$\mathbf{U}_X\left(\frac{\delta\psi_2}{2} + \frac{\delta\psi_1}{2}\right) = \left(\mathbf{1} + \mathrm{i}\frac{\delta\psi_2}{2}\boldsymbol{\sigma}_3\right)\left(\mathbf{1} + \mathrm{i}\frac{\delta\psi_1}{2}\boldsymbol{\sigma}_3\right).$$

If we put $\delta\psi = \psi/N$ and if $N \to \infty$, then

$$(10)\quad \mathbf{U}_X\left(\frac{\psi}{2}\right) = \lim_{N\to\infty}\left[\mathbf{1} + \mathrm{i}\frac{\delta\psi}{2}\boldsymbol{\sigma}_3\right]^N = \exp\left(\mathrm{i}\frac{\psi}{2}\boldsymbol{\sigma}_3\right).$$

Similarly

$$(10)\quad \mathbf{U}_Y\left(\frac{\psi}{2}\right) = \exp\left(\mathrm{i}\frac{\psi}{2}\boldsymbol{\sigma}_1\right), \quad \mathbf{U}_Z\left(\frac{\psi}{2}\right) = \exp\left(\mathrm{i}\frac{\psi}{2}\boldsymbol{\sigma}_2\right).$$

The spin matrices can be o btained by the differentiation of equations (8). If the differentiation of a matrix means a matrix of the differentiation, then

$$(11)\quad \left(\frac{\mathrm{d}\mathbf{U}_X}{\mathrm{d}\psi}\right)_{\psi=0} = \mathrm{i}\boldsymbol{\sigma}_3 , \quad \left(\frac{\mathrm{d}\mathbf{U}_Y}{\mathrm{d}\psi}\right)_{\psi=0} = \mathrm{i}\boldsymbol{\sigma}_1 ,$$

$$\left(\frac{\mathrm{d}\mathbf{U}_Z}{\mathrm{d}\psi}\right)_{\psi=0} = \mathrm{i}\boldsymbol{\sigma}_2 .$$

The validity of these equations depends on the differentiability of the Lie group.

The generators (10) can be developed by MacLaurin development and we get finally :

$$\mathbf{U}_X\left(\frac{\psi}{2}\right) = \exp\left(\frac{1}{2}\,\mathrm{i}\psi\boldsymbol{\sigma}_3\right) = \mathbf{1}\cos\frac{\psi}{2} + \mathrm{i}\boldsymbol{\sigma}_3\sin\frac{\psi}{2},$$

$$(12)\quad \mathbf{U}_Y\left(\frac{\psi}{2}\right) = \exp\left(\frac{1}{2}\,\mathrm{i}\psi\boldsymbol{\sigma}_1\right) = \mathbf{1}\cos\frac{\psi}{2} + \mathrm{i}\boldsymbol{\sigma}_1\sin\frac{\psi}{2},$$

$$\mathbf{U}_Z\left(\frac{\psi}{2}\right) = \exp\left(\frac{1}{2}\,\mathrm{i}\psi\boldsymbol{\sigma}_2\right) = \mathbf{1}\cos\frac{\psi}{2} + \mathrm{i}\boldsymbol{\sigma}_2\sin\frac{\psi}{2}.$$

And we get the generator representation corresponding to equation (9)

$$(13)\quad \mathbf{U}(\alpha, \beta, \gamma) = \exp(\tfrac{1}{2}\,\mathrm{i}\gamma\boldsymbol{\sigma}_2)\exp(\tfrac{1}{2}\,\mathrm{i}\beta\boldsymbol{\sigma}_3)\exp(\tfrac{1}{2}\,\mathrm{i}\alpha\boldsymbol{\sigma}_2) .$$

7. — Generators for O_3^+

A matrix which describes a finite rotation $\mathbf{R}_Z(\psi)$ around the Z axis comes from equation (8)

$$(14)\quad \mathbf{R}_Z(\psi) = \begin{pmatrix} \cos\psi & \sin\psi & 0 \\ -\sin\psi & \cos\psi & 0 \\ 0 & 0 & 1 \end{pmatrix}.$$

Let $\delta\psi$ be an infinitesimal angle, then

$$\mathbf{R}_Z(\delta\psi) = \begin{pmatrix} 1 & \delta\psi & 0 \\ -\delta\psi & 1 & 0 \\ 0 & 0 & 1 \end{pmatrix} = \mathbf{1} + \mathrm{i}\,\delta\psi\mathbf{M}_Z$$

where $\mathbf{1}$ and $\mathbf{M}_Z$ are

$$\mathbf{1} = \begin{pmatrix} 1 & 0 & 0 \\ 0 & 1 & 0 \\ 0 & 0 & 1 \end{pmatrix}, \quad \mathbf{M}_Z = \begin{pmatrix} 0 & -\mathrm{i} & 0 \\ \mathrm{i} & 0 & 0 \\ 0 & 0 & 0 \end{pmatrix}.$$

$\mathbf{M}_Z$ corresponds to the infinitesimal rotation around the Z axis. $\mathbf{M}_Z$ can be obtained also by the differentiation equation (14),

$$\left(\frac{\mathrm{d}\mathbf{R}_Z}{\mathrm{d}\psi}\right)_{\psi=0} = \mathrm{i}\mathbf{M}_Z .$$

It is the same as in the case of SU(2) that a finite rotation ψ can be obtained by successive infinitesimal rotation $\delta\psi$ and

$$(15)\quad \mathbf{R}_Z(\psi) = \exp(\mathrm{i}\psi\mathbf{M}_Z), \mathbf{M}_Z = \begin{pmatrix} 0 & -\mathrm{i} & 0 \\ \mathrm{i} & 0 & 0 \\ 0 & 0 & 0 \end{pmatrix}.$$

This is the generator of $\mathbf{R}_Z(\psi)$. Similarly

$$\left(\frac{\mathrm{d}\mathbf{R}_X}{\mathrm{d}\psi}\right)_{\psi=0} = \mathrm{i}\mathbf{M}_X , \quad \left(\frac{\mathrm{d}\mathbf{R}_Y}{\mathrm{d}\psi}\right)_{\psi=0} = \mathrm{i}\mathbf{M}_Y ,$$

$$\mathbf{R}_X(\psi) = \exp(\mathrm{i}\psi\mathbf{M}_X) , \quad \mathbf{R}_Y(\psi) = \exp(\mathrm{i}\psi\mathbf{M}_Y) ,$$

$$\mathbf{M}_X = \begin{pmatrix} 0 & 0 & 0 \\ 0 & 0 & -\mathrm{i} \\ 0 & \mathrm{i} & 0 \end{pmatrix}, \quad \mathbf{M}_Y = \begin{pmatrix} 0 & 0 & \mathrm{i} \\ 0 & 0 & 0 \\ -\mathrm{i} & 0 & 0 \end{pmatrix}.$$

By using these generator representations, we get the expression corresponding to equation (2)

$$(16)\quad \mathbf{R}(\alpha, \beta, \gamma) = \exp(\mathrm{i}\gamma\mathbf{M}_Z)\exp(\mathrm{i}\beta\mathbf{M}_X)\exp(\mathrm{i}\alpha\mathbf{M}_Z) .$$

There is the correspondence between the Pauli's spin matrices and the generators of O_3^+

$$\mathrm{SU(2)} \quad \begin{matrix} \boldsymbol{\sigma}_3 \leftrightarrow \mathbf{M}_X \\ \boldsymbol{\sigma}_1 \leftrightarrow \mathbf{M}_Y \\ \boldsymbol{\sigma}_2 \leftrightarrow \mathbf{M}_Z \end{matrix} \quad O_3^+ .$$

It is not difficult to show that the commutation relation of the generators of O_3^+ is

$$[\mathbf{M}_i, \mathbf{M}_j] = \mathrm{i}\varepsilon_{ijk}\mathbf{M}_k$$

and that of the Pauli's spin matrices is

$$[\boldsymbol{\sigma}_i, \boldsymbol{\sigma}_j] = 2\,\mathrm{i}\varepsilon_{ijk}\boldsymbol{\sigma}_k$$

where ε_{ijk} is the Levi-Civita symbol.

If we take into consideration $\mathbf{M}_k^3 = \mathbf{M}_k$ $(k = X, Y, Z)$, it is not difficult to develop equations (15) and we get finally, corresponding to equations (12)

$$\mathbf{R}_X(\psi) = \mathbf{1} + \mathrm{i}\mathbf{M}_X\sin\psi + \mathbf{M}_X^2(\cos\psi - 1) ,$$

$$(17)\quad \mathbf{R}_Y(\psi) = \mathbf{1} + \mathrm{i}\mathbf{M}_Y\sin\psi + \mathbf{M}_Y^2(\cos\psi - 1) ,$$

$$\mathbf{R}_Z(\psi) = \mathbf{1} + \mathrm{i}\mathbf{M}_Z\sin\psi + \mathbf{M}_Z^2(\cos\psi - 1) .$$

PARTIALLY TRANSPARENT SYSTEM

In the case of the totally transparent system, from the fact that the intensity is invariant, the analysis of the four dimensional space reduces to that of the three dimensional space. In the case of the partially transparent system, we must treat the four dimensional space itself. But as the four dimensional space can be obtained by extending the three dimensional space, we can discuss this case in the same way as the three dimensional case. So we discuss briefly in this case.

1. — Unimodular group and Lorentz group

The 2×2 matrix with determinant $+1$ forms a group. This group is called the unimodular group. On the other hand, the 4×4 orthogonal matrix which represents the rotation in the four dimensional space (X_0, X_1, X_2, X_3), that is the Lorentz transformation, also forms a group. This group is called the Lorentz group.

The unimodular group and the Lorentz group are homomorphic. The correspondence is two to one.

The optical interpretation is straightforward. The Jones matrix of dichroism or circular dichroism represents the operation in the unimodular group and the Mueller matrix of dichroism or circular dichroism represents the operation in the Lorentz group. The correspondence between the Jones and Mueller matrices is two to one.

These are expressed as :

$$\mathbf{U}_1\left(\frac{\delta}{2}\right) = \begin{pmatrix} e^{\delta/2} & 0 \\ 0 & e^{-\delta/2} \end{pmatrix} \leftrightarrow$$

$$\leftrightarrow \mathbf{L}_1(\delta) = \begin{pmatrix} \cosh\delta & \sinh\delta & 0 & 0 \\ \sinh\delta & \cosh\delta & 0 & 0 \\ 0 & 0 & 1 & 0 \\ 0 & 0 & 0 & 1 \end{pmatrix}$$

$$(18) \quad \mathbf{U}_2\left(\frac{\delta}{2}\right) = \begin{pmatrix} \cosh\frac{\delta}{2} & \sinh\frac{\delta}{2} \\ \sinh\frac{\delta}{2} & \cosh\frac{\delta}{2} \end{pmatrix} \leftrightarrow$$

$$\leftrightarrow \mathbf{L}_2(\delta) = \begin{pmatrix} \cosh\delta & 0 & \sinh\delta & 0 \\ 0 & 1 & 0 & 0 \\ \sinh\delta & 0 & \cosh\delta & 0 \\ 0 & 0 & 0 & 1 \end{pmatrix}$$

$$\mathbf{U}_3\left(\frac{\delta}{2}\right) = \begin{pmatrix} \cosh\frac{\delta}{2} & -\mathrm{i}\sinh\frac{\delta}{2} \\ \mathrm{i}\sinh\frac{\delta}{2} & \cosh\frac{\delta}{2} \end{pmatrix} \leftrightarrow$$

$$\leftrightarrow \mathbf{L}_3(\delta) = \begin{pmatrix} \cosh\delta & 0 & 0 & \sinh\delta \\ 0 & 1 & 0 & 0 \\ 0 & 0 & 1 & 0 \\ \sinh\delta & 0 & 0 & \cosh\delta \end{pmatrix}.$$

Here let us consider the meaning of the matrix $\mathbf{L}_k(\delta)$, $(k = 1, 2, 3)$. As an example we take $\mathbf{L}_1(\delta)$. If we replace X_0 by $\mathrm{i}X_0$, then we get

$$\begin{pmatrix} \mathrm{i}X'_0 \\ X'_1 \\ X'_2 \\ X'_3 \end{pmatrix} = \begin{pmatrix} \cosh\delta & \mathrm{i}\sinh\delta & 0 & 0 \\ -\mathrm{i}\sinh\delta & \cosh\delta & 0 & 0 \\ 0 & 0 & 1 & 0 \\ 0 & 0 & 0 & 1 \end{pmatrix} \begin{pmatrix} \mathrm{i}X_0 \\ X_1 \\ X_2 \\ X_3 \end{pmatrix}.$$

Taking into consideration

$$\cosh\delta = \cos(\mathrm{i}\delta)\,, \quad \mathrm{i}\sinh\delta = \sin(\mathrm{i}\delta)\,,$$

then the above equation becomes

$$\begin{pmatrix} \mathrm{i}X'_0 \\ X'_1 \\ X'_2 \\ X'_3 \end{pmatrix} = \begin{pmatrix} \cos(\mathrm{i}\delta) & \sin(\mathrm{i}\delta) & 0 & 0 \\ -\sin(\mathrm{i}\delta) & \cos(\mathrm{i}\delta) & 0 & 0 \\ 0 & 0 & 1 & 0 \\ 0 & 0 & 0 & 1 \end{pmatrix} \begin{pmatrix} \mathrm{i}X_0 \\ X_1 \\ X_2 \\ X_3 \end{pmatrix}$$

and the Stokes vectors

$$(X'_0, X'_1, X'_2, X'_3) \quad \text{and} \quad (X_0, X_1, X_2, X_3)$$

satisfy the equation

$$(\mathrm{i}X'_0)^2 + X'^2_1 + X'^2_2 + X'^2_3 = (\mathrm{i}X_0)^2 + X_1^2 + X_2^2 + X_3^2\,.$$

Therefore if we consider the $\mathrm{i}X_0$ axis as a real axis, the modified matrix of $\mathbf{L}_1(\delta)$ represents the rotation of a complex angle $\mathrm{i}\delta$ in the $\mathrm{i}X_0 - X_1$ plane.

Similarly $\mathbf{L}_2(\delta)$ and $\mathbf{L}_3(\delta)$ represent the rotations of complex angles $\mathrm{i}\delta$ in the $\mathrm{i}X_0 - X_2$ plane and $\mathrm{i}X_0 - X_3$ plane, respectively. This is the extension of the Poincaré sphere into the four dimensional space.

2. — Generators for the unimodular group and the Lorentz group

It is not difficult to show that

$$(19) \quad \begin{aligned} \mathbf{U}_1\left(\frac{\delta}{2}\right) &= \exp\left(\frac{1}{2}\delta\boldsymbol{\sigma}_3\right) = \mathbf{1}\cosh\frac{\delta}{2} + \boldsymbol{\sigma}_3\sinh\frac{\delta}{2}\,, \\ \mathbf{U}_2\left(\frac{\delta}{2}\right) &= \exp\left(\frac{1}{2}\delta\boldsymbol{\sigma}_1\right) = \mathbf{1}\cosh\frac{\delta}{2} + \boldsymbol{\sigma}_1\sinh\frac{\delta}{2}\,, \\ \mathbf{U}_3\left(\frac{\delta}{2}\right) &= \exp\left(\frac{1}{2}\delta\boldsymbol{\sigma}_2\right) = \mathbf{1}\cosh\frac{\delta}{2} + \boldsymbol{\sigma}_2\sinh\frac{\delta}{2}\,, \end{aligned}$$

and

$$(20) \quad \begin{aligned} \mathbf{L}_1(\delta) &= \exp(\delta\mathbf{N}_1) = \mathbf{1} + \mathbf{N}_1\sinh\delta + \mathbf{N}_1^2(\cosh\delta - 1)\,, \\ \mathbf{L}_2(\delta) &= \exp(\delta\mathbf{N}_2) = \mathbf{1} + \mathbf{N}_2\sinh\delta + \mathbf{N}_2^2(\cosh\delta - 1)\,, \\ \mathbf{L}_3(\delta) &= \exp(\delta\mathbf{N}_3) = \mathbf{1} + \mathbf{N}_3\sinh\delta + \mathbf{N}_3^2(\cosh\delta - 1) \end{aligned}$$

where $\mathbf{N}_k$ ($k = 1, 2, 3$) and $\mathbf{1}$ are

$$\mathbf{N}_1 = \begin{pmatrix} 0 & 1 & 0 & 0 \\ 1 & 0 & 0 & 0 \\ 0 & 0 & 0 & 0 \\ 0 & 0 & 0 & 0 \end{pmatrix}, \quad \mathbf{N}_2 = \begin{pmatrix} 0 & 0 & 1 & 0 \\ 0 & 0 & 0 & 0 \\ 1 & 0 & 0 & 0 \\ 0 & 0 & 0 & 0 \end{pmatrix},$$

$$\mathbf{N}_3 = \begin{pmatrix} 0 & 0 & 0 & 1 \\ 0 & 0 & 0 & 0 \\ 0 & 0 & 0 & 0 \\ 1 & 0 & 0 & 0 \end{pmatrix}, \quad \mathbf{1} = \begin{pmatrix} 1 & 0 & 0 & 0 \\ 0 & 1 & 0 & 0 \\ 0 & 0 & 1 & 0 \\ 0 & 0 & 0 & 1 \end{pmatrix}.$$

These are the generator representations, and this corresponds to the Galilean transformation. When we compare equation (12) and (19), equations (17) and (20), we recognize the very close similarities.

CONCLUSION

We have explained the structure and relationship of three polarization calculus methods using group theory.

— The generators for the Jones matrix and the Mueller matrix are introduced and considered their meanings and used extensively.

— The geometrical meaning of the matrix of dichroism or circular dichroism is shown and we see that it is related to the Lorentz transformation.

— The similarity between the matrix of birefringence and that of dichroism is pointed out.

— The relationship between the Jones matrix and the Mueller matrix is examined.

ACKNOWLEDGEMENTS. — The author wishes to express his thanks to Dr. T. Yamamoto, Nikon Research Laboratory, who read this paper and gave fruitful discussions, and Professor M. Goto, the Mathematics Department of the University of Pennsylvania, who gave useful suggestions.

REFERENCES

[1] SCHRCLIFF (W. A.). — Polarized light, *Harvard Univ. Press*, Cambridge, 1962.
[2] SCHMIEDER (R. W.). — *J. Opt. Soc. Amer.*, 1969, **59**, 297.
[3] MARATHY (A. A.). — *J. Opt. Soc. Amer.*, 1965, **55**, 969.
[4] TAKENAKA (H.). — *Japan J. appl. Phys.* (to be published).
[5] HIGMAN (B.). — Applied group-theoretic and matrix methods, *Dover*, New York 1964.
[6] GOLDSTEIN (H.). — Classical mechanics, *Addison-Wesley*. Reading, Mass. 1950.

(Manuscrit reçu le 20 décembre 1972.)

VI
Synthetic Polarizers

Editor's Comments on Papers 31 Through 33

Tourmaline, a naturally occurring mineral, possessing the ability to polarize light by selective absorption (dichroism), was discovered by Biot in 1815. In 1852 W. B. Herapath and his student discovered, by accident, a crystal that had similar properties to tourmaline. The crystal, to be known later as Herapathite, had been grown synthetically, so there was interest in attempting to grow much larger crystals than those originally found. Paper 31, by Herapath, is a detailed account of the discovery, and of the chemical and optical properties of the new material. It was not until eighty years later that Herapathite was successfully used in the production of the first large-aperture synthetic polarizer.

The development of the sheet polarizer is one of the great achievements in the field of polarized light. Paper 32 is a review paper that describes the early work leading to the commercial production of various types of sheet polarizers. Paper 33 is an earlier, more comprehensive discussion of dichroism and dichroic polarizers. It contains an excellent bibliography.

Edwin H. Land has been President and Director of Research of the Polaroid Corporation since its founding in 1935. He is also well known for his work in color vision and one-step photography. Land has been awarded many honors for his work. These include the Cresson Medal, the Potts Medal, the National Modern Pioneer Award, the Holly Medal, and the Duddell Medal.

31

Reprinted from *Phil. Mag.*, Ser. 4, **3**(17), 161–173 (1852)

XXVI. *On the Optical Properties of a newly-discovered Salt of Quinine, which crystalline substance possesses the power of polarizing a ray of Light, like Tourmaline, and at certain angles of Rotation of depolarizing it, like Selenite.* By William Bird Herapath, *M.D. London University, M.R.C.S. Engl., Member of the Bristol Microscopical Society, &c.**

[With a Plate.]

SOME short time since my pupil called my attention to some peculiarly brilliant emerald-green crystals, which had formed by accident in a solution of the disulphate of quinine. He could give me no account of their formation. Some experiments made upon them convinced me of their importance, both chemically and optically, and led me to suspect that iodine was in some way necessary to their composition.

Upon dropping tincture of iodine into the solution of disulphate of quinine in diluted sulphuric acid, an abundant deposition of similar crystals immediately occurred. However, it was found exceedingly difficult to experiment in a satisfactory manner upon the crystals thus formed, as it was almost impossible to isolate them from their mother-liquid. It was subsequently found, that by dissolving the disulphates of quinine and cinchonine of commerce in concentrated acetic acid, upon warming the solution, and dropping into it a spirituous solution of iodine carefully by small quantities at a time, and placing the mixture aside for some hours, large brilliant plates of this substance were produced. These could be readily separated from their mother-liquid, and by frequent recrystallization, purified.

The crystals of this salt, when examined by reflected light,

* Communicated by the Author; to whose liberality we are likewise indebted for the beautiful plate which accompanies this paper.

have a brilliant emerald-green colour, with almost a metallic lustre; they appear like portions of the elytra of cantharides, and are also very similar to murexide in appearance. When examined by transmitted light, they scarcely possess any colour, there is only a slightly olive-green tinge; but if two crystals crossing at right angles be examined, the spot where they intersect appears as black as midnight, even if the crystals are not $\frac{1}{500}$dth of an inch in thickness. (See Plate IV. fig. 1.) If the light used in this experiment be in the slightest degree polarized, as by reflexion from a cloud, or by the blue sky, or from the glass surface of the mirror of the microscope placed at the polarizing angle, 56° 45′, these little prisms immediately assume complementary colours. One appears green and the other pink; and the part at which they cross is a chocolate or deep chestnut-brown instead of black.

Their optical properties will be more minutely examined hereafter.

Their *chemical characters* are the following:—

They are immediately redissolved upon heating the acid liquid to 180°, and recrystallize on cooling; those formed in the sulphuric acid solution, if exposed to the air on a narrow slip of glass, upon the concentration of the mother-liquid by evaporation, will slowly disintegrate by dissection, and at length dissolve completely. They are also altered by diluting the solution with distilled water, appearing to become disintegrated. The only mode of preparing these crystals as microscopic objects, is cautiously to neutralize the excess of acid of the mother-liquid by the addition of liquid ammonia, but to take care that it be added in successive small quantities, short of precipitation of the excess of disulphates of quinine and cinchonine; then depositing upon a glass slide with a dropping tube a portion of the fluid charged with these crystals, allowing the crystals to subside, gradually removing the fluid by the capillary attraction of bibulous paper, and immediately drying the crystals by a current of cold air. They may then be mounted in Canada balsam in the usual way; taking care, however, to use no heat in liquefying the balsam; otherwise the crystals would be immediately destroyed.

Boiling alcohol readily dissolves these crystals; a clear orange-yellow solution results; this on cooling deposits crystals in abundance, having the same optical and chemical characters; but they have lost the prismatic form, and now appear as rosettes of minute hexagonal plates, or forms derived from the hexagon by truncation of the angles. Cold alcohol does not dissolve them.

Sulphuric æther and chloroform appear to have no solvent power over them.

Ammonia rapidly decomposes them; this power is greatly increased by heat. A colourless solution results, and an opake Naples-yellow precipitate remains, which is fusible at the boiling temperature of the ammoniacal liquid. A deep brownish-yellow resinous mass results; this is a compound of iodine and the alkaloid.

Liquor potassæ has the same action on these crystals; but the resulting resin is deeper in colour, being now a chocolate-brown.

The alkaline solutions in both instances contain sulphuric and hydriodic acids.

Analysis.—About ten grains of the mixed disulphates of quinine and cinchonine were dissolved in half an ounce of pyroligneous or acetic acid; into the hot solution was dropped a spirituous solution of iodine (without iodide of potassium); as the mixture cooled, these little gems gradually formed; they were carefully separated on a filter, and again dissolved in acetic acid by the aid of heat; a few drops of the tincture of iodine, as before, added, and on cooling they were again deposited; a second recrystallization removed all traces of impurity; they were collected on a filter and dried.

(A.) About three grains of these purified acetic crystals were redissolved in acetic acid, and whilst the solution was hot, acetate of baryta was added; a white precipitate was immediately produced; this was insoluble in concentrated nitric acid, and proved to be sulphate of baryta. [The acetic acid used in this experiment gave no trace of sulphuric acid when tested in the same way.]

(B.) Into the filtered portion from A (the excess of baryta having been removed by a solution of sulphate of ammonia and again filtered) were dropped nitric acid and then granules of starch; an abundant indication of the presence of iodine was instantly made evident.

(C.) Into the fluid filtered from B was dropped solution of ammonia; a flocculent, white, gelatinous precipitate was produced; this separated and fell to the bottom after some delay. Upon agitating the mixture with sulphuric æther, the precipitate was dissolved, and separated by decanting the æthereal solution from the watery fluid; upon slowly evaporating the æther by exposure to the air, a gummy resinous mass remained, nearly destitute of colour and crystalline appearance; it was probably quinine, as it did not crystallize.

Other experiments were instituted to decide the question whether the alkaloid was quinine or cinchonine.

1st. By taking advantage of the different solubility of the two disulphates, they were separated, and at length rendered perfectly pure.

2nd. It was found that by treating solutions of each alkaloid in a precisely similar manner with iodine, crystals possessing the peculiar properties were only produced in the solution of the pure disulphate of quinine.

The disulphate of cinchonine solution was merely slightly reddened upon the addition of iodine· becoming turbid, a cinnamon-brown precipitate falling, which upon heating in contact with the mother-liquid became indigo-coloured, and did not redissolve. No crystals were produced.

Therefore it became probable that iodine, sulphuric acid, and quinine, were absolutely necessary for the production of these crystals.

1st. To decide the question whether the sulphuric acid was absolutely essential, a portion of the crystals was dissolved in acetic acid, and whilst hot, acetate of baryta was dropped in until no further precipitation occurred; the solution was filtered whilst hot, and upon cooling there was no appearance of any crystallization after remaining several days.

2nd. Another portion of these crystals was dissolved by boiling n rectified spirit; a sherry wine-coloured fluid resulted; it was divided into two portions.

(A.) The first was allowed to cool spontaneously; the crystals were deposited again, but in rosettes, as before described.

(B.) The second portion of the alcoholic solution was treated whilst hot with acetate of baryta; the sulphate precipitated directly; the supernatant fluid on cooling remained perfectly transparent, and no crystals formed after some days.

3rd. An alcoholic solution of the pure alkaloid quinine was carefully prepared, and an alcoholic solution of iodine added; a sherry wine-coloured fluid resulted; no crystals were deposited; and upon spontaneous evaporation an ochry-yellow precipitate remained, without crystalline form, and having a very resinous appearance.

Therefore it now became evident that iodine, sulphuric acid, and quinine, were the constituent elements of this peculiar body. How associated, it is difficult to say; but it is probable that they are arranged as a binary compound, the disulphate of quinine acting as a feebly electro-positive base to the iodine as an electro-negative. It is conjectured therefore to be an iodide of the disulphate of quinine.

It now became an interesting question to decide whether any other of the vegetable alkaloids would act in a similar manner with iodine. The same experiments were tried with the salts of morphine, brucine, strychnine, salicine and cinchonine, but without success; we may therefore almost confidently depend on the production of these crystals being indicative of the presence of quinine in a given solution. It is to be regretted that the atomic weight has not yet been determined, but hitherto time has not permitted the necessary experiments to be undertaken.

This substance presents itself under a variety of crystalline forms; a slight change in the manner of producing them will occasion an alteration in their shape. (See fig. 2.)

When formed from a solution of the disulphates of quinine and cinchonine in diluted sulphuric acid, they present the form of parallelopipeds, exceedingly slender and elongated; the terminal planes are rectangular, the thickness being scarcely appreciable, even less than $\frac{1}{1000}$dth part of an inch, the breadth and length being variable.

The transition from this form to the square plate is very easy, and frequently observed.

By truncating the angles of the square plate we derive the octagonal plate; very common.

Under other circumstances, the aciculæ change the form of their terminal planes and become acutely pointed.

By shortening the length and increasing the breadth we obtain the half hexagon.

By joining two of these, base to base, we obtain the hexagonal plate. Very frequently found in crystals deposited from the acetic acid solution.

The rhomboidal plate is a very common form.

When a quantity of the disulphates is dissolved in acetic acid, a very few drops of a spirituous solution of iodine employed (say four or five), and the mixture left some hours in perfect repose to cool and crystallize, very large broad plates are produced, apparently formed of many aciculæ cohering by their elongated edges. These plates, by very careful manipulation indeed, may be transferred to a thin plate of microscopic glass and dried; when set up in Canada balsam and properly mounted, this becomes available as a polarizer; and in this way a crystal has been mounted by the author, and adapted to his microscope in place of a tourmaline. Frequently these crystals assume a form derived from the cuboid plate, several of which joined edge to edge produce a compound plate, the angles being at the same time more or less truncated.

Occasionally the constituent rhombic or square plates cohere by their flat surfaces instead of by their edges. They are all arranged in the same optical plane; and are not merely superimposed by accident in this peculiar position, all the crystals formed in the solution having the same extraordinary shape.

Under other circumstances we obtain this substance in the form of most beautiful compound rosettes, the component crystal being either the minute hexagon or a form derived from it; or the lozenge, like lithic acid. The crystals deposited from alcohol are of this character.

At other times it forms small stellæ, composed of acicular crystals radiating from a centre like the spokes of a wheel.

A more cautious crystallization will produce short pyramids like the ammoniaco-magnesian phosphate from alkaline urine. This is the case when a solid plate of iodine is suspended in a solution of the disulphates in acetic acid. Some days elapse ere they form, in consequence of the very slow solution of the iodine. (See fig. 2.)

The primary form of these crystals appears to be derived from the rhombic prism, but it is very possible that the substance may be dimorphous.

One remarkable fact is evident throughout the whole of this crystalline metamorphosis,—the optical properties remain the same; and the merest film of this remarkable substance possesses decided power over the rays of light.

In the following examination of their optical properties I made use of Oberhauser's achromatic microscope, with half an inch object-glass and No. 2 eye-piece; a low power, certainly under 100 diameters.

A. Their brilliant emerald-green colour reflected to the eye has been already noticed. This beam of green light produced by reflexion is decidedly a polarized ray when the plane of the crystal is inclined 41° to the plane of the incident ray.

B. Their perfectly transparent and almost colourless appearance when examined by transmitted light has also been noticed.

C. The production of complementary colours, when examined by means of a slightly polarized light, has also been spoken of.

D. The action of a single tourmaline upon them is very decided.

E. The action of two tourmalines must also be investigated.

F. The action of one tourmaline and a selenite stage is also very peculiar, and will be minutely examined.

G. The action of two tourmalines and a selenite stage must also be explained.

H. The phænomena exhibited by these crystals, when used as polarizers and analysers, is also worthy of remark, and of course permit of various crystalline substances being used as tests of their remarkable polarizing properties.

I. The phænomena of depolarization by these crystals will be touched upon under sections C. and E. &c.

(B.) The perfect polarizing powers which these crystals exhibit must now be proved and illustrated.

When two crystals of the prismatic form are examined in a superimposed condition, the following effects will be apparent:—

1st. If the two prisms are perfectly parallel in their long diameters, the ray of light will pass through unaltered. (Figs. 3 and 4.)

2nd. When they cross each other at a right angle, the small square spot where they cross will be as black as midnight. The two rays are both obstructed; that is, about half the incident ray of ordinary light is stopped or absorbed by the first or lowest crystal, the other half transmitted by it in a polarized state; this impinging upon the superior crystal is stopped by it effectually.

3rd. When the two crystals intersect each other at an angle of 45°, polarization also occurs, but not to the same extent; the spot where they are superposed is decidedly darkened.

4th. There is a perceptible polarizing effect produced at an angle of 30°; below this there does not appear to be any effect on the transmitted light.

5th. Similar effects are equally well observed in the superposition of the hexagonal plates and other forms. (See figs. 4 and 5.)

(C.) When three crystals are examined in a superimposed condition, two being crossed at right angles, and therefore dark, and a third introduced between them, the phænomena of depolarization are produced: the interposed crystal permits the light to pass through, and at the same time communicates to it the order of colour, equivalent to its thickness, in the same manner as a plate of selenite would do if interposed between two tourmalines at right angles to each other.

The angle of depolarization appears to be 45° to the plane of either polarizing or analysing crystal; but the phænomenon will take place in a minor degree at other angles. (Vide fig. 5.)

Similar phænomena are produced by the hexagonal plates and other crystalline forms. (Vide fig. 5.)

(D.) The action of a single tourmaline or Nicol's prism is very marked, and proves beyond a doubt that these crystals possess both the polarizing and the depolarizing powers.

In the first place, upon examining two of these crystals placed at right angles the one to the other, with a single tourmaline or Nicol's prism, one crystal is perfectly black and obstructs all the light, the other is as transparent as ever. Upon more closely analysing this experiment, it will be found that the crystal whose length crosses the plane of the tourmaline at right angles is the dark one, whilst that one whose long diameter is parallel to the plane of the tourmaline is transparent. (Fig. 6.)

Upon rotating the tourmaline 90°, or one quarter of a circle, the crystal which was before transparent becomes dark, and that which was dark now becomes transparent. (Figs. 6 and 7.)

Their polarizing power may now be considered to be established.

It remains to prove their depolarizing power with equal certainty.

If three crystals, *a*, *b* and *c*, arranged as in fig. 8 be examined with one tourmaline, the latter (*c*) is inferior to the two former, which of course cross it at an angle of 45°, and at 90° to each other respectively. Upon placing a tourmaline over the eye-piece of the microscope at right angles to the plane of *c*, the phænomenon of polarization will be exhibited by *c*; it will appear black.

The crystals *a* and *b* are of course at 45° respectively to both the tourmaline and to *c*; they are therefore at the angle of depolarization as in section B, and will consequently exhibit coloured images where they cross the polarizing crystal *c*, one being complementary in colour to the other; and as they intersect each other at right angles, they there exhibit the appearance due to polarization, and darkness is the result.

Similar phænomena are exhibited by the hexagonal plates, &c.

(E.) The next phænomena to be described will be the result of examining these new polarizing and depolarizing crystals by means of two tourmalines, a polarizing and an analysing plate as they are commonly called; in fact, they would be submitted to the ordinary arrangement of the polarizing microscope.

Select two crystals superimposed and crossing at right angles, and the whole object capable of being revolved horizontally on its own axis.

Let the crystals as a cross coincide with the planes of the tourmalines. (Fig. 9, *a* and *b*.)

The field of the microscope will be dark, as the tourmalines are at right angles, and consequently nearly the whole of the incident light will be obstructed or polarized.

The crystal (*a*) being at right angles to tourmaline (*d*), of course produces an increase to the polarizing effect.

The crystal (*b*) being at right angles to the tourmaline (*c*), also polarizes and increases the depth of darkness.

And at the centre (*e*) we have the combined influence of the two tourmalines and the two crystals also; we consequently have the maximum polarizing effect which it is possible to produce with this combination.

In the second place, we will rotate the object through an arc of 45° whilst the tourmalines remain stationary. The crystals are now in the position most favourable for exhibiting depolarization: they compel the light to pass through, and at the same time communicate colour to the beam, unless their thickness be too great, when of course white light will be transmitted. (Fig. 9, *e*, *f*.) Similar results follow in the examination of hexagonal plates.

(F.) We will now proceed to examine these crystals by means of a single tourmaline and the selenite stage.

This arrangement consists in placing a tourmaline in the

centre of the field of the microscope on the stage, and superimposing upon it a plate of selenite, of such a thickness that it will give a brilliant wine colour, or the complementary green, when examined by the second or analysing tourmaline placed over the eye-piece. But in the experiment now to be described the *superior* plate of tourmaline is *not employed.*

In fig. 10 four prismatic crystals of the iodide of disulphate of quinine are supposed to be arranged at various angles of rotation. (*a*) is placed in the position from 0° to 180°, and at right angles to the plane of the tourmaline. This crystal, acting as a tourmaline in the field of the microscope, developes the colour of the selenite stage, and of course appears wine-coloured.

(*b*) is across the field at 90° to the former one; it shows the complementary green.

The crystals (*c*) and (*d*) are across the field at 45° to the plane of the tourmaline; they are therefore in a position to exert but a minimum of polarizing power: an olive-green tint is produced.

The force of the argument may not be apparent at first sight, but upon experimenting with the hexagonal plates we are soon convinced of the fact. Here *a* is at right angles to the tourmaline below the stage, and therefore appears wine-coloured; whilst *b* is parallel to the polarizing plate, and of course is complementary.

In the centre of the field the crystals cross at 90°, and therefore polarize. But any lingering doubt we may yet have of the truth of this position is most certainly removed upon proceeding to the following experiment, in which we simply substitute for the pink selenite stage a plate of the same substance of a different thickness, one which developes the sky-blue tint in polarized light.

We now perceive that the crystal which is at right angles to the tourmaline is a beautiful blue, whilst that crystal which is parallel to it is the complementary yellow. The two intermediate crystals are of a slight neutral tint, as they produce but a minor degree of polarizing power at these angles.

The hexagonal crystals show the same phænomenon, but in a more marked degree.

The phænomena exhibited by this substance in these experiments were so remarkable, and so different from those of any crystals I had previously examined, that I was induced to make a comparative series of experiments upon some other crystalline compounds, as I felt convinced that the single tourmaline and the selenite stage would become a very delicate test of the power which any substance may possess of polarizing a ray of light.

Upon submitting disulphate of cinchonine to this experiment, I found it to possess a decided power of polarizing light.

This substance crystallizes as a tuft of minute radiating aciculæ, sometimes arranged in a perfect circle. (Fig. 11.) Upon placing such a tuft above the *red* selenite stage, having a tourmaline beneath it, one-half the circle appears red, the other half green. But there are four segments to the circle; one quarter red, one green, one red, and the fourth green.

Now all those prisms which are arranged at right angles to the plane of the tourmaline are red; all those parallel to it are the complementary green; but as the power exists in a minor degree on each side of this line through an arc of 45°, of course we get the whole half-circle so coloured, but in two quarter-segments placed apex to apex.

The same phænomenon is also to be found with the blue selenite stage, but it requires a better light of illumination to discover it: the segments are respectively blue and yellow alternately.

Upon examining pure cinchonine in a crystalline state as deposited from its hot alcoholic solution, the evidence of the same power exhibits itself.

The oxalurate of magnesia (discovered in the urine of a patient afflicted with the oxalic acid diathesis, after having administered the bicarbonate of magnesia for some time) possesses the same power to a considerable extent. The colours are shown in these dumb-bell crystals with tolerable splendour.

Taurine possesses this power in a very slight degree only.

Some radiating crystals of carbonate of lime or magnesia found between the tegumentary layers of the shrimp are also capable of polarizing, or rather analysing the ray to a slight degree.

The nitrate of urea, the oxalate of urea, nitrate of potassa, and nitrate of soda (rhombic), possess this power in great splendour.

There is very little doubt that more time spent in the investigation of this phænomenon would considerably enlarge the catalogue of those substances which possess the faculty of polarizing light; but none of those enumerated possess it to the extent of the iodide of the disulphate of quinine.

The disulphate of quinine does not exert the slightest influence upon the ray of light under these circumstances.

(G.) To return from our digression to the point from which we started, our next mode of examining the properties of the new crystals will be by the *two* tourmalines and by the selenite stage.

Fig. 12 shows four crystals arranged at various angles as before.

It will be understood that in this experiment the two tourmalines are arranged at right angles; one being on the stage, the other upon the eye-piece of the microscope; the plate of selenite superimposed on the inferior tourmaline. In fact, it is the same as the last arrangement, but with the addition of the superior tourmaline. We will first employ the pink selenite stage.

The prism (*a*) is at right angles to the inferior tourmaline and parallel to the superior; it developes the red colour of the stage.

The prism (*b*), being parallel with the inferior tourmaline, is at *right angles* to the superior tourmaline; it consequently obstructs the whole of the light.

But the crystals *c* and *d* are at 45° to either tourmaline, and therefore at that angle which is most favourable for showing the phænomena of depolarization. They are coloured green and yellow respectively, as they now add the influence of their own thickness to that of the selenite stage.

The experiment being varied by employing the blue selenite plate, all the other arrangements being as before, of course the field will be blue.

The prism (*a*) becomes blue from the same cause, the superior tourmaline having no influence upon it.

The crystal (*b*) is dark, as before, the superior tourmaline obstructing the beam polarized by it.

The crystals *c* and *d* are now violet and orange respectively, being complementary in colour the one to the other. They act as depolarizing crystals to the light polarized by the inferior tourmaline, and analysed by the superior tourmaline, and add their thickness to the selenite stage; in this position they exert the same influence upon polarized light that any other crystalline substance belonging to the rhombic prismatic series would do under similar circumstances.

Upon revolving the superior tourmaline, the whole appearance changes; the field passes to green with one stage and yellow with the other. The crystals pass through various changes in colour and appearance, each in its turn becoming a polarizer in action with the superior tourmaline.

(H.) It has already been stated that the author has succeeded in adapting one of these artificial tourmalines to the stage of his microscope; it is sufficiently large to give an uniform tint to the whole field, and covers a surface of an eighth of an inch in diameter. This crystal will bear magnifying to any extent, and he has been enabled to illuminate a field of eleven inches in diameter with light polarized by its means.

It is at once evident that such a crystalline plate would be far too small to be serviceable as the analysing plate above the eye-piece; a crystal of at least half an inch in diameter would be necessary for this purpose. There is frequently found in the mother-liquid a crystal sufficiently large to be so used, were it possible to transfer it safely from the fluid to a plate of glass in order to mount it: the difficulty consists in the extreme fragility of these microscopically thin compound plates; the slightest touch is sufficient to disrupt them. The slightest movement in the liquid will at times destroy the connexion existing between the edges of the component prisms; they thus lose their uniform and parallel arrangement. Wherever they cross, polarization and obstruction of the rays of light necessarily occur, and the plate is of course useless for the purpose designed.

But although it is not possible to obtain one large enough to surmount the eye-piece, yet it is perfectly easy to procure plates of sufficient size to act as analysing crystals upon the field or stage of the instrument, of course used superimposed on the tourmaline, or artificial tourmaline attached to the stage; and when these are placed at right angles, the phænomena of polarization may be exhibited with great splendour. (Vide fig. 13.)

When these crystals have been thus arranged, and the selenite stage interposed, the field becomes coloured according to the thickness of the plate of selenite; and the extent of the field so coloured will depend on the magnitude and breadth of the superior artificial tourmaline employed; frequently the whole field of seven or eight inches, or even eleven inches in diameter, has been coloured with an uniform tint.

Fig. 13 exhibits a polyhedral compound crystal of the new substance employed as an analysing plate, the artificial tourmaline being placed beneath it. The radiating crystals are those of disulphate of quinine, which crystallized upon the plate of glass used to mount the analysing crystal, in consequence of the evaporation of the mother-liquid from which the plate was formed, and, depositing the excess of disulphate beneath the plate, thus produced the splendid specimen now attempted to be depicted. The crystals of course depolarize the light, and it is transmitted by the superior or analysing plate; and if the crystals are thin enough, the prismatic colours are shown. But when the selenite stage is placed upon the polarizing plate on the stage of the microscope, and therefore inferior to the analysing plate with the disulphate of quinine beneath it, the appearances exhibited are of the most gorgeous character—it is in vain to attempt to depict them. The analysing plate of course developes the colour of the stage employed; its whole breadth therefore assumes the colour of the stage if at right angles to the plane of the plate below, or the complementary tint if parallel to it. The radiating crystals of disulphate of quinine assume every hue and tint of the spectrum: the experiment must be witnessed to be fully understood and properly appreciated.

It will be recollected that in this experiment it is not at all necessary to employ a tourmaline; the whole phænomenon may be exhibited with equal brilliancy by using the two plates of iodide of the disulphate of quinine; one as a polarizer, the other as an analyser, the selenite and disulphate of quinine being interposed. This will fully establish the fact of this substance possessing optical properties precisely equivalent to those of the tourmaline, or of a Nicol's prism, and will be sufficient to show that all the phænomena capable of being produced by the one may be exhibited by the other.

Upon submitting these artificial polarizing plates to micrometrical admeasurement, it was found that those which possessed sufficient thickness to adhere together in clusters, and to raise themselves on their edges so as to show their thickness, were none of them more than $\frac{1}{300}$dth of an inch; many were about one-half or one-third of this thickness—$\frac{1}{600}$ or $\frac{1}{900}$dth of an inch. But even these were much larger than any of those thin broad plates so readily broken; and some of which, after great trouble, were mounted and experimented with. The tourmalines commonly sold and employed for optical purposes are from $\frac{1}{100}$dth to $\frac{1}{200}$dth of an inch thick; such a one as the latter size was employed in the comparative experiment above related, whence it follows that this newly-discovered substance possesses the power of polarizing a ray of light with at least *five times* the intensity that the best tourmaline is capable of. It must consequently be the most powerful polarizing substance known, and it has been proved to be a new salt of a vegetable alkaloid.

32 Old Market Street, Bristol,
Nov. 30, 1851.

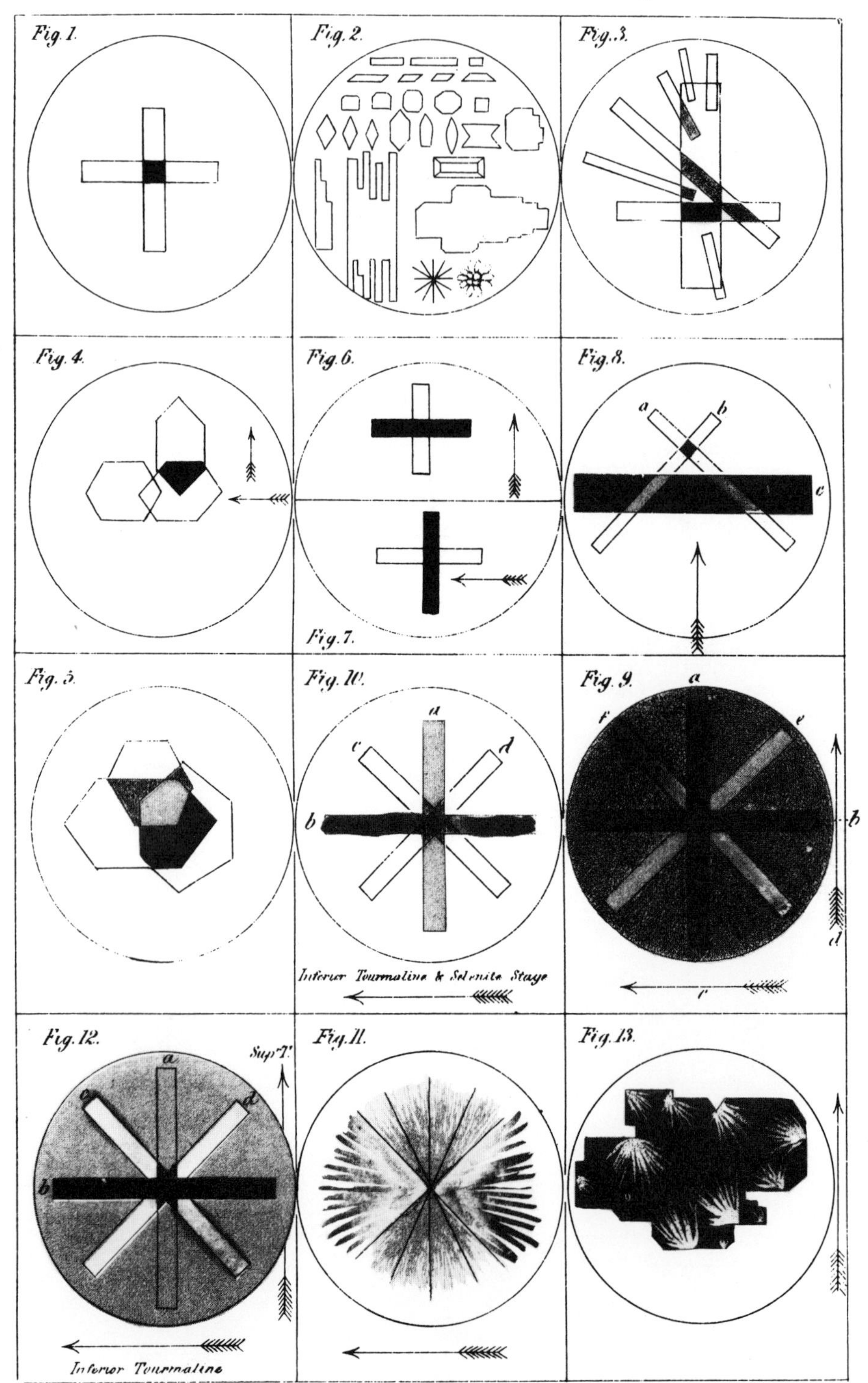

The arrows in all the figures (excepting 4 & 13) mark the planes of the Tourmalines

J. Basire, Lith.

32

Reprinted from *J. Opt. Soc. Am.*, **41** (12), 957–963 (1951)

Some Aspects of the Development of Sheet Polarizers*

Edwin H. Land
Polaroid Corporation, Cambridge, Massachusetts
(Received October 10, 1951)

This paper reviews the 25 years of activity, by the author and his co-workers, in the development of synthetic sheet polarizers. The early work during the nineteenth century is described briefly, and then the various stages of the modern development in the author's laboratory are chronicled. A description is given of the nature and the optical properties of the currently-available sheet polarizers of the Polaroid *J*, *H*, *K*, and *L* types, and of the quantitative methods used in characterizing them. The reasons are given for developing special polarizers for each of a variety of applications such as optical instruments, vectographs, and headlights.

INTRODUCTION

I SHOULD like to begin by describing some work that I did about twenty-five years ago and which should have been reported at that time. But the days slipped by and the months slipped by and now twenty-five years have slipped by, and, although my co-workers and I have reported from time to time on some of the applications of that work, I was surprised to find that the work itself has not yet been reported to the Optical Society of America.† I want to take this occasion to review with you the field to which a number of us have devoted a large part of our professional careers, in order to see what our early purposes were, and to what extent we have fulfilled them.

In about 1926 I became interested in the question of whether or not one could make a material in the form of a synthetic sheet which could polarize light in the same way that tourmaline does, only more efficiently—a material that might have the desirable properties of a Nicol prism and yet be free from some of its undesirable properties such as small area and limitation in angular aperture.

HERAPATH'S WORK

In the literature there are a few pertinent high spots in the development of polarizers, particularly the work of William Bird Herapath,[1] a physician in Bristol, England, whose pupil, a Mr. Phelps, had found that when he dropped iodine into the urine of a dog that had been fed quinine, little scintillating green crystals formed in the reaction liquid. Phelps went to his teacher, and Herapath then did something which I think was curious under the circumstances; he looked at the crystals under a microscope and noticed that in some places they were light where they overlapped and in some places they were dark. He was shrewd enough to recognize that here was a remarkable phenomenon, a new polarizing material.

Doctor Herapath spent about ten years trying to grow these green crystals large enough to be useful in covering the eyepiece of a microscope. He did get a few, but they were extremely thin and fragile—for it is very hard to grow them. We find that they were not successfully grown as large single crystals until at the Zeiss Works, Bernauer[2] succeeded about ten or twelve years ago in growing some that were several inches in diameter.

Herapath's work caught the attention of Sir David Brewster, who was working in those happy days on the kaleidoscope. Brewster thought that it would be more interesting to have interference colors in his kaleidoscope than it would to have just different-colored pieces of glass. The kaleidoscope was the television of the 1850's and no respectable home would be without a kaleidoscope in the middle of the library. Brewster, who invented the kaleidoscope, wrote a book[3] about it, and in that book he mentioned that he would like to use the herapathite crystals for the eyepiece. When I was reading this book, back in 1926 and 1927, I came across his reference to these remarkable crystals, and that started my interest in herapathite.

AMBRONN'S WORK

In the work of Herapath, iodine was seen to be definitely associated with polarizers. Iodine appears in another place in the early literature, in the work of Ambronn,[4] who noticed that when he stained membranes of cells, living cells of animal tissue, he obtained an elevation of the absorption coefficient of the iodine. As he treated certain natural tissues, they became darker, and he observed that these darkened tissues were dichroic. In a sense he "missed the boat" on making a polarizer. In fact he said flatly that he saw no prospect of making a polarizer by using this phenomenon, a statement which I find very hard to understand. Ambronn was a competent worker, but he was

* Presented as an invited paper in the Symposium on Polarization at the March, 1951, meeting of the Optical Society of America, Washington, D. C.

† The development of the *J* polarizer was reported to the Harvard Physics Colloquium on February 8, 1932, under the title "A new polarizer for light, in the form of an extensive synthetic sheet." See reference 6 for many references to polarizers and their applications.

[1] W. B. Herapath, Phil. Mag. (4th ser.) 3, 161 (1852).

[2] F. Bernauer, Fortschr. d. Mineralogie **19**, 22 (1935).

[3] D. Brewster, *The Kaleidoscope, Its History, Theory, and Construction* (John Murray, London, 1858), second edition.

[4] H. Ambronn, Ann. Physik u. Chem. **34**, 340 (1888).

one of those persons who are more interested in conducting scientific experiments than in the utilization of the results obtained. If Ambronn had been working in America, the outcome might have been different!

One of the concepts that has been emphasized thus far is the important place of iodine in the history of the development of synthetic polarizers. Herapathite is a crystalline form of quinine sulfate per-iodide.

USE OF MICROSCOPIC CRYSTALS

My immediate inspiration was the work of Herapath. It was apparent that, since his ten years of effort had not sufficed to make a large polarizing crystal, this was probably a hard way to make a polarizer. I wondered if, instead of trying to grow one large crystal, one might not use a multiplicity of small crystals, all similarly oriented. However, there are several difficulties: (1) A number of little crystals will scatter light if they are not carefully prepared; (2) There is a problem of orientation; (3) How does one prepare very small crystals; and (4) In what form does one handle them? In visualizing this problem—how would one go about making a synthetic sheet containing oriented herapathite crystals—the first elementary concept is not to make big crystals, but to make microscopic ones. If one is then able to turn all of the microscopic crystals the same way, the remaining difficulty is that it is not sufficient to have them microscopic, but because they will scatter light, they must be submicroscopic.

THE FIRST SYNTHETIC POLARIZER

In my then youthful innocence it seemed to me that these problems could be solved in a rather short time, perhaps in a few months. I have preserved as an exhibition piece the first synthetic polarizer which I made. This first polarizer was made by grinding herapathite crystals in a ball mill for a month; the mill contained a solution of nitrocellulose lacquer.

Then came the most exciting single event in my life. The suspension of herapathite crystals was placed in a small cell—a cylinder of glass about a half-inch in diameter and a quarter-inch in length. The cell was placed in the gap of a magnet which could produce about 10,000 gauss. Before the magnet was turned on, the Brownian motion caused the particles to be oriented randomly so that the liquid was opaque and reddish black in color. When the field was turned on—and this was the big moment—slowly and somewhat sluggishly the cell became lighter and quite transparent; when we examined the transmitted light with a Nicol prism, it went from white to black as the prism was turned. This first polarizer experiment was a success, but in the twenty-five years between then and now, there have been many technical details which required solution.

In making a solid polarizer from a liquid polarizer, the following steps were taken: The same suspension was placed in a test tube and a plastic sheet was dipped into it; the test tube as a whole was placed in the magnetic field and then the tube was withdrawn so as to leave a coating of the dispersion of polarizing crystals on the sheet; the sheet was then allowed to dry in the magnetic field, and a few moments later we had a polarizer.

OTHER METHODS OF ORIENTATION

One can also orient the particles with an electrostatic rather than a magnetic field; however, I would like to mention a quite different orientation method. A special sheet of rubber was procured, a sheet about eight inches square, and this sheet was stretched in one direction to a length of several feet. A very viscous suspension of crystals was coated on the sheet in the stretched condition. The sheet was then permitted to contract, and was then stretched in the perpendicular direction. The stretch ratio obtained in this way is the square of the stretch ratio that can be obtained by a simple stretching operation.

All of this, while exciting at the time and indicative of the potential of this approach, did not provide a good way to make polarizers in large quantities. We abandoned use of electrostatic and magnetic fields and put about a year's work into making a colloidal dispersion of submicroscopic needles of herapathite, this dispersion being formed into a viscous mass which was then extruded between long narrow slits. As the sheet was extruded, the needles were oriented parallel to one another. This is the method employed in making the *J* polarizer.

SIZE OF THE CRYSTALS

These needles were thought to be about one micron long and a small fraction of a micron in diameter. It was quite important to have the diameter that small, because this minimized the light scattered by this two-phase system. In the recent decade it became feasible to study these needles with the electron microscope, and thus a good estimate could be made of their actual size. The technique was this: The polarizing sheet made with the small crystals of quinine sulfate per-iodide was treated with silver nitrate to form silver iodide in the place where the herapathite crystals had been. This was done in the solid state, the liquid being introduced without dissolving the sheet. Then the plastic sheet itself was dissolved and the crystals of silver iodide were examined with the electron microscope. Figure 1 is one of the pictures obtained in this manner. In this figure the needles are indeed of about the size originally suspected; these needles were prepared from the Polaroid *J* polarizer. An interesting fact which will be discussed in another part of this presentation is that the sheet is still a polarizer when one replaces the herapathite with metallic silver.

THE PRINCIPAL ABSORPTION COEFFICIENTS

Dichroic polarizers have two different principal absorption coefficients which are effective for normal incidence, so that the absorption of normally incident linearly polarized light is markedly different for vibration directions parallel to the two principal axes of the polarizer. In a good dichroic polarizer the two absorption coefficients must be very different; in an ideal polarizer one linear component is transmitted without absorption and the other is absorbed completely. Thus an ideal polarizer involves an infinite ratio for the two absorption coefficients. As will appear later, we are theoretically unable to obtain an infinite ratio, but in good modern polarizers the ratio can run as high as 100 or more. It is not necessary for the ratio of absorption coefficients to be constant throughout the visible spectrum; indeed, there are some applications where one does not want the ratio to be constant. It is desirable, however, in an ordinary polarizer that the one absorption coefficient be so high throughout the visible spectrum that substantially none of the light of that orientation passes through the sheet, and that the other absorption coefficient be so low that substantially all of the light goes through. Such a polarizing material is remarkable in that its transmittance for unpolarized light is quite constant throughout the spectrum, with the result that a good visual polarizer is sometimes a better neutral-density filter than filters which are made with nonpolarizing absorbing materials.

QUANTITATIVE DESCRIPTION OF POLARIZERS

It is useful to be able to describe a polarizer in several different ways. The first way is in terms of the two principal transmittances. In Fig. 2, for one density-type of *H* polarizer, k_z is the transmittance for linearly-polarized incident light whose electric vector is parallel to the "machine direction" z, and similarly k_y is the transmittance when the electric vector of the incident light is perpendicular to the machine direction. This method of describing a polarizer permits one to show in a graph the value of k_y with fair precision, but the actual value of k_z is not indicated well. Figure 3 shows the same information described now in terms of the principal optical densities. In Fig. 3 the transmitted component is at the bottom and the nearly extinguished component appears as the upper curve. In the density representation the behavior of the z component is shown well, but the transmittance for the y component is not shown very accurately.

THE STRETCH RATIO

It is of interest to inquire as to what actually determines the density ratio of the polarizing sheet—i.e., the ratio of the principal optical densities. There are two parameters; obviously one is the efficiency of the crystals (the intrinsic density ratio of the dichroic absorbing material), and the other is the degree of orientation of the crystals.

The degree of orientation is measured in terms of the stretch ratio; the method of determining the stretch ratio in the laboratory is very simple. A circle is drawn on the sheet; then after stretching, one measures the axial ratio of the ellipse into which the circle has been deformed. Figure 4 shows the relation between the density ratio of the final polarizer and the stretch ratio, on the basis of the assumption that the dichroic crystals are needle-like and that the absorption axis is parallel with the needle axis.[5] In Fig. 4 it is apparent that even if the density ratio of the crystals is infinite, that is, even if the crystal itself is a perfect polarizer, the sheet cannot have a density ratio greater than about 1.3 times the stretch ratio. We have found, however, that this limitation on the density ratio which holds for the microcrystalline *J* sheet does not apply to the more modern *H* and *K* polarizers.

THE SCATTERING POLARIZER

We have developed an entirely different kind of polarizer, which very few people indeed have ever

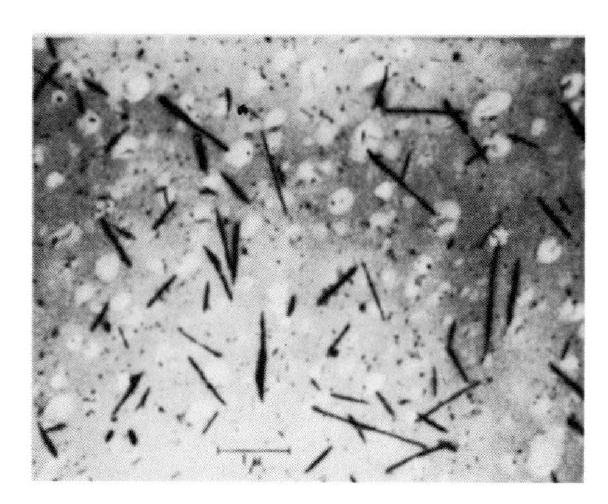

FIG. 1. Silver iodide replicas of (disoriented) herapathite needles in the *J* polarizer. (Electron micrography by C. E. Hall, Massachusetts Institute of Technology, 1949.)

seen, and for which there seems to be no obvious use. In this polarizer we use crystals of a doubly-refracting material instead of dichroic crystals, and we deliberately make the crystals with a size several times the wavelength of light. Furthermore, the refractive index of the plastic matrix is made equal to one of the two principal indexes of the birefringent crystals. What is then obtained is quite a beautiful product: when the sheet is examined with an analyzer, it is as transparent as glass for one position of the analyzer, but when the analyzer is turned through 90°, the sheet is turbid and highly diffusing. It is a variable diffuser, and one would think that there would be a use in optical instrumentation for such a component.

[5] R. Clark Jones, J. Opt. Soc. Am. 35, 803(A) (1945).

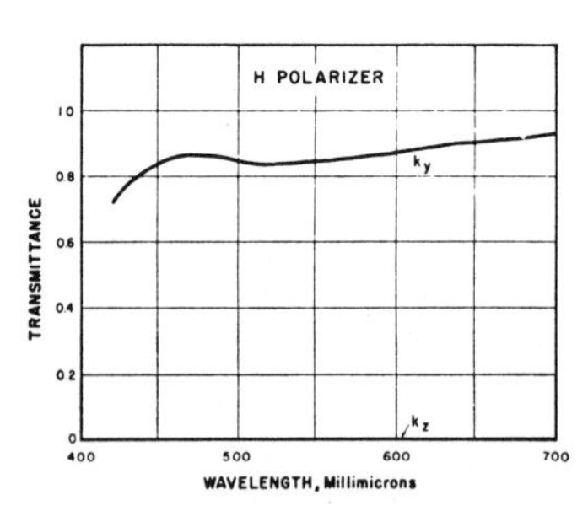

FIG. 2. Principal transmittances *versus* wavelength, for a typical *H* polarizer. (Transmittance values have been corrected for reflection at the surfaces.)

THE METAL POLARIZERS

Still different are the metal polarizers, like the one whose properties are illustrated in Fig. 5 which utilizes very small needles of tellurium, all oriented parallel with one another. This polarizer works, probably not so much because of the dichroism of the individual needles as because the absorption of light by the lattice or grating of needles depends on the direction of the electric vector with respect to the needle axis. The tellurium crystals were prepared and oriented in a way quite different from that used in making the *J* polarizer. Instead of being dispersed mechanically in a plastic matrix before extrusion, the reagents were introduced into a different plastic, namely polyvinyl alcohol. Either before or after the sheet is stretched, the tellurium needles are precipitated chemically. Figure 6 is an electron micrograph which indicates the actual size of these needles. Because of the method of preparation of the pellicle, the needles are disoriented and are much fewer in number per unit area than in the original polarizer. The preceding sentence applies also to Fig. 1.

We have also prepared metal polarizers of silver, gold, and mercury; the optical density curves of these have been published.[6] The particles of all three of these metals are optically isotropic.

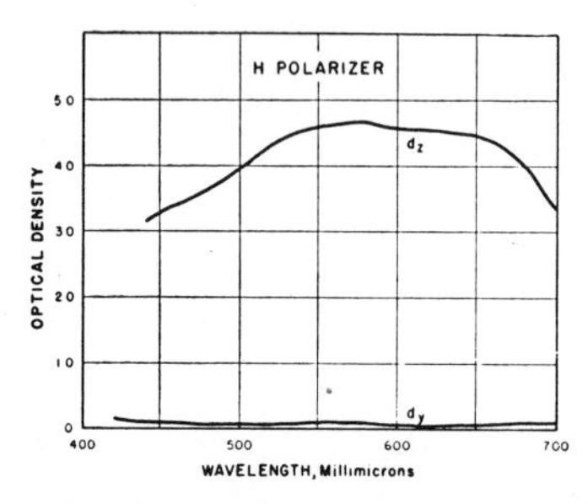

FIG. 3. Principal optical densities *versus* wavelength, for the typical *H* polarizer of Fig. 2. (The data have been corrected for surface reflection.)

POLYVINYL ALCOHOL

A new material of considerable importance in the field of polarizers is polyvinyl alcohol, which can be fabricated into a beautiful, water-clear sheet which lends itself to stretching when heated. In the stretched state it is the matrix for a series of important and useful polarizers.

The *H* polarizer is made by absorbing iodine in a stretched sheet of polyvinyl alcohol. The various dye polarizers are made by absorbing dyes in stretched polyvinyl alcohol. These dye polarizers, which we call *L* polarizers, have the property of polarizing only portions of the spectrum and of transmitting both components in the remainder of the spectrum. A very important new polarizer is the *K* polarizer, whose dichromophore is polyvinylene. The *K* polarizer is made by dehydrating stretched polyvinyl alcohol.

Figure 7 indicates the approximate relative shape and size of the molecular structures of polyvinyl alcohol,

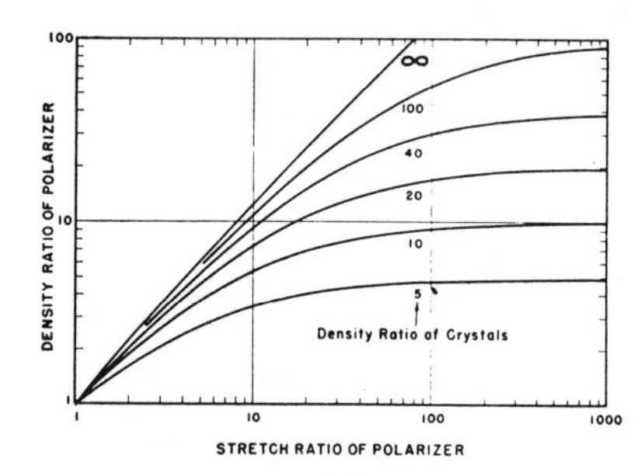

FIG. 4. Optical density ratio of a microcrystalline polarizer (calculated) *versus* the stretch ratio of the polarizer sheet, for needle-shaped crystals of different intrinsic density ratios. The crystals are assumed to be uniaxial in optical properties; the unique axis is the axis with the greater absorption coefficient, and is parallel with the geometrical axis of the needle.

polyvinylene (*K* polarizer), and the polymeric iodine of the *H* polarizer.

Blake, Makas, and West showed that when dichromophores of the *H* and *K* types are combined in a single layer of polyvinyl alcohol, contrary to expectation a new dichromophore termed *HR* is formed which has a definite infrared absorption maximum at about 1.5μ. This dichromophore makes possible the preparation of efficient dichroic polarizers for the broad spectral range 0.75 to 2.8μ. (See U. S. Patent 2494686, 1950.)

THE VECTOGRAPH

We have found one rather interesting application of the iodine-polyvinyl alcohol polarizer in the vectograph, and this is one of the applications which has

[6] E. H. Land and C. D. West, "Dichroism and Dichroic Polarizers." *Colloid Chemistry*, J. Alexander, Ed. (Reinhold Publishing Corporation, New York, 1946), Vol. 6, pp. 160–190. This paper contains much original data, and lists many references to articles and patents concerning polarized light, including those of Käsemann and Marks. Dreyer's work is first described in U. S. Patent 2400877, 1946.

been reported previously to the Optical Society.[7] Prior to this it was not realized that three-dimensional pictures could be obtained if the two images of the stereoscopic pair were made in terms of percent polarization. We then undertook to develop methods for obtaining good "polarization anaglyphs." If two vectographs are superposed with their stretch directions at right angles, one will see one image and then the other, if one views the pair with an analyzer which is rotated slowly. In order to prepare such images, the same procedure is used as that employed in making color separations by the dye transfer process, except that iodine is used instead of a dye. The procedure involves making what is called a wash-off relief, in which the thickness of the gelatin is proportional to the image density that is desired. The wash-off relief is then dipped into an iodine bath and is then placed in contact with a sheet of stretched polyvinyl alcohol. Images formed in this way with

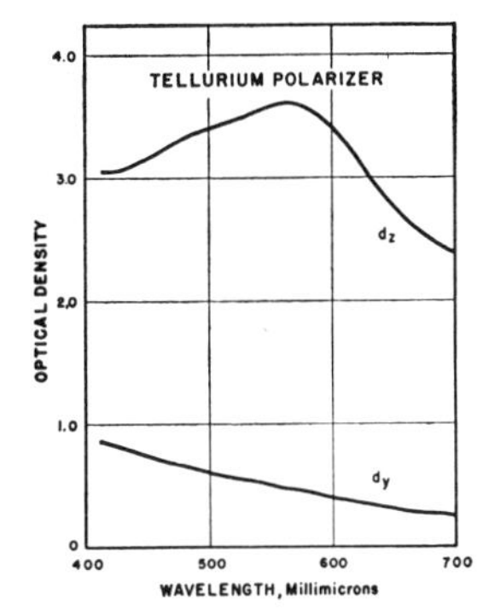

FIG. 5. Principal optical densities *versus* wavelength, for a tellurium polarizer. (The data have been corrected for surface reflection.)

iodine in polyvinyl alcohol are really images formed in terms of degree of polarization.

In particular, the two images one observes can be members of a stereoscopic pair, so that one can use an ordinary lantern slide projector or an ordinary movie projector with a nondepolarizing screen, to obtain pictures which are threedimensional when viewed with polarizing spectacles. A method has also been worked out by which one can make vectograph prints of ordinary size that can be held in the hand.

By using the dye polarizers mentioned above, instead of the iodine polarizers, vectographs can be prepared in color. This provides the only method of projecting full-color stereoscopic pictures in standard motion-picture and lantern-slide projectors without any alteration of the instrument. No polarizers are used on the projector, so that all the light from the source is utilized. This application illustrates a requirement for

[7] E. H. Land, J. Opt. Soc. Am. **30**, 230 (1940); see also p. 184 of reference 6 regarding the work of Land and Mahler.

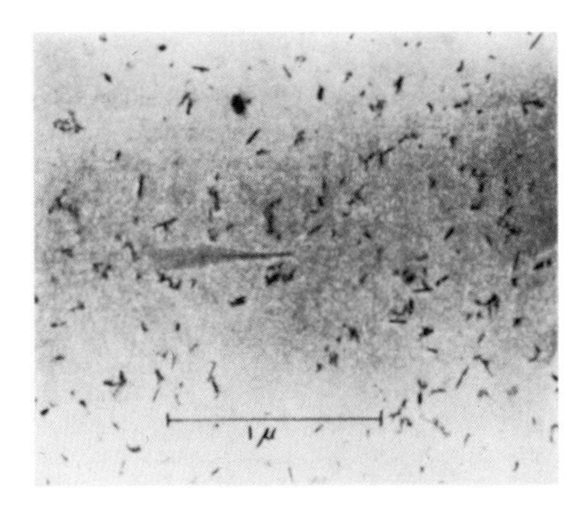

FIG. 6. Electron micrograph of tellurium needles in polyvinyl alcohol, precipitated *in situ* in the oriented state. (The matrix was re-dissolved in making the pellicle, which fact accounts for the disorientation.) It is interesting to speculate that needles like these may possibly account for the polarizing action of interstellar space.

polarizers which have high absorption for one component over only a portion of the visible spectrum.

THE *H* POLARIZER

The *H* polarizer is the one that is now in most common use. Its principal transmittances and principal optical densities were shown in Figs. 2 and 3 respectively, and are shown also in Fig. 8, which is a combination graph designed to exhibit both k_y and k_z effectively. The optical density d_z is high in the middle of the spectrum where the eye is most sensitive; even at 680 mμ the optical density is greater than three. Also, the transmittance k_y is quite high and stays high throughout the visible spectrum. This polarizer looks neutral in color, when viewed in unpolarized light, and it is remarkably free of light scattering because the dichromophore is of molecular dimensions. The extinction color is ordinarily a deep blue, but a very bright light is required to see the extinction color in the denser varieties of the *H* polarizer.

THE *K* POLARIZER

Figure 8 also shows the optical behavior of a typical *K* polarizer. The density d_z of this polarizer is greater than that of the typical *H* polarizer just discussed for

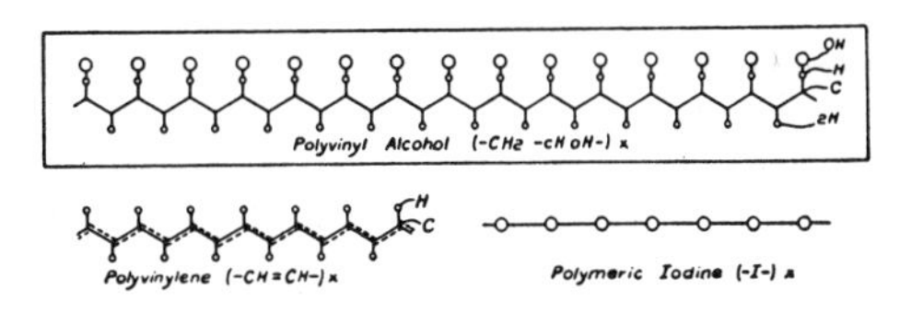

FIG. 7. The approximate molecular structure of polyvinyl alcohol, polyvinylene (*K* polarizer), and polymeric iodine (*H* polarizer).

all wavelengths shorter than about 650 mμ, but the density d_z is less than that of the *H* polarizer for longer wavelengths. As a result, the *K* polarizer has a red extinction color instead of the blue extinction of the *H* polarizer.

The *K* polarizer is particularly interesting for two reasons: first, the chemical simplicity of the simple straight-chain hydrocarbon, polyvinylene, employed as the dichromophore; second, its extraordinary stability with respect to heat and radiation.

The *K* polarizer was developed especially for use on automobile headlamps in connection with our system for eliminating the glare of opposing headlamps in night driving while still providing drivers with excellent visibility of the road ahead. In this system a polarizer is laminated to the front surface of the headlamp; the requirements placed upon this polarizer are severe: it must stand the intense heat of the radiation during the periods when the car is stationary with the lights turned on, and it should last for the life of the car, even though its owner subjects it to continuous weathering in the tropics.

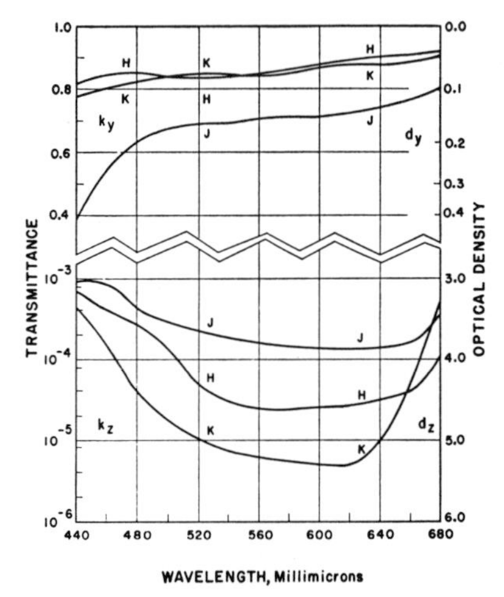

FIG. 8. The optical behavior of typical *J*, *H*, and *K* polarizers *versus* wavelength. The curves represent transmittances if referred to the ordinate scale on the left, and represent optical densities if referred to the scale on the right. The principal transmittance k_y is plotted on a linear ordinate scale in the upper part of the diagram, and k_z, the transmittance of the highly absorbed component, is plotted on a logarithmic scale in the lower part. (The data have been corrected for surface reflection.)

The motivation back in the early days when I was making the polarizers in the magnetic field was threefold: First, there was the tremendous excitement of trying to make a polarizer the size of a large piece of window glass such as you see here on the stage; second, there was a good deal of scientific interest in such things as Kerr cells, the Faraday effect, and polarized-light microscopy, and all of these interests required polarizers with an optical performance that could be met at that time only by the Nicol prism, which of course had a severely limited aperture. Incidentally, Kerr cells were a lively item in the early television research; Alexanderson[8] was using them in his researches, and it looked as though Kerr cells might play an important part in television. But the third and most important motivation was, and still is, the field of polarizing headlights.

HEADLIGHTS

The glareless headlight system involves placing on the headlights polarizers with their axes tilted in the vertical plane at about 45° from the horizontal. A similar polarizer is incorporated in the windshield or otherwise placed before the driver, and its axis is substantially parallel with the polarizers on the headlights. If you now consider two such cars, and if you imagine that one is turned around so that it faces the other, you see that the headlight polarizers of the approaching car are "crossed" with the one before the driver's eyes. The performance of this arrangement is such that the driver's viewer selects the opposing headlamps from among all other objects in the field of view and extinguishes them. Meanwhile, the same viewer causes little reduction in the driver's visibility of his own roadway because of the parallel relationship between the viewer and his own headlight polarizers. With this system in universal use, every driver will be able to drive at night with light at least comparable to the *upper* beam of your present headlamps, and approaching cars will seem to have headlights whose brightness is equal to only a fraction of that of the present *lower* beams.

At the present time the status of our glareless headlight system might be summed up as follows: We have available for the system polarizers of such stability and high efficiency that little would be gained by making them better; and 125-watt headlamps have been developed by General Electric for meeting the requirements of the system. These headlamps, three times as powerful as present day headlamps, are, when in use in the polarizing system, about 50 percent brighter than today's. With this matched system, including the polarizing viewer, large-scale quantitative tests involving careful measurements[9] have been conducted. These tests show that through the viewer one has sufficient light for at least as good visibility as one has today with the high beam for open road driving, and

[8] E. F. W. Alexanderson and R. D. Kell, U. S. Patent 1783031, 1930.

[9] V. J. Roper, Traffic Engineering **19**, 151 (1949), and E. H. Land and L. W. Chubb, Jr., Traffic Engineering **20**, 265 and 384, 399 (1950).

that this good visibility is maintained while a similarly-equipped car is being met and passed.

CONCLUSION

In closing I would like to acknowledge the assistance received from the many colleagues, co-workers, and assistants who have been associated in this program. Many of them are known from their own publications, and others are named in the references that have been cited. One of the great pleasures in reviewing this field has been the recalling to mind of these many happy relationships.

33

Reprinted from *Colloid Chemistry*, Vol. 6, Jerome Alexander, ed., Van Nostrand Reinhold Co., New York, 1946, pp. 160–190

Dichroism and Dichroic Polarizers

EDWIN H. LAND AND C. D. WEST
Polaroid Corporation, Cambridge, Mass.

Historical Introduction

The phenomenon of linear dichroism was first observed in the mineral tourmaline by Biot in 1815. Since that time numerous observations, but relatively few quantitative measurements, of linear dichroism have been made on the widest variety of materials; these include, besides natural and artificial crystals, dichroic products obtained by the staining, dyeing or coloring of crystals, fibers, and other biological tissues, films, glass surfaces, and the like. This subject, although faithfully cultivated by a few devotees, never aroused a wide interest in the literature of science and hardly passed beyond a preliminary descriptive stage of development. The data of dichroism remained diffusely scattered through the journals of physics, chemistry, crystallography and mineralogy, biology, and dye and textile technology. This may explain why examples of later rediscoveries of the results of much of the earlier work are fairly numerous in the literature of this subject.

The phenomenon of circular dichroism was first reliably observed in alkaline chromium tartrate solution by another Frenchman, Cotton, in 1895. This property has been carefully measured for a wide variety of optically active molecules in the liquid and vapor states and in solutions; and these measurements have contributed substantially to the development of the theory of optical activity in asymmetric molecules, as related in several books on this subject.

The first satisfactory light polarizer was found in 1828 by W. Nicol, who devised the type of calcite prism which bears his name. The superiority of this type of polarizer needs no other testimony than its leadership of the field for over one hundred years; and it is still the preferred polarizer, when suitably constructed, for giving a completely polarized beam over the spectral range for which calcite transmits the extraordinary ray without absorption, namely about $0.23 - 3.0\mu$. The limitations of this class of polarizer are equally well recognized and have effectively prevented their application outside the laboratory. These are an angular aperture limited by the optical constants of calcite, and a linear aperture limited by the scarcity of large pieces of calcite of optical quality; great thickness; and occasionally the astigmatism of images transmitted through such prisms.

It was probably recognized quite early that physically the most satisfactory substitute for a calcite polarizer would be some kind of dichroic polarizer. To be specific, tourmaline sections were employed in the simple form of polariscope known as the tourmaline tongs, and also as polarizers for microscopes. The Englishman Herapath in 1852 discovered the intense dichroism of crystals of a sulfate periodide of the cinchona alkaloid quinine. He described in detail his attempts to grow and mount these crystals, which were named herapathite after him, for use as polarizers; and later worked with a similar crystal from cinchonidine alkaloid, a relative of quinine. Later attempts to realize Herapath's conception were described by Zimmern and Coutin in France (1926);[1] and about 1936 A. M. Marks in the United States began commercial production of macrocrystalline cinchonidine herapathite polarizers, while the Carl Zeiss firm in Germany was doing the same with quinine herapathite polarizers developed by F. Bernauer. In 1935 he [2] had previously outlined general methods for making large area polarizers, but none of these apparently was ever described in any detail. E. H. Land had already been occupied with the same problem for several years, and in February 1932 he had described and demonstrated before the Physical Colloquium of Harvard University a form of dichroic polarizer that was entirely new in that its structure was microcrystalline rather than macrocrystalline. In 1934 Land was in commercial production of this type of polarizer, using quinine herapathite as the crystal component.

The last six years have witnessed a rapid application of dichroic polarizers in a diversity of popular and scientific uses, and an equally rapid development of new types of dichroic polarizers. The most notable of these are not crystalline in the ordinary sense of this term, but would better be described as molecular in structure.

In this brief account we shall treat the general phenomenon of linear dichroism and its measurement, describe the structures of several kinds of useful dichroic polarizers, and point out some of their practical applications.

Dichroism: Definition and Measurement

Absorption. For ordinary homogeneous absorbing media, the absorption of light is conveniently expressed by the optical density of each wave length: $\frac{J}{J_0} = k = 10^{-d}$, or $\log \frac{J_0}{J} = \log \frac{1}{k} = d$. When different thicknesses of the same medium are compared, the law becomes $\log \frac{J_0}{J} = d = Kt$. When the volume concentration of the absorbing agent may be varied as well as the thickness, then in the ideal case the law becomes $\log \frac{J_0}{J} = d = Kt = ect$. It is sometimes more convenient to determine the area concentration c' than the volume concentration c, and in this case by a suitable choice of units $ect = ec'$. The proportionality constants K and e are constants of the absorbing agent, and like d they vary with wave length; when they are employed, the units for c and t must always be stated. We shall take t in cm, c in g per liter, and c' in g per 1000 cm^2; thus our e is a specific extinction coefficient. For ordinary media it is indifferent whether natural, linear, or circular polarized light is used to measure the absorption constants.

Dichroism. It is sometimes overlooked that there exist homogeneous media whose absorption of natural light does not follow the simple thickness law. If we let natural light of intensity J_0 fall on such a medium, and measure successively the emergent intensities J_1 and J_2 after traversing thicknesses t_1 and $t_2 = 2t_1$, we find after making any necessary corrections that $\frac{J_2}{J_0}$ is not equal to $\left[\frac{J_1}{J_0}\right]^2$ as required by $\frac{J}{J_0} = 10^{-Kt}$, but is somewhat greater than $\left[\frac{J_1}{J_0}\right]^2$. If now we analyze the emerging light we find that it has experienced a change in its vibration state; instead of being natural light, it is partially polarized, either circularly or linearly as the case may be. Accordingly, caution must always be exercised in calculating the results of a transmission measurement with natural light to an absorption coefficient if the specimen has this property to an appreciable degree.

Such media are called dichroic, and are characterized by unequal absorptions for two oppositely polarized beams of incident light; in one case the two beams of light are linear-polarized at right angles to each other; in the other case the two beams are circular-polarized in opposite senses. For such media *two* absorption constants are required for a complete description; for example, for linear dichroism $d_{\parallel}$ and $d_{\perp}$, or $K_{\parallel}$ and $K_{\perp}$ or $e_{\parallel}$ and $e_{\perp}$. Then the conventional mathematical expression for dichroism is $\Delta d = d_{\parallel} - d_{\perp}$, or ΔK, or Δe.

The constant $d_{\parallel} = \log \frac{1}{k_{\parallel}} = K_{\parallel}t$ of a dichroic specimen denotes the optical density of the specimen for linear-polarized light vibrating parallel to one of the two vibration directions of the specimen, and similarly for $d_{\perp}$. For a linear dichroic specimen $d = \log \frac{1}{k} = Kt$ may be defined and measured in two ways; it is either the optical density for natural or circular-polarized light, or it is the optical density for linear-polarized light vibrating at 45° to the vibration directions of the specimen.

The optical constants, for natural light, of a combination of ideal polarizer plus a dichroic specimen characterized by the constants $d_{\parallel}$, d, and $d_{\perp}$, are found as follows: since the d for an ideal polarizer is log 2 = 0.301, the constants $d_{\parallel}$, d, $d_{\perp}$ are each to be increased by 0.301 to give the constants of the combination; and similarly the k constants are to be multiplied by 0.5 to give the constants of the combination.

Linear dichroism is in fact more complicated than this, for in the most general linear dichroic crystal the absorption constants $K_{\parallel}$ and $K_{\perp}$ vary with the orientation of the crystal section being measured, with respect to the orthogonal principal axes X, Y, Z, of the triaxial absorption ellipsoid. In the dichroic materials treated here, we shall avoid ambiguity by considering only what amounts to crystal sections of fixed orientation, and only that light which traverses such sections at normal incidence.

Detection of Dichroism. While a suitably mounted calcite cleavage rhomb was formerly employed as a dichroscope for detecting linear dichroism, it is now easier to construct an equivalent device by juxtaposing two linear film polarizers (plate P in Fig. 1). If this divided field plate be superposed with a quarter-wave sheet (plate

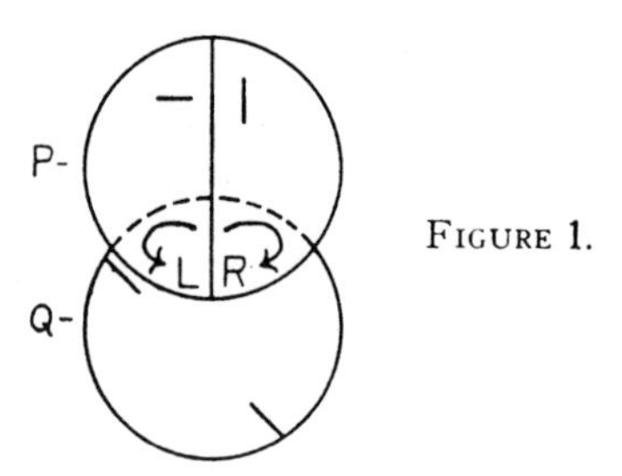

FIGURE 1.

Q), the combination serves as a circular dichroscope when the polarizer layer is held next to the eye, and as a linear dichroscope when it is turned over and the quarter-wave layer is next to the eye of the observer. With dichroscopes of this kind, in fact in the general use of dichroic polarizers, it is a great convenience to have the vibration direction or sense of the transmitted light plainly indicated by a reference mark.

For detecting linear dichroism in extra-visual portions of the spectrum, C. D. West has made use of the interference bands (Müller bands) given by a thin crystal section such as quartz cut parallel to the optic axis. This section is placed in the diagonal position between a known polarizer, such as a Nicol prism, and the unknown specimen, in front of the slit of a prism or grating spectrograph; and varying exposures are made with a light source giving a continuous spectrum. Then the interference bands, which occur at approximately equal intervals on a frequency scale, are registered only in those portions of the spectrum for which the unknown specimen is dichroic.

The Savart plate in conjunction with a Nicol prism is the familiar device for detecting traces of linear polarization in a light source. A convenient and easily made equivalent device results, when a film polarizer whose vibration direction is marked is cemented to a basal section of a strongly birefringent crystal such as calcite. The appearance of isochromatic circles will reveal the presence of polarization, while the azimuth at which the black isogyres join in the center will indicate the vibration direction of the polarized component.

Measurement of Absorption Constants of Linear Dichroic Media. The absorption constants of a dichroic medium can be measured with any kind of a photometer; but in practice the spectral variation of these constants, as measured with a spectrophotometer, is of primary interest.

It is to be recognized that the measurement of the absorption constants of strongly dichroic samples such as the present polarizing layers is equivalent to the measurement of relatively high and relatively low optical densities; if done with accuracy, these are difficult tasks by any spectrophotometric methods. In contrast to the more usual practice with isotropic liquid solutions, the operator here is generally not free to choose for measurement density ranges over which a given instrument yields results of maximum accuracy.

With a non-polarizing spectrophotometer one measures k, the transmission of a single dichroic layer; k_2, the transmission of two superposed identical layers in the parallel position; and k_x, the transmission of the same two layers crossed. In the measurement of k and k_2, polarization effects in the instrument which would affect the results must be guarded against. The three transmissions should satisfy the relations

$$k_2 > k_x,$$
$$2k^2 = k_2 + k_x.$$

The constants k_2 and k_x are related to the transmissions of a single layer for each of two oppositely polarized rays, $k_{\parallel}$ and $k_{\perp}$, by the equations (when $k_x = 0$, $k_{\perp} = 2k = \sqrt{2k_2}$)

$$k_2 = \tfrac{1}{2}(k_{\parallel}{}^2 + k_{\perp}{}^2), \qquad k_{\perp} = \frac{\sqrt{2}(\sqrt{k_2 + k_x} + \sqrt{k_2 - k_x})}{2},$$

$$k_x = k_{\parallel} \cdot k_{\perp}; \qquad k_{\parallel} = \frac{\sqrt{2}(\sqrt{k_2 + k_x} - \sqrt{k_2 - k_x})}{2}.$$

Any non-polarizing spectrophotometer may be adapted to measure directly the constants of dichroic layers by inserting a fixed polarizer, preferably a calcite polarizer, in the optical system. By such means one can determine $k_{\parallel}$ and $k_{\perp}$ with only a single specimen, which obviously must be rotated through a right angle between the two measurements; k also may be measured in the diagonal position. This method is suggested for the photoelectric spectrophotometers which have in recent years become readily available, and which cover portions of the ultraviolet and infrared spectra as well as the visible.

Polarization spectrophotometers are also available for measurements over the visible spectrum. Such subjective instruments can be employed to measure independently four different constants of a dichroic specimen, namely k, $k_{\parallel}$, $k_{\perp}$, and the ratio of the last two or $R = \frac{k_{\parallel}}{k_{\perp}}$. For the measurement of the first three of these, the dichroic specimen covers only one window of the photometer, whereas in measuring R the specimen must be large enough to cover both windows. In terms of density the corresponding constants are d, $d_{\parallel}$, $d_{\perp}$, and $\Delta d = d_{\parallel} - d_{\perp}$. With such an instrument the logical practice is to measure whichever pair of these constants can be obtained with the greater accuracy, for example, Δd and $d_{\parallel}$, and calculate the other two constants therefrom. Incidentally, these four constants may be readily measured to a fair approximation for a not too strongly absorbing dichroic specimen (*e.g.*, a piece of commercial dyed cellophane) by means of an improvised photometer consisting of the dichroscope plate described above, in combination with an analyser and a divided circle (such as the stage of a polarizing microscope) for reading off the angular settings.

From a measured R the degree of polarization may be calculated as $V = \frac{1-R}{1+R}$, where V is the per cent of linear polarized light, $1 - V$ the per cent of unpolarized light in the total light transmitted.

The k, V and R constants are related to $k_{\parallel}$ and $k_{\perp}$ by the equations:

$$k_{\parallel} = k(1-V) = \frac{2kR}{1+R}, \qquad k = \tfrac{1}{2}(k_{\parallel} + k_{\perp}),$$

$$k_{\perp} = k(1+V) = \frac{2k}{1+R}; \qquad V = \frac{k_{\perp} - k_{\parallel}}{k_{\perp} + k_{\parallel}}.$$

All calculations are simplified by freeing the measured densities ($d_{\parallel}$, $d_{\perp}$, d) or transmissions from reflection losses. All the measurements in this paper have been corrected in this manner.

The effect of adding an isotropic absorber to a dichroic absorber is to decrease both the transmission k and the dichroic ratio $d_{\parallel}/d_{\perp} > 1$, while the dichroism Δd, the ratio R and the degree of polarization V characteristic of the dichroic absorber are left unchanged.

The average density $\bar{d}$ of a dichroic medium will enter into some of our calculations. It is defined as

$$\begin{aligned} \bar{d} &= \tfrac{1}{3}(d_x + d_y + d_z) \text{ for biaxial media;} \\ &= \tfrac{1}{3}(2d_{\perp} + d_{\parallel}) \quad \text{for uniaxial media;} \\ &\sim \tfrac{1}{3}d_{\parallel} \text{ when } d_{\perp} \sim 0. \end{aligned}$$

This is distinctly different from the density for natural light of a section of a dichroic medium

$$\begin{aligned} d = -\log k &= -\log \tfrac{1}{2}(k_{\parallel} + k_{\perp}) \\ &\sim \log 2 + d_{\perp} \text{ when } k_{\parallel} \sim 0. \end{aligned}$$

The spectral curves of $d_{\parallel}$, $d_{\perp}$, $\bar{d}$ and d for a strongly dichroic medium are illustrated in Fig. 7; it is seen here how the pronounced absorption band shown by the $d_{\parallel}$ and $\bar{d}$ curves vanishes completely when the data are calculated to the d curve for this medium. As has already been stated, the constants $d_{\parallel}$, d, and $d_{\perp}$ are measures of the intensity of the emergent light when a linear dichroic specimen is traversed by a beam of linear-polarized light vibrating at the respective azimuths 0, 45° and 90° to one of the vibration directions of the specimen. For a better idea of the relations between these three constants the reader is referred to Fig. 12.

Presentation of Results. In comparing a wide variety of dichroic media it is desirable to reduce measured quantities to a uniform system of constants. In the past the practice has been almost unanimous to report values of $d_{\parallel}$ and $d_{\perp}$, or absorption constants differing from these by a simple factor. But in describing the current dichroic polarizers one notes a tendency to give other constants, for example k_2 and k_z (Strong, 1936) [3] or k and V (Grabau, 1937).[4] This last system is doubtless quite satisfactory from the point of view of the person who expects to use a dichroic preparation as a polarizer; but for the purposes of the present treatment we prefer to retain the older $d_{\parallel}$, $d_{\perp}$ system. One advantage here is that not only the dichroism, Δd, but also the dichroic ratio $\frac{d_{\parallel}}{d_{\perp}}$ is easily obtained in this system; and this ratio is seen to be, for a given absorbing agent, a constant invariant with respect to changes in concentration and thickness, and thus it serves as a convenient figure of merit, other things being equal. This ratio was early employed in reporting quantitative measurements of dichroism by Pulfrich (1882),[5] and it was more recently resurrected by Preston (1931).[6]

Example of Linear Dichroism. Flow Dichroism of a Hydrosol. Previous measurements of flow dichroism and flow birefringence of colloidal solutions of dyes and of vanadium pentoxide in water were given in a valuable paper on linear dichroism of Zocher and Jacoby (1927).[7] These measurements give only the difference of the absorption, Δd, but not the value of either $d_{\parallel}$ or $d_{\perp}$ for the systems studied.

West has found that the strong blue color resulting from mixing extremely dilute solutions of α-naphthoflavone and iodine, first observed by Barger and Starling (1915) [8] and lately employed as an iodometric indicator, is characterized by a remarkable flow dichroism and flow birefringence not previously reported. This flow dichroism is positive, which means that the component of the incident light vibrating parallel to the streamlines is the more strongly absorbed; but the flow birefringence is negative (the slow ray is transverse to the streamlines), as would be expected if it were governed by the dichroic absorption band in the far red.

A glass cell was constructed in which the solution could be made to flow with a slow nearly uniform motion about a closed path between stationary walls. Measurements were made of $d_{\parallel}$ and $d_{\perp}$ for the liquid in motion, and of $\bar{d}$ for the stationary system, by means of a Gaertner polarization spectrophotometer for the visual spectrum.

The accompanying curves (Fig. 2) show the course of the measured $\bar{d}$, $d_{\parallel}$, and $d_{\perp}$ curves; and the $\Delta d = d_{\parallel} - d_{\perp}$ and $\bar{d}$ calc. $= \frac{1}{3}(d_{\parallel} + 2d_{\perp})$ curves calculated therefrom. It is seen that the calculated $\bar{d}$ points fall on the measured $\bar{d}$ curve to a good approximation; they would be expected to coincide exactly for a system in uniform uniaxial flow, a condition somewhat difficult to realize experimentally.

Dichroic Polarizers: Structure and Properties

The following list is representative of recent patents disclosing dichroic polarizers or processes of preparing them.*

Linear dichroism is a general property of absorbing anisotropic matter. It is not difficult to state the characteristics of dichroic materials potentially suitable for preparation of dichroic polarizers. Such materials should have:

(1) An absorption band covering the portion of the spectrum it is desired to polarize, ordinarily the visible spectrum 400 — 700 mμ.

(2) An adequately high dichroism throughout this band.

(3) Stability and permanence under the conditions to which it will be exposed in use—or generally, exposure to heat, to visible and ultraviolet radiation, to humidity, and to mechanical stresses.

(4) Freedom from scattering or diffusion of the transmitted component.

Under (2), we might arbitrarily take a dichroic ratio of about 7 as the threshold

* While this paper is intended to represent the state of the art as of January, 1943, this list includes patents up to June, 1944.

E. D. Bailey and M. M. Brubaker. U. S. 2,246,087 (1941).
R. P. Blake. U. S. 2,256,108 (1941).
C. H. Brown. U. S. 2,224,214 (1940); 2,287,598 (1942).
E. Käsemann. U. S. 2, 236,972 (1941).
L. A. Keim. U. S. 2,274,706 (1942); 2,340,476 (1944).
E. H. Land. U. S. 1,918,848 (1933) (Land and Friedman); 1,951,664 (1934); 1,956,867 (1934); 1,989,371 (1935); 2,011,553 (1935); 2,041,138 (1936); 2,078,254 (1937); 2,165,973 (1939); 2,173,304 (1939) (Land and Rogers); 2,178,996 (1939); 2,237,567 (1941); 2,281,100 (1942); 2,289,712 (1942) (Land and West); 2,289,713 (1942); 2,289,-714 (1942); 2,298,058 (1942); 2,306,108 (1942) (Land and Rogers); 2,319,816 (1943); 2,328,219 (1943); 2,343,775 (1944); 2,346,766 (1944).
H. Lapp. German 674,840 (1939).
A. M. Marks. U. S. 2,104,949 (1938); 2,167,899 (1939); 2,199,227 (1940); 2,344,514 (1944).
K. Meyer and others. German 675,217 (1939); 681,237 (1939); 681,347 (1940).
L. Pollack. U. S. 2,286,569 (1942); 2,346,784 (1944).
H. G. Rogers. U. S. 2,255,940 (1941); 2,284,590 (1942).
O. Vierling and P. G. Gänswein. U. S. 2,344,117 (1944).
G. Wilmanns and W. Schneider. U. S. 2,176,516 (1939).
H. Zocher. U. S. 1,873,951 (1932).

of usefulness. A polarizer with $k_{\|} = 0.01$, $d_{\|} = 2.0$; $k_{\perp} = 0.50$, $d_{\perp} = 0.30$; and $k = 0.25$, $d = 0.61$, $V = 96.1$ per cent, would approximate this figure. It is of interest that colored tourmalines so far reported have a dichroic ratio of this order of magnitude, irrespective of their color. An ideal polarizer would have, for comparison, $k_{\|} = 0$, $d_{\|} = \infty$; $k_{\perp} = 1.00$, $d_{\perp} = 0$; $k = 0.50$, $d = 0.30$, $V = 100$ per cent, giving an infinite dichroic ratio.

Linear dichroic polarizers are conveniently treated in three classes according to structure: macrocrystalline, microcrystalline, and those associated with oriented linear high polymers. In the following, dichroic specimens of these classes will be illustrated by $d_{\|}$, $d_{\perp}$ curves, all taken in these laboratories by means of the Gaertner spectrophotometer (Martens type) previously mentioned, and all corrected for reflection losses (Figs. 2-18); these are chosen to show the nature of the absorption bands

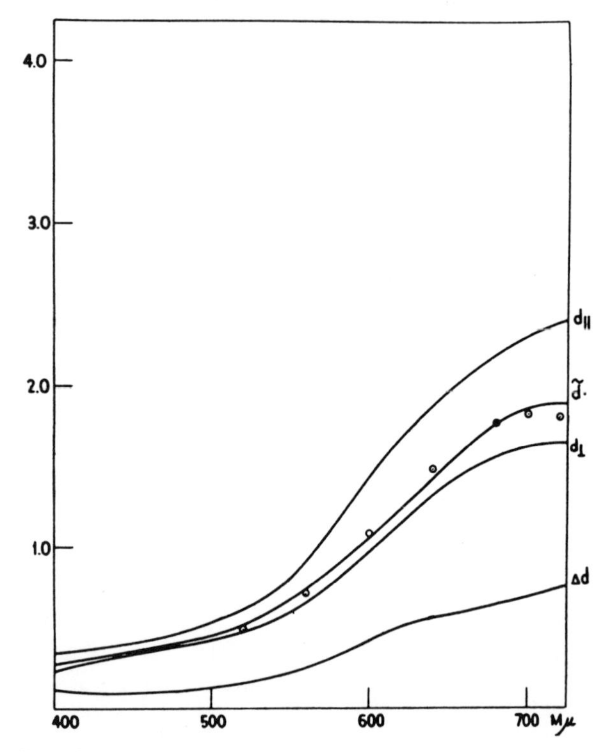

FIGURE 2. Flow dichroism of α-naphthoflavone product with iodine. The curve of average density $\bar{d}$ is for the fluid at rest, the circles are values of $\bar{d}$ calculated from $d_{\|}$ and $d_{\perp}$ for the fluid in motion. Cell thickness 2 cm.

in question rather than to be representative of the commercial polarizers which are or may be prepared from these materials.

In considering these curves it is to be kept in mind that optical densities greater than 3 are in general subject to increasingly large errors of measurement as the density increases above 3.

We wish to mention, among other members of the research laboratory of Polaroid Corporation, especially W. F. Amon, Jr., who prepared many photometric samples, and A. S. Makas, who took many spectrophotometric curves, which were the necessary preliminary to the discussion of the following section.

Macrocrystalline Polarizers. These are prepared by cutting, or by growing, sections of dichroic crystals of suitable orientation and thickness to give useful polarizers. When the section is thin or mechanically weak, it is cemented between transparent protecting covers such as glass plates.

Tourmaline Polarizers. This relatively abundant silicate mineral crystallizes in the trigonal system, and when colored always shows stronger absorption of the ordinary ray (negative dichroism). As with all uniaxial crystals, the maximum dichroism is exhibited by sections parallel to the optic axis, whereas sections perpendicular to the axis are isotropic. In the past the absorption constants $K_{\parallel}$ and $K_{\perp}$ have been repeatedly measured through the visible spectrum for specimens of various colors, and while they are found to vary quite widely, the dichroic ratio is substantially constant. This suggests that the absorbing agent is of the same nature in all tourmalines, but is present in varying concentrations. No explanation of the dichroism has yet been undertaken; it is recognized only that colored tourmalines generally contain cations of the transition metals, for example manganese in rose tourmaline and iron in the brown or black varieties. The dichroism of a black tourmaline is shown in the accompanying density curves (Fig. 3).

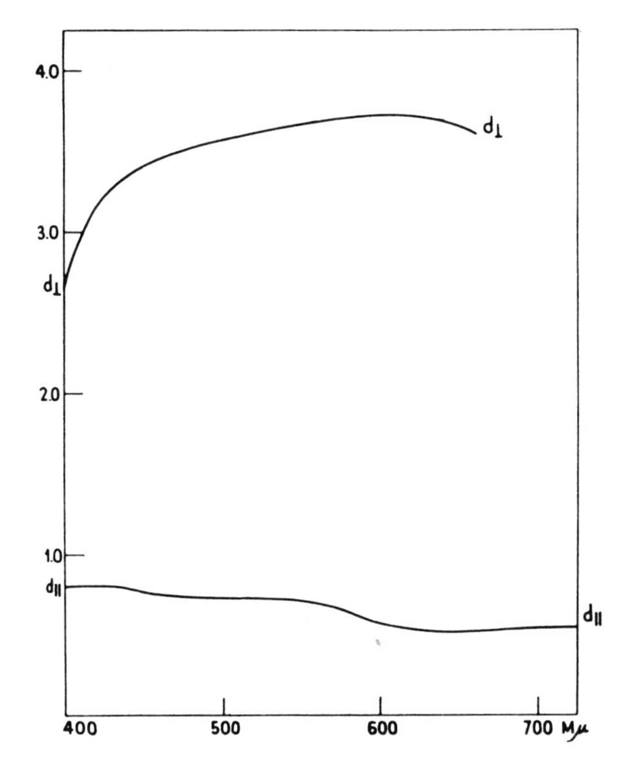

FIGURE 3. Dichroism of black tourmaline, section about 0.2 mm thick, cut 24° from parallelism with the optic axis, for light traveling perpendicular to the optic axis.

A mineral which, when colored, has a very strong uniaxial positive dichroism colorless-deep blue, but whose absorption constants have not been measured, is the rare silicate, benitoite, $BaTiSi_3O_9$ or $BaO{\cdot}TiO_2{\cdot}3\ SiO_2$. It is interesting to speculate whether the absorption here might be due to the presence of another transition metal, trivalent titanium, which is known in water solutions to have a narrow absorption band with a peak at 744 mμ.

Quinine Herapathite Polarizers. This quinine sulfate tri-iodide crystallizes in thin pinacoidal plates belonging to the orthorhombic system, which absorb one vibration in their plane and transmit the other two vibrations (positive dichroism). Its formula, according to Jörgensen (1876),[9] is $4C_{20}H_{24}N_2O_2 \cdot 3H_2SO_4 \cdot 2HI \cdot 2I_2 \cdot xH_2O$, and it has been observed in these laboratories to begin to melt at about 165°. The crystallography of this material has been described by West (1937).[10] The absorption constants $K_{\parallel}$ and $K_{\perp}$ of this crystal are not yet accurately known, since $d_{\parallel}$ and $d_{\perp}$ have not been measured for specimens of known thickness; however, a thickness of about 0.0002 inch (5 microns) is probably enough to give a good polarizer. Plates that are too thin have red extinction, while those that are too thick have red-brown transmission when viewed against a white light source. The absorption curves for this crystal are illustrated by the data for a Zeiss Herotar filter (Fig. 4).

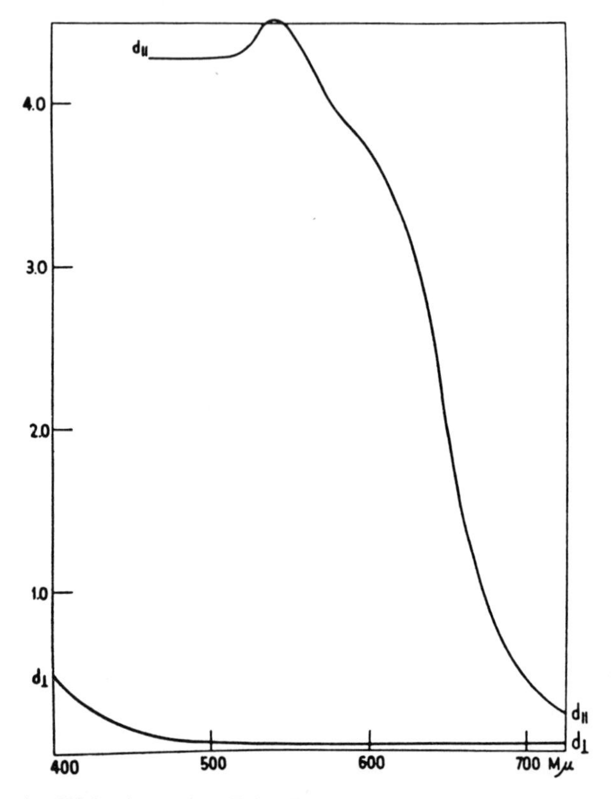

FIGURE 4. Dichroism of a Zeiss Herotar filter (quinine herapathite).

Cinchonidine Herapathite Polarizers. The crystal habit and optical properties of cinchonidine herapathite are similar to but not identical with those of quinine herapathite; it grows in orthorhombic pinacoidal plates, which begin to melt at about 154°. The formula given by Jorgensen (1876),[11] $12C_{20}H_{24}N_2O \cdot 9H_2SO_4 \cdot 8HI \cdot 12I_2 \cdot 8H_2O$, requires revision, since the formula now accepted for cinchonidine is $C_{19}H_{22}N_2O$. This crystal is superior to herapathite in that its positive dichroic absorption band persists into the near infrared, while that of herapathite cuts off in the visible far red; but as in the preceding example the absorption constants have not yet been published for this crystal. Presumably they will be of the same order as the constants of herapathite. Measurements on a Marks polarizing plate may be taken as representative of the dichroism of this crystal (Fig. 5).

Microcrystalline Polarizers. Polarizers of this type, invented by Land, are made by first preparing a suspension of dichroic crystal needles in a viscous medium, and then subjecting the suspension to a uniform flow process which will yield a sheet containing the crystal needles all oriented parallel to a common direction, namely the direction of the streamlines in the flow process. After the flow process the sheet is hardened, as by the escape of the volatile solvent used in preparing the viscous medium. Practically the process differs from the example of flow dichroism described in a previous section in respect to the higher concentrations, viscosities, and velocity gradients employed. Suitable crystal needles may have one of two main kinds of dichroism: they may absorb light vibrating parallel to the needle length, and transmit the two light vibrations perpendicular to this direction (positive dichroism), or they may transmit the parallel vibrations and absorb the two perpendicular vibrations (negative dichroism); a third possibility would be offered by a crystal needle which transmitted the parallel vibration and absorbed only one of the two perpendicular vibrations. In practice, only crystal needles of the first class, those characterized by positive dichroism, have so far been employed. If one starts with a strongly dichroic crystal species, the quality of the resulting polarizer depends chiefly on the perfection of the orientation attained, and this is conditioned by two factors: first, all the crystals in the suspension should have the correct shape and should fall in a certain size range; second, the flow conditions should be correctly chosen to set them all in parallelism. There are further obvious precautions to be observed in making these polarizers: the viscous medium should not have an appreciable solvent power for the crystal; temperatures must be avoided which melt, decompose or dissolve the crystals; and mechanical stresses must be avoided which break the crystals.

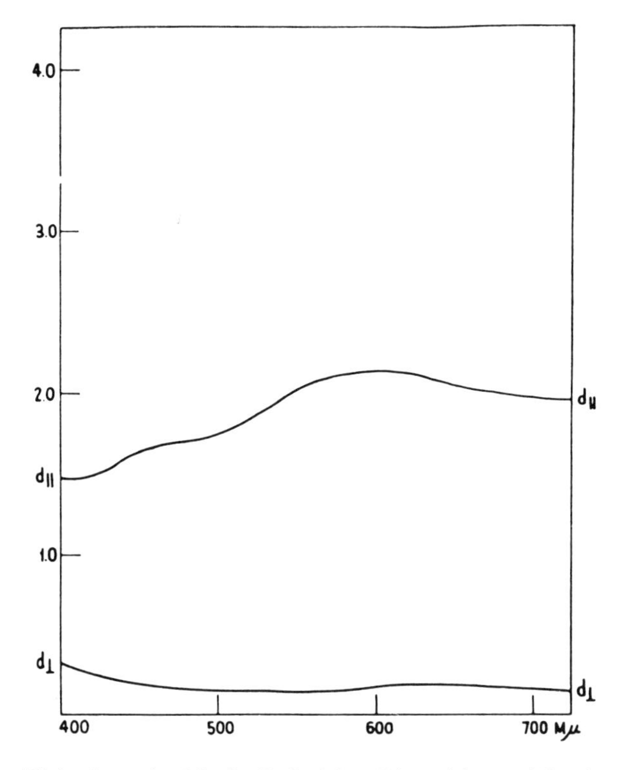

FIGURE 5. Dichroism of a Marks Polarizing Plate (cinchonidine herapathite).

West has observed that a section of Polaroid J-sheet cut across the needle axes is found to be not isotropic but weakly dichroic, which indicates a departure of its optical properties from strict uniaxiality. If we denote the film normal by X, the direction of the needle axes in the plane of the film by Z, and the transverse direction in the plane of the film by Y, then the absorptions are in the order $Z \gg Y > X$. Consistent with this is the observation that when a film is immersed in mineral oil and tilted about *Y*, two positions (corresponding to the optic axes of a biaxial dichroic crystal) are found where the dichroism vanishes. The angle between these two positions measured in the *XZ* plane over *X* might be termed the "monochroic angle." The simplest interpretation is that the long axes of the needles have a spatial distribution such that they are more nearly parallel to the plane of the film (*YZ* plane) than they are to the *XZ* plane.

Dichroic polarizers of this class are often characterized by a slight scattering of the transmitted light; this scattered light is far from having a uniform distribution, rather it reaches a maximum in what we have just called the *XY* plane. This subject was studied by Farwell (1938).[12] While this sometimes undesirable property would be minimized by having the refractive index of the medium as close as possible to the refractive index of the crystals, practically it is found easier to control it by keeping the largest crystals below a limiting size. It would be even more effectively eliminated, as Land has suggested, by using negative dichroic crystals if such were available.

The conclusion from experience so far is that microcrystalline polarizers are technically easier to produce than macrocrystalline polarizers, over any useful range of areas.

Useful dichroic polarizers have been made by this process from both quinine and cinchonidine herapathites. The preparation of quinine herapathite suspensions for this purpose was described in the Land patents and in a recent paper by Godina and Faerman (1941).[13] The commercial polarizer known as Polaroid J-sheet is of this type and consists of oriented quinine herapathite needles in a cellulose acetate film. It is of interest that overheating under suitable conditions will shift the dichroic absorption band of this film towards the blue. The effect of this shift is to change the extinction color of crossed films from purple to red, when viewed against a white light source. Absorption curves for a dichroic film made by the Polaroid J-sheet process before and after such heating are shown in the accompanying figure (Fig. 6).

Data are also given for a dichroic film prepared from microcrystalline cinchonidine herapathite (Fig. 7). A general similarity of Figs. 7 and 5 and likewise of Figs. 6 and 4, is to be observed.

A third microcrystalline dichroic polarizer was developed by Polaroid Corporation in 1942 from an acid periodide of 3, 4, 5, 6 dibenzacridin. The somewhat lower dichroism of this polarizer (see Fig. 8) as compared to the two preceding ones, is to some extent offset by the higher melting point of this crystal, which is in the neighborhood of 200°. The colored sorption compound of dibenzacridin with iodine was described in a paper by Kermack, Slater and Spragg (1930).[14]

Dichroic Polarizers Associated with Oriented Linear High Polymers. Linear high polymeric materials initially in an isotropic state are quite generally capable of being put into a uniformly oriented state by subjection to a relatively large uniform tensile strain or deformation under suitable conditions. The deformation is generally at least partly recoverable, likewise under suitable conditions. The process of orienting a linear high polymer may be formulated as a rather complex case of combined elastic and viscous deformation, the resistance to which is quite large; much larger, apparently than the viscous resistance encountered in the manufacture of the microcrystalline polarizing films treated in the preceding section.

Evidences of the oriented state of a linear high polymer are the anisotropy of mechanical properties (increase of modulus and tensile strength in the direction of the original tension); the anisotropy of optical properties (birefringence which cannot be explained in terms of internal stresses, uniaxial about the direction of the original tension); and frequently but not always the property of giving an x-ray fiber diffraction pattern, when the primary beam travels perpendicular to the direction of the original tension. Some natural fibers such as ramie (cellulose) and silk

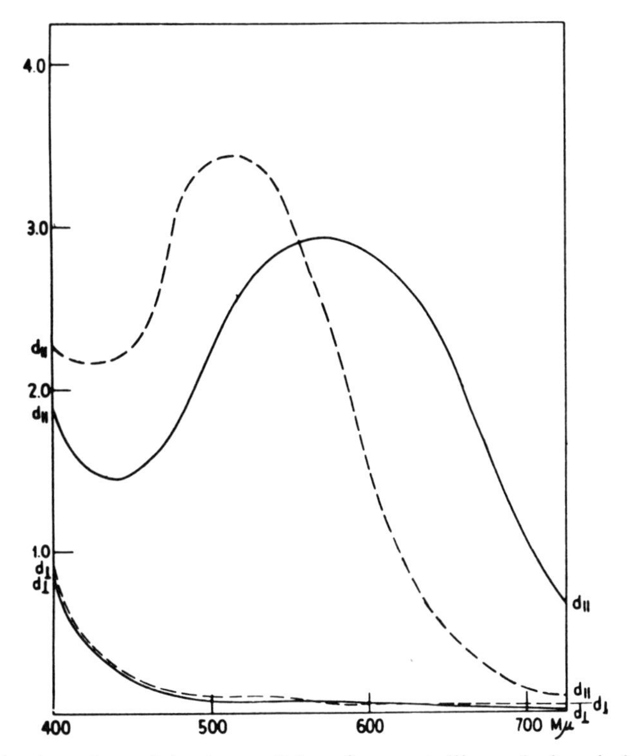

FIGURE 6. Dichroism of a quinine herapathite microcrystalline polarizer before (full lines) and after (broken lines) heat treatment.

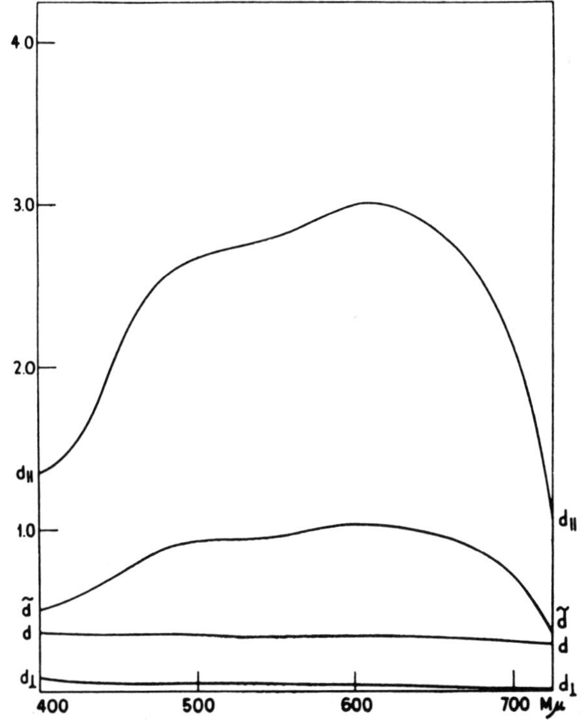

FIGURE 7. Dichroism of a cinchonidine herapathite microcrystalline polarizer. Curves of average density $\tilde{d}$ and density for natural light d are also shown.

represent quite perfectly oriented states of linear high polymers, but these are not very suitable for preparing dichroic polarizers.

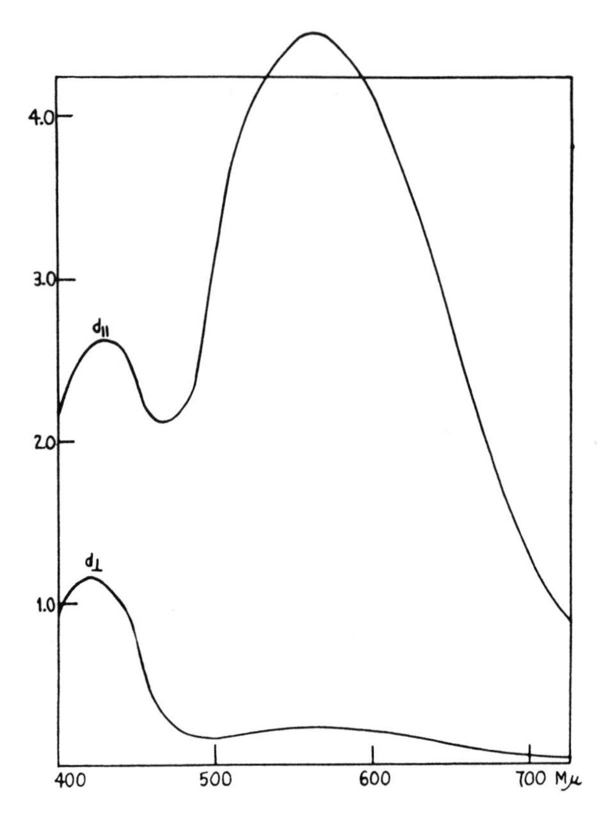

FIGURE 8. Dichroism of a dibenzacridine periodide polarizer.

A polymer which Land has found to be well suited for preparing the present class of dichroic polarizers is the synthetic product polyvinyl alcohol $(-CH_2-CHOH-)_x$, manufactured commercially by hydrolysis of polyvinyl acetate $(-CH_2-CH(OOCCH_3)-)_x$. This polymer, which has an extensibility of the same order as that of rubber, is formed into sheets which are oriented by uniform tensile deformations. The fully oriented sheet has a definitely fibrous structure and is characterized by high tensile strength in the fiber direction, of the order of 10^5 psi; by uniaxial positive birefringence ($\varepsilon = 1.560$, $\omega = 1.526$, $\varepsilon - \omega = 0.034$ are representative saturation values); and by a sharp x-ray fiber diffraction pattern. The atomic structure of the oriented polyvinyl alcohol fiber as worked out by x-ray methods by Mooney (1941)[15] has a plane zigzag of carbon atoms (*trans, trans, . . .*) with hydroxyls attached to alternate carbons (1, 3 diol groups) and all lying on the same side of the plane of the carbon chain. As is true of cellulose films, the transparency of polyvinyl alcohol films extends over considerable ranges into the ultraviolet and infrared. The weak ultraviolet absorption band reported for polyvinyl alcohol at 280 mμ by Marvel and Denoon (1938)[16] does not exist, according to our measurements.

As with the microcrystalline polarizers, all the useful dichroic polarizers so far made from oriented high polymers are characterized by positive dichroism.

Dye Polarizers. Preston (1933)[17] found that dichroic products resulted from the application of direct dyes to films or fibers of oriented cellulose, and published quantitative data on the dichroism; he also referred to earlier observations of this effect.

Other dichroic polarizers can also be made from oriented polyvinyl alcohol films, and these are found to have useful dichroic ratios.

An outstanding characteristic of such preparations is that the light absorption by each dye molecule (more strictly, ion) does not change very much, whether it is measured in solution, in the unoriented film, or in the oriented film. This observation suggests that the dye goes onto the oriented film in single molecules, and that the dichroism of the film generally may be taken as a direct measure of the dichroism of the individual dye molecules. In making this comparison, one measures the e of the solution in the customary manner for comparison with the average extinction coefficient $\tilde{e}$ of the dichroic film, as we have already done in the example of flow dichroism. This line of investigation is a relatively easy one, first because it is possible to cast the dye directly into the polyvinyl alcohol film by the aid of mutual solvents; secondly, because it is possible to calculate the e's of the dye in the film from the measured d's through the area concentration of the dye, taken as the product of the weight per cent of the dye in the film times the area density of the film: $c' =$ (g dye per g of dyed film) $\times$ (g dyed film per 1000 cm^2); then $\tilde{e} = \bar{d}/c' = \frac{1}{3c'}(2d_{\perp} + d_{\parallel})$. In this way the labor of a chemical analysis becomes unnecessary.

Another indication that dichroic direct dyes on oriented high polymers are present as single ions, and not as polymeric forms, is afforded by the observation that x-ray diffraction photographs of such preparations show only amorphous scattering even when the dye is present in high concentrations. Our unpublished data confirm Valko (1941)[18] on this point.

Figs. 9 and 10 illustrate the constancy of the absorption by the Congo Red anion; curves are also given for dichroic preparations of both the red and the blue (acid) forms of this dye.

The curves for Congo Red are representative of the generality of direct dyes: the absorption bands of single dyes are too narrow to cover the entire visible spectrum satisfactorily at once. For this purpose one may combine two or more dyes in separate parallel films, or in the same oriented film (Fig. 11). Also it is seen that

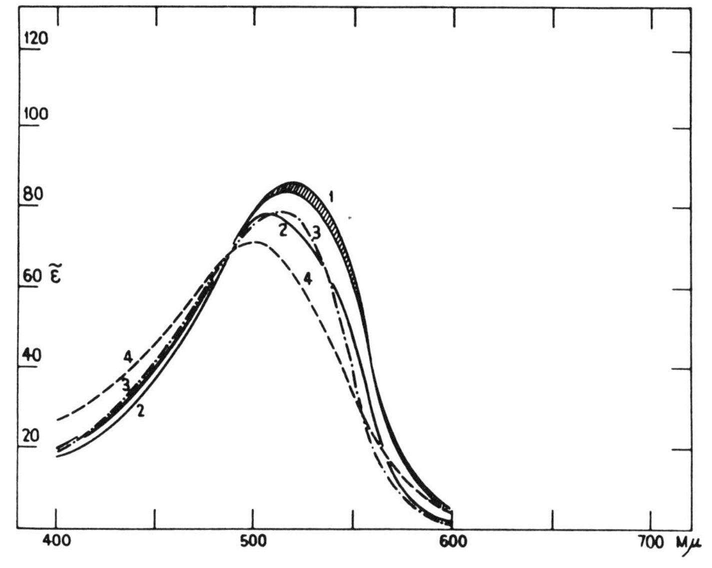

FIGURE 9. Absorption curves of Congo Red in varying environments. (1) Two dichroic preparations on polyvinyl alcohol, $c' = 0.00566$ and 0.01187 respectively. (2) Isotropic preparation on unoriented polyvinyl alcohol, $c' = 0.01155$. (3) Methyl "Cellosolve" solution. (4) Water solution. For (3) and (4), $c = 0.0174$, thickness $= 1$ cm.

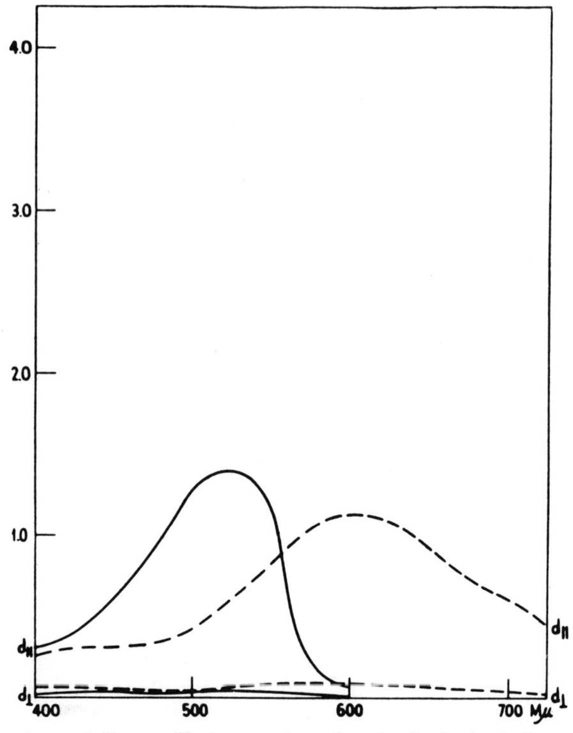

FIGURE 10. Dichroism of Congo Red on oriented polyvinyl alcohol, red (full lines) and blue (dotted lines) forms.

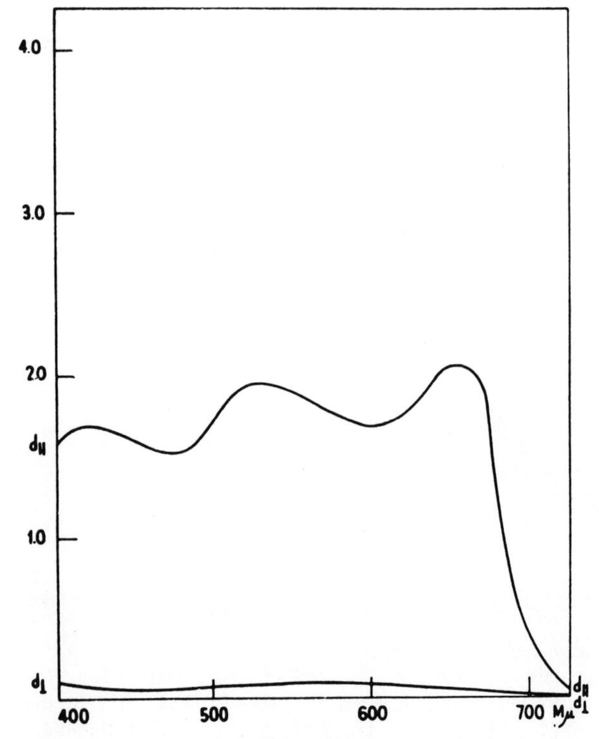

FIGURE 11. Dichroism of a mixture of dyes on oriented polyvinyl alcohol: yellow (Color Index No. 622), red (C. I. 278), and blue (C. I. 518).

quite striking effects are observed when two dichroic films, made with dyes whose absorption bands do not overlap, are crossed and examined with a dichroscope. Absorption curves of such a combination are shown in Fig. 12. When the two dyes in

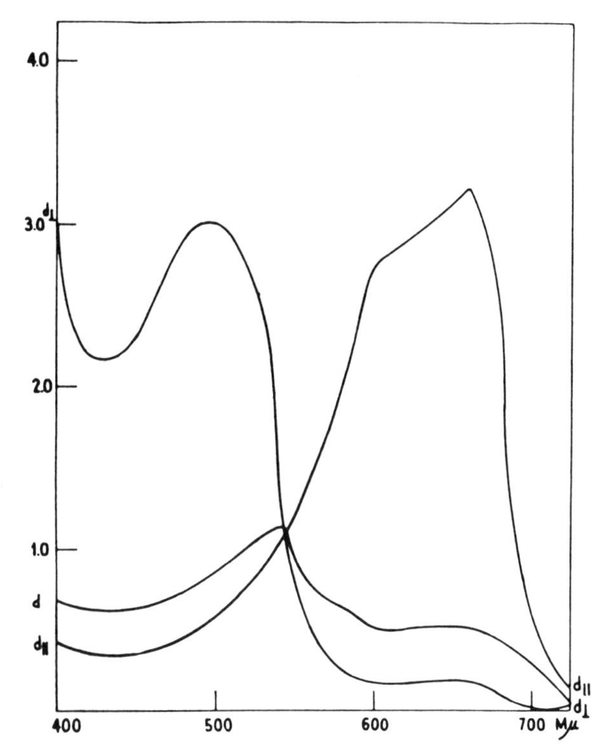

FIGURE 12. Dichroism of crossed dyed films of polyvinyl alcohol, orange (C. I. 374) and blue (C. I. 518).

a crossed combination of this kind are complementary, such combinations when taken in pairs have the interesting property of extinguishing white light when rotated to the crossed position.

It should not be inferred from the foregoing that only direct dyes give useful dichroism on oriented polyvinyl alcohol. The basic dye methylene blue, in which the dye ion is a cation, is an example of a dye with a dichroic ratio of about ten on the oriented polymer.

For dyes having disc-shaped molecules one would expect to find a maximum dichroic ratio not greater than 2, namely $\frac{d_\parallel}{d_\perp} = \frac{e_x}{\frac{1}{2}(e_x + e_z)}$, which equals 2 when $e_z \sim 0$ (z denotes the normal to the molecule plane, x a direction in the molecule plane). This is confirmed by observations of such dye cations and anions as crystal violet, aurin, and picric acid on oriented polyvinyl alcohol.

From the complete absence of dichroism of other dyes on oriented polyvinyl alcohol, one concludes that their structures are substantially spherically symmetrical. An example is the coördination compound of trivalent iron, Naphthol Green B.

From evidence of this sort one is justified in drawing conclusions concerning proposed structural formulas for dye molecules. The stilbene dye Color Index 620 has a dichroic ratio of about 15; therefore, the ring structure assigned to it seems highly improbable. Similarly, the observation that aniline black applied to oriented polyvinyl alcohol in a variety of ways always exhibits negligible dichroism apparently precludes the beautifully linear structures ascribed to this molecule in its various stages of oxidation.

The present method also has some value in helping to identify the several absorption maxima of a dye molecule with geometrical directions or axes in the molecule. Our experience has been that when a given molecule has positive dichroism for its strongest absorption maximum in the visible, then a second maximum will generally exhibit dichroism of the same sign. For example, when polyvinyl alcohol sheet containing ¼ per cent by weight of the much studied basic dye 1, 1′-diethyl-2, 2′-cyanine chloride is stretched and its dichroism is measured, twin maxima having the same dichroism (dichroic ratio about 5) and the same sign are found at 527 and 495 mμ; from which we conclude that both these maxima are identified with electronic transitions along the long direction of the planar molecule. Such a conclusion is at variance with the views of Scheibe, and of Lewis and Calvin, who ascribe the 495 maximum to an electronic transition transverse to the long direction of the molecule and in its plane.

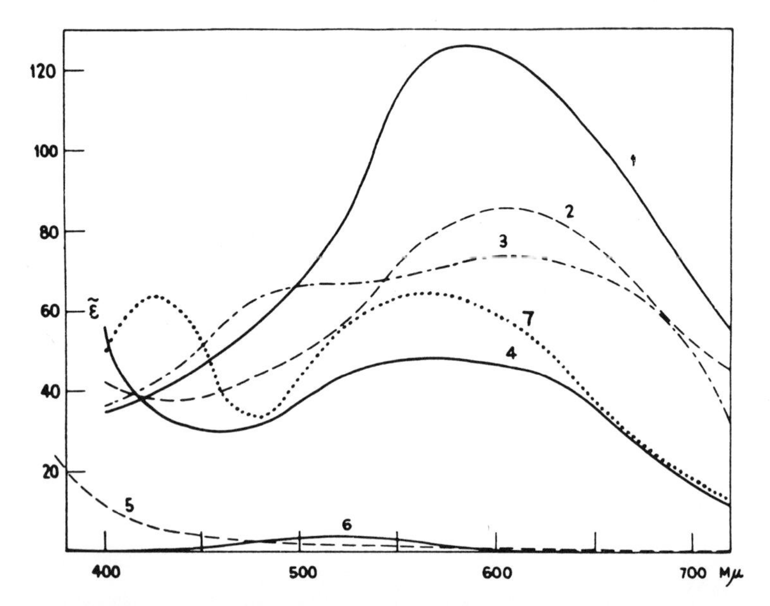

FIGURE 13. Absorption curves of iodine in varying environments. (1) Starch-iodine (2.54 mg I, 50 mg starch per 100 cc water; $c' = 0.0254$). (2) Iodine in preparation of Fig. 14, $c' = 0.012$ by titration. (3) Iodine in preparation of Fig. 7, $c' = 0.014$ calc. (4) Iodine in preparation of Fig. 6, $c' = 0.02115$ calc. (5) Iodine in aqueous KI, after Brode, *J. Am. Chem. Soc.*, **48**, 1877 (1926). (6) Iodine in hexane, after Groh, *Z. physik. Chem.*, **149**, 153 (1930). (7) Iodine in preparation of Fig. 8, $c' = 0.02625$ calc.

Iodine Polarizers. The ability of iodine to give dichroic dark colors on oriented cellulose has been known for some time. That a water solution of polyvinyl alcohol gives a blue coloration with iodine, when the concentrations are relatively high as compared for example with the starch-iodine system, was shown by Staudinger (1927)[19] and by Gallay (1936).[20] Land has prepared films of extraordinarily high dichroism from iodine and the oriented polymers of vinyl alcohol and vinyl butyral; and this reaction seems to be a fairly general property of oriented polyvinyl oxy-compounds, represented by the formula $(-CH_2-CHOR-)_x$, where OR is taken to indicate hydroxyl, acetal, ketal, ether or ester groups, etc., or mixed polymers thereof.

These colored products differ distinctly from the foregoing products made with dye molecules. For all simple iodine solutions, whether brown or violet, are characterized by relatively weak absorption in the visible spectrum, and especially in the red, whereas iodine in association with water solutions of suitable polymeric materials gives very intense absorption bands in the visible, as for example the well-known blue color with starch solution or the red color with dextrine solution. The state of the iodine in these water solutions is evidently similar to the state of the iodine in the dichroic preparations from oriented films of cellulose or polyvinyl alcohol, but must be radically different from the state of the iodine in the brown or violet solutions in simple solvents, as shown by an examination of the quantitative absorption curves (Fig. 13).

A vast amount of work has been done and published in the past with the object of elucidating the structure of this strongly absorbing state of iodine, which is called forth by the presence of suitable high polymers and water; but even the recent activity in this field has not given the final word.

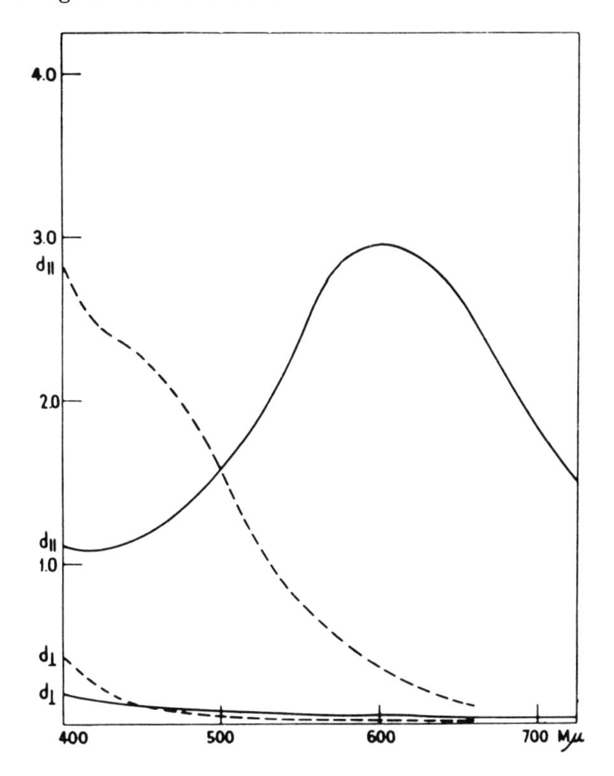

FIGURE 14. Dichroism of iodine on oriented polyvinyl alcohol, blue (full lines) and brown (broken lines) forms.

West has obtained definite evidence, not yet published, from the x-ray diffraction study of dichroic films of iodine on oriented polyvinyl alcohol that the iodine is present in a *polymeric* form, namely as independent long strings of iodine atoms all lying parallel to the fiber axis, with a periodicity in this direction of about 3.10Å. A slightly smaller interatomic distance is known for the tri-iodide ion in crystals.

The commercial polarizer known as Polaroid H-sheet is of this type.

When oriented polyvinyl alcohol film is stained with iodine in the absence of water, a dichroic brown product results. An x-ray photograph of one such preparation indicates the presence of polymeric iodine here also.

Absorption curves of brown and blue dichroic preparations of iodine on oriented polyvinyl alcohol are shown in Fig. 14.

Polyvinylene Polarizers. Land and Rogers found that when an oriented transparent film of polyvinyl alcohol is heated in the presence of an active dehydration catalyst such as HCl, the film darkens slightly and acquires a strong positive dirchroism. It is also observed that the darkened film, which has undergone only a slight decrease in weight of the order of 1 per cent in the process, can still be strongly swelled but no longer dissolved in hot water.

The chemistry of this change in light absorption obviously involves a partial dehydration of the transparent polyol $(-CH_2-CHOH-)_x$ to form an unsaturated compound or polyene $(-CH = CH-)_x$. The polyene must be pictured as a conjugated polyene, with the conjugation extending through long chains, in order to provide an acceptable explanation for the absorption of visible light. To account for the insolubility of the product, it seems quite probable that some dehydration takes place between chains; this would have the effect both of interrupting the conjugation and of forming a three-dimensional structure from the original linear polymer.

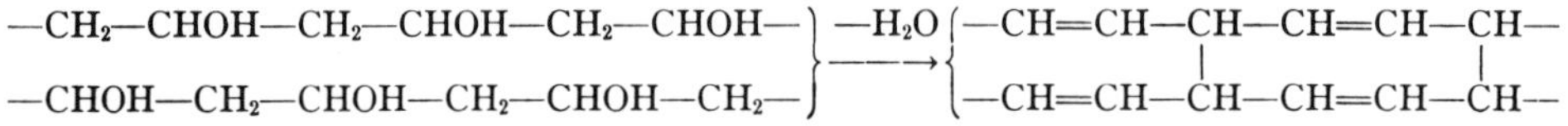

$$\left.\begin{array}{l}-CH_2-CHOH-CH_2-CHOH-CH_2-CHOH- \\ -CHOH-CH_2-CHOH-CH_2-CHOH-CH_2-\end{array}\right\} \xrightarrow{-H_2O} \left\{\begin{array}{l}-CH{=}CH-CH-CH{=}CH-CH- \\ \qquad\qquad\quad | \qquad\qquad\qquad\quad | \\ -CH{=}CH-CH-CH{=}CH-CH-\end{array}\right.$$

There might also be some oxygen bridge formation by elimination of water from hydroxyl pairs in adjacent chains.

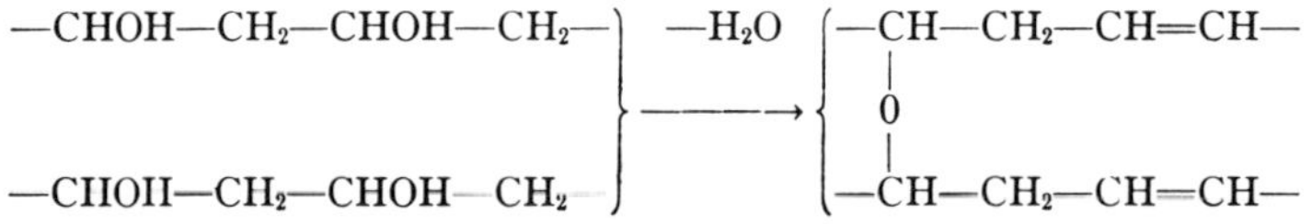

$$\left.\begin{array}{l}-CHOH-CH_2-CHOH-CH_2- \\ \\ \\ -CHOH-CH_2-CHOH-CH_2\end{array}\right\} \xrightarrow{-H_2O} \left\{\begin{array}{l}-CH-CH_2-CH{=}CH- \\ \quad | \\ \quad O \\ \quad | \\ -CH-CH_2-CH{=}CH-\end{array}\right.$$

As required by this conjugated polyene structure, which was first proposed by West, the product is found to be bleached by common oxidizing agents with relative ease, but it is very stable indeed when oxidizing agents are excluded.

A quantitative determination of the number of double bonds in a photometered specimen of this polarizer would allow a comparison with the known absorption of light by the low molecular polyenes, a subject extensively studied by Kuhn and his associates (1935,[21] 1938 [22]). Qualitatively the foregoing picture is consistent with both Kuhn's experimental work and Mulliken's theoretical treatment of this subject. We quote from a summary by the latter (1939) : [23] "The longest wave length electronic transition of appreciable intensity in absorption (in a conjugated polyene $CH_2(=CH-CH=)_nCH_2$) is then $N \rightarrow V_1$. This shifts to longer wave lengths with increasing *n*. The calculations showed that if the polyene chain has the most elongated possible form (*trans, trans, trans,* . . .), the $N \rightarrow V_1$ transition should be far stronger than all the other $N \rightarrow V$ transitions together, and its strength should increase with the length of the chain. This calculated result is paralleled by experimental data on the polyenes, especially the carotinoids. The calculations showed that for less elongated (more *cis*-like) molecules, the intensity of $N \rightarrow V_1$ should be reduced in favor of $N \rightarrow V_2$, $N \rightarrow V_3$, and so on. A further point of interest was that the $N \rightarrow V_1$ transition in molecules of the most elongated form should be polarized approximately along the long axis of the molecules."

When oriented polyvinyl alcohol films are heated in the absence of active dehydration catalysts, dichroic yellow, orange or brown products result; the structure of these products has not been more closely investigated. The absorption curves for such a preparation are also shown in Fig. 15.

To give a better picture of the spatial relationships involved in the foregoing linear polymer polarizers, the approximate atomic dimensions of polyvinyl alcohol, of the Congo Red molecule and the iodine and polyvinylene polymers, are shown in the accompanying figures (Fig. 16).

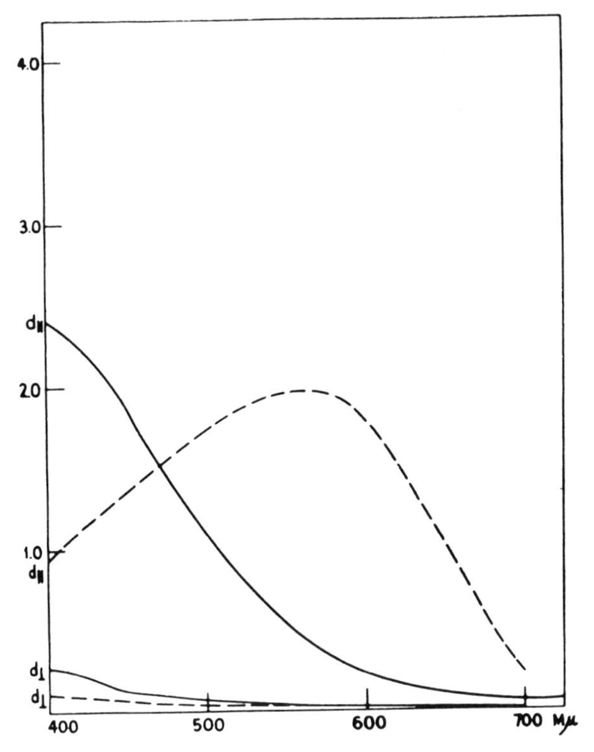

FIGURE 15. Dichroism of oriented polyvinylene (dotted lines) and of yellow product from polyvinyl alcohol (full lines).

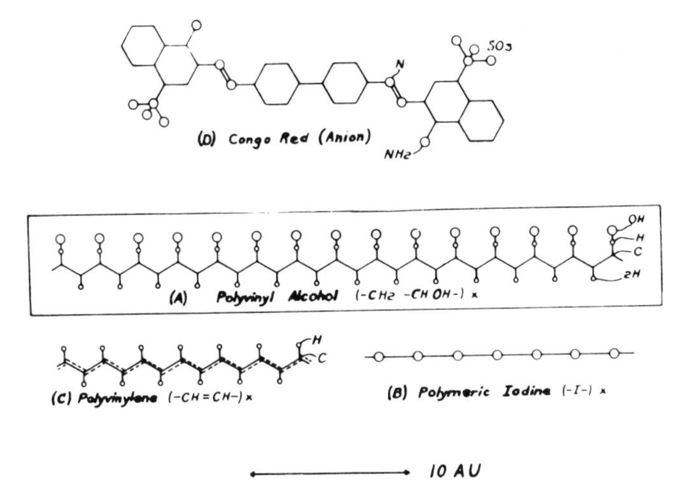

FIGURE 16. Schematic drawings, approximately to scale, of polyvinyl alcohol, Congo Red, polyiodine and polyvinylene molecules.

Metallic Polarizers. When metallic salts are incorporated in oriented linear high polymers and are subsequently reduced to the metals, the product is often characterized by a strong dichroism. The preparation and the structure investigation of such products has been described at some length by Frey-Wyssling (1937),[24] although the writers are not aware that quantitative measurements of dichroism have yet been published. For example, Frey-Wyssling finds for dichroic preparations made by depositing gold or mercury on oriented cellulose, that the gold is microcrystalline with its well known cubic structure and the mercury is liquid; the metallic particles are extremely small, widely separated rods oriented parallel to the cellulose fibers. Particle dimensions of the order of 10mμ diameter, 100mμ in length, are mentioned by Frey-Wyssling. Thus these particles are plainly much larger than the dichroic molecules treated earlier in this section. Furthermore, in contrast to all the polarizers so far treated, the oriented absorbing particles here are not of themselves dichroic, since every cubic crystal or liquid particle is of necessity an optically isotropic system.

An earlier review of this subject was given by Berkman, Boehm and Zocher (1926) [25]; these writers show how some of the optical properties of dichroic metal preparations may be predicted from O. Wiener's theory, but point out the failure, important from practical considerations, of this theory to relate the optical properties of a given preparation to the dimensions of the metallic particles. Figs. 17, 18, and 19

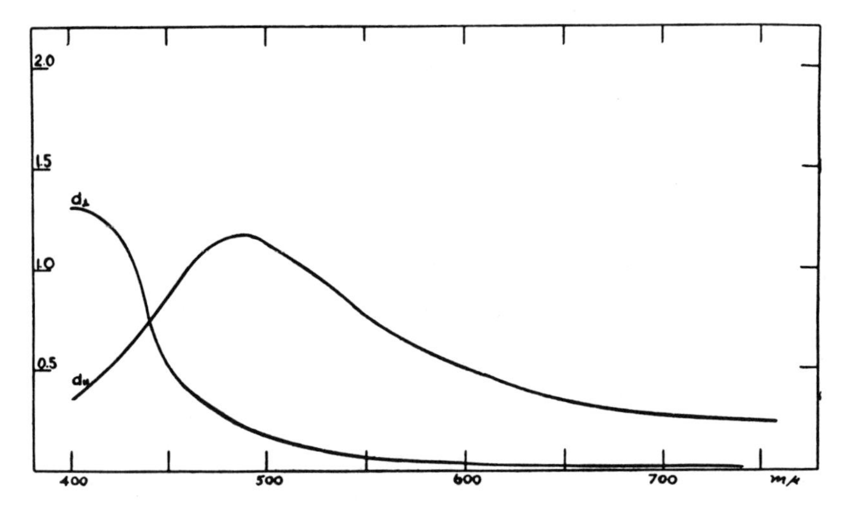

FIGURE 17. Dichroism of silver on oriented polyvinyl alcohol.

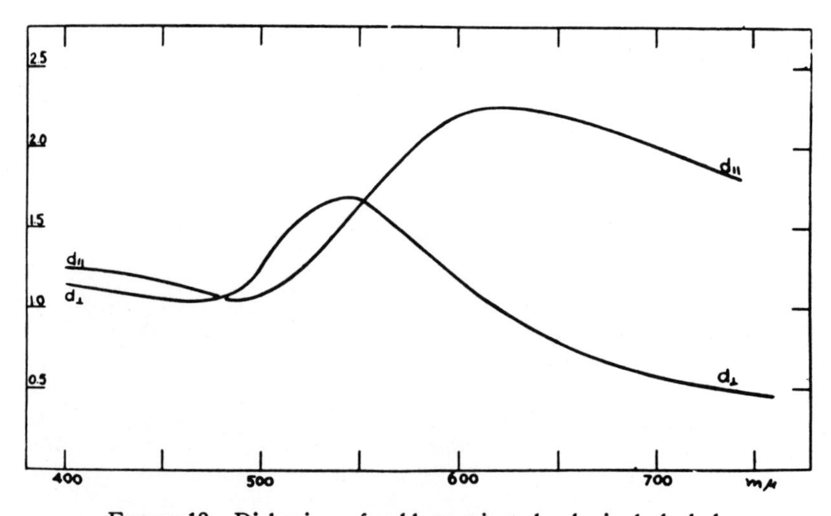

FIGURE 18. Dichroism of gold on oriented polyvinyl alcohol.

give data for dichroic preparations made from silver, mercury and gold on oriented polyvinyl alcohol. It is of interest that for the gold preparation, which has a red-green dichroism, the curves show a suggestive resemblance to some of Lange's (1928) [26] absorption curves for gold hydrosols of measured particle diameters; and namely the $d_{\parallel}$ curve resembles in shape Lange's curve for the 56mμ hydrosol, while the $d_{\perp}$ curve is close to that of the 7mμ hydrosol, even if their relative heights are out of order. It is seen that these dimensions are of the same order as those obtained by Frey-Wyssling by entirely different methods.

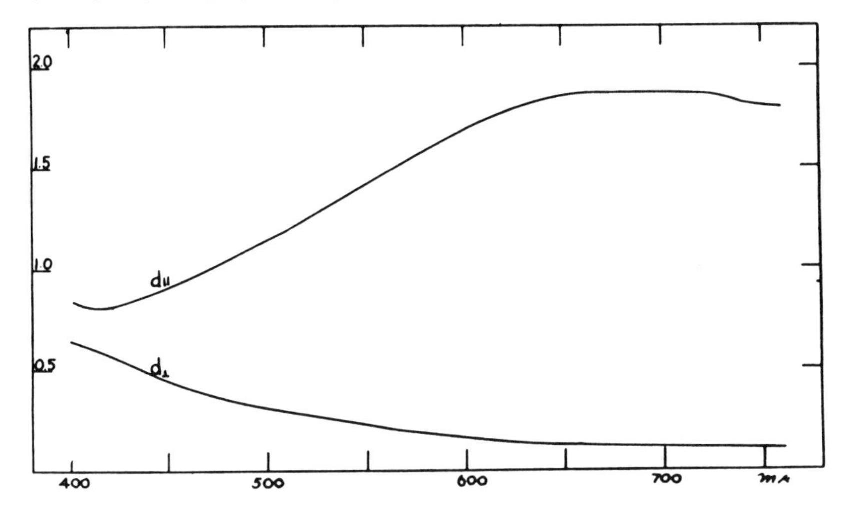

FIGURE 19. Dichroism of mercury on oriented polyvinyl alcohol.

Circular Polarizers. It appears doubtful that useful polarizers will ever be made from homogeneous circular dichroic materials, because the dichroism here is so far of such a low order of magnitude. The circular polarizers widely used in polariscopes designed for photoelastic investigations are customarily made by combining linear dichroic polarizers with quarter-wave birefringent films. However, if anyone ever discovers a useful circular dichroic polarizing material, this could obviously be combined with a quarter-wave film to make a useful linear polarizer also.

Summary. The useful artificial dichroic polarizers treated here, setting an arbitrary threshold of usefulness at a dichroic ratio of about 7:1, all have linear positive dichroism, and are made either by growing dichroic macrocrystals (herapathite), or by orienting dichroic microparticles. These particles may be dichroic microcrystalline needles (herapathite), dichroic simple molecules (dyes), or dichroic polymeric molecules (iodine, polyvinylene). The preferred methods of orienting the absorbing microparticles all involve some kind of a flow process in a viscous fluid medium.

SOME PRACTICAL APPLICATIONS OF DICHROIC POLARIZERS

Introduction

Since the applications of dichroic polarizers have in recent years been repeatedly written up in readily available sources, as shown by the following list of general papers, a brief recapitulation will suffice here.

W. E. Forsythe and E. Q. Adams, "Polarized Light." *Gen. Elec. Rev.*, **42**, 346-52 (1939).

H. Freundlich, "Polaroid Films." *Chemistry & Industry*, **56**, 698-9 (1937).

M. Grabau, "Polarized Light Enters the World of Everyday Life." *J. Applied Phys.*, **9**, 215-25 (1938).

—, "Introduction to Polarized Light and Its Applications." 46 p., Cambridge, Mass., Polaroid Corporation, 1940.

M. Haase, "New Polarization Filters Using Dichroic Crystals." *Z. tech. Physik,* **18,** 69-72 (1937).
H. A. Levey, "New Polarizing Media and Their Applications." *Ind. Eng. Chem., News Ed.,* **17,** 527-8 (1939).
E. Nähring, "Polarization Filters," *Phot. Ind.,* **38,** 599-601; 629-30 (1940).
A. Pollard, "Polarization of Light and Some Technical Applications." *Nature,* **138,** 311-14 (1936).
S. Rösch, "Some Properties and Applications of Dichroic Polarizers." *Z. Instrumentenk.,* **58,** 181-92 (1938).
L. W. Taylor, "Physics, the Pioneer Science." Chapter 38, pp. 549-75, Polarized Light. Boston, Houghton Mifflin Co., 1941.

For our purposes we may consider applications of dichroic polarizers for the man in the street or in the home, and applications for the man in the laboratory or in the lecture hall. In the former class, the numerous uses of dichroic polarizers are characterized throughout by novelty, even though some of the underlying ideas were proposed before the advent of practical dichroic polarizers, and by a practically exclusive concern with visual processes, either directly or at second hand through photography. In the latter class of applications, dichroic polarizers have widely replaced calcite polarizers for many laboratory uses. In some of the more expensive precision polarizing instruments, the instrument maker has been understandably slow in making the change. The period of active exploration in both classes of applications is by no means closed. Military applications, a third class which has now outstripped the other two, are not treated here.

Dichroic Polarizers in Applied Science

As Aids in the Visual Process. The widest application of dichroic polarizers so far has been in aiding the visual process. This end is achieved by cutting out, by means of one or more polarizers between the eye and the light source, some of the visual radiation, so that one may see more clearly or more comfortably with what radiation is left. This field of applied physical optics is primarily subject to physiological, and to some extent to psychological considerations.

There are three general ways of using polarizers to aid vision:

(1) A single polarizer is used over the eye, to cut out undesired light partially polarized by natural reflection and scattering.

(2) A single polarizer is used over the source for the same purpose as (1).

(3) The polarizers are used in pairs to cut out undesired light, and either superposed in space, or else separated in space, so that one of the pair goes with the eye, the other with the source.

Sun Glasses. By observation as with a dichroscope it is quickly found, first, that when sunlight is reflected or scattered from flat surfaces, it thereby generally becomes partially polarized with the vibrations normal to the plane of incidence predominating; secondly, that this component of the light is a hindrance rather than an aid in perceiving detail in or possibly under the flat surface. The degree of polarization varies widely with the angle of incidence, reaching a maximum for angles intermediate between normal and grazing incidence, and with the nature of the flat surface.

The function of the sun glasses is to absorb the component of the light vibrating normal to the plane of incidence, and to transmit only the component vibrating in the plane of incidence. Since flat surfaces are more often horizontal, for example the surface of a body of water, the pavement of a road, etc., polarizing sun glasses commonly have a fixed orientation so as to transmit only vertical vibrations.

It is of interest that when the sun is in the zenith on a clear day the skylight just above the horizon is partially polarized by scattering with the horizontal vibrations predominating, so that polarizing sun glasses "selectively absorb," so to speak, this horizon light as well as the ground light.

Reading Lamps. The light scattered or reflected from printed pages, by which we read, is also found to be partially polarized, and here too perception of detail under the optimum conditions is greatly aided when only vibrations in the plane of incidence are admitted to the eye. As would be expected, some improvement may be noted when one reads with sun glasses under high levels of illumination. It is generally more efficient and more convenient to employ a polarized source of light, that is, a lamp provided with a polarizer and designed so that its radiation is incident on the printed page at an angle to give a maximum effect. The improvement is especially noticeable with material printed on glossy paper.

Variable Density Filters. Under some conditions it is desired to attain a wider range of illumination control than is possible by absorbing a single component of the light. There is then the possibility of using dichroic polarizers in pairs, and either superposed, or separated in space.

Large polarizers have been used superposed in pairs to provide variable daylight illumination through windows, smaller polarizers similarly in variable-density sun glasses. The illumination in both cases is varied by rotating one polarizer of a pair with respect to the other.

When the problem arises of completely eliminating the mirror reflection from any flat surface, regardless of the angle of incidence of the light and the nature of the mirror, the solution is found in polarizing the light source in or normal to the plane of incidence, and in viewing the reflected ray with an analyser crossed with the polarizer. Then the flat surface can be seen only by that light, if any, which is depolarized as by scattering or by defects in the mirror. By suitable rotations of one polarizer with respect to the other more or less of the mirror component of the reflected light may be transmitted to the eye.

Automobile Headlights. Here the polarizers are to be used in crossed pairs to eliminate dazzle, with the components of the crossed pairs necessarily separated in space. The polarizers have a fixed orientation so that the analyzer in front of a driver's eyes is always automatically parallel to the polarizers over the headlights on his car and on other cars driving in his direction, while it is always crossed with the polarizers over the headlights of all approaching cars. This is effected by orienting the vibration directions of all polarizers at a uniform angle of 45° from the horizontal, always measured in the same sense.

Of all the applications of dichroic polarizers this probably has been the one most thoroughly discussed in the patent and technical literature.[27-34]

From the manufacturing point of view, the polarized headlight that fills the rather exacting requirements of this application is best made by starting with a headlamp of the sealed beam type and applying to the outer surface of its lens a thin layer containing dichroic polyvinylene.

Photographic Images. Many of the principles which apply in improving visual images through the use of polarizing filters can be utilized equally well in photography. Here, however, the situation is somewhat simpler in that ordinary cameras have only one objective, and filter holders which can be set at any desired azimuth are readily available for most cameras. In using a polarizing filter one must take into consideration the curve of its transmission for natural light as a function of wave length, since the sensitivity curve of the photographic layer may be quite different from the sensitivity curve of the eye.

There is already a considerable literature on the photographic uses of polarizing filters. Filters over the objective are employed to cut out vibrations perpendicular to the plane of incidence when there is undesired illumination from reflecting surfaces,—horizontal mirrors such as bodies of water, vertical mirrors such as window panes, etc. Such filters too are advantageously employed to darken a background that is illuminated by light partially linear-polarized by scattering, such as the skylight 90° from the sun on a clear day; their use in underwater photography has been proposed on the same principle.

In photography too there is the already mentioned possibility of using polarizers in pairs—together as with both components over the objective or over the source, or more commonly separated as with one polarizer over the source and one over the objective, to vary the illumination over wider ranges than is possible with a single rotatable polarizer. Polarizers have been used in pairs over the objectives of motion picture cameras for taking fades and lap dissolves.

As Aids in Viewing Stereoscopic Pairs of Images. Polarizing devices have already been widely used for viewing stereoscopic pairs of images, either directly or after projection. Three different systems may be cited here to illustrate some of the possibilities.

In all these systems, the observer employs polarizing spectacles to admit the one image of the stereo pair to the eye to which it belongs.

Projection of juxtaposed stereo images through twin projectors.[35, 36] This system is the exact analog of the anaglyph system, except that polarizing filters are employed in place of complementary (*e.g.*, red and green) color filters.

Projection or viewing of vectograph stereo images.[37, 38] In this strikingly novel application, due to Land and Mahler, each image of the stereo pair is formed directly as a positive image in a polarizing layer; for the high lights, $d_{\parallel} = d_{\perp} = 0$, while for the shadows $d_{\perp} = 0$, $d_{\parallel} > 0$. For viewing or projection, the two images are superposed with their axes crossed; thus only a single ordinary projector is required in contrast to current practice for the preceding case. The vectograph may also be viewed directly, and either as a diapositive, or by backing with a suitable non-depolarizing diffusing layer. An aluminum lacquer has been used for this purpose as well as for preparing projection screens for reflecting polarized images without depolarization.

In addition to their uses in stereoscopic viewing or projection, vectographic images have other novel applications.

A third system,[39] also due to Land, has been used for stereoscopic viewing of the large diapositives encountered in x-ray radiographic work.

Display Devices. The full range of interference colors obtainable with doubly refracting materials of the normal dispersion of birefringence, when viewed between polarizers against a standard white light source, has been analytically determined, tabulated and plotted in a valuable paper by Buchwald (1940).[40] This writer starts with the I.C.I. (International Commission on Illumination) data, using the illuminant B and gypsum as bases for calculation, and ends up with the first five orders of interference colors (gypsum thicknesses 0-0.30 mm) tabulated according to the color triangle coördinates x, y, and the dominant wave length λ and saturation σ derivable therefrom. The color curves for the first three orders of such a gypsum wedge between crossed and parallel polarizers respectively are shown as spirals around the white point in Figs. 20 and 21; the colors of the fourth and fifth orders between crossed polarizers are shown in Fig. 22; and the brightness, Hx, for the first five orders between crossed polarizers is shown in Fig. 23 (the brightness for parallel polarizers is $H\| = 100 - Hx$). These data for $(x, y, H)_x$ and $(x, y, H)\|$ could advantageously be presented in three-dimensional figures. Fig. 24 shows the transition of the color in a 0.10-mm section of gypsum to its complementary as the analyzer is rotated from the crossed position ($\beta = 90°$) to the parallel position ($\beta = 0°$).

Buchwald's results would be of scientific interest only, were it not for the fact that some common transparent sheet materials such as cellophane are characterized by birefringence and normal dispersion thereof. This means that any point on Buchwald's color curves can be reproduced by selecting a suitable piece of birefringent film and viewing it between dichroic polarizers. Still other colors are obtainable by preparing transparent films having abnormal dispersion of birefringence.

An application of such polarization colors, either alone or in combination with ordinary isotropic colors or the dichroic colors described earlier, has been made in various display devices. Common to many of these is the play of colors resulting from continuous rotation of one or more members of the optical system, ordinarily one of the polarizers.

Dichroic Polarizers in Scientific Research

The papers of the following two lists have in common that they all include some mention of dichroic polarizers. Somewhat in the manner of *Science Abstracts,* the

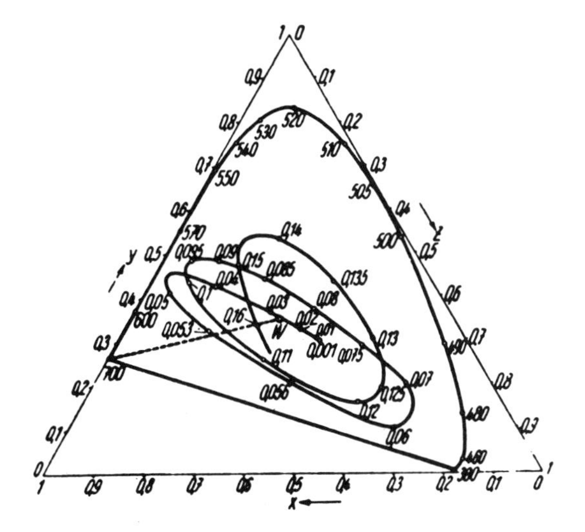

FIGURE 20. Metric of normal linear-birefringent interference colors, reproduced from E. Buchwald. Gypsum sections 0 to 0.30 mm thick, I.C.I. illuminant B. Crossed polarizers, orders I-III.

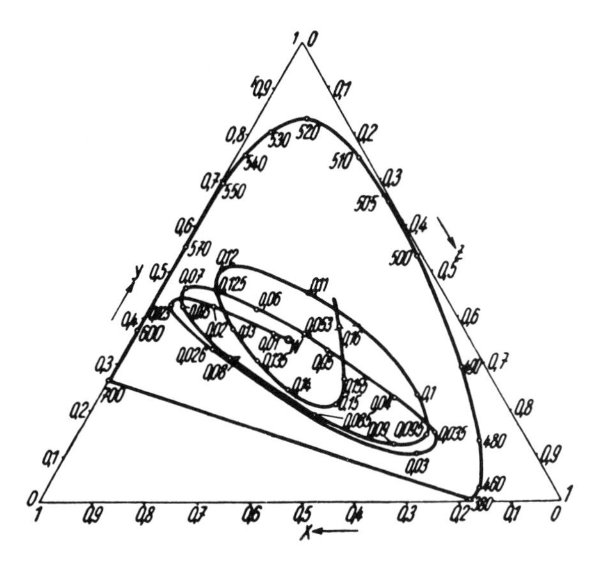

FIGURE 21. Same as Figure 20, with parallel polarizers, orders I-III.

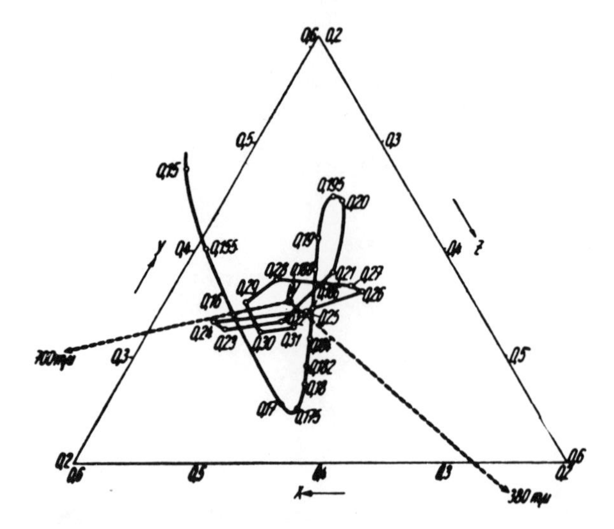

FIGURE 22. Same as Figure 20, with crossed polarizers, orders IV and V.

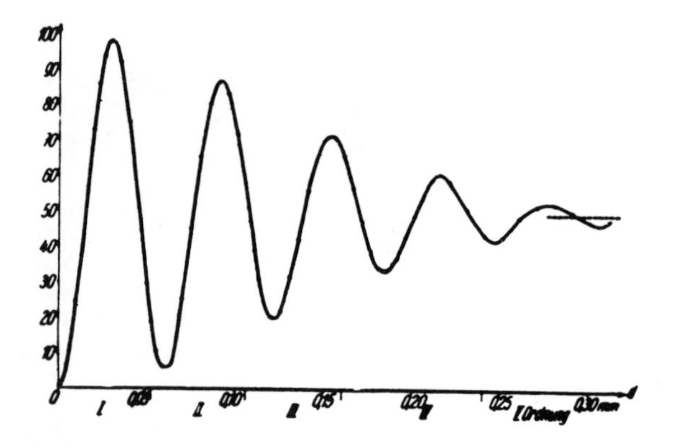

FIGURE 23. Same as Figure 20, with crossed polarizers, brightness for orders I-V.

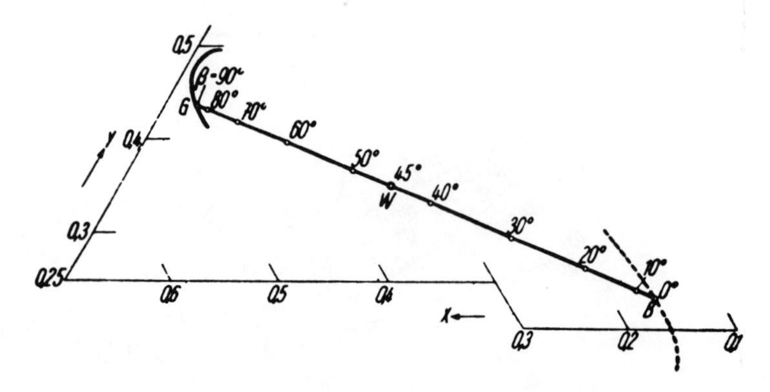

FIGURE 24. Same as Figure 20. Transition of interference color of 0.10 mm section to its complementary as analyzer is rotated from crossed ($\beta = 90$) to parallel position ($\beta = 0$).

first is by subjects and the second deals with apparatus and instruments. While these lists are not offered as exhaustive, it is hoped that they will be fairly representative of the uses to which dichroic polarizers have been put in the laboratory up to January 1943.

For colloid chemists, there are several fields of direct interest in which polarization effects are observed and measured:

(1) The study of the size and shape of colloid particles from the state of polarization of the light scattered thereby. S. Bhagavantam (1940) has treated this subject in his book "Scattering of Light and the Raman Effect."

(2) The study of certain fluid systems by the optical anisotropy of flow. This term is used broadly to include both birefringence and dichroism, and to include matter in states of both actual flow and "frozen" flow. A valuable review of streaming birefringence has been written by J. T. Edsall (1942), as cited below.

(3) The study of very thin films by methods involving the reflection of polarized light.

List of Subjects

Mechanics:

Photoelasticity of solids under steady stress

Frocht, M. M., "Photoelasticity," Vol. I. New York, Wiley, 1941.

Weller, R., "Three-Dimensional Photoelasticity Using Scattered Light," *J. Applied Phys.*, **12**, 610 (1941).

Photoelasticity of vibrating systems

Ferry, J. D., "Photoelastic Method for Transverse Vibrations in Gels," *Rev. Sci. Instruments*, **12**, 79 (1941).

—, "Rigidities of System Polystyrene-Xylene and Their Dependence on Temperature and Frequency," *J. Am. Chem. Soc.*, **64**, 1323 (1942).

Photoviscosity of fluids in steady flow

Astronomical optics:

Allen, C. W., "Polarization of Solar Corona," *Monthly Notices Roy. Astron. Soc.*, **101**, 281 (1941).

Meteorological optics:

Byram, G. M., "Polarization of Atmospheric Haze," *Science*, **94**, 192 (1941).

Physiological optics:

Bannon, R. E., "Technique of Measuring and Correcting Aniseikonia," *Am. J. Optometry*, **18**, 145 (1941).

Gehrcke, E., "New Phenomena in Physiological Optics," *Physik. Z.*, **41**, 540 (1940).

Land, E. H., and Hunt, W. A., "Use of Polarized Light in Simultaneous Comparison of Retinally and Cortically Fused Colors," *Science*, **83**, 309 (1936).

Physical optics:

Optical anisotropy in various systems

Biological materials

Tartar, V., and Richards, O. W., "Use of Polarized Light in Biological Science," *Collecting Net*, **12**, 4 p. (1937).

Crystals, etc.

Burbage, E. J., and Anderson, B. W., "Analysis of Movements of Shadow-Edges on the Refractometer in the Case of Biaxial Gemstones," *Mineralog. Mag.*, **26**, 246 (1942).

West, C. D., "Optical Properties and Polymorphism of Paraffins," *J. Am. Chem. Soc.*, **59**, 742 (1937).

Fluids in flow

Edsall, J. T., and Mehl, J. W., "Effect of Denaturing Agents on Myosin. II. Viscosity and Double Refraction of Flow," *J. Biol. Chem.*, **133**, 409 (1940).

Farwell, H. W., "Double Refraction in Colloidal Systems," *J. Applied Phys.*, **8**, 416 (1937).

Hauser, E. R., and Dewey, D. R., "Visual Studies in Flow Patterns," *J. Phys. Chem.*, **46**, 212 (1942).
Kraemer, E. O., ed., "Advances in Colloid Science." Vol. 1, p. 269. Article by Edsall, J. T.: "Streaming Birefringence and Its Relation to Particle Size and Shape," New York: Interscience Publishers, Inc., 1942.
Langmuir, I., "Role of Attractive and Repulsive Forces in Formation of Tactoids, Thixotropic Gels, Protein Crystals and Coacervates," *J. Chem. Phys.*, **6**, 873 (1938).
Lauffer, M. A., "Optical Properties of Solutions of Tobacco Mosaic Virus," *J. Phys. Chem.*, **42**, 935 (1938).
Lauffer, M. A., and Stanley, W. M., "Stream Double Refraction of Virus Proteins," *J. Biol. Chem.*, **123**, 507 (1938).
Stanley, W. M., "Biophysics and Biochemistry of Viruses," *J. Applied Phys.*, **9**, 148 (1938).

Glass

Edgerton, H. E., and Barstow, "Further Studies of Glass Fracture with High Speed Photography," *J. Am. Ceram. Soc.*, **24**, 131 (1941).
Schardin, H., Elle, D., and Stroth, W., "Velocity of the Fracture Process in Glass and Artificial Resins," *Z. tech. Physik*, **21**, 393 (1940).

Plastics

Farwell, H. W., "Double Refraction and Change of Length of Certain Plastics," *J. Applied Phys.*, **10**, 109 (1939).
—, "Tension and Birefringence in a Vinylite Plastic," *J. Applied Phys.*, **11**, 271 (1940).
Middlehurst, D., and Weller, R., "Polarizing Properties of Cellophane," *Rev. Sci. Instruments*, **11**, 108 (1940).
Spence, J., "Optical Anisotropy and the Structure of Cellulosic Sheet Materials," *J. Phys. Chem.*, **43**, 865 (1939).

Electro-optics

Mueller, H., "Electro-optical Field Mapping," *J. Optical Soc. Am.*, **31**, 286 (1941).
Mueller, H., and Sakmann, B. W., "Electro-optical Properties of Colloids," *J. Optical Soc. Am.*, **32**, 309 (1942).

Reflection

Laurence, J., "Reflection Characteristics with Polarized Light," *J. Optical Soc. Am.*, **31**, 9 (1941).

Scattering, Raman

Bernstein, H. J., and Martin, W. H., "Depolarization of Raman Lines and Structure of Complex Ions," *Trans. Roy. Soc. Can., Sect. III*, **32**, 43 (1938).
Nedungadi, T. M. K., "Effect of Crystal Orientation on Raman Spectrum of Sodium Nitrate," *Proc. Indian Acad. Sci.*, **10**, 197 (1939).

Scattering, Rayleigh or Tyndall

Gehman, S. D., and Field, J. E., "Colloidal Structure of Rubber in Solution," *Ind. Eng. Chem.*, **29**, 793 (1937).
Hoover, C. R., Putnam, F. W., and Wittenberg, E. G., "Depolarization of Tyndall Light of Bentonite and Iron Oxide Sols," *J. Phys. Chem.*, **46**, 81 (1942).

Thin Films

Bishop, F. L., Jr., "Surface Layers of Sheet Glass," *Phys. Rev.*, **62**, 295 (Abstract) (1942).
Blodgett, K. B., and Langmuir, I., "Built-up Films of Barium Stearate and Their Optical Properties," *Phys. Rev.*, **51**, 964 (1937).
Pfund, A. H., "Haidinger Interferences in Silvered Mica," *J. Optical Soc. Am.*, **32**, 383 (1942).

Photochemistry:

Rama Char, T. L., "Photochemical Stability of Mixtures of Vanadic and Tartaric Acids. II . . . Induced Optical Activity by Circularly Polarized Light," *J. Indian Chem. Soc.*, **18**, 563 (1941).

List of Apparatus and Instruments

Color filters

Lyot, B., "A Monochromatic Filter Especially Practical for Investigations of the Sun," *Compt. rend.*, **212**, 1013-17, (1941).
Öhman, Y., "A New Monochromator," *Nature*, **141**, 157 (1938).
Pettit, E., "The Interference Polarizing Monochromator." Pub. of the Astronomical Soc. of the Pacific, **53**, 313 17-27 (1941).

Compensators and retardation plates

Piecewicz, C. T., "Plastic Wedge Compensator," *Rev. Sci. Instruments,* **13,** 296 (1942).
Tuzi, Z. and Oosima, H., "Artificial Quarter Wave Plate and Its Theory," *Sci. Papers Inst. Phys. Chem. Research (Tokyo),* **905,** 72 (1939).
West, C. D., "Substitute for the Quartz Wedge Used with the Polarizing Microscope," *Am. Mineral.,* **23,** 531 (1938).

Dichroscopes, divided field devices

Pfund, A. H., "Polaroid Half Shade Analyzer," *J. Optical Soc. Am.,* **26,** 453 (1936).
Thibault, N. W., "A Simple Dichroscope," *Am. Mineral.,* **25,** 88 (1940).

Ellipticity and retardation, measurement of

Ingersoll, L. R., "Improved Spectroradiometric Method and Apparatus for Measuring Small Polarization Ellipticities," *J. Optical Soc. Am.,* **27,** 411 (1937).
Kent, C. V., and Lawson, J., "Photoelectric Method for the Determination of Elliptically Polarized Light," *J. Optical Soc. Am.,* **27,** 116 (1937).
v. Look, Th., "New Method for Measuring Retardations; the Temperature Dependence of Birefringence of Quartz, Beryl and Adular," *Z. angew. Mineral.,* **1,** 229 (1938).

Microscopes

Plato, W., "Determination of the Optical Crystallographic Properties of Polished Crystalline Stones," *Z. wiss. Mikroskop.,* **57,** 137 (1940).
West, C. D., "Polarizing Accessories for Microscopes," *J. Chem. Education,* **19,** 66 (1942).

Optometry and ophthalmology, instruments for

American Optical Company, "Eikonometer," *Rev. Sci. Instruments,* **12,** 282 (1941).
Bausch and Lomb Company, "Ortho-Fusor," *Am. J. Optometry,* **19,** 39 (1942).

Photometers, densitometers, spectrophotometers, etc.

Adam Hilger, Ltd., "Photoelectric Color Comparator (for Reflected Light)," *Rev. Sci. Instruments,* **12,** 337 (1941).
Moon, P., and Laurence, J., "Construction and Test of a Goniophotometer," *J. Optical Soc. Am.,* **31,** 130 (1941).
Öhman, Y., "New Method of Standardization in (Photographic) Spectrophotometry," *Stockholms Obs. Ann.,* **13,** 6 16 p. (1940).
Sweet, M. H., "Physical Density Comparator," *J. Optical Soc. Am.,* **28,** 349 (1938).

Polarimeters and Saccharimeters

Estey, R. S., "The Use of the Spencer Polarimeter," 32 p. Buffalo, Spencer Lens Company, 1941.
Karush, F., "Photoelectric Polarimeter," 16 p. Thesis, Univ. of Chicago, 1941.
Marion, A., "Adapting Polarizing Microscopes for Use as Polarimeters," *Ind. Eng. Chem., Anal. Ed.,* **12,** 777 (1940).
Spengler, O., and Hirschmüller, H., "New Photoelectric Saccharimeter," *Z. Wirtschaftsgruppe Zuckerind.,* **90,** 26 (1940).

Polariscopes, for crystals, etc.

Arshinov, V. V., "Pocket Mineralogical or Polarizing Magnifier," *Am. Mineral.,* **23,** 287 (1938).
Buckley, H. E., "Demonstration of Optical Interference Figures," *Nature,* **143,** 801 (1939).
Quirke, T. T., "Direct Projection of Optical Figures," *Am. Mineral.,* **23,** 595 (1938).
Rösch, S., "Polariscope for Physical Chemistry, Mineralogy and Examination of Gems," *Fortschr. Mineral. Krist. Petrog.,* **22,** li-lii (1937).
Schaub, B. M., "On the Use of Polaroid for Photographing Large Thin Sections in Crossed Polarized Light," *Am. Mineral.,* **21,** 384 (1936).
Zuloaga, G., "Photomicrography with an Enlarger," *Phototechnique,* p. 53 (Nov. 1941).

Polariscopes, for glass products

Douglas, R. W., and Breadner, B. L., "Strain Viewer for Glass Articles," *J. Sci. Instruments,* **17,** 187 (1940).
Hull, A. W., and Burger, E. E., "Simple Strain Tester for Glass Seals," *Rev. Sci. Instruments,* **7,** 98 (1936).
Swicker, V. C., "The Polariscope as a Glass Factory Instrument," *Glass Ind.,* **23,** 13 (1942).

Polariscopes, photoelastic

Föppl, L., and Müller-Lufft, E., "Stress-Optical Apparatus with Polarization Filters," *Arch. tech. Messen,* **100,** T125-6 (1939).

Frocht, M. M., "Photoelasticity," Vol. I, p. 382-9. New York, Wiley, 1941.

Tuzi, Z., and Nisida, M., "New Apparatus for Photoelastic Stress Type. Quarter Wave Plate of Large Field," *Sci. Papers Inst. Phys. Chem. Research (Tokyo)*, **31,** 676, 99 (1937).

Refractometers

West, C. D., "Measurement of Refractive Indices of Resins and Plastics," *Ind. Eng. Chem., Anal. Ed.,* **10,** 627 (1938).

Sextants, telescopes, etc.

Hardy, A. C., "Automatic Telescope Control," *J. Optical Soc. Am.,* **30,** 654 (Abstract) (1940).

Hulburt, E. O., "Sextant with Improved Filters," *J. Optical Soc. Am.,* **26,** 216 (1936). **26,** 216 (1936).

Spectrographs for detecting polarization of spectral lines

Cleveland, F., and Murray, M. J., "Depolarization Measurements on Raman Lines by an Easy Accurate Method," *J. Chem. Phys.,* **7,** 396 (1939).

Edsall, J. T., and Wilson, E. B., Jr., "Simple Method of Determining Polarization of Raman Lines," *J. Chem. Phys.,* **6,** 124 (1938).

Glockler, G., and Baker, H. T., "Polarization Measurements of Raman Lines," *J. Chem. Phys.,* **10,** 404 (1942).

Spectrometers, goniometers, etc.

Anderson, B. W., "Refractive Index Measurement by Brewster's Angle," *Gemnologist (London)*, **10,** 61 (1941).

Pfund, A. H., "Brewsterian Angle and Refractive Indices," *J. Optical Soc. Am.,* **31,** 679 (1941).

References

1. Zimmern, A., and Coutin, M., *Compt. rend.,* **182,** 1214 (1926).
2. Bernauer, F., *Fortschr. Mineral. Krist. Petrog.,* **19,** 22 (1935).
3. Strong, J., *J. Optical Soc. Am.,* **26,** 256 (1936).
4. Grabau, M., *J. Optical Soc. Am.,* **27,** 420 (1937).
5. Pulfrich, C., *Z. Krist. Mineral. Petrog.,* **6,** 142 (1882).
6. Preston, J., *J. Soc. Dyers Colourists,* **47,** 312 (1931).
7. Zocher, H., and Jacoby, F., *Kolloidchem. Beihefts,* **24,** 365 (1927).
8. Barger, G., and Starling, *J. Chem. Soc.,* **107,** 411 (1915).
9. Jörgensen, S., *J. prakt. Chem.,* **14,** 213 (1876).
10. West, C., *Am. Mineral.* **22,** 731 (1937).
11. Jörgensen, S., *J. prakt. Chem.,* **14,** 356 (1876).
12. Farwell, H., *J. Optical Soc. Am.,* **28,** 460 (1938).
13. Godina, D., and Faerman, G., *J. Applied Chem. (U.S.S.R.)*, **14,** 362 (1941).
14. Kermack, Slater and Spragg, *Proc. Roy. Soc. Edinburgh,* **50,** 243-61 (1930).
15. Mooney, R., *J. Am. Chem. Soc.,* **63,** 2828 (1941).
16. Marvel, C., and Denoon, C., *J. Am. Chem. Soc.,* **60,** 1045 (1938).
17. Preston, J., see reference 6. See also Silverman, S., *Rev. Sci. Instruments,* **12,** 212 (1941).
18. Valkò, E., *J. Am. Chem. Soc.,* **63,** 1433 (1941).
19. Staudinger, H., Frey, K., and Starck, W., *Ber.,* **60,** 1782 (1927).
20. Gallay, W., *Can. J. Research,* **14B,** 105 (1936).
21. Hausser, K., Kuhn, R., and others, *Z. physik. Chem.,* **29,** 363 (1935).
22. Kuhn, R., *J. Chem. Soc.,* p. 605 (1938).
23. Mulliken, R., *J. Chem. Phys.,* **7,** 570 (1939).
24. Frey-Wyssling, A., *Protoplasma,* **27,** 372-411; 563-571 (1937).
25. Berkman, S., Boehm, J., Zocher, H., *Z. physik. Chem.,* **124,** 83 (1926).
26. Lange, B., *Z. physik. Chem.,* **132,** 27-46 (1928).
27. Christoph, W., and Neugebauer, H., *Z. Tech. Physik.,* **20,** 257 (1939).
28. Chubb, L., *Trans. Illum. Eng. Soc. (N.Y.)*, **32,** 505 (1937).
29. Gamble, A., *Bull. soc. franc. élec.,* **7,** 511 (1937).
30. Gibbs, H., *J. Franklin Inst.,* **228,** 719 (1939).
31. Land E., *J. Franklin Inst.,* **224,** 269 (1937).
32. Land, E., *Univ. Mich. Official Pub.,* **42,** No. 42 (1940): "Michigan-Life Conference on New Technologies in Transportation," pp. 151-68.
33. Roper, V., and Scott, K., *Illum. Eng. Soc. (N.Y.)*, **36,** 1205 (1941).
34. Sauer, H., *Z. Ver. deut. Ing.,* **82,** 201 (1938)

35, 36. Norling, J., *J. Soc. Motion Picture Eng.,* **33,** 612 (1939); **37,** 516 (1941).

37. Land, E., *J. Optical Soc. Am.,* **30,** 230 (1940).
38. Dudley, B., *Phototechnique,* **3,** 30 (May, 1941).
39. Stamm, R., *Am. J. Roentgenol. and Radium Therapy,* **45,** 744 (1941).
40. Buchwald, E., *Ann. Physik,* **38,** 325 (1940).

VII
Bibliography

Editor's Comments on Papers 34 and 35

34 Anonymous: *Bibliography from* The Proceedings of the Optical Convention

35 Shurcliff: *Bibliography from* Polarized Light

The Optical Convention was held at the Northampton Institute, London, in 1905, and the proceedings were published by Messrs. Norgate and Williams of Covent Garden, London, in that year. The last of the thirty-one articles, included in this volume as Paper 34, is a bibliography on double refraction and polarizing prisms. It is one of the most comprehensive listings of pre-twentieth-century literature on this subject.

One of the very few textbooks devoted entirely to the subject of polarized light is Schurcliff 's book *Polarized Light.* A valuable feature of the book is the bibliography; it is an excellent source of references, especially in the practical and device-oriented areas. Although by now somewhat dated, it is included here because it is still the most complete bibliography on the subject.

William A. Shurcliff (1909–) was senior scientist at the Polaroid Corporation (1948–1959) before becoming senior research associate at the Cambridge Electron Accelerator. His research interests have included emission spectroscopy, absorption spectrophotometry, gamma radiation dosimeters, and color vision. He is now hard at work on schemes for the solar heating of houses in New England. He is active in fighting noise pollution.

34

Reprinted from *Proceedings of the Optical Convention,* Norgate and Williams, London, 1905, pp. 236–240

Proceedings of the Optical Convention

BIBLIOGRAPHY.

REFRACTION OF ICELAND SPAR, THEORY OF DOUBLE REFRACTION, AND CONSTRUCTION OF POLARIZING PRISMS.

1670. ERASMUS BARTHOLINUS. Experimenta Crystalli Islandici disdiaclastici quibus mira et insolita refractio detegitur. Amstelodami, 1670.

1690. HUYGENS. Traité de la Lumière. Leyden, 1690.

1704. NEWTON. Opticks. Book III., Question 35.

1739. DU FAY. Observations sur la double réfraction du spath d'Islande. *Mém. de l'anc. Acad. des Sciences*, 1739, p. 81.

1762. BECCARIA. Account of the Double Refraction in Crystals. *Phil. Trans.*, 1762, p. 486.

1777. ROCHON. (In a paper relating to his micrometer.) *Nova Acta Academiæ Petropolitanæ*, vi.; see *Recueil de mémoires sur la mécanique et sur la physique*, Brest, 1783; also *Gilb. Ann.*, xl. p. 141; and Marbach's *Physikalisches Lexicon* 2 (Ed. iv.), p. 1039.

1788. HAÜY. Sur la double réfraction du spath d'Islande. *Mém. de l'anc. Acad. des Sciences*, 1788, p. 34; *Ann. de Chim.* (1), xvii., p. 140.

1797. BROUGHAM. Further experiments on light. *Phil. Trans.*, p. 352.

1801. HAÜY. Traité de Minéralogie, i. p. 271, and ii. p. 132. Paris. (Second Edition, Paris, 1822, contains fuller notes on calcite.)

1802. WOLLASTON. On the Oblique Refraction of Iceland Crystal. *Phil. Trans.*, 1802, p. 381; *Gilb. Ann.*, xxxi. (1809), p. 252.

1808. BOURNON, LE COMTE DE. Traité complet de la chaux carbonatée (with many plates of crystal forms). London.

1809. LAPLACE. Sur le mouvement de la lumière dans les corps diaphanes. *Mém. de l'Inst.* x. p. 306.

1809. YOUNG. Review of Laplace's Mémoire sur les lois de la réfraction extraordinaire dans les milieux diaphanes. *Quarterly Review*, Nov. 1809; *Miscell. Works*, i. p. 228.

1810. MALUS. Théorie de la double réfraction. *Mém. des sav. étrang*, ii. 303.

1813. BIOT. Examen comparé de l'intensité d'action que la force répulsive extraordinaire du spath d'Islande exerce sur les molécules des diverses couleurs. *Ann. de Chim.* (1), xciv., p. 281.

1813. MONTEIRO. Sur la détermination directe d'une nouvelle variété de forme crystalline de chaux carbonatée, et sur les propriétés rémarquables qu'elle présente. *Journal des Mines.*, ii. p. 161.

1815. AMPÈRE. Démonstration d'un théorème d'où l'on peut déduire toutes les lois de la réfraction ordinaire et extraordinaire. *Mém. de l'Inst.*, xiv. p. 235.

1817. YOUNG. Theoretical Investigation intended to illustrate the Phenomena of Polarisation. *Supplt. to Encyclop. Brit.*

1818. BREWSTER. On the Laws of Polarisation and Double Refraction in Crystallised Bodies. *Phil. Trans.*, 1818, p. 199.

1819. BREWSTER. On the action of Crystalline Surfaces on Light. *Phil. Trans.*, 1819, p. 146; *Edinb. Encyclop.*, Art. *Optics*, p. 600.

1820. MONTEIRO. Mémoire sur un nouveau problème crystallographique . . . avec des applications à une nouvelle variété déterminable de chaux carbonatée. *Annales des Mines.*, v. p. 3.

1821. FRESNEL. Mémoire sur la double réfraction. *Mém. de l'Acad. des sc.*, vii. p. 45 ; *Ann. de Chim.* (2), xxviii. 263 ; *Pogg. Ann.*, xxiii. (1847), pp. 372, 424.

1823. MARTIN. An Essay on the Nature and Wonderful Properties of Iceland Crystal Respecting its Unusual Refraction of Light. *Edin. Philos. Jour.*, viii. p. 150.

1826. WACKERNAGEL. See *Kastner's Archiv. für die gesammte Naturlehre*, ix. A criticism of Count Bournon's work.

1828. RUDBERG. Untersuchungen über die Brechung des farbigen Lichts im Bergcrystall und im Kalkspath. *Pogg. Ann.*, xiv. p. 45. (Determination of refractive indices.)

1828. NICOL. On a Method of so far increasing the Divergency of the two Rays in Calcareous Spar that only one Image may be seen at a time. *Edin. New Philos. Journal* (Jameson's), vi. p. 83; referred to in *Pogg. Ann.*, xxix. p. 182 (1833).

1828. NAUMANN. Kristallographische Notiz. *Pogg. Ann.*, xiv. p. 235.

1831. SEEBECK. Ueber die Polarisationswinkel am Kalkspath. *Pogg. Ann.*, xxi. 290, 299 ; *Pogg. Ann.*, xxii. p. 126 ; see also *Pogg. Ann.*, xxxviii. p. 277.

1832. NEUMANN. Theorie der doppelten Strahlenbrechung abgeleitet aus den Gleichungen der Mechanik. *Pogg. Ann.*, xxvi. 291.

1834. FOX-TALBOT. Facts relating to Optical Science. No. II. on Mr Nicol's Polarizing Eyepiece. *Philos. Magaz.* (3), iv. p. 289.

1836. WEISS. Neue Bestimmung einer Rhomboëder - Fläche am Kalkspath. *Berl. Akad.*, 1836, p. 207.

1836. MACCULLAGH. On the Laws of Crystalline Reflection and Refraction. *Trans. Roy. Irish Academy*, xviii. Pt. i. ; *Philos. Magaz.* (3), vii. p. 295 ; viii. p. 103 ; x. p. 42.

1837. KELLAND. On the Transmission of Light in Crystallized Media. *Cambr. Phil. Trans.*, vi. p. 323 ; *Philos. Magaz.* (3), x. p. 336.

1837. LEVY. Description d'une collection de Mineraux formée par M. Heuland. London.

1838. SPASKY. Note über das Nicol'sche Prisma. *Pogg. Ann.*, xliv. p. 138.

1839. NICOL. Notice concerning an Improvement in the Construction of the Single Vision Prism of Calcareous Spar. *Edin. New Philos. Jour.* xxvii. p. 332 ; referred to in *Pogg. Ann.*, xlix. (1840), p. 238.

1839. RADICKE. De Phænomenis quibusdam quæ prismata Nicoliana offerunt, de subsidiisque quibus quam optima construantur . . . commentatio. Berlin. (A pamphlet suggesting use of square ends, and use of balsam of copaiba instead of Canada balsam.)

1840. RADICKE. Ueber Vervollkommnung der Nicol'schen Polarisationsprisma. *Pogg. Ann.*, l. p. 25.

1841. GREEN. On the Propagation of Light in Crystallized Media. *Cambr. Phil. Trans.*, vii. pt. ii. p. 120.

1843. BREWSTER. On the Ordinary Refraction in Iceland Crystal. *Report* (13th) *of Brit. Assoc.*, p. 7.

1846. STOKES. On a Formula for determining the Optical Constants of Doubly Refracting Crystals. *Camb. Math. Jour.*, i. 183.

1847. SWAN. Experiments on the Ordinary Refraction of Iceland Spar. *Edin. Trans.*, xvi. p. 375.

1848. HAIDINGER. Dichroscopische Loupe. *Ber. Wien. Akad.*, xlvii. p. 70.

1850. DE SENARMONT. Sur un nouveau Polariscope. *Ann de Chimie*, xxviii. p. 279 ; *Pogg. Ann.*, lxxx. p. 293.

1851. SALM-HORSTMAR, Der Fürst zu. Ueber das Verhalten einiger Krystalle gegen polarisirtes Licht. *Pogg. Ann.*, lxxxiv. p. 515.

1851. ZIPPE. Uebersicht der Krystallgestalten des rhomboëdrischen Kalkhaloides. *Denkschrift der Wien. Akad.*, iii.

1853. BEER. Beiträge zur Dioptrik und Katoptrik kristallinischer Mittel mit einer optischen Axe. *Pogg. Ann.*, lxxxix. p. 56.

1854. HOCHSTETTER. Das Krystallsystem des rhomboëdrischen Kalkhaloides. *Denkschrift der Wien. Akad.*, vi., p. 89.

1855. BILLET. Sur une nouvelle manière d'étudier la marche du rayon extraordinaire dans le spath d'Islande. *C.R.*, xli. p. 514; *Ann. di Chim.* (3), iv. p. 250; *Pogg. Ann.* xcvii. (1856), p. 148.

1856. DE SENARMONT. Sur la réflexion totale de la lumière extérieurement à la surface des cristaux biréfringents. *Journal de Liouville*, 1856, p. 305.

1856. DE SENARMONT. Recherches sur la double réfraction. *C.R.*, xlii. p. 65.

1856. SELLA. Studi sulla mineralogica Sarda. *Mem. d. R. Accad. d. scienzi di Torino* (2), xvii. p. 289.

1856. KENNGOTT. Mittheilung über einige besondere exemplare des Calcit. *Pogg. Ann.*, xcvii. p. 310.

1857. DE SENARMONT. Note sur la construction d'un prisme biréfringent propre à servir de polariseur. *Ann. de Chim.* (3), l. p. 480.

1857. FOUCAULT. Nouveau polariseur en spath d'Islande. *C.R.*, xlv. p. 238; *Pogg. Ann.*, cii. p. 642.
See also *Ann. de Chim.* (5), xvii. (1879), p. 429; *Beiblätter*, 1879, p. 47.

1857. POTTER. On the Principle of Nicol's Rhomb, and on some Improved Forms of Rhombs for Procuring Beams of Plano-polarized Light. *Philos. Magaz.* (5), xiv. p. 482;

1858. POTTER. On the Properties of Compound Double - Refracting Rhombs. *Philos. Magaz.* (5), xvi. p. 419.

1860. JELLETT. Researches in Chemical Optics. *Rep. Brit. Assoc.*, 1860, ii. p. 13; *Proc. Roy. Irish. Acad.*, viii. (1863), p. 279; *Trans. Roy. Irish Acad.*, xxv. (1875), p. 371.

1861. PICHOT. Note sur la vérification expérimentale des lois de la double réfraction. *C.R.*, lii. p. 356.

1861. HASERT. Verbesserte Konstruktion des Nicol'schen Prismas. (Recommends balsam to have same index as extraordinary ray, and end faces to make 81° with the film.) *Pogg. Ann.*, cxiii. p. 188.

1862. STOKES. Report on Double Refraction, *Report* (22nd) *of Brit. Assoc.*, p. 253.

1863. VAN CALKER. De Phænomenis Opticis quæ præbent Crystalli Spathi Calcarii. *Univ. Diss.*, Bonn.

1864. DOVE. Ueber ein neues polarisirendes Prisma. *Pogg. Ann.*, cxxii. p. 18, and p. 456.

1864. BREWSTER. On the Influence of the Refracting face of Calcareous Spar on the Polarization, the Intensity and the Colour of Light which is Reflected. *Proc. Roy. Soc. Edin.*, v. p. 174.

1864. MASCART. Recherches sur le spectre solaire ultra-violet. (Contains his determinations of refractive indices of calcite for different Wave-lengths.) *Annales de l'Ecole Normale*, i. p. 238.

1866. HARTNACK and PRASMOUSKI. Prisme polarisateur de M. M. Hartnack et Prasmouski. *Ann. de Chim.* (4), vii. p. 181; *Pogg. Ann.*, cxxvii. p. 494; *Carl Rep.*, ii. p. 217; *Pogg. Ann.*, cxxviii. p. 336; *Jour. of the Roy. Micros. Soc.* (2), iii. (1883), p. 428.

1867-76. VOM RATH. Mineralogische Mittheilungen, in *Pogg. Ann.* Amongst these notes are a number on calcite:—1867 in vol. cxxxii. p. 517 and p. 549; 1868, in vol. cxxxv. p. 572; in Erganz. Bd. v. (1871), p. 438; in 1874, vol. clii. p. 17; in 1875, vol. clv. p. 481; in 1876, vol. clviii; p. 414. Also by same author in Groth's *Zeitschrift für Krystallographie*, vol. i. No. 6, p. 604.

1869 JAMIN. Sur un nouveau prisme polarisant. *C.R.*, lxviii. p. 221; *Pogg. Ann.*, cxxxvii. p. 174.

1870. VAN DER WILLIGEN. Les indices de réfraction du quartz et du spath d'Islande. Deuxième mémoire. *Archives du Musée Teyler.* (Haarlem), iii p. 34.

1870. GLAN. Ueber die Absorption des Lichtes. *Pogg. Ann.*, cxli. p. 58.

1872. VON DE SANDE BOKHUYZEN. (Remarks on Displacement of Images by Nicol Prisms.) *Pogg. Ann.*, cxlv. p. 258.

1872. STOKES. On the Law of Extraordinary Refraction in Iceland Spar. *Philos. Magaz.* (4), xliv. p. 316; *C.R.*, lxxvii. p. 1150.

1873. ABRIA. Études de double réfraction: vérification de la loi d'Huyghens. *C.R.*, lxxvii., p. 814; *Annales de Chim.* (5), i. p. 289.

1873. CORNU. Prisme polarisant à pénombres. *Bulletin Soc. Chim.* (2), xiv. p. 140.

1875. MACH. Ueber einen Polarisations-Apparat mit rotirendem Zerleger. *Pogg. Ann.*, clvi. p. 169; *Journal de Physique*, v. p. 71.

1875. SPOTTISWOODE. On a new Revolving Polariscope. *Philos. Magaz.* (4) xlix. p. 472; *Pogg. Ann.*, clvi. p. 654.

1875-1893. HESSENBERG. Mineralogischen Notizen. A number of these mineralogical notes relating to calcite from different sources, are to be found in vols. iii., iv., vi., vii., viii., and x. of the *Abhandlungen der Senckenbergischen Naturforschenden Gesellschaft zu Frankfurt am Maine.*

1876. SCHARFF. Ueber den inneren Zusammenhang der Krystallgestalten des Kalkspaths. Frankfurt-a-M.

1877. THOMPSON. On Interference Fringes within the Nicol Prism. *Proc. Phys. Soc.*, vol. ii. p. 157.

1878. IRBY. On the Crystallography of Calcite. (Inaugural Dissertation, Univ. Göttingen.) Bonn.

1878. LAURENT. Sur l' orientation précise de la section principale des Nicols dans les appareils de polarisation. *C.R.*, lxxxvi. p. 662.

1879. KOHLRAUSCH, W. Ueber die experimentelle Bestimmung von Lichtgeschwindigkeiten in Krystallen. *Wied. Ann.*, vi. p. 86; *Journal de Physique*, viii. p. 287.

1879. LANDOLT. Die Polarisationsapparate (A Report on Polarimeters, dealing specially on p. 366 with the Half-Shadow Prism of Schmidt and Haensch.) *Bericht über die wissenschaftlichen Instrumente auf der Berliner Gewerbeausstellung im Jahre*, 1879, p. 357. Berlin, 1880.

1879. CROVA. Étude des prismes polariseurs employés dans les observations photométriques. *Journal de Physique*, ix. p. 152.

1880. GLAN. Ueber einen Polarisator. *Carl. Rep.* xvi. p. 570; *Journal de Physique*, x. p. 175; Nachtrag zum Polarisator. *Carl Rep.*, xvii. p. 195.

1880. GLAZEBROOK. Double Refraction and Dispersion in Iceland Spar. An experimental Investigation, with a comparison with Huyghens' construction for the extraordinary wave. *Phil. Trans.*, clxxi. p. 421; *Beiblätter*, iv. p. 469.

1880. LIPPICH. Ueber ein Halbschattenpolarimeter. *Naturwiss. Jahrbuch, Lotos.* ii.

1881. THOMPSON. On a new Polarizing Prism. *Rep. Brit. Assoc.* (1881), ii. p. 563; *Philos. Magaz.* (5), xii. p. 349; *Journal de Physique* (2), i. p. 200; *Centralzeitung für Optik und Mechanik* (1882), p. 121.

1882. SARASIN. Indices de Réfraction du Spath d'Islande. *Archives des Sciences phys. et nat. de Genève* (3), viii. p. 392; *Journal de Physique* (2), ii. p. 369.

1882. LIPPICH. Ueber ein neues Halbschattenpolarimeter. *Zeitschrift für Instrumentenkunde*, 1882, p. 167. Ueber polaristrobometrische Methoden, *Wien. Ber.*, lxxxv. p. 268; *Beiblätter*, xi. (1887), p. 443

1883. GLAZEBROOK. On Polarizing Prisms. *Philos. Magaz.* (5), xv. p. 352; *Proc. Phys. Soc.*, v. p. 204.

1883. THOMPSON. Polarizing Prisms. *Philos. Magaz.* (5), xv. p. 435; *Jour. of the Roy. Micros. Soc.* (2), iii. p. 575; *English Mechanic*, No. 962, 31st August 1883, p. 596.

1884. BERTRAND, E. Sur différents prismes polarisateurs. *Bull. de la Soc. Mineralogique de France* (1884), p. 339; *C.R.*, xcix. p. 538; *Beiblätter*, ix. (1885), p. 169.

1884. FEUSSNER. Ueber die Prismen zur Polarisation des Lichtes. *Zeitschrift für Instrumentenkunde*, iv. p. 41; *Nature*, 27th March 1884, p. 515.

1884. AHRENS. On a new Polarizing Prism. *Journ. of the Roy. Microscop. Soc.* (2) iv. p. 533; *Phils. Magaz.* (5), xix. p. 69.

1885. THOMPSON. *Rep. Brit. Assoc.* (1885), p. 912.

1885. LIPPICH. Ueber polaristrobometrische Methoden, inbesondere über Halbschattenapparate. *Wien. Ber.*, xci. p. 1059; *Beiblätter*, xi. (1887), p. 443.

1885. ZENKER. (Remarks on Recent Work on Prisms.) *Zeitschrift für Instrumentenkunde*, iv. p. 135.

1886. MADAN. Modification of Polarizing Prisms of Foucault and of Ahrens. *Nature*, 1886, p. 371.

1886. ABBE. (An Improved Double-Image Prism of Glass and Spar), see Grosse's pamphlet. *Die gebräuchlichen Polarisationsprismen*, 1887.

1886. AHRENS. New Polarizing Prism. *Jour. of the Roy. Micros. Soc.* (2), vi. p. 397; see also *Philos. Magaz.* (5), xxi. p. 476; *Zeitschrift für Instrumentenkunde*, vi. p. 310.

1886. THOMPSON. Notes on some New Polarizing Prisms. *Philos. Magaz.* (5), xxi. p. 476; *Jour. of the Roy. Micros. Soc.* (2), vi. p. 1054.

1887. KETTELER. (Discussion of Indices of Ordinary Refraction in Calc - Spar) *Wied. Ann.*, xxx. p. 313.

1887. LAURENT. Méthode pratique pour l'exécution des prismes de Nicol et de Foucault. *Journal de Physique* (2), vi. p. 38.

1887. THOMPSON. Twin Prisms for Polarimeters. *Philos. Magaz.* (5), xxiv. p. 399.

1887. GROSSE. Die gebräuchlichen Polarisationsprismen. Clausthal, 1887.

1888. WENHAM. Polishing Calc-Spar. *English Mechanic*, 29th June 1888.

1888. NORRENBERG. Ueber Totalreflexion an doppeltbrechenden Krystallen. *Wied. Ann.*, xxxv. p. 843.

1889. HASTINGS. The Law of Double Refraction in Iceland Spar. *Amer. Jour. of Science* (3), xxxv. p. 60.

1889. THOMPSON. Note on Polarizing Apparatus for the Miscroscope. *Journ. of the Roy. Micros. Soc.* (2), ix. p. 617.

1891. THOMPSON. On the Use of Fluor-Spar in Optical Instruments [Spectropolariscope]. *Philos. Magaz.* (5), xxxi. p. 120.

1895. WEINSCHENK. Eine Methode zur genauen Justierung der Nicol'schen Prismen. *Zeitschrift für Instrumentenkunde*, xxiv. p. 81; *Beiblätter*, xx. (1896), p. 43.

1896. LIPPICH. Dreitheiliger Halbschattenpolarisator. *Wien. Ber.*, cv. p. 317.

1896. CATHREIN. Vervollkommnung des Dichroscops. *Zeitschrift für Instumentenkunde*, xvi. p. 225; *Beiblätter*, xx. p. 620.

1897. THOMPSON. [An Economical Method of cutting Improved Nicol Prisms, described on p. 122 of *Light, Visible and Invisible*, London, 1897.]

1897. LEISS. Ueber ein neues, aus Kalkspath und Glas zusammengesetztes Nicol'sches Prisma. *Ber. Akad.* (Oct. 1897), p. 901; *Beiblätter*, xxii. (1898), p. 104.

1897. STEWART. On the Absorption of the Extraordinary Ray in Uniaxial Crystals. *Phys. Rev.*, iv. p. 433.

1898. VON LOMMEL. Ueber aus Kalkspat und Glas zusammengesetzte Nicol'sche Prisma. *Muenchener Akad.* (Feb. 1898), p. 111; *Beiblätter*, xxii. p. 404.

1898. CARVALLO. Recherches de précision sur la dispersion infra-rouge du spath d'Islande. *C.R.*, cxxvi. p. 950.

1899. WILLIAMS. On Double Refraction in Naphtalin, etc. *Chem. News*, lxxx. p. 305.

1900. MACÉ DE LÉPINAY. Sur un nouveau polarimètre à pénombres. *Journal de Physique* (3), x. p. 585. *C.R.*, cxxxi. p. 832; *Beiblätter*, xxv. p. 444.

1901. MADAN. The Colloid Form of Piperine with Especial Reference to its Refractive and Dispersive Powers. *Trans. Chem. Soc.*, lxxix. p. 922.

1902. GIFFORD. Refractive Indices of Fluorite, Quartz, and Calcite. *Proc. Roy. Soc.*, cxx. p. 329.

35

Reprinted from *Polarized Light,* Harvard University Press, Cambridge, Mass., 1966, 173–202

Polarized Light

WILLIAM A. SHURCLIFF

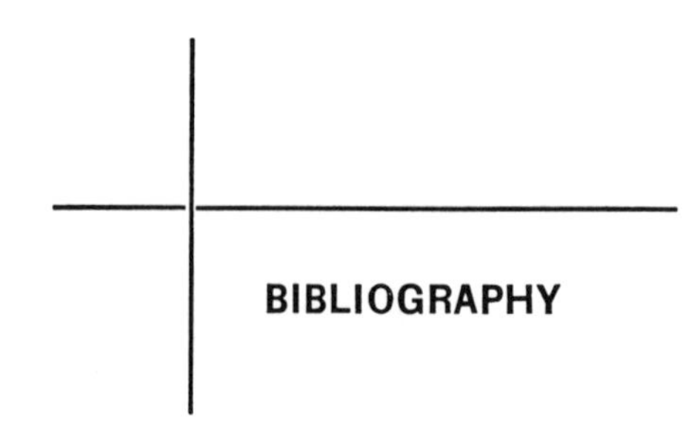

BIBLIOGRAPHY

A-1 Abraham, M., and R. Becker, *Classical theory of electricity and magnetism* (Blackie, London, 1937).

A-2 Alexander, N., *Photoelasticity* (Rhode Island State College, Kingston, 1936).

A-3 Allen, P. J., "An instantaneous microwave polarimeter," *Proc. IRE 47*, 1231 (1959).

A-4 Ambronn, H., "Über den Dichroismus pflanzlicher Zellmembranen," *Ann. Physik 34*, 344 (1888).

A-5 Ambronn, H., and A. Frey, *Das Polarisationsmikroskop, seine Verwendung in der Kolloidforschung und in der Farberei* (Akademische Verlag, Leipzig, 1926).

A-6 Ambrose, E. J., A. Elliott, and R. B. Temple, "Use of polarized infrared radiation in the study of crystal structures," *Proc. Roy. Soc. (London) A 206*, 192 (1951).

A-7 *American Institute of Physics handbook* (McGraw-Hill, New York, 1957). Presents data on sheet polarizers, crystals, etc.

A-8 Anderson, S., "Orientation of methylene blue molecules adsorbed on solids," *J. Opt. Soc. Amer. 39*, 49 (1949).

A-9 Angenetter, H., and H. Verleger, "Polarization of Light Emitted from Canal Rays," *Physik. Z. 39*, 328 (1938).

A-10 Anonymous historical account of discoveries on double refraction and polarization, *Edinburgh Phil. J. 1*, 289 (1819).

A-11 Anon., *Lectures on polarized light*, lectures delivered before the Pharmaceutical Society of Great Britain and in the Medical School of the London Hospital (Longman, Brown, Green, and Longman, London, 1843) 110 pp.

A-12 Anonymous article on National Bureau of Standards NBS Sky Compass, *Rev. Sci. Instr. 20*, 460 (1949).

A-13 Anonymous, "Polaroid's vectograph: the system that puts 3-D on one film," *Ideal Kinema 20*, 14 (1954).

A-14 Anonymous article on use of polarizers in three-dimension television system worked out by Pye, Ltd. *British Kinematography 26*, 46 (1955).

A-15 Anonymous article on use of polarizers in 3-D TV system used at Harwell, England, *Nucleonics 15*, 117 (July 1957).

A-16 Anonymous, "Polarization of light of moon and planets," *Sky and Telescope 19*, 17 (1959).

A-17 Appel, A. V., and D. A. Pontarelli, "Infrared polariscope for photoelastic measurements of semiconductors," talk given at meeting of American Optical Society, March 29, 1958; employed Glan-Thompson polarizers, quartz retarders.

A-18 Arago, F. J., *Oeuvres completes*, vol. 10, p. 36 (1812).

— Archambault, C., U.S. patent 2,813,459. Variable-density sunglasses.

A-19 Archard, J. F., and A. M. Taylor, "Improved Glan-Foucault prism," *J. Sci. Instr. 25*, 407 (1948).

A-20 Archard, J. F., "Performance and testing of polarizing prisms," *J. Sci. Instr. 26*, 188 (1949).

A-21 Archard, J. F., P. L. Clegg, and A. M. Taylor, "Photoelectric analysis of elliptically polarized light," *Proc. Phys. Soc. (London) B, 65*, 758 (1952).

A-22 Automobile Manufacturers Association. *Public side of the problem of polarized automobile lights*, pamphlet issued in May 1952, 12 pp. Lists drawbacks to polarized headlights.

A-23 Autrum, H., and H. Stumpf, "Das Bienenauge als Analysator für Polarisiertes Licht," *Z. Naturforsch., Pt. b, 5*, 116 (1950).

— Baerwald, H. G., U.S. patent 2,766,659 (1956). An electro-optic modulator of great complexity.

— Bailey, E. D., and M. M. Brubaker, U.S. patent 2,246,087 (1941). Describes methods of stretching various materials as for improving the performance of the resulting polarizer.

B-1 Baird Associates, Technical Circular RD 505 (1951). Describes a modified Lyot-Öhman filter priced at $3000.

B-2 Barer, R., "Variable colour-amplitude phase-contrast microscopy," *Nature 164*, 1087 (1949).

B-3 Barer, R., "Some experiments with polarizing films in the ultraviolet," *J. Sci. Instr. and Phys. in Ind. 26*, 325 (1949). Describes use of a PVA-iodine polarizer in the near ultraviolet.

B-4 Barer, R., "Phase contrast and interference microscopy," in D. A. Spencer *Progress in photography* (Focal Press, London), Vol. 2, 1951–1954.

B-5 Bartholinus, E., *Experimenta crystalli Islandici disdiaclastici quibus mira & insolita refractio detegetur* (Hafnia [Copenhagen], 1670). Announces discovery of double refraction.

B-6 Bates, F. J., *Polarimetry, saccharimetry and the sugars*, National Bureau of Standards. Circular C-440 (1942).

B-7 Baxter, L., private communication.

B-8 Baxter, L., A. S. Makas, and W. A. Shurcliff, "Measuring spectral properties of high-extinction polarizers," *J. Opt. Soc. Amer. 46*, 229 (1956). Describes rotating holder, and H and K polarizers for which G_{90} exceeds 5.0.

B-9 Baxter, L., "On the properties of polarization elements as used in optical instruments. III. Angular aperture functions of a positive dichroic film polarizer," *J. Opt. Soc. Amer. 46*, 435 (1956). Discusses obliquity effects in HN-22.

B-10 Bayley, H. G., "Gelatin as a photo-elastic material," *Nature 183*, 1757 (1959).

B-11 Baylor, E. R., and F. E. Smith, "Orientation of *Cladocera* to polarized light," *Am. Naturalist 87*, 97 (1953).

B-12 Bennett, H. E., J. M. Bennett, and M. R. Nagel, "Question of the polarization of infrared radiation from the clear sky," *J. Opt. Soc. Amer. 51*, 237, (1961).

B-13 Bennett, H. S., "Microscopical investigation of biological materials with polarized light," in Ruth McClung, ed., *McClung's handbook of microscopical technique* (Hoeber, New York, 1949).

B-14 Berkman, S., J. Boehm, and H. Zocher, "Anisotropes Kupfer, Silber, und Gold," *Z. Physik. Chem. 124*, 83 (1926), article on dichroism of aligned microcrystals of gold, silver, mercury, etc.

B-15 Bernauer, F., "Neue Wege zur Herstellung von Polarisatoren," *Fortschr. Mineral. Krist. Petrog. 19*, 22 (1935).

B-16 Bertrand, E., "Optique — sur un nouveau prisme polarisateur," *Compt. rend. 96*, 538 (1884).

B-17 Besse, A., and F. Desvignes, "An infrared polariscope," *Rev. opt. 38*, 344 (1959). Device employing silicon mirrors at Brewster's angle.

B-18 Beth, R. A., "Mechanical detection and measurement of the angular momentum of light," *Phys. Rev. 50*, 115 (1936). Deals with torque produced by circularly polarized light.

B-19 Bhagavantam, S., "Polarization of Raman lines in liquids," *Indian J. Phys. 7*, 79 (1932).

B-20 Billardon, M., and J. Badoz, "Spectropolarimètre photoélectrique destiné à l'étude de la dispersion du pouvoir rotatoire naturel," *Compt. rend. 248*, 2466 (1959).

B-21 Billings, B. H., "Tunable narrow-band optical filter," *J. Opt. Soc. Amer. 37*, 738 (1947).

B-22 Billings, B. H., and E. H. Land, "Comparative survey of some possible systems of polarized headlights," *J. Opt. Soc. Amer. 38*, 819 (1948).

B-23 Billings, B. H., "Monochromatic depolarizer," *J. Opt. Soc. Amer. 41*, 966 (1951).

B-24 Billings, B. H., "Un depolarizador monocromatico," *Ciencia e invest. (Buenos Aires) 8*, 99 (1952).

B-25 Biot, J. R., "Sur un mode particulier de polarisation qui s'observe dans la tourmaline, "*Bull. soc. philomath. Paris 6*, 26 (1815).

B-26 Bird, G. R., M. Parrish, Jr., and E. R. Blout, "Apparatus for the observation of infrared streaming dichroism of polymer solutions," *Rev. Sci. Instr. 29*, 305 (1958).

B-27 Bird, G. R., and M. Parrish, Jr., "The wire grid as a near-infrared polarizer," *J. Opt. Soc. Amer. 50*, 886 (1960).

B-28 Bird, G. R., and W. A. Shurcliff, "Pile-of-plates polarizers for the infrared: improvement in analysis and design," *J. Opt. Soc. Amer. 49*, 235 (1959).

B-29 Birge, R. T., and L. A. DuBridge, "Nature of unpolarized light," *J. Opt. Soc. Amer. 25*, 179 (1935).

B-30 Blackwell, H. R., "Effect of tinted optical media upon visual efficiency at low luminance," *J. Opt. Soc. Amer. 43*, 815 (1953). Points out harm of tinted windshields and tinted glasses.

B-31 Blake, R. P., A. S. Makas, and C. D. West, "Molecular-type dichroic film polarizers for 0.75 to 2.8 micron radiations," *J. Opt. Soc. Amer. 39*, 1054 (1949). Describes HR polarizer.

— Blake, R. P., U.S. patent 2,494,686. Infrared polarizer.

B-32 Blout, E. R., and R. Karplus, "Infrared spectrum of polyvinyl alcohol," *J. Am. Chem. Soc.* *70*, 862 (1948).
B-33 Blout, E. R., G. R. Bird, and D. S. Grey, "Infrared microspectroscopy," *J. Opt. Soc. Amer.* *39*, 1052A (1949).
B-34 Blout, E. R., G. R. Bird, and D. S. Grey, "Infrared microspectroscopy," *J. Opt. Soc. Amer.* *40*, 304 (1950).
B-35 Blout, E. R., and G. R. Bird, "Infrared microspectroscopy, II," *J. Opt. Soc. Amer.* *41*, 547 (1951).
B-36 Blout, E. R., talk on "Chemical optics" to New England Section of the Optical Society of America, May 20, 1954.
B-37 Boehm, G., and R. Signer, "Über die Strömungsdoppelbrechung von Eiweisslösungen," *Helv. Chim. Acta.* *14*, 1370 (1931). Article on flow birefringence.
B-38 Boehm, G., "Über maculare (Haidinger'sche) Polarisationbüschel und über einen polarisationoptischen Fehler des Auges," *Acta Ophthalmol.* *18*, 109 (1940).
B-39 Bolla, G. V., "Polarisationseffekte in Quarzspektrographen," *Z. Physik* *103*, 756 (1936).
B-40 Bond, W. L., and J. Andrus, "Photograph of the stress field around edge dislocations," *Phys. Rev.* *101*, 1211 (1956).
— Bond, W. L., U.S. patent 2,768,557. Electro-optic shutter compensated for divergent rays.
B-41 Born, M., *Optik* (Springer, Berlin, 1933).
B-42 Born, M., "Theory of optical activity," *Proc. Roy. Soc.* (*London*) *A 150*, 84 (1935).
B-43 Born, M., and E. Wolf, *Principles of optics* (Pergamon, New York, 1959).
B-44 Bouasse, H., *Optique cristalline double refraction, polarisation rectiligne et elliptique* (Librairie Delegrave, Paris, 1925).
B-45 Bouhet, C., and Lafont, "New normal field polarizers," *Rev. opt.* *28*, 490 (1949). Describes a square-ended prism-type polarizer.
B-46 Bouriau, Y., and J. Lenoble, "Etudé des polariseurs pour l'ultraviolet," *Rev. opt.* *36*, 531 (1957). Describes various cemented-prism-type, ultraviolet polarizers.
B-47 Bovis, P., "Spectres d'absorption et pléochroïsme de l'iode et de l'hérapathite," *Compt. rend.* *184*, 1237 (1927). States the dichroic ratio of individual crystals of iodine.
B-48 Brewster, D., "On the laws which regulate the polarisation of light by reflexion from transparent bodies," paper read on March 16, 1815, published in *Phil. Trans.* *105*, 125 (1815).
— Bridges, S. W., and J. R. Roehrig, U.S. patent 2,824,488. Polarizing-microscope apparatus for grading fibers.
B-49 Brode, W. R., and C. H. Jones, "Recording spectrophotometer and spectropolarimeter," *J. Opt. Soc. Amer.* *31*, 743 (1941).
B-50 Brode, W. R., "Optical rotation of polarized light by chemical compounds," *J. Opt. Soc. Amer.* *41*, 987 (1951).
— Brown, C. H., U.S. patents 2,224,214 and 2,287,598. Polarizer employing wires.
B-51 Brown, T. B., "Elliptic polarimeter for the student laboratory: specimens of elliptically polarized light," *Am. J. Phys.* *26*, 183 (1958).

B-52 Browne, C. A., and F. W. Zerban, *Physical and chemical methods of sugar analysis* (Wiley, New York, 1941).

B-53 Bruhat, G., "Le dichroïsme circulaire," *Rev. opt. 8*, 365, 413 (1929).

B-54 Bruhat, G., *Traité de polarimetrie* (Editions Revue d'Optique, Paris, 1930).

B-55 Bruhat, G., and P. Guenard, "Étude du dichroïsme circulaire de solutions de camphre dans des solvants organiques," *Compt. rend. 203*, 784 (1936).

B-56 Buijs, K., "Preparation of selenium polarizers for the near infrared region," *Appl. Spectroscopy 14*, 81 (1960).

B-57 Burri, C., *Das Polarisationsmikroskop* (Birkhauser, Basel, 1950).

C-1 Cabannes, J., and P. Daure, "Sur le spectre Raman du benzène en lumière circulaire," *Compt. rend. 208*, 1700 (1939).

C-2 Campbell, C. D., "Sky compass for field use," *Northwest Sci. 28*, 43 (1954).

C-3 Carpenter, R. O., "Electro-optic effect in uniaxial crystals of the dihydrogen phosphate type, III. Measurements of coefficients," *J. Opt. Soc. Amer. 40*, 225 (1950). Discusses smallest measurable changes.

C-4 Cayrel, R., and L. Schatzman, "Nouveaux polariseurs et leur procédé de fabrication," French patent 1,109,170 (July 1954).

C-5 Central Scientific Co., *Cenco News Chats* (No. 86, 1959). Discussion of equipment for demonstrating polarization of microwaves.

C-6 Cerf, R., and H. A. Scheraga, "Flow birefringence in solutions of macromolecules," *Chem. Revs. 51*, 185 (1952).

C-7 Chandrasekhar, S., "On the radiative equilibrium of a stellar atmosphere. X," *Astrophys. J.*, *103*, 350 (1946). Prediction on polarization of starlight.

C-8 Chandrasekhar, S., *Radiative transfer* (Clarendon Press, Oxford, 1950), 393 pp.

C-9 Chandrasekhar, S., "The optical rotatory power of quartz and its variation with temperature," *Proc. Indian Acad. Sci. 35*, 103 (1952).

C-10 Chandrasekhar, S., and D. D. Elbert, "Illumination and polarization of the sunlit sky on Rayleigh scattering," *Trans. Am. Phil. Soc. 44*, Pt. 6, 88 (1954).

C-11 Chandrasekaran, K. S., "Perfect polarization of x-rays by crystal reflection," *Proc. Indian Acad. Sci. A 44*, 387 (1956).

C-12 Chapman, J. A., "Device for visualizing the pattern of plane polarized light from blue sky," *Nature 181*, 1393 (1958).

C-13 Charney, E., "Dichroic ratio measurements in the infrared region," *J. Opt. Soc. Amer. 45*, 980 (1955). Discusses errors due to polarization by the spectrophotometer prism.

C-14 Charru, M. A., "Expériences à 3000 MHz sur les aériens hélicoïdaux; réalisation et analyse d'un dichroïsme circulaire," *J. phys. radium 21*, 93 S (1960). Describes experiments in which right and left helical antennas were employed.

C-15 Choudhuri, K., "On the polarized fluorescence-dyestuffs in solution," *Indian J. Phys. 18*, 74 (1944).

— Chubb, L. W., U.S. patent 2,087,795. Headlight system employing polarizers at 45°.

C-16 Chubb, L. W., Jr., D. S. Grey, E. R. Blout, and E. H. Land, "Properties of polarizers for filters and viewers for 3-D motion pictures," *J. Soc. Motion Picture Television Engrs. 62*, 120 (1954).

C-17 Clarke, J. T., and E. R. Blout, "Nature of the carbonyl groups in polyvinyl alcohol," *J. Polymer Sci. 1*, 419 (1946).

C-18 Cohen, M. H., "Radio astronomy polarization measurements," *Proc. I.R.E. 46*, 172 (1958).

C-19 Cohen, S. G., H. C. Haas, and H. Slotnick, "Studies on hydroxyethylpolyvinyl alcohol," *J. Polymer Sci. 11*, 193 (1953). Shows x-ray diffraction pattern for PVA and related materials.

C-20 Coker, E. G., and L. N. G. Filon, *A treatise on photo-elasticity* (Cambridge University Press, New York, ed. 2, rev. by H. T. Jessop, 1957), 720 pp.

C-21 Colman, K. W., D. Courtney, J. B. Freeman, and R. Bernstein, "The control of specular reflections from bright tube radar displays," Nov. 15, 1958, Report No. 23 by Courtney & Co., Philadelphia, for Engineering Psychology Branch of Office of Naval Research, contract Nonr-2346(00).

C-22 Compton, W. D., "Polarized light as a tool for studying the symmetry properties of color centers," talk at March 1958 meeting of American Physical Society.

C-23 Condon, E. U., "Theories of optical rotatory power," *Revs. Mod. Phys. 9*, 432 (1937).

C-24 Conn, G. K. T., and F. J. Bradshaw, *Polarized light in metallography* (Butterworth, London, 1952).

C-25 Conn, G. K. T., and G. D. Eaton, "On the use of a rotating polarizer to measure optical constants in the infrared," *J. Opt. Soc. Amer. 44*, 484 (1954). Claim 99.7-percent polarizance for an eight-film selenium pile of plates.

C-26 Conn, G. K. T., and G. D. Eaton, "On polarization by transmission with particular reference to selenium films in the infrared," *J. Opt. Soc. Amer. 44*, 553 (1954).

C-27 Conroy, J., "Polarization of light by crystals of iodine," *Proc. Roy. Soc. (London) 25*, 51 (1876).

C-28 Cords, O., "Das Haidingersche Büschel und seine Erklärung als Beitrag für eine physikalische Deutung des Sehvorganges," *Optik 2*, 423 (1947).

C-29 Cotton, A., "Absorption inégale des rayons circulaires droit et gauches dans certains corps actifs," *Compt. rend. 120*, 989 (1895). Discovery of circular dichroism in solutions.

C-30 Cotton, A., "Prismes polariseurs a champ normal fondés sur la reflexion crystalline interne," *Compt. rend. 193*, 268 (1931). Describes the Cotton polarizer.

C-31 Coulson, K. L., D. Deirmendjian, R. S. Fraser, C. Seaman, and Z. Sekera, *Investigation of polarization of skylight* (University of California, Department of Meteorology, June 1955), 198 pp.

C-32 Culver, W. H., "The maser," *Science 126*, 810 (1957).

D-1 D'Agostino, J., D. C. Drucker, C. K. Liu, and C. Mylonas, "Analysis of plastic behavior of metals with bonded birefringent plastics," *Proc. Soc. Exp. Stress Anal. 12*, 115 (1955).

D-2 D'Agostino, J., D. C. Drucker, C. K. Liu, and C. Mylonas, "Epoxy adhesives and casting resins as photoelastic plastics," *Proc. Soc. Exp. Stress Anal. 12*, 123 (1955).

D-3 Dawson, E. F., and N. O. Young, "Helical Kerr cell," *J. Opt. Soc. Amer. 50*, 170 (1960). Confirms calculations easily made with the aid of the Jones calculus that a skewed series of linear retarders is equivalent to a single circular retarder.

D-4 Dehmelt, H. G., "Slow spin relaxation of optically polarized sodium atoms," *Phys. Rev. 105*, 1487 (1957). Deals with absorption of circularly polarized light in masers.

D-5 Demon, L., "Propriétés optiques des cristaux de bleu de méthylène," *Ann. phys. 1*, 101 (1946). Describes rubbing glass surfaces and applying dyes.

D-6 de Vaucouleurs, G., *Physics of the planet Mars* (Macmillan, New York, 1954). Reviews use of polarizers in determining the amount of atmosphere on Mars.

D-7 Dévé, C., *Optical workshop principles* (Hilger and Watts, London, 1954). Discusses manufacture of polarizing prisms.

D-8 de Vries, H., R. Jielof, and A. Spoor, "Properties of the human eye with respect to linearly and circularly polarized light," *Nature 166*, 958 (1950).

D-9 Dietze, G., *Einführung in die Optik der Atmosphäre* (Akademisch Verlag, Leipzig, 1957), 263 pp. Discusses polarization.

D-10 Ditchburn, R. W., *Light* (*Interscience*, New York, 1953), 680 pp.

D-11 Ditchburn, R. W., "Some new formulas for determining the optical constants from measurements on reflected light," *J. Opt. Soc. Amer. 45*, 743 (1955).

D-12 Dmitri, I., "Two new tools for color," *Photography 9*, 68 (October 1954). Describes 1954 General Electric variable color filters.

D-13 Dobrowolski, J. A., "Mica interference filters with transmission bands of very narrow half-width," *J. Opt. Soc. Amer. 49*, 794 (1959).

D-14 Dollfus, A., "Détermination de la pression atmosphérique sur la planète Mars," *Compt. rend. 232*, 1066 (1951).

D-15 Downie, A. R., "Removal of signal fluctuations in a photoelectric polarimeter," *J. Sci. Instr. 35*, 114 (1958).

D-16 Dreyer, J. F., "A variable light polarizing coating," *J. Opt. Soc. Amer. 37*, 983 (1947). Discusses Beilby layers.

D-17 Dreyer, J. F., and C. W. Ertel, "Orientation of the surface of glass," *Glass Ind. 29*, 197 (1948).

D-18 Dreyer, J. F., "Some of the applications of polarized light to photography," *PSA Journal 20*, 28 (June 1954).

— Dreyer, J. F., U.S. patents:
2,400,877 (1946). Dichroic polarizer.
2,432,867 (1947). Light-polarizing coating on curved surface such as that of a lamp.
2,481,830 (1949). Method of preparing a dichroic polarizer; the method employs an intermediate dichroic film.
2,484,818 (1949). Polarizing mirror involving nematic state.
2,544,659 (1951). Dichroic polarizer.
2,776,598 (1957). Polarizing mirror involving nematic state.

D-19 Dreyfus, M., "Measurement of birefringence," *J. Opt. Soc. Amer. 46*, 142 (1956).

D-20 du Bois, H., and H. Rubens, "Polarisation ungebeugter langwelliger Wärmstrahlen durch Drahtgitter," *Ann. Physik 35*, 243 (1911).

D-21 Dudley, B., "Vectograph stereograms," *Photo Tech. 3*, 30 (May 1941).

D-22 Dufay, J., *Galactic nebulae and interstellar matter* (Philosophical Library, New York, about 1957). Chapter 15 discusses polarization of the light from stars by interstellar matter.

D-23 Dunsmuir, P., "Use of polarized light for the examination of etched metal crystals and their orientation," *Brit. J. Appl. Phys. 3*, 264 (1952).

D-24 DuPont (E. I. duPont de Nemours and Company), *DuPont ELVANOL polyvinyl alcohols* (DuPont, Wilmington, Delaware, 1953).

D-25 Duverney, R., and A. M. Vergnoux, "Polarimétrie dans l'infra-rouge," *J. phys. radium 18*, 527 (1957). A long review article comparing many types of infrared polarizers.

D-26 Dyson, J., "Interferometer for straightness measurement," *Nature 175*, 559 (1955). Describes use of polarizers and concave mirrors in measuring straightness as a probe (Wollaston prism) is moved along the alleged straight line.

E-1 Eastman Kodak Co., *Photography by polarized light* (Pamphlet 4-36-CH-10; April 1936), 32 pp.

E-2 Eastman Kodak Co., *Photography by polarized light* (booklet, 1938), 40 pp.

E-3 Eastman Kodak Co., *Eastman Pola-Screens* (Pamphlet 8-38-CH-25, 1938), 6 pp.

E-4 Eastman Kodak Co., *Kodak filters and Pola-Screens for black-and-white photography* (booklet, ed. 4, 1956), 54 pp.

E-5 Edgerton, H. E., and C. W. Wyckoff, "A rapid-action shutter with no mechanical moving part," *J. Soc. Motion Picture Engrs. 56*, 398 (1951).

E-6 Edsall, J. T., A. Rich, and M. Goldstein, "An instrument for the study of double refraction of flow at low and intermediate velocity gradients," *Rev. Sci. Instr. 23*, 695 (1952).

E-7 Edwards, D. F., and M. J. Bruemmer, "Polarization of infrared radiation by reflection from germanium surfaces," *J. Opt. Soc. Amer. 49*, 860 (1959).

E-8 Elliott, A., and E. J. Ambrose, "Polarization of infra-red radiation," *Nature 159*, 641 (1947). Discusses a transmission-type polarizer using three selenium films.

E-9 Elliott, A., E. J. Ambrose, and R. B. Temple, "Polarization of infra-red radiation," *J. Opt. Soc. Amer. 38*, 212 (1948). Describes a device employing selenium.

E-10 Elliott, A., E. J. Ambrose, and R. B. Temple, "Double orientation and infra-red dichroism in polymers," *Nature 163*, 567 (1949). Discusses polyvinyl alcohol, nylon, etc.

E-11 Elliott, A., "Infra-red dichroism and chain orientation in crystalline ribonuclease," *Proc. Roy. Soc. (London) A 211*, 490 (1952).

E-12 Elliott, A., "Infra-red dichroism in synthetic polypeptides," *Nature 172*, 359 (1953).

E-13 Ellis, J. W., and J. Bath, "The near infra-red absorption spectrum of sucrose crystals in polarized light," *J. Chem. Phys. 6*, 221 (1938). Discusses polarizing prisms suitable for use in near infrared.

E-14 Ellis, J. W., and L. Glatt, "Channeled infra-red spectra produced by birefringent crystals," *J. Opt. Soc. Amer. 40*, 141 (1950). Discusses inadvertent polarization by prisms in infrared spectrophotometers.

E-15 Eshbach, J. R., and M. W. P. Strandberg, "Apparatus for Zeeman effect measurement of microwave spectra," *Rev. Sci. Instr. 23*, 623 (1952).

E-16 Evans, J. W., "The birefringent filter," *J. Opt. Soc. Amer. 39*, 229 (1949). Discusses the Lyot-Öhman filter.

E-17 Evans, J. W., "Šolc birefringent filter," *J. Opt. Soc. Amer. 48*, 142 (1958).

F-1 Fagg, L. W., and S. S. Hanna, "Polarization measurements on nuclear gamma rays," *Revs. Mod. Phys. 31*, 711 (1959). Comprehensive article on the production and detection of polarized gamma rays.

F-2 Fahy, E. F., and M. A. MacConaill, "Optical properties of cellophane," *Nature 178*, 1072 (1956).

F-3 Falkoff, D. L., and J. E. McDonald, "On the Stokes parameters for polarized radiation," *J. Opt. Soc. Amer. 41*, 861 (1951).

F-4 Fallon, J., unpublished manuscript written in about 1950.

F-5 Fano, U., "Remarks on the classical and quantum-mechanical treatment of partial polarization," *J. Opt. Soc. Amer. 39*, 859 (1949). Relates Stokes parameters to quantum-mechanical properties of photons.

F-6 Fano, U., "Description of states in quantum mechanics by density matrix and operator techniques," *Revs. Mod. Phys. 29*, 74 (1957).

F-7 Farwell, H. W., "Scattered light from Polaroid plates," *J. Opt. Soc. Amer. 28*, 460 (1938).

F-8 Federov, F. I., "On the theory of optical activity in crystals," *Optics and Spectroscopy 6*, 49 (1959).

F-9 Filon, L. N. G., *Manual of photo-elasticity for engineers* (Cambridge University Press, Cambridge, 1936), 140 pp.

F-10 Finch, D. M., J. M. Chorlton, and H. F. Davidson, "The effect of specular reflection on visibility," in *Proceedings of the Fourteenth Session of the International Commission on Illumination* (Illuminating Engineering Society, New York, 1959).

F-11 Foitzik, L., and K. Lenz, "Einfluss des Aerosols auf die Himmelslichtpolarisation," *Optik 17*, 554 (1960).

F-12 Försterling, K., *Lehrbuch der Optik* (Hirzel, Leipzig, 1928). Contains long section on crystals, birefringence, etc.

F-13 Fox, A. G., S. E. Miller, and M. T. Weiss, "Behavior and applications of ferrites in the microwave region," *Bell System Tech. J. 34*, 5 (1955).

F-14 Françon, M., "Polarization apparatus for interference microscopy and macroscopy of isotropic transparent objects," *J. Opt. Soc. Amer. 47*, 528 (1957).

F-15 Fresnel, A. J., "Sur la diffraction de la lumière, où l'on examine particulièrement le phénomène des franges colorées que présentent les ombres des corps éclairés par un point lumineux," *Ann. chim. (Paris)* [2] *1*, 239 (1816). Describes noninterference of orthogonally polarized rays.

F-16 Friedman, G. H., "An elliptical polarization synthesizer," Communication and Electronics, July, 1955.

F-17 Frisch, K. von, *Bees: their vision, chemical sense, and language* (Cornell University Press, Ithaca, 1950).

F-18 Frocht, M. M., *Photoelasticity* (Wiley, New York, 1941).

F-19 Frocht, M. M., and R. Guernsey, "Studies in three-dimensional photoelasticity," *Proceedings of the First U.S. National Congress of Applied Mechanics* (American Society of Mechanical Engineers, New York, 1951), p. 301.

— Frost, R. H., U.S. patent 2,856,810. Automobile sun visor.

F-20 Fuessner, K., "Ueber die Prismen zur Polarisation des Lichtes," *Z. Instrumentenk. 4*, 47 (1884).

G-1 Gabler, F., and P. Sokob, "Senarmont compensator," *Z. Instrumentenk. 58*, 301 (1938).

G-2 Gänge, C., *Polarisation des Lichtes* (Quandt & Händel, Leipzig, 1894).

G-3 Ganguly, S. C., and N. K. Chaudhury, "Anisotropy of fluorescence of some organic crystals," *Phys. Rev. 95*, 1148 (1954).

G-4 General Electric Company, *How to use your variable color filter* (Pamphlet GE-J-2411A, 11-54), 16 pp.

G-5 George, W. H., "Production of polarized x-rays," *Proc. Roy. Soc. (London) A 156*, 96 (1936).

G-6 Gillham, E. J., and R. J. King, "New design of spectropolarimeter," *J. Sci. Instr. 38*, 21 (1961).

G-7 Godina, D. A., and G. P. Faerman, "Colloidal suspensions of herapathite for the construction of polarizing luminous filters," *J. Appl. Chem. (U.S.S.R.) 14*, 362 (1941). Describes making suspensions of herapathite crystals.

G-8 Godina, D. A., "Optical properties of polarizing filters made of polyvinyl alcohol," *J. Tech. Phys. (U.S.S.R.) 18*, 1317 (1948).

G-9 Goetze, R., "La ley de Malus y su comprabación experimental," paper presented in Caracas, Venezuela, at the February 1955 meeting of the Venezuelan Association for the Advancement of Science (AVAC). Presents data on the degree of polarization in light emitted at grazing angles from tungsten filaments.

G-10 Grabau, M., "Optical properties of Polaroid for visible light," *J. Opt. Soc. Amer. 27*, 420 (1937).

G-11 Grabau, M., "Polarized light enters the world of everyday life," *J. Appl. Phys. 9*, 215 (1938).

G-12 Grabau, M., *Introduction to polarized light and its applications* (Pamphlet, Polaroid Corporation, Cambridge, Massachusetts, 1940), 46 pp. (A revised version was prepared by the Polaroid Corporation in 1945; see P-22).

G-13 Groosmuller, J. T., "Das Polarisationsfeld Nicolscher Prismen," *Z. Instrumentenk. 46*, 563 (1926). Excellent discussion of obliquity effects in one nicol or two crossed nicols.

G-14 Gurnee, E. F., "Theory of orientation and double refraction in polymers," *J. Appl. Phys. 25*, 1232 (1954).

H-1 Haas, H. C., "Note on the infrared absorption spectrum of polyvinyl alcohol," *J. Polymer Sci. 26*, 391 (1957).

H-2 Haase, M., "Dichroitische Kristalle und ihre Verwendung für Polarisationsfilter," *Zeiss Nachr. 2*, 55 (August 1936).

H-3 Haase, M., "Beispiele zur Wirkungsweise der Polarisationsfilter," *Zeiss Nachr. 2*, 55 (August 1936).

H-4 Haase, H. H., "Neue Polarizations filter und der Verwendung Dichroitischer Kristalle," *Z. tech. Phys. 18*, 69, 1937. Mentions Zeiss Herotare and Mipolare linearly polarizing filters and discusses obliquity effects in certain types of polarizers.

H-5 Haber, H., "Safety hazard of tinted automobile windshields at night," *J. Opt. Soc. Amer. 45*, 413 (1955). States that use of isotropic dyes in windshields or "night-glasses" is harmful in night driving.

H-6 Haidinger, W., "Über das direkte Erkennen des polarisierten Lichts und der Lage der Polarisationsebene," *Ann. Physik 63*, 29 (1844).

H-7 Haidinger, W., "Ueber den Pleochroismus des Amethysts," *Ann. Physik 70*, 531 (1847). Discovery of circular dichroism.

H-8 Hall, J. S., and A. H. Mikesell, *Polarization of light in the galaxy as determined from observations of 551 early-type stars* (Government Printing Office, Washington, 1950; U.S. Naval Observatory Publications, vol. 17, pt. 1), p. 16.

H-9 Hall, J. S., "Some polarization measurements in astronomy," *J. Opt. Soc. Amer. 41*, 963 (1951).

H-10 Hallimond, A. F., *Manual of the polarizing microscope* (Cooke Troughton and Simms, Ltd., York, England, n. d.).

H-11 Hallimond, A. F., "Use of Polaroid for the microscope," *Nature 154*, 369 (1944).

H-12 Hardy, A. C., "A new recording spectrophotometer," *J. Opt. Soc. Amer. 25*, 305 (1935). Mentions spectropolarimeter.

H-13 Hariharan, P., "Accurate measurements of phase differences with the Babinet compensator," *J. Sci. Instr. 37*, 278 (1960). Device produces fringes by a double-pass method.

H-14 Harrick, N. J., "Reflection of infrared radiation from a germanium-mercury interface," *J. Opt. Soc. Amer. 49*, 376 (1959).

H-15 Harrick, N. J., "Infrared polarizer," *J. Opt. Soc. Amer. 49*, 379 (1959). Employs germanium-mercury interface. Effective from 2 to 200 μ.

H-16 Hartshorne, N. H., and A. Stuart, *Crystals and the polarizing microscope* (Arnold Press, London, ed. 2, 1950).

H-17 Hartwig, G., and H. Schopper, "Circular polarization of internal bremsstrahlung emitted in the *K* capture of A^{37}," *Bull. Am. Phys. Soc.* [2] *4*, 77 (1959). Abstract.

H-18 Heller, W., "Polarimetry," in A. Weissberger, ed., *Physical methods in organic chemistry*, vol. 1, part 3 (Interscience, New York, ed. 3, 1960), chap. 33.

H-19 Helmholtz, H. von, *Physiological optics*, ed. J. P. C. Southall (Optical Society of America, 1924), 3 vols.

H-20 Helwich, O., *Wissenschaftliche Photographie* (Helwich, Darmstadt, 1958).

H-21 Henroit, E., "Les couples exercés par la lumière polarisée circulairement," *Compt. rend. 198*, 1146 (1934).

H-22 Henry, F. G., "Improved lighting in the Bendix and Gilfillan GCA operations trailers," Technical Memorandum No. TM-293, June 25, 1958, issued by U.S. Navy Electronics Laboratory, San Diego.

H-23 Herapath, W. B., "On the optical properties of a newly-discovered salt of quinine which crystalline substance possesses the power of polarizing a ray of light, like tourmaline, and at certain angles of rotation of depolarizing it, like selenite," *Phil. Mag.* [4] *3*, 161 (1852). Reports discovery of the crystal herapathite.

H-24 Herapath, W. B., "Further researches into the properties of the sulphate of iodo-quinone or herapathite," *Phil. Mag. 9*, 366 (1855).

H-25 Hetényi, M., *Handbook of experimental stress analysis* (Wiley, New York, 1950), 1080 pp.

H-26 Hiltner, W. A., "On the presence of polarization in the continuous radiation of early-type stars," *Astrophys. J. 106*, 231 (1947).

H-27 Hiltner, W. A., "Polarization of light from distant stars by interstellar medium," *Science 109*, 165 (1949).

H-28 Hiltner, W. A., "On the presence of polarization in the continuous radiation of stars. II," *Astrophys. J. 109*, 471 (1949).

H-29 Hiltner, W. A., "On polarization of radiation by interstellar medium," *Phys. Rev. 78*, 170 (1950).

H-30 Hiltner, W. A., "Polarization of stellar radiation, III. The polarization of 841 stars," *Astrophys. J. 114*, 241 (1951).

H-30a Hiltner, W. A., "Polarization of the Crab Nebula," *Astrophys. J. 125*, 300 (1957). Attributes the polarization to a synchrotron mechanism.

H-31 Hiltner, W. A., "Photoelectric polarization observations of the jet in M87," *Astrophys. J. 130*, 340 (1959).

H-32 Holbourn, A. H. S., "Angular momentum of circularly polarized light," *Nature 137*, 31 (1936).

H-33 Howard, F. J., J. M. Hood, and S. S. Ballard, "A three-polarizer calibration unit for photometric instruments," *J. Opt. Soc. Amer. 45*, 904 (1955).

H-34 Hsu, H., M. Richartz, and Y. Liang, "A generalized intensity formula for a system of retardation plates," *J. Opt. Soc. Amer. 37*, 99 (1947).

H-35 Hughes, R. H., "Modified Wollaston prism for spectral polarization studies," *Rev. Sci. Instr. 31*, 1156 (1960). Use of thin wedges of quartz as pseudo depolarizers.

H-36 Hulburt, E. O., "Polarization of light at sea," *J. Opt. Soc. Amer. 24*, 35 (1934).

H-37 Hulburt, E. O., "Sextant with improved filters," *J. Opt. Soc. Amer. 26*, 216 (1936).

H-38 Hull, G. F., Jr., "Microwave experiments and their optical analogues," appendix in J. Strong, S-31.

H-39 Hurlbut, C. S., Jr., and J. L. Rosenfeld, "Monochromator utilizing the rotary power of quartz," *Am. Mineralogist 37*, 158 (1952).

— Hurlbut, C. S., Jr., U.S. patent 2,742,818. Monochromator using several quartz plates of unlike thickness and several linear polarizers.

H-40 Hurwitz, H., Jr., and R. C. Jones, "A new calculus for the treatment of optical systems, II. Proof of three general equivalence theorems," *J. Opt. Soc. Amer. 31*, 493 (1941).

H-41 Hurwitz, H., Jr., "The statistical properties of unpolarized light," *J. Opt. Soc. Amer. 35*, 525 (1945).

H-42 Huyghens, C., *Traité de la lumière* (Leyden, 1690). Announces the discovery of polarized light.

H-43 Hyde, W. L., "Polarization techniques in the infrared," *J. Opt. Soc. Amer. 38*, 663 (1948).

H-44 Hyde, W. L., E. F. Tubbs, and C. J. Koester, "An automatic photoelectric polarimeter," *J. Opt. Soc. Amer. 49*, 513 (1959).

I-1 Ingersoll, L. R., and D. H. Liebenberg, "Faraday effect in gases and vapors. I," *J. Opt. Soc. Amer. 44*, 566 (1954).

I-2 Ingersoll, L. R., and D. H. Liebenberg, "Faraday effect in gases and vapors. II," *J. Opt. Soc. Amer. 46*, 538 (1956).

I-3 Ingersoll, L. R., and D. H. Liebenberg, "Faraday effect in gases and vapors. III," *J. Opt. Soc. Amer. 48*, 339 (1958).

I-4 Inoué, S., "Polarization-optical studies of the mitotic spindle," *Chromosoma 5*, 487 (1953). Describes an improved polarizing microscope.

I-5 Inoué, S., and W. L. Hyde, "A device to obtain high extinction at high apertures in polarizing microscopes," *J. Opt. Soc. Amer. 46*, 372 (1956).

I-6 Inoué, S., and W. L. Hyde, "Studies on depolarization of light at microscope lens surfaces. II. The simultaneous realization of high resolution and high sensitivity with the polarizing microscope," *J. Biophys. Biochem. Cytol. 3*, 831 (1957).

I-7 Inoué, S., and C. J. Koester, "Optimum half-shade angle in polarizing instruments," *J. Opt. Soc. Amer. 49*, 556 (1959).

I-8 Insley, H., and V. D. Frechette, *Microscopy of ceramics and cements* (Academic Press, New York, 1955), 286 pp.

I-9 Ivanoff, A., "Degree of polarization of submarine illumination," *J. Opt. Soc. Amer. 46*, 362 (1956).

J-1 Jaffe, L., "Effect of polarized light on polarity of *Fucus*," *Science 123*, 1081 (1956).

J-2 Jaffe, L. F., "Tropistic responses of zygotes of the Fucaceae to polarized light," *Exptl. Cell Research 15*, 282 (1958).

J-3 Jamnik, D., and P. Axel, "Plane polarization of 15.1 Mev bremsstrahlung from 25-Mev electrons," *Phys. Rev. 117*, 194 (1960). A degree of polarization of 21 percent was found.

J-4 Jander, R., and T. H. Waterman, "Sensory discrimination between polarized light and light intensity patterns by arthropods," *J. Cellular Comp. Physiol. 56*, 137 (1960).

J-5 Janeschitz-Kriegl, H., "New apparatus for measuring flow-birefringence," *Rev. Sci. Instr. 31*, 119 (1960). Employs Glan-Thompson polarizer.

J-6 Jauch, J. M., and F. Rohrlich, *Theory of photons and electrons* (Addison-Wesley, Reading, Massachusetts, 1955). Presents quantum-mechanical approach to polarized light and the Stokes parameters.

J-7 Jehu, V. J., "Assessment of polarized headlighting," *Intern. Road Safety Traffic Rev. 4*, 26 (1956). Describes German and American systems; proposes use of mixed (polarized and unpolarized) beams.

J-8 Jelley, J. V., *Čerenkov radiation and its applications* (Pergamon, London, 1958).

J-9 Jenkins, F. A., and H. E. White, *Fundamentals of optics* (McGraw-Hill, New York, ed. 2, 1950).

J-10 Jensen, C., "Die Himmelsstrahlung," in H. Geiger and K. Scheel, ed., *Handbuch der Physik*, vol. 19 (Springer, Berlin, 1928), chap. 4.

J-11 Jerrard, H. G., "Optical compensators for measurement of elliptical polarization," *J. Opt. Soc. Amer. 38*, 35 (1948).

J-12 Jerrard, H. G., "Use of a half-shadow plate with uniform field compensators," *J. Sci. Instr. 28*, 10 (1951).

J-13 Jerrard, H. G., "The calibration of quarter-wave plates," *J. Opt. Soc. Amer. 42*, 159 (1952).

J-14 Jerrard, H. G., "Transmission of light through birefringent and optically active media: the Poincaré sphere," *J. Opt. Soc. Amer. 44*, 634 (1954).

J-15 Jessop, H. T., and F. C. Harris, *Photoelasticity principles and methods* (Dover, New York, 1950). 180 pp.

J-16 Jessop, H. T., "On the Tardy and Senarmont methods of measuring fractional relative retardation," *Brit. J. Appl. Phys. 4*, 138 (1953).

J-17 Jessop, H. T., "Photoelasticity," in S. Flügge, ed., *Encyclopedia of physics*, vol. 6 (Springer, Berlin, 1958).

J-18 Johannsen, A., *Manual of petrographic methods* (McGraw-Hill, New York, 1918). Contains descriptions of many kinds of birefringence polarizers.

J-19 Jones, R. C., "New calculus for the treatment of optical systems. I. Description and discussion of the calculus," *J. Opt. Soc. Amer. 31*, 488 (1941). For Part II, see Hurwitz and Jones, H-40.

J-20 Jones, R. C., "New calculus for the treatment of optical systems. III. The Sohncke theory of optical activity," *J. Opt. Soc. Amer. 31*, 500 (1941).

J-21 Jones, R. C., "New calculus for the treatment of optical systems. IV," *J. Opt. Soc. Amer. 32*, 486 (1942).

J-22 Jones, R. C., unpublished notes used in talk abstracted in J-23.

J-23 Jones, R. C., "Theory of sheet polarizers," *J. Opt. Soc. Amer. 35*, 803 (1945). Deals with density ratio, extent of parallelism of dichroic needles; a brief abstract.

J-24 Jones, R. C., "Theory of sheet polarizers." Unpublished manuscript no. 539 of Oct. 1945 showing relation between axial ratio and density ratio.

J-25 Jones, R. C., "New calculus for the treatment of optical systems. V. A more general formulation and description of another calculus," *J. Opt. Soc. Amer. 37*, 107 (1947).

J-26 Jones, R. C., "New calculus for the treatment of optical systems. VI. Experimental determination of the matrix," *J. Opt. Soc. Amer. 37*, 110 (1947).

J-27 Jones, R. C., "New calculus for the treatment of optical systems. VII. Properties of the N-matrices," *J. Opt. Soc. Amer. 38*, 671 (1948).

J-28 Jones, R. C., "On the possibility of a spathic polarizer which transmits more than one-half of the incident unpolarized light," *J. Opt. Soc. Amer. 39*, 1058 (1949).

J-29 Jones, R. C., and C. D. West, "On the properties of polarization elements as used in optical instruments. II. Sinusoidal modulators," *J. Opt. Soc. Amer. 41*, 982 (1951).

J-30 Jones, R. C., "On reversibility and irreversibility in optics," *J. Opt. Soc. Amer. 43*, 138 (1953).

J-31 Jones, R. C., and W. A. Shurcliff, "Equipment to measure and control synchronization errors in 3-D projection," *J. Soc. Motion Picture Television Engrs. 62*, 134 (1954).

J-32 Jones, R. C., "Transmittance of a train of three polarizers," *J. Opt. Soc. Amer. 46*, 528 (1956).

J-33 Jones, R. C., "New calculus for the treatment of optical systems. VIII. Electromagnetic theory," *J. Opt. Soc. Amer. 46*, 126 (1956). Deals with N-matrices.

J-34 Jones, R. V., and J. C. S. Richards, "Polarization of light by narrow slits," *Proc. Roy. Soc. (London) A 225*, 122 (1954).

K-1 Kalmus, H., "Sun navigation by animals," *Nature 173*, 657 (1954).

K-2 Kalmus, H., "The sun navigation of animals," *Sci. Am.* (Oct. 1954), p. 74.

K-3 Käsemann, E., "Die Dichroismus des Zellulosefarbstoffkomplexes und seine technische Anwendung als Polarisationsfilter," *Optik 3*, 521 (1948).

K-4 Kaufman, I., "Band between microwave and infrared regions," *Proc. IRE 47*, 381 (1959).

K-5 Kaye, W., "Near-infrared spectroscopy. II. Instrumentation and technique," *Spectrochim. Acta 7*, 181 (1955).

K-6 Kelly, R. L., "Shift of photoconductive peak with polarization in CdS," *Bull. Am. Phys. Soc.* [2] *2*, 387 (1957). Abstract.

K-7 Kennedy, D., and E. R. Baylor, "Analysis of polarized light by the bee's eye," *Nature 191*, 34 (1961). Casts doubt on Autrum's hypothesis that analysis of the light occurs in the receptors themselves.

K-8 Kerker, M., "Use of white light in determining particle radius by the polarization ratio of the scattered light," *J. Colloid Sci. 5*, 165 (1950).

See also M. Kerker, and M. I. Hampton, article with same title, *J. Opt. Soc. Amer. 43*, 370 (1953).

K-9 Kerr, J., "A new relation between electricity and light: dielectrified media birefringent," *Phil. Mag.* [4] *50*, 337 (1875).

— Keston, A. S., U.S. patent 2,829,555 (1958). Polarimeter employing two beams (polarized in slightly different azimuths) and two photocells.

K-10 Keussler, V., and P. Manogg, "Über die Emission polarisierten Lichtes durch glühende Metalloberflächen und die darbei vorhandene räumliche Intensitätsverteilung," *Optik 17*, 602 (1960).

K-11 Klyne, W., and A. C. Parker, "Optical rotatory dispersion," in A. Weissberger, ed., *Physical methods in organic chemistry*, vol. 1, part 3 (Interscience, New York, ed. 3, 1960), chap. 34.

K-12 Koester, C. J., "Achromatic combinations of half-wave plates," *J. Opt. Soc. Amer. 49*, 405 (1959).

K-13 Koester, C. J., "Half-shade eyepieces for the A. O. Baker interference microscope," *J. Opt. Soc. Amer. 49*, 560 (1959).

K-14 Koester, C. J., H. Osterberg, and H. E. Willman, Jr., "Transmittance measurements with an interference microscope," *J. Opt. Soc. Amer. 50*, 477 (1960). Describes use of polarizers and retarders in interference microscopes.

K-15 Kondo, T., "Photoanisotropic effects in dyes," *Z. wiss. Phot. 31*, 153 (1932). Says that Weigert effect was found in 450 out of 1700 dyes.

K-16 Kossel, D., U.S. patent 2,809,555 (1957). "Light rays dividing system" for binocular microscope to provide equal intensity for both eyes, even for a polarizing specimen.

K-17 Kraft, C. L., "Broad band blue lighting system for radar approach control centers: Evaluations and refinements based on three years of operational use," Wright Air Development Center Tech. Rept. 56–71; Armed Services Technical Information Agency document No. AD 118090 (1956), 96 pp.

K-18 Kremers, H. C., "Optical silver chloride," *J. Opt. Soc. Amer. 37*, 337 (1947).

K-19 Kriebel, R. T., "Stereoscopic photography," *Complete Photographer 9*, 3308 (1943).

K-20 Krimm, S., C. Y. Liang, and G. B. B. M. Sutherland, "Infrared spectra of high polymers, V. Polyvinyl alcohol," *J. Polymer Sci. 22*, 227 (1956). An excellent review of the structure of PVA; deals with polarized infrared spectra and x-ray diffraction patterns.

K-21 Kubota, H., and K. Shimizu, "Experiment on the sensitive color," *J. Opt. Soc. Amer. 47*, 1121 (1957). Proposes means for getting a more sensitive color effect than is afforded by a conventional full-wave retarder between crossed polarizers. Employs a half-wave retarder between parallel polarizers.

K-22 Kuhn, R., "Synthesis of polyenes," *J. Chem. Soc. 1*, 605 (1938).

K-23 Kuscer, I., and M. Ribaric, "Matrix formalism in the theory of diffusion of light," *Optica Acta 6*, 42 (1959).

L-1 Lagemann, R. T., and T. G. Miller, "Thallium bromide-iodide (KRS-5) as an infra-red polarizer," *J. Opt. Soc. Amer. 41*, 1063 (1951).

L-2 Laine, P., "Sur les erreurs entraînées par l'inexactitude des lames demi-onde dans l'analyse des vibrations faiblement elliptiques et sur l'étalonnages des lames demi-onde et quart d'onde," *Compt. rend. 192*, 1215 (1931).

L-3 Lambe, J., and W. D. Compton, "Luminescence and symmetry properties of color centers," *Phys. Rev. 106*, 684 (1957). Discusses dichroism in KBr.

L-4 Land, E. H., "A new polarizer for light in the form of an extensive synthetic sheet," talk given at the Harvard Physics Colloquium, Cambridge, Massachusetts, Feb. 8, 1932.

L-5 Land, E. H., "Polaroid and the headlight problem," *J. Franklin Inst. 224*, 269 (1937).

L-6 Land, E. H., "Vectographs: Images in terms of vectorial inequality and their application to three-dimensional representation," *J. Opt. Soc. Amer. 30*, 230 (1940).

L-7 Land, E. H., "Polarized light in the transportation industries," University of Michigan Official Publication *42*, No. 42, 1940. (Michigan-Life Conference on New Technologies in Transportation, p. 151.)

L-8 Land, E. H., and C. D. West, "Dichroism and dichroic polarizers," in J. Alexander, ed., *Colloid chemistry* (Reinhold, New York, 1946), vol. 6, chap. 6.

L-9 Land, E. H., *The completion of the technical development stage of the Polaroid glare-eliminating headlight system* (pamphlet prepared for presentation at a Nov. 10, 1947, meeting of the American Association of Motor Vehicles Administrators), 23 pages, 17 illustr.

L-10 Land, E. H., *The Polaroid headlight system* (Highway Research Board Bulletin No. 11, Division of Engineering and Industrial Research, National Research Council, Washington, 1948), 20 pp.

L-11 Land, E. H., "Polarized headlights for safe night driving," *Traffic Quarterly* (Eno Foundation for Highway Traffic Control, Saugatuck, Connecticut; October 1948), 11 pp.

L-12 Land, E. H., and L. W. Chubb, Jr., "Polarized light for auto headlights," *Traffic Eng. Mag.* (April and July 1950).

L-13 Land, E. H., "Some aspects of the development of sheet polarizers," *J. Opt. Soc. Amer. 41*, 957 (1951).

— Land, E. H., U.S. patents:

1,918,848 (1933), with J. S. Friedman. Polarizer containing aligned crystals of herapathite.

1,951,664 (1934). Suspensions of herapathite, etc.

1,955,923 (1939). Light valve employing controllable herapathite crystals.

1,956,867 (1934). Polarizer employing a periodide.

1,963,496 (1934). Light valve employing controllable particles.

1,989,371 (1935). Extrusion method of orienting objects.

2,005,426 (1935). Variable-density sunglasses.

2,011,553 (1935). Stretching method of orienting objects.

2,018,214 (1935). Advertising display employing polarizers, etc.

2,018,963 (1935). Suppression of specularly reflected glare by means of linear or circular polarizers.

2,031,045 (1936). Headlight system.

2,041,138 (1936). Flow method of orienting objects.

2,078,181 (1937). Polarizer system for microscope.

2,078,254 (1937). Polarizer containing fine, nonscattering crystals.

2,079,621 (1937). Polarizer system for microscope used in examining opaque objects.

2,084,350 (1937). 3-D viewer for viewing a pair of nearby, large-area pictures; employs a large semitransparent mirror at 45°.
2,096,696 (1937). Reading lamp employing two polarizers.
2,099,694 (1937). Circular polarizer of two-layer type, and use thereof in visors or viewers for headlight systems or 3-D projection systems.
2,102,632 (1937). Automobile visor adjustable (by interposing a retarder) to either of two unlike functions, namely reducing reflection from road or dimming of coded light from oncoming car's headlight.
2,106,752 (1938). Prism-type beam splitter used with polarizers in taking or projecting 3-D photographs.
2,122,178 (1938). Diffusing polarizer.
2,123,901 (1938). Diffusing polarizer producing two useful components.
2,123,902 (1938). Diffusing polarizer producing a specular component and a cylindrically spread component.
2,146,962 (1938). Dynamic display device employing polarizers and retarders.
2,158,129 (1939). Dynamic display device to be affixed to opaque surface.
2,158,130 (1939). Diffusing polarizer.
2,165,973 (1939). Polarizer containing very small crystals of herapathite in very low volume concentration.
2,168,220 (1939). Polarizing safety glass.
2,168,221 (1939). Laminated polarizer to be used close to a hot body.
2,173,304 (1939), with H. G. Rogers. Relates to K-sheet.
2,174,269 (1939). Photoelastic analysis apparatus employing polarizers and a variable achromatic retarder.
2,174,270 (1939). Display device employing reflection from surface of birefringent sheet.
2,178,996 (1939). Polarizer employing small crystals of a sulfate of an alkaloid.
2,180,113 (1939). Translucent screen employing immiscible polymers having different indices.
2,180,114 (1939). Headlight polarizer assembly having efficiency exceeding 50 percent.
2,184,999 (1939). Color filter employing several polarizers and retarders.
2,185,000 (1939). Headlight polarizer system making use of a specular and a diffuse component.
2,200,959 (1940). Display system involving showcase windows.
2,203,687 (1940), with J. Mahler. Vectograph system.
2,204,604 (1940). Images in terms of polarizance, with the aid of "resists."
2,212,880 (1940). Prism-type beam splitter used in 3-D projection.
2,237,565 (1941). Polarizing visor containing adjustable retarder to compensate for windshield's retardance.
2,237,566 (1941). Variable-density window employing polarizers and an adjustable retarder.
2,237,567 (1941). Relates to H-sheet.
2,252,324 (1941). Incandescent lamp coated with polarizing material.
2,255,933 (1941). Variable-density window employing polarizers and an adjustable retarder consisting of a rubbery layer in shear between the polarizers.
2,256,093 (1941). Method of fabricating a retarding coating.

2,270,323 (1942), with C. D. West. PVA film having high birefringence.
2,270,535 (1942), with C. J. T. Young. Automobile-headlight polarizer employing oblique surfaces and having efficiency exceeding 50 percent.
2,281,100 (1942). Orienting polarizing particles on a softened surface.
2,281,101 (1942). Reflection-type vectograph.
2,287,556 (1942). Diffusing screen employing incompatible polymers and conserving polarization.
2,289,712 (1942), with C. D. West. Discusses making herapathite crystals and polarizers employing them.
2,289,713 (1942). Method of making J-type polarizer on a curved support by means of stroking.
2,289,714 (1942). Colored vectograph images.
2,289,715 (1942). Composite films for receiving vectograph images.
2,298,058 (1942). Variable-hue sunglasses.
2,298,059 (1942). Variable-hue filter for camera.
2,299,906 (1942). Light-sensitive, layered material for vectograph.
2,302,613 (1942). Polarizing desk lamp.
2,306,108 (1942), with H. G. Rogers. Method of making K-sheet.
2,311,840 (1943). Variable-density window employing two polarizers, one of which can be rotated or moved out of the way.
2,313,349 (1943). Variable-density window employing adjustable retarder.
2,315,373 (1943). Process of making a vectograph print.
2,319,816 (1943). Polarizer consisting of stretched glass that contains a reduced metal.
2,323,059 (1943). Colored wall panels employing polarizers and retarders.
2,328,219 (1943). H-sheet based on vinyl compounds.
2,329,543 (1943). Polarizing images produced by destroying the polarizance in certain regions.
2,334,418 (1943). Use of circular polarizers in traffic-signal light.
2,343,775 (1944). Polarizer manufacture employing extrusion and evaporation.
2,346,766 (1944). Polarizer containing aligned fibers in which aligned dichroic molecules are present.
2,348,912 (1944). Self-analyzing dichroic image.
2,356,250 (1944). Adhesive for use in polarizers.
2,356,251 (1944). Polarizer employing birefringent needles.
2,356,252 (1944). Shatterproof polarizer.
2,359,428 (1944). Aligning needle-like crystals by means of stretching, smearing, or rolling operations.
2,362,832 (1944). Remote-communication or -control system employing light beams of varying polarization form.
2,373,035 (1945). Polarizing image, as in vectograph.
2,376,493 (1945), with M. Grabau. Apparatus for generating sound waves controllable by means of polarizers.
2,380,363 (1945), with R. P. Blake. Means of orienting the surface of a PVA sheet.
2,397,149 (1946). Producing opposite orientations on opposite faces of a PVA sheet to be used in vectograph.
2,397,272 (1946). Vectograph identification badge.

2,397,273 (1946). Range finder employing polarizers.
2,397,276 (1946). Vectograph system employing materials having opposite sign of refraction indicatrix.
2,402,166 (1946). Vectograph sheet.
2,407,306 (1946). Range finder employing polarizers.
2,416,528 (1947). 3-D motion-picture viewers integral with theater ticket.
2,420,252 (1947). Optical ring sight.
2,420,253 (1947). Optical ring sight that includes a biaxial element.
2,423,503 (1947). Vectograph sheet.
2,423,504 (1947). Vectograph photography employing silver and iodine.
2,431,942 (1947). 3-D motion-picture viewers employing retarders and usable either side around.
2,431,943 (1947), with J. R. Swanton and J. W. Gibson. Press used in manufacturing polarizer sheet.
2,440,102 (1948). Producing two-tone polarizing images.
2,440,103 (1948). Polarizer protected by laminated glass plates.
2,440,104 (1948). 3-D motion-picture viewers made by a folding process.
2,440,105 (1948). Producing polarizing and nonpolarizing images in a single picture.
2,440,106 (1948). Producing polarizing images.
2,445,581 (1948). H-sheet production method employing boric acid.
2,454,515 (1948). H-sheet basic patent. Calls for a rubber-elastic base and an added dichromophore.
2,458,179 (1949). Headlight system employing polarizer at 35°.
2,493,200 (1950). Variable-hue filter for use in producing color television. Employs several spectrally selective dichroic polarizers and three electro-optic retarders.
2,547,763 (1951). Method of stretching sheet.
2,788,707 (1957). Calcite parfocalizing plate for vectograph.

L-14 Langsdorf, A., Jr., and L. A. DuBridge, "Optical rotation of unpolarized light," *J. Opt. Soc. Amer. 24*, 1 (1934).

L-15 Laurence, J., "Reflection characteristics with polarized light," *J. Opt. Soc. Amer. 31*, 9 (1941).

L-16 Lawrence, A. S. C., in *Thorpe's dictionary of applied chemistry* (Longmans, Green, New York, vol. 7, 1946), p. 350. Discusses mesomorphic (smectic and nematic) states.

L-17 Le Fèvre, C. G., and R. J. W. Le Fèvre, "The Kerr effect," in A. Weissberger, ed., *Physical methods in organic chemistry*, vol. 1, part 3 (Interscience, New York, ed. 3, 1960), chap. 36.

L-18 Lenoble, J., "État de polarisation du rayonnement diffusé dans les milieux naturels (mer et atmosphere)," *J. phys. radium 18*, 47 S (1957). Employs the Stokes vector and the matrices of scatterers.

L-19 LeRoux, P., "Étude du pléochroïsime du spath d'Islande dans le spectre infrarouge," *Compt. rend. 196*, 394 (1933).

L-20 Lester, H. M., and O. W. Richards, "Stereo photomicrography with cameras of fixed interocular distance," *Photo. Eng. 5*, 149 (1954). Discusses not only 3-D photography but also the adapting of mono-objective, dual-ocular microscopes to 3-D.

L-21 Leven, M. M., "Quantitative three-dimensional photoelasticity," *Proc. Soc. Exp. Stress Anal. 12*, 157 (1955).

L-22 Linhart, J. G., "Cherenkov radiation," *Research 8*, 402 (1955).

L-23 Lodge, A. S., "Network theory of flow birefringence and stress in concentrated polymer solutions," *Trans. Faraday Soc. 52*, 120 (1956).

L-24 Loferski, J. J., "Optical polarization in single crystal of tellurium," *Phys. Rev. 87*, 905 (1952).

L-25 Locquin, M., *Bull. microscop. appl. 6*, 33 (1956). Reviews refractoanisotropy of biological materials.

L-26 Lostis, M. P., "Étude et réalisation d'une lame demi-onde en utilisant les propriétés des couches minces," *J. phys. radium 18*, 51 S (1957). Achromatic retarder.

L-27 Lowry, T. M., *Optical rotatory power* (Longmans, Green, London, 1935).

L-28 Lyot, B. F., "Un monochromator à grand champ utilisant les interférences en lumière polarisée," *Compt. rend. 197*, 1593 (1933).

L-29 Lyot, B. F., "Le filtre monochromatique polarisant et ses applications en physique solaire," *Ann. astrophys. 7*, 31 (1944).

— Lyot, B. F., U.S. patent 2,718,170. Slitless spectrophotometer.

M-1 McDermott, M. N., and R. Novick, "Large-aperture polarizers and retardation plates for use in the far ultraviolet," *J. Opt. Soc. Amer. 51*, 1008 (1961). Deals with dichroic polarizers useful in the range from 215 to 400 mμ, and with retarders of polyvinyl alcohol and mica.

M-2 McMaster, W. H., "Polarization and the Stokes parameters," *Am. J. Phys. 22*, 351 (1954).

M-3 McMaster, W. H., "Matrix representation of polarization," *Revs. Mod. Phys. 33*, 8 (1961).

M-4 McMaster, W. H., and F. L. Hereford, "Angular distribution of photoelectrons produced by 0.4–0.8-Mev polarized photons," *Phys. Rev. 95*, 723 (1954).

M-5 McMaster, W. H., *Matrix representation of polarization* (Report UCRL 5496, University of California Radiation Laboratory, 1959); also *Revs. Mod. Phys. 33*, 8 (1961).

M-6 McNally, J. G., and S. E. Sheppard, "Double refraction in cellulose acetate and nitrate films," *J. Phys. Chem. 34*, 165 (1930). Describes improved polariscope. Says cellulose acetate, for example, can be isotropic, uniaxial, or biaxial depending on how it was supported during drying.

M-7 Madden, R. P., "10 to 15 micron thick evaporated silver films for infrared gratings," paper delivered to the spring meeting of the Optical Society of America, April 8, 1955. Dealt with polarization produced by reflection gratings made of silver.

M-8 Maden, H. G., "On a modification of Foucault's and Ahrens's prisms," *Nature 31*, 371 (1884–85). Polarizing prism suitable for transmitting polarized-light images that are almost free of deviation and dispersion.

— Mahler, J., U.S. patent 2,674,156. Scheme for reducing ghosts in 3-D images.

M-9 Makas, A. S., and W. A. Shurcliff, "New arrangement of silver chloride polarizer for the infrared," *J. Opt. Soc. Amer. 45*, 998 (1955).

M-10 Malcolm, B. R., and A. Elliott, "Sensitive photoelectric polarimeter," *J. Sci. Instr. 34*, 48 (1957). Precision of 0.001° is claimed.

M-11 Mallemann, R. de., "Constantes selectionnées pouver rotatoire magnétique (effet Faraday)," in *Tables de constantes et données numériques* (Herman, Paris, 1951; published for the International Union for Pure and Applied Chemistry), pt. 3.

M-12 Malus, E., *Mém soc. Arcueil 1*, 113 (1808). Announces discovery of Malus's cosine-squared law.

M-13 Manchester, H., "The magic crystal," a chapter from *New world of machines* (Random House, New York, 1945). Popular discussion of polarizers and their applications.

M-14 Mariot, M. L., "Polarisation de la lumière en relativité générale," *J. phys. radium 21*, 80 S (1960).

M-15 Marks, A. M., "Multilayer polarizers and their application to general polarized lighting," *Illum. Eng. 54*, 123 (1959).

— Marks, A. M., U.S. patents:
2,104,949 (1938). Optical sheet on which a continuous layer of optically active substance has been formed.
2,199,227 (1940). Evaporation method of producing crystalline films.
2,777,011 (1957). Three-dimensional display system employing a special thick screen.

M-16 Meecham, W. C., and C. W. Peters, "Reflection of plane-polarized, electromagnetic radiation from an echelette diffraction grating," *J. Appl. Phys. 28*, 216 (1957).

M-17 Meier, R., and H. H. Günthard, "Germanium polarizers for the infrared," *J. Opt. Soc. Amer. 49*, 1122 (1959).

M-18 Mesnager, M., "Sur la determination optique des tensions intérieures dans les solides a trois dimensions," *Compt. rend. 190*, 1249 (1930).

M-19 Mielenz, K. D., and R. C. Jones, "Die Eignung von Polarisations-filtern für photometrische Messungen," *Optik 15*, 656 (1958).

M-20 Mindlin, R. D., "A reflection polariscope for photoelastic analysis," *Rev. Sci. Instr. 5*, 224 (1934).

M-21 Mitchell, S., and J. Veitch, "Rotary dispersion measurements with Unicam spectrophotometer," *Nature 168*, 662 (1951). Describes a spectropolarimeter.

M-22 Mitchell, S., "Accessories for measuring circular dichroism and rotatory dispersion with a spectrophotometer," *J. Sci. Inst. 34*, 89 (1957). Covers 3000–10,000 A range.

M-23 Mitsuishi, A., Y. Yamada, S. Fujita, and H. Yoshinaga, "Polarizer for the far-infrared region," *J. Opt. Soc. Amer. 50*, 433 (1960). Device consists of 15 layers of thin polyethylene films and performs well in the range from 3 to 200 μ.

M-24 Mooney, F., "A modification of the Fresnel rhomb," *J. Opt. Soc. Amer. 42*, 181 (1952).

M-25 Motz, H., W. Thon, and R. N. Whitehurst, "Experiments on radiation by fast electron beams," *J. Appl. Phys. 24*, 826 (1953). Describes the Motz generator.

M-26 Mueller, H., "Memorandum on the polarization optics of the photoelastic shutter," Report No. 2 of the OSRD project OEMsr-576, Nov. 15, 1943. A declassified report.

M-27 Mueller, H., informal notes of about 1943 on Course 8.26 at Massachusetts Institute of Technology.

M-28 Mueller, H., "The foundations of optics," *J. Opt. Soc. Amer.* *38*, 661, 1948. Deals with Stokes parameters and matrices.

— Mueller, H., U.S. patent 2,707,749. System of light beam communication. Involves two beams modulated differently in terms of polarization.

M-29 Muralt, A. von, and J. T. Edsall, "Studies in the physical chemistry of muscle globulin. III. The anisotropy of myosin and the angle of isocline. IV. The anisotropy of myosin and double refraction of flow," *J. Biol. Chem.* *89*, 315, 351 (1930).

M-30 Mussett, E. A., "On refractive index determinations by Brewster angle measurements," *J. Opt. Soc. Amer.* *46*, 369 (1956).

N-1 Nathan, A. M., "Polarization technique for improving visibility in fog and haze," *J. Opt. Soc. Amer.* *48*, 285 (1958). Linear or circular polarizers can cause five- to fiftyfold improvement.

N-2 Newman, R., and R. S. Halford, "An efficient, convenient polarizer for infra-red radiation," *Rev. Sci. Instr.* *19*, 270 (1948).

N-3 Ney, E. P., W. F. Huch, R. W. Maas, and R. B. Thorness, "Eclipse polarimeter," *Astrophys. J.* *132*, 812 (1960). Employs a rotating disk containing many small windows covered with HN-32 and HR polarizers at a variety of azimuths.

N-4 Nicholson, W. Q., and I. Ross, "Kerr-cell shutter has submicrosecond speed," *Electronics* *28*, 171 (June 1955).

N-5 Nicol, W., "On a method of so far increasing the divergence of the two rays in calcareous-spar that only one image may be seen at a time," *Edinburgh New Phil. J.* *6*, 83 (1828, 1829, 1833).

N-6 Nikitine, S., "Généralisation de la théorie du photodichroïsme," *Compt. rend.* *207*, 331 (1938).

N-7 Nikitine, S., "Sur l'anistropie d'absorption de différentes radiations pour les molécules de quelques colorants photosensibles," *Compt. rend.* *208*, 805 (1939).

O-1 Öhman, Y., "A new monochromator," *Nature* *141*, 291 (1938).

O-2 Optical Society of America, Committee on Colorimetry, *The science of color* (Crowell, New York, 1953).

O-3 Ore, A., "Entropy of radiation," *Phys. Rev.* *98*, 887 (1955). Recalls Planck's derivation of an entropy equation for radiation.

O-4 Osipov-King, W. A., "Über eine neue Konstruktion von Polarisationsprismen," *Compt. rend. acad. sci. U.R.S.S.* *4* (No. 2), 53 (1936).

O-5 Oster, G., and A. W. Pollister, *Physical techniques in biological research* (Academic Press, New York, 1955).

P-1 Pancharatnam, S., "Achromatic combinations of birefringent plates. Part I. An achromatic circular polarizer," *Proc. Indian Acad. Sci. A* *41*, 130 (1955).

P-2 Pancharatnam, S., "Achromatic combinations of birefringent plates. Part II. An achromatic quarter-wave plate," *Proc. Indian Acad. Sci. A* *41*, 137 (1955).

P-3 Pancharatnam, S., "Propagation of light in absorbing biaxial crystals: II. Experimental," *Proc. Ind. Acad. Sci.* *42*, 235 (1955).

P-4 Pancharatnam, S., "Generalized theory of interference and its applications: Part 2. Partially coherent pencils," *Proc. Ind. Acad. Sci.* *44*, 398 (1956).

P-5 Pancharatnam, S., "Light propagation in absorbing crystals possessing optical activity," *Proc. Indian Acad. Sci.* *46*, 280 (1957). Discusses non-orthogonality of eigenvectors of absorbing, retarding bodies.

P-6 Parke, N. G., III, "Matrix optics," Ph.D. thesis, Department of Physics, Massachusetts Institute of Technology (May 1, 1948), 181 pp.

P-7 Parke, N. G., III, "Matrix algebra of electromagnetic waves," Technical Report No. 70, Research Laboratory of Electronics, Massachusetts Institute of Technology (June 30, 1948), 28 pp. Discusses Jones calculus, Mueller calculus, and Wiener algebra.

P-8 Parke, N. G., III, "Statistical optics. I. Radiation," Technical Report No. 95, Research Laboratory of Electronics, Massachusetts Institute of Technology (January 31, 1949), 15 pp. Shows how to proceed from the Jones calculus to the Mueller calculus.

P-9 Parke, N. G., III, "Statistical optics. II. Mueller phenomenological algebra," Technical Report No. 119, Research Laboratory of Electronics, Massachusetts Institute of Technology (June 15, 1949). Describes and relates the Jones calculus and the Mueller calculus.

P-10 Partington, J. R., *An advanced treatise on physical chemistry*, vol. 4, "Physico-chemical optics" (Longmans, Green, New York, 1953), 688 pp. Detailed discussion of birefringence, optical activity, dichroism.

P-11 Perrin, F., "La fluorescence des solutions," *Ann. Phys. 12*, 169 (1929). Describes a method of calculating fluorescence decay time in terms of the loss in degree of polarization, the viscosity, etc., assuming that the exciting radiation was 100-percent linearly polarized.

P-12 Perrin, Francis, "Polarization of light scattered by isotropic opalescent media," *J. Chem. Phys. 10*, 415 (1942). Discusses the Stokes vector and a new calculus.

P-13 Perry, C. C., "Visual flow analysis," *Product Eng. 26*, 154 (1955). Shows photographs of "flow through a convergent-divergent nozzle as indicated by the streaming double refraction system using a synthetic dye solution."

P-14 Peterlin, A., and H. A. Stuart, *Doppelbrechung insbesonders kunstliche Doppelbrechung* (Akademische Verlag, Leipzig, 1943), 115 pp.; republished in 1948 by J. W. Edwards, Ann Arbor, Michigan. Discusses birefringence produced by flow, electric fields, magnetic fields, etc.

P-15 Pfund, A. H., "Improved infra-red polarizer," *J. Opt. Soc. Amer. 37*, 558 (1947).

P-16 Piddington, J. H., "Cosmical electrodynamics," *Proc. I.R.E. 46*, 349 (1958). Discusses polarized light from Crab Nebula.

P-17 Planck, M., *Theorie der Warmestrahlung* (Barth, Leipzig, ed. 1, 1906).

P-18 Planck, M., *Theory of heat radiation* (Blakiston, Philadelphia, 1914).

P-19 Pockels, F., *Lehrbuch der Kristalloptik* (Teubner, Leipzig and Berlin, 1906).

P-20 Poincaré, H., Théorie mathématique de la lumière (Gauthiers-Villars, Paris, 1892).

P-21 Polacoat, Inc., Bulletin p105 of Nov. 1, 1957.

P-22 Polaroid Corporation, *Polarized light and its applications* (booklet, 1945), 48 pp. Revision of M. Grabau's earlier version; see G-12.

P-23 Polaroid Corporation, *How to make Polaroid 3-D vectographs; preliminary instruction manual* F54 (pamphlet issued by Polaroid Corp. in 1940 or early 1941).

P-24 Polaroid Corporation, "Show me the way to go home by polarized light," *Polaroid Reporter*, No. 1 (1951).

P-25 Polaroid Corporation, *Polaroid vectograph film, supplementary information* (Pamphlet F-1380, May 1954), 4 pp.

P-26 Polaroid Corporation, *Photometric specifications of Polaroid Corporation sheet polarizers* (pamphlet dated Oct. 10, 1954).

P-27 Polaroid Corporation, *The application of polarized light to basic design problems* (Pamphlet F-2165, 1959), 12 pp.

P-28 Pollak, L. W., and H. Wilhelm, "Über die Verwendung von Flächenpolarisatoren in der meteorologischen Optik," *Zeiss Nachr.*, p. 307 (January 1939).

P-29 Porter, C. S., E. G. Spencer, and R. C. LeCraw, "Transparent ferromagnetic light modulator using yttrium iron garnet," *J. Appl. Phys. 29*, 495 (1958).

P-30 Preston, T., *The theory of light* (Macmillan, London, ed. 5, 1928).

P-31 Prishivalko, A. P., "Determination of the optical constants of absorbing substances from the measurements of Stokes' parameters of reflected light," *Optics and Spectroscopy 9*, 256 (1960).

P-32 Pritchard, B. S., and H. R. Blackwell, "Preliminary studies of visibility on the highway in fog," University of Michigan Engineering Research Institute, Report 2557-2-F, July 1957.

P-33 Provostaye, F., and P. Desains, "Mémoire sur la polarisation de la chaleur par réfraction simple," *Ann. chim. et phys. 30*, 158 (1850).

R-1 Ramachandran, G. N., and S. Ramaseshan, "Magneto-optic rotation in birefringent media. Application of the Poincaré sphere," *J. Opt. Soc. Amer. 42*, 49 (1952).

R-2 Ramachandran, G. N., and S. Ramaseshan, "Crystal optics," a section in S. Flügge, ed., *Encyclopedia of Physics*, vol. 25/1 (Springer, Berlin, 1961). Includes discussions of the Poincaré sphere and the Jones calculus.

R-3 Randall, D. D., "New photoelectric method for the calibration of retardation plates," *J. Opt. Soc. Amer. 44*, 600 (1954).

R-4 Ravilious, C. F., R. T. Farrar, and S. H. Liebson, "Measurement of organic fluorescence decay times," *J. Opt. Soc. Amer. 44*, 238 (1954). Determines fluorescence decay time in terms of polarization defect of the fluorescent light, assuming that the exciting light is 100-percent polarized.

R-5 Rich, A., "Use of the Senarmont compensator for measuring double refraction of flow," *J. Opt. Soc. Amer. 45*, 393 (1955).

R-6 Richards, O. W., *A. O. Baker interference microscope Model 7; Reference manual* (pamphlet; American Optical Co., 1958), 32 pp.

R-7 Richartz, M., and H. Hsu, "Analysis of elliptical polarization," *J. Opt. Soc. Amer. 39*, 136 (1949).

R-8 Richartz, M., "On the measurement of skylight polarization," *J. Opt. Soc. Amer. 50*, 302 (1960). Compares the visual methods available.

R-9 Rinne, F., and M. Berek, *Anleitung zu optischen Untersuchungen mit dem Polarisationsmikroskop* (Schweizerbart'sche Verlagsbuchhandlung, Stuttgart, ed. 2, 1953), 366 pp.

R-10 Roberts, S., "Interpretation of the optical properties of metal surfaces," *Phys. Rev. 100*, 1667 (1955).

— Roehrig, J. R., U.S. patent 2,824,487. Apparatus for grading anisotropic fibers.

— Rogers, H. G., U.S. patent 2,255,940. Orientation method involving shrinking and restretching, as of K-sheet.

R-11 Roper, V., and K. D. Scott, "Seeing with polarized headlamps," *Illum. Eng. 36*, 1205 (1941).

R-12 Rosen, P., "Entropy of radiation," *Phys. Rev. 96*, 555 (1954).
R-13 Rossi, B., *Optics* (Addison-Wesley, Reading, Massachusetts, 1957), 510 pp.
R-14 Rothen, A., and M. Hanson, "Optical properties of surface films, II," *Rev. Sci. Instr. 20*, 66 (1949).
R-15 Rothen, A., "Improved method to measure the thickness of thin films with a photoelectric ellipsometer," *Rev. Sci. Instr. 28*, 283 (1957).
R-16 Rudolph, H., "Photoelectric polarimeter attachment," *J. Opt. Soc. Amer. 45*, 50 (1955). Contains an excellent discussion of accuracy, and an extensive bibliography.
— Ryan, W. H., U.S. patents:
2,263,684 (1941). Polarizers in variable-hue filter for darkroom.
2,811,893 (1957). Deals with methods of compensating for ghost images in 3-D presentations.
— Sage, S. J., U.S. patent 2,834,254. Electronic color filter.
S-1 Schawlow, A. L., and C. H. Townes, "Infrared and optical masers," *Phys. Rev. 112*, 1940 (1958). Mentions production of polarized light by masers.
S-2 Scheraga, H., and R. Signer, "Streaming birefringence," in A. Weissberger, ed., *Physical methods of organic chemistry*, vol. 1, part 3 (Interscience, New York, ed. 3, 1960), chap. 35.
S-3 Scherer, H., "Dichroitisch angefärbte Polarisatoren," *Z. Naturforsch. 6a*, 440 (1951).
S-4 Schick, L. H., and S. C. Miller, "Compton cross section for circular photon polarization and arbitrary electron spin orientation," *Bull. Am. Phys. Soc.* [2] *2*, 312 (1957).
S-5 Schmidt, W. J., "Polarisationsoptische analyse des Submikroskopischen Baues von Zellen und Geweben," in *Handbuch der biologischen Arbeitsmethoden*, Abt. 5, Teil 10, Heft 3 (1934), pp. 435–665.
S-6 Schöne, Herman, and Hedwig Schöne, "Eyestalk movements induced by polarized light in the ghost crab, *Ocypode quadrata*," *Science 134*, 675 (1961).
S-7 Schubert, F., "Optische Eigenschaften glühender Metalle," *Ann. Physik 29*, 473 (1937).
S-8 Sekera, Z., "Polarization of skylight," in S. Flügge, ed., *Encyclopedia of physics*, vol. 48 (Springer, Berlin, 1957).
S-9 Sekera, Z., "Light scattering in the atmosphere and the polarization of sky light," *J. Opt. Soc. Amer. 47*, 484 (1957).
S-10 Shklovsky, I. S., *Cosmic radio waves* (Harvard University Press, Cambridge, Mass., 1960). Includes discussion of polarization of light from Crab nebula.
— Short, F., U.S. patent 1,734,022 (1921). Polarizing headlight system using reflection polarizers.
S-11 Shurcliff, W. A., "Screens for 3-D and their effect on polarization," *J. Soc. Motion Picture Television Engrs. 62*, 125–133 (1954).
S-12 Shurcliff, W. A., "Polarized light," *Encyclopedia Americana* (Americana, New York, 1956).
S-13 Shurcliff, W. A., "Haidinger's brushes and circularly polarized light," *J. Opt. Soc. Amer. 45*, 399 (1955).
S-14 Shurcliff, W. A., "Circular polarizer improves viewing," *Electronics Design 4*, April 1, 1956.
S-15 Shurcliff, W. A., section on polarizing filters in *American Institute of Physics handbook* (McGraw-Hill, New York, 1957).

S-16 Simpson, J. A., H. W. Babcock, and H. D. Babcock, "Association of a unipolar magnetic region on the sun with changes of primary cosmic ray intensity," *Phys. Rev. 98*, 1402 (1955).

S-17 Skinner, C. A., "A universal polarimeter," *J. Opt. Soc. Amer. 10*, 491 (1925).

S-18 Sloan, L. L., "The Haidinger brush phenomenon," *J. Opt. Soc. Amer. 45*, 402 (1955).

S-19 Smartt, R. N., and W. H. Steel, "Birefringence of quartz and calcite," *J. Opt. Soc. Amer. 49*, 710 (1959).

S-20 Smith, F. H., "Microscopic interferometry," *Research 8*, 385 (1955).

S-21 Smith, S. J., and E. M. Purcell, "Visible light from localized surface charges moving across a grating," *Phys. Rev. 92*, 1069 (1953).

S-22 Šolc, I., "Further investigations on the birefracting filter" [in Russian], *Czechoslov. J. Phys. 5*, 80 (1955). Describes narrow-passband filter employing a large number of linear retarders.

S-23 Šolc, I., "Chain birefringent filters," *Czechoslov. J. Phys. 9*, 237 (1959). Describes a 14-plate filter having a 2-A transmission band.

S-24 Soleillet, P., "Sur les paramètres caractérisant la polarisation partielle de la lumière dans les phénomènes de fluorescence," *Ann. phys. 12*, 23 (1929). Discusses Stokes parameters and matrices.

S-25 Spence, J., "Optical anisotropy and the structure of cellulosic sheet material," *J. Phys. Chem. 43*, 865 (1939).

— Stadler, A. E. K., U.S. patent 2,527,593 (1950). Variable-hue filter employing two polarizers, a 90° retarder, and a 180° retarder.

S-26 Stamm, R. W., "The Polaroid stereoscope," *Am. J. Roentgenol. Radium Therapy 45*, 744 (1941). Shows how to view two x-ray photographs to obtain a 3-D effect.

S-26a Steel, W. H., R. N. Smartt, and R. G. Giovanelli, "A 1/8 Å birefringent filter for solar research," *Australian J. Phys. 14*, 201 (1961).

S-27 Stein, R. S., "Optical properties of oriented polystyrene," *J. Appl. Phys. 32*, 1280 (1961). Describes methods of calculating the birefringence and infrared dichroism.

S-28 Stokes, G. G., "On Haidinger's brushes" (1850), reprinted in *Mathematical and physical papers*, vol. 2 (Cambridge University Press, Cambridge, England, 1883), p. 362.

S-29 Stokes, G. G., "On the composition and resolution of streams of polarized light from different sources," *Trans. Cambridge Phil. Soc. 9*, 399 (1852); *Mathematical and physical papers*, vol. 3 (Cambridge University Press, Cambridge, England, 1901), p. 233.

S-30 Stokes, G. G., "On the intensity of the light reflected from or transmitted through a pile of plates," *Proc. Roy. Soc. (London) 11*, 545 (1862); *Mathematical and physical papers*, vol. 4 (Cambridge University Press, Cambridge, England, 1904), p. 145.

S-31 Strong, J., *Concepts of classical optics* (Freeman, San Francisco, 1958), 692 pp. Contains appendixes by G. F. Hull, Jr., on microwaves and by A. C. S. van Heel on Savart plates.

S-32 Sultanoff, M., "A 0.1 microsecond Kerr-cell shutter," *Phot. Eng. 5*, 80 (1954).

S-33 Swann, M. M., and J. M. Mitchison, "Refinements in polarized light microscopy," *J. Exptl. Biol. 27*, 226 (1950).

T-1 Tadokoro, H., S. Seki, and I. Nitta, "Infrared absorption spectrum of deuterated polyvinyl alcohol film," *J. Chem. Phys. 23*, 1351 (1955).

T-2 Takasaki, H., "Photoelectric measurement of polarized light by means of an ADP polarization modulator. I. Photoelectric polarimeter," *J. Opt. Soc. Amer. 51*, 462 (1961). Device employs a Senarmont compensator in which an ADP crystal excited by alternating current is incorporated.

T-3 Theophanis, G. A., "A Kerr-cell camera with synchronized light source for millimicrosecond reflected light photography," *J. Soc. Motion Picture and Television Engrs. 70*, 522 (1961). Employs three polarizers and a 60,000-v square-wave pulse to achieve 0.05-μs exposure time.

T-4 Thiessen, G. von, "Polarization des Lichtes beim Durchgang durch Metallspalte," *Optik 2*, 266 (1947).

T-5 Thompson, S. P., "On the Nicol prism and its modern varieties," in *Proceedings of the optical convention of 1905* (Williams and Norgate, London, 1905), pp. 216–235.

T-6 Thorpe, W. H., "Orientation and methods of communication of the honey bee and its sensitivity to the polarization of the light," *Nature 164*, 11 (1949).

T-7 Tinkham, M., and M. W. P. Strandberg, "The excitation of circular polarization in microwave cavities," *Proc. I.R.E. 43*, 734 (1955).

T-8 Townes, C. H., "Microwave spectroscopy," in M. H. Shamos and G. M. Murphy, eds., *Recent advances in science* (Interscience, New York, 1956).

T-9 Trentini, G. V., "Maximum transmission of electromagnetic waves by a pair of wire gratings," *J. Opt. Soc. Amer. 45*, 883 (1955).

T-10 Tsurumi, I., "Optical method for determining the small phase retardation with white light," *J. Opt. Soc. Amer. 45*, 1021 (1955).

T-11 Tuckerman, L. B., "On the intensity of the light reflected from or transmitted through a pile of plates," *J. Opt. Soc. Amer. 37*, 818 (1947).

T-12 Tutton, A. E. H., *Crystallography and practical crystal measurement* (Macmillan, London, 1911). Contains an excellent section on birefringence polarizers.

T-13 Twyman, F., *Optical glassworking* (Hilger and Watts, London, ed. 2, 1955), 288 pp. Contains an appendix on "Making polarizing prisms."

U-1 United States District Court for the District of Massachusetts. Civil Action No. 53-168. Opinion by Chief Justice Sweeney, issued Feb. 28, 1955.

V-1 Van de Hulst, H. C., *Light scattering by small particles* (Wiley, New York, 1957), 470 pp. Chapter 5 discusses Stokes parameters and matrices of Perrin, Mueller, and Jones.

V-2 Van Doorn, C. Z., and Y. Haven, "Dichroism of the F and M absorption bands in KCl," *Phys. Rev. 100*, 753 (1955).

V-3 Van Heel, A. C. S., "Interferometry with Savart's plate," appendix in J. Strong, S-31.

V-4 Vickers, A. E. J., "The polarizing microscope in organic chemistry and biology," *Research 9*, 67 (1956).

W-1 Wahlstrom, E. E., *Optical crystallography* (Wiley, New York, ed. 2, 1951).

W-2 Walker, J., *The analytical theory of light* (Cambridge University Press, Cambridge, England, 1904). Deals at length with many types of compensator.

W-3 Walker, M. J., "Matrix calculus and the Stokes parameters of polarized radiation," *Am. J. Phys. 22*, 170 (1954).

W-4 Waring, C. E., and R. L. Custer, "Determination of the Faraday effect," in A. Weissberger, ed., *Physical methods in organic chemistry*, vol. 1, part 3 (Interscience, New York, ed. 3, 1960), chap. 37.

W-5 Waterman, T. H., "Polarization patterns in submarine illumination," *Science 120*, 927 (1954).

W-6 Waterman, T. H., "Polarized light and animal navigation," *Sci. Am.* (July 1955), p. 88.

W-7 Waterman, T. H., and W. E. Westell, "Quantitative effect of the sun's position on submarine light polarization," *J. Marine Research (Sears Foundation) 15*, 149 (1956).

W-8 Waterman, T. H., "Interaction of polarized light and turbidity in the orientation of *Daphnia* and *Mysidium*," *Z. vergleich. Physiol. 43*, 149 (1960).

W-9 Wayland, H., "Quantitative fluid flow visualization with streaming birefringence," *Phys. Rev. 98*, 255 (1955). Abstract.

W-10 Weber, G., "Polarization of the fluorescence of macromolecules," *Biochem. J. 51*, 145 (1952).

W-11 Weeks, D. W., "A study of sixteen coherency matrices," *J. Math. and Phys. 13*, 380 (1957).

W-12 Weigel, R. G., "Die Anwendung polarisierten Lichtes zur Verhinderung der Blendung im Kraftverkehr," *Optik 5*, 169 (1949).

W-13 Weigert, F., "New time phenomenon in photographic emulsions," *Trans. Faraday Soc. 34*, 927 (1938). Discusses photodichroism in gelatin emulsions that contain AgCl.

W-14 Weill, M. G., "Un appareil de mesure de la dépolarisation de la lumière diffusée," *J. phys. radium 18*, 78 S (1957). Employs Wollaston and Glazebrook polarizers in the photoelectric measurement of depolarization produced by liquids and gases.

W-15 Wellington, W. G., "Motor responses, etc., to plane polarized light," *Nature 172*, 1177 (1953).

W-16 West, C. D., "Crystallography of herapathite," *Am. Mineralogist 22*, 731 (1937).

W-17 West, C. D., "Polarizing accessories for microscopes," *J. Chem. Ed. 19*, 66 (1942).

W-18 West, C. D., "Structure-optical studies. I. X-ray diffraction by addition compounds of halogens with hydrophyllic organic polymers," *J. Chem. Phys. 15*, 689 (1947).

W-19 West, C. D., "On rendering surfaces anisotropic," *Glass Ind. 30*, 272 (May 1949). Discusses Beilby layer polarizers and work by Zocher.

W-20 West, C. D., and A. S. Makas, "The spectral dispersion of birefringence, especially of birefringent plastic sheets," *J. Opt. Soc. Amer. 39*, 791 (1949). Discusses achromatic retarders.

W-21 West, C. D., "Polariscopic and polarimetric examination of materials by transmitted light," in W. G. Berl, *Physical methods in chemical analysis* (Academic Press, New York, vol. 1, 1950), pp. 425–483.

W-22 West, C. D., and R. C. Jones, "On the properties of polarization elements as used in optical instruments. I. Fundamental considerations," *J. Opt. Soc. Amer. 41*, 976 (1951).

— West, C. D , U.S. patents:
2,420,273 (1947). Achromatic optical ring sight.
2,441,049 (1948). Achromatic retarder of organic plastic.

2,447,828 (1948). Prism-type polarizer employing isotropic block and thin birefringent sheet.
2,788,710 (1957). Electro-optical shutter employing cubic crystals of the class T_d, for example, cuprous halide.

W-23 Westfold, K. C., "New analysis of the polarization of radiation and the Faraday effect in terms of complex vectors," *J. Opt. Soc. Amer.* *49*, 717 (1959).

W-24 White, C. T., "Polarized-light illumination of radar and sonar spaces and comparison with limited spectrum methods," Research and Development Report No. 669, U.S. Navy Electronics Laboratory, San Diego, California, Feb. 21, 1956; 12 pp.

— White, C. T., U.S. patent 2,793,361. Use of polarizers in combat information centers, etc.

W-25 Wilkinson, D. H., "A source of plane polarized gamma-rays of variable energy above 5.5 Mev," *Phil. Mag.* *43*, 659 (1952). A degree of polarization of nearly 100 percent is achieved in the $^2H(p,\gamma)^3He$ reaction.

W-26 Wood, R. W., *Physical optics* (Macmillan, New York, ed. 3, 1934).

W-27 Worthing, A. G., "Deviation from lambert's law and polarization of light emitted by incandescent tungsten, tantalum, and molybdenum, and changes in the optical constants of tungsten with temperature," *J. Opt. Soc. Amer.* *13*, 635 (1926).

W-28 Wright, F. E., "A spherical projection chart for use in the study of elliptically polarized light," *J. Opt. Soc. Amer.* *20*, 529 (1930). Uses Poincaré sphere.

W-29 Wright, N., "A transmitting polarizer for infra-red radiation," *J. Opt. Soc. Amer.* *38*, 69 (1948).

Y-1 Yamaguti, T., "On a sodium nitrate polarization plate of scattering type," *J. Opt. Soc. Amer.* *45*, 891 (1955).

Y-2 Young, T., "Miscellaneous Works," Vol. 1, 1855.

Z-1 Zandman, F., and M. R. Wood, "Photostress: a new technique for photoelastic stress analysis for observing and measuring surface strains on actual structures and parts," *Product Eng.* *27*, 167 (1956).

Z-2 Zandman, F., "Make strain visible with photostress analysis," *Product Eng.* *30*, 43 (1959).

Z-3 Zarem, A. M., F. R. Marshall, and S. M. Hauser, "Millimicrosecond Kerr cell camera shutter," *Rev. Sci. Instr.* *29*, 1041 (1958). Describes shutter having an exposure time of 5 mμs.

— Zeiss (Carl Zeiss, Inc.), German patent 1,015,236 (1957). Infrared dichroic polarizer by Drechsel.

Z-4 Zimm, B. H., "Photoelectric flow birefringence instrument of high sensitivity," *Rev. Sci. Instr.* *29*, 360 (1958).

Z-5 Zimmern, A., Belgian and Austrian patents mentioned in U-1. An article entitled "Sur une nouvelle méthode de production de l'hérapathite," *Compt. rend.* *182*, 1082 (1926). These papers deal with methods for producing large area, crystalline, polarizing layers on specially prepared surfaces.

Z-6 Zocher, H., and F. C. Jacoby, "Über die optische Anisotropie selektiv absorbierender Farbstoffe," *Kolloidchem. Beih.* *24*, 365 (1927). Tabulates dichroism data on about 100 organic dyes. Discusses streaming dichroism.

Z-7 Zocher, H., and K. Coper, "Über die Erzeugung der Anisotropie von

Oberflächen," *Z. physik. Chem.* *132*, 295 (1928). Describes polarizers made by rubbing a glass surface and applying methylene blue. (Translated and presented with W-19 by West.)

Z-8 Zocher, H., and K. Coper, "Über die durch den Weigerteffekt in Photochlorid erzeugte Anisotropie," *Z. physik. Chem.* *132*, 303 (1928).

Z-9 Zocher, H., and K. Coper, "Über die Erzeugung optischer Aktivität durch zirkulares Licht," *Z. physik. Chem.* *132*, 313 (1928).

Author Citation Index

Subject Index